AF685538

SOLAR PHYSICS RESEARCH TRENDS

SOLAR PHYSICS RESEARCH TRENDS

PINGZHI WANG
EDITOR

Nova Science Publishers, Inc.
New York

Copyright © 2008 by Nova Science Publishers, Inc.

All rights reserved. No part of this book may be reproduced, stored in a retrieval system or transmitted in any form or by any means: electronic, electrostatic, magnetic, tape, mechanical photocopying, recording or otherwise without the written permission of the Publisher.

For permission to use material from this book please contact us:
Telephone 631-231-7269; Fax 631-231-8175
Web Site: http://www.novapublishers.com

NOTICE TO THE READER

The Publisher has taken reasonable care in the preparation of this book, but makes no expressed or implied warranty of any kind and assumes no responsibility for any errors or omissions. No liability is assumed for incidental or consequential damages in connection with or arising out of information contained in this book. The Publisher shall not be liable for any special, consequential, or exemplary damages resulting, in whole or in part, from the readers' use of, or reliance upon, this material.

Independent verification should be sought for any data, advice or recommendations contained in this book. In addition, no responsibility is assumed by the publisher for any injury and/or damage to persons or property arising from any methods, products, instructions, ideas or otherwise contained in this publication.

This publication is designed to provide accurate and authoritative information with regard to the subject matter covered herein. It is sold with the clear understanding that the Publisher is not engaged in rendering legal or any other professional services. If legal or any other expert assistance is required, the services of a competent person should be sought. FROM A DECLARATION OF PARTICIPANTS JOINTLY ADOPTED BY A COMMITTEE OF THE AMERICAN BAR ASSOCIATION AND A COMMITTEE OF PUBLISHERS.

Library of Congress Cataloging-in-Publication Data

Solar physics research trends / Pingzhi Wang (editor).
p. cm.
ISBN-13: 978-1-60021-987-0 (hardcover)
ISBN-10: 1-60021-987-X (hardcover)
1. Sun. 2. Solar activity. 3. Solar atmosphere. 4. Astrophysics. I. Wang, Pingzhi.
QB521.S627 2007
523.7'2--dc22

2007047053

Published by Nova Science Publishers, Inc. ✢ New York

Contents

Preface vii

Expert Commentary 1

The Quasi-biennial Oscillation in Time Series of Solar Activity Parameters 3
Ana G. Elias and Marta Zossi de Artigas

Short Communication 15

Remote Radio Sounding of Perturbed Plasma Formations at Small Distances from the Sun with a Spacecraft 17
A. N. Afanasiev and N. T. Afanasiev

Research and Review Studies 29

Chapter 1 Solar Filaments and Their Eruptions 31
Boris Filippov

Chapter 2 Multifractal Nature of Solar Phenomena 95
Valentyna I. Abramenko

Chapter 3 Time-Distance Helioseismology of Sunspots 137
Sébastien Couvidat

Chapter 4 The Quasi-biennial Oscillations in the Equatorial Stratosphere: Seasonal Regularities, Dependence on the Solar UV Flux, and Relation to Ozone Depletion in Antarctica 165
I. Gabis and O. Troshichev

Chapter 5 Some Peculiarities of Research into Multi-scale Ionospheric Irregularities by Radio Methods 195
M. V. Tinin

Chapter 6 A Fresh Insight into the Energy Source and Transport for Large Solar Flares **233**
G. M. Simnett

Chapter 7 Application of Machine Learning in Solar Physics **259**
Frank Y. Shih, Ju Jing, Ming Qu and Haimin Wang

Chapter 8 Solar Type IV Bursts at Frequencies 10-30 MHZ **287**
V. N. Melnik, H. O. Rucker, A. A. Konovalenko, V. V. Dorovskyy, E. P. Abranin, A. I. Brazhenko, B. Thide and A. A. Stanislavskyy

Chapter 9 Forcing of the Earth's Atmosphere by Solar Radiation and Solar Wind **327**
Klemens Hocke and Niklaus Käampfer

Chapter 10 Numerical Study of Formation and Propagation of Giant Electric Jets between a Thundercloud and the Lower Ionosphere **345**
Lizhu Tong and Kenichi Nanbu

Chapter 11 Sunspot Influence on Photometric and Morphological Properties of Some Photospheric Fine Structures **357**
Ernesto R. Rodríguez-Flores and Ramón E. Rodríguez Taboada

Chapter 12 Propagation of Radiowaves at Frequencies Below the Electron Gyrofrequency from the Earth Surface to the Higher Atmosphere **381**
V. M. Krasnov

Index **401**

PREFACE

This new book presents the latest results on solar-stellar research and the study of solar terrestrial physics. It includes all aspects of solar physics, ranging from the internal structure of the Sun and its evolution, to outer corona and solar wind in interplanetary space.

Chapter 1 - In the hot and rarefied plasma of the solar corona, relatively dense and cold clouds are encountered, with the parameters more typical for the chromosphere, referred to as solar prominences. Some of them are very dynamic and have a short lifetime (loop prominences, surges, sprays) while others are observed over a long period of time and change rather slowly. The latter are usually visible within the chromospheric spectral lines against the background of the solar disk as dark filaments. It is well known that the filaments are elongated along the polarity inversion lines of the radial component of magnetic field in the photosphere. It is the magnetic field that holds the dense substance of filaments high in the corona and prevents it from imminent falling into the chromosphere. Resent research suggests that the magnetic field in the filament environment has a structure of a flux rope, which consists of the helical lines of force. The lower parts of the helixes serve as magnetic traps in the gravitational field, where the dense plasma of prominences can be accumulated. The equilibrium of a filament is determined by the equilibrium of the integral electric current of the flux rope. Its loss of equilibrium or stability leads to the rapid ascent of the filament named eruption. Further development of the process in the corona manifests itself as a coronal mass ejection (CME). Thus, the filaments serve as indicators of the presence of the magnetic flux ropes in the solar atmosphere and on the basis of analysis of their state (the height above the chromosphere, the intensity of internal motions, the upward motion, etc.) it is possible to judge the prospects for sudden disturbances in the solar wind.

Chapter 2 - Last decades, Solar Physics research demonstrated a considerable progress in understanding of macro-scale processes (e.g., the magnetohydrodynamic modeling of the heliosphere, modeling of large-scale coronal magnetic structures, etc.), on one hand and micro-scale phenomena (e.g., turbulence of solar plasma) on the other hand. Further progress seems to be associated with the realization of how various micro-scale processes are involved and manifested in the macro-scale behavior of the entire Sun. A similar problem unavoidably arises in studies of any, other than the Sun, non-linear dynamical dissipative system present in nature. Such systems, that can be placed in between a chaos and a completely determined structure, display a scale-invariant behavior. A broad class of models for scale-invariant structures is now elaborated based on the notion of a fractal, which can be mathematically described as an object with fractional dimensions. The goal of this chapter is to show how the concept of fractality and multifractality is currently elaborated in Solar Physics and how these ideas help us to understand the unpredictable behavior of our closest star. The first section provides introduction to multifractality and outlines physical connections between various aspects of this paradigm. The second section is devoted to reviewing the multifractality approach in the study of the Sun: the authors will discuss how high-order statistical moments, fractal geometry, multifractality, percolation theory, and the concept of the self-organized criticality are now applied to a broad range of solar phenomena: from explosive events of various scales to geometrical properties of observed filigree structures. The last section outlines new ideas and results based on the concept of multifractality.

Chapter 3 - Time-distance helioseismology is a recently developed method in solar physics giving us unprecedented access to the local structure and dynamics of subsurface solar layers. Since its inception in 1993, seminal and widely publicized results have been achieved: for instance, the inversion of the sound-speed subsurface structure of a sunspot, of the subsurface flows below and around an active region, the precise measurement of meridional and zonal flows, the mapping of large scale subsurface flows, the measurement of supergranular flows, and even the precise determination of the solar rotation axis. In this chapter, the authors focus on the results related to sunspots and the authors describe how recent works on surface effects and the magnetic field interaction with the acoustic wavefield question the relevance of these major breakthroughs.

Chapter 4 - The term Quasi-Biennial Oscillation (QBO) was designated to describe the regular reversals of zonal winds in the equatorial stratosphere occurred with periodicity about 28 months. The phase of QBO cycle, East (E) or West (W) is determined by direction of the zonal winds. Although the wind QBO is mainly a tropical circulation feature, its effects are displayed well beyond the equatorial region. Indeed, the year-to-year fluctuations in the extratropical atmosphere are controlled by QBO, which influence is observed in such parameters as temperature, the total ozone distribution, and others. The most important thing is the well known at present QBO property to modulate the solar activity impact on the stratosphere and troposphere.

Today the dissipation of upward propagating equatorial waves usually examined as a theoretical mechanism for QBO generation. So, it is generally accepted that the proximity of the average QBO period to the double annual one being merely statistical result, whereas the actual period varies from 19 to 36 months. As a result of the high variability and elusive relation to the seasonal cycles, the QBO phase duration seemed to be unpredictable.

A new method for the investigation of the QBO cycle evolution is presented in this chapter. The analysis of the height wind profiles makes it possible to reveal new regularities in the equatorial stratosphere wind reversals. The stagnation stage of descending easterlies always starts in solstice, either in December–January or in June–July, and always completes by the equinox being of quantized duration in different QBO-cycles (about 3, 9, or 15 months). So the length of the complete QBO-cycle, if defined as a period between the inceptions of the successive stagnation stages, depends on their durations and may be equal to 24, 30, or 36 months. Consequently, the full QBO cycle length can be predicted as soon as end of the stagnation stage is found while observing the zonal wind transformation.

In turn the duration of the stagnation stage depends on the intensity of solar UV flux. The short stalling period (3–4 months) is observed under condition of the high level or steady increase of the UV irradiance during the first equinox in course of the partial QBO cycle. If the UV flux is low or decreases during the proper equinox, the easterly winds do not descend from the upper layer to the lower stratosphere, and the stalling period length at about 20-40 hPa increases to 9–10 or 15–16 months. The intensity of the spring ozone depletion within Antarctic polar vortex is thought to be also dependent on the phase of the equatorial QBO cycle. It is shown the relation of the season regularities of the equatorial QBO-circulation to the interannual fluctuations and the intensity of total ozone (TOZ) destruction over Antarctica during "ozone hole" phenomenon.

Chapter 5 - A characteristic feature of ionospheric plasma is the presence of irregularities with different sizes. There is a background irregularity in the form of ionospheric layers with the scales of order 100-400 km along the vertical. Smaller-size irregularities are observed at this background. These ionospheric irregularities have different scales between 100 km and lower. Furthermore, the presence of the geomagnetic field, gravity, wind in the upper atmosphere, and of other factors leads to the fact that ionospheric irregularities are different in size along different directions. Such irregularities are said to be anisomeric. This paper presents some of the latest results derived from developing new methods of describing the propagation of radio waves in a multi-scale inhomogeneous ionosphere and analyzes the possible approaches in perfecting them. In addition to a conventional description of the singly scattered wave from an elongated irregularity smaller than the Fresnel radius in size, the authors consider the single scattering from a strongly anisomeric irregularity in which the major axis exceeds the corresponding Fresnel radius. Furthermore, the authors examine the single scattering from a flattened irregularity. This situation is investigated both for the case of a weakly anisomeric irregularity, all scales of which are less than the Fresnel radius, and for the case of a strongly anisomeric irregularity, two axes of which exceed the Fresnel radius. The authors point to a number of characteristics of the scattering from anisomeric irregularities which, when taken into account, are useful for their diagnostics. With increasing irregularity amplitude, the scattering ceases to be a weak one, and the single scattering approximation becomes invalid. The geometrical optics approximation and its generalizations are used in investigating the scattering in this case. In this paper, the authors examine some generalizations of the geometrical optics approximation which furnish a means to investigate strong fluctuations in the field, and effects of random irregularities in the neighborhood of the background caustic. The authors explore the possibility of extending the results obtained in the neighborhood of a regular caustic that is produced in the ionosphere, to similar investigations of the effects of random irregularities under conditions of the solar corona. It is pointed out that mixed integral representations obtained using a double (over the coordinates of the source and observer) weighted Fourier transform hold much promise for the description of the characteristics of radio wave propagation in such strongly regularly and randomly inhomogeneous media as the ionosphere, solar plasma, and fusion plasma.

Chapter 6 - The physical processes involved in the total energy release and transport in a large solar flare are not well understood. The energy is derived from magnetic fields and the magnetic free energy arises through convective motions of plasma at levels below the photosphere. In studying flare energetics the goal is to establish a comprehensive and self-consistent picture of the sources and transport of energy. The first objective must be to define the relevant volume encompassing the important energy sources and sinks and it is believed that most previous studies of flare energetics have grossly underestimated this. A second objective is to identify the principal energy transport mechanisms. Much of the work to date has concentrated on mass motions, Joule heating, radiative transport and thermal conduction. Energy transport by non-thermal ions has been given a secondary, often minor role, although transport by non-thermal electrons has featured in an unsatisfactory way in some models. In this paper the authors re-examine flare energetics by postulating the relevant volume for the energy source to be more like the volume of the corona involved in the initiation of the coronal mass ejection and eruption of coronal magnetic fields that almost invariably accompany a major flare. The authors then examine how the energy transport might proceed if the initial result of the release of magnetic energy is to create a population of non-thermal charged particles in which the ions contain the bulk of the energy released. An inevitable outcome of this approach is that the active region flare becomes primarily a secondary phenomenon to energy release elsewhere. The processes occurring in the volume of the corona containing the magnetic structures that erupt into the interplanetary medium are the primary driver of the energetic transient, which includes the flare. Often these erupting structures contain excess mass over the ambient corona, in which case they are identified as coronal mass ejections. The photospheric/chromospheric active region does play an important role, including provision of some energy, in forming the spectacular dynamic structures that for over a century have been the epitome of the optical flare, and the authors incorporate this into the global picture.

Chapter 7 - The authors present the application of machine learning techniques in solar feature detection and solar event classification. The focus of automatic solar feature detection is on the development of efficient feature-based classifiers. The authors have conducted experiments using different classifiers for the detection of solar flare on the solar H-α (Hydrogen-Alpha) images. These images were obtained from the Big Bear Solar Observatory in California, USA. Series of H-α full-disk images are taken in regular time intervals to observe the changes of solar disk features. In addition, the authors have proposed the classifier for detection of Coronal Mass Ejections (CMEs) in Large Angle Spectrometric Coronagraph (LASCO) C2 and C3 images, which were obtained from the Solar and Heliospheric Observatory (SOHO). The authors use image filtering and enhancement techniques to preprocess the images and extract a set of critical features for event classification. The proposed method can detect and track solar events as well as can measure their properties. This automatic process is valuable for the forecasting and studies of solar events since it dramatically improves efficiency and accuracy.

Chapter 8 - The results of the first observations of Type IV bursts at frequencies 10-30 MHz are presented. These observations were carried out at radio telescopes UTR-2 (Kharkov, Ukraine) and URAN-2 (Poltava, Ukraine) during the period 2003-2006. Detection of Type IV bursts in wide band from 10 to 30MHz with high sensitivity and time resolution allowed to study their properties in details. These bursts have fluxes 10-2000s.f.u. at maximum phase. Their durations are about 1-2 hours and even more. Some of Type IV bursts drift from high to low frequencies with drift rates about 10kHz/s. All observed Type IV bursts have fine structures in the form of sub-bursts with durations from 2s to 20s and frequency drift rates in a majority of 1-2MHz/s. In most cases, sub-bursts with negative drift rates were registered. Sometimes sub-bursts in absorption with durations 10-200s against Type IV burst background have been observed. The Type IV burst observed on July 22, 2004 had zebra structure, in which single zebra stripes had positive, negative and infinite drift rates.

Chapter 9 - If you imagine the Earth's atmosphere without insolation, you may see a quiet, dark and cold atmosphere with small atmospheric fluctuations due to the Earth's rotation and gravitational forcing by the Moon, Sun, and planets. There is no doubt that the electromagnetic radiation from the Sun is the main driver of the climate, the circulation, and the wave flux of the Earth's atmosphere.

In the first part of the authors' study, the authors investigate how the annual variation of the total solar irradiance (TSI at the position of the Earth) influences the global mean temperature of the lower troposphere. The annual variation of TSI is caused by the changing Sun-Earth distance during the year, and its amplitude amounts 45.7 W/m^2. The annual variation of TSI is much larger than the other variations of TSI, e.g., the 11-year solar cycle corresponds to an amplitude of around 1 W/m^2. The authors estimate the impact of the annual variation of TSI on the time series of global mean temperature (from NCEP/NCAR reanalysis). Additionally the authors derive the influence of the inter-hemispherical asymmetry of the Earth on the time series of the global mean temperature. Since the Northern Hemisphere (NH) has more land surfaces and mountains than the Southern Hemisphere (SH), the characteristics of wind circulation, water vapor convection, atmospheric waves, concentration of greenhouse gases, oceanic currents, surface albedo, etc. are different and lead to a different thermal equilibrium for the NH than for the SH. Particularly the warmest sea surface temperatures are observed in the NH which can be explained by complex ocean-atmosphere interactions depending on surface wind, cloud cover, coastal geometry of the continents, the depth of the oceanic thermocline, and other factors.

The time series of the global and hemispheric mean temperature, humidity, and geopotential height of the lower troposphere reveal a linear, positive trend which is mainly caused by the strong increase of anthropogenic emissions of greenhouse gases such as CO_2 The feedback of atmospheric water vapor to the increase of temperature is derived from the time series of the time interval 1996-2006.

In the second part of the authors' study, the relationship between the active Sun evidenced by solar flares, solar wind speed, and geomagnetic activity, and tropospheric dynamics is investigated. Strong solar flares are often associated with interplanetary coronal mass ejections (ICME) which travel at high speeds through the heliosphere and cause severe disturbances of the Earth's magnetosphere, ionosphere, and upper atmosphere. The strongest flares of the solar cycle-23 are selected, and a superposed epoch analysis is performed by using the time series of solar X-ray flux, solar wind speed, planetary geomagnetic activity A_p, and geopotential height. The typical sequence of the processes is: 1) sun-spot darkening in TSI, 2) X-ray burst and faculae brightening in EUV/UV, 3) ICME arrival at the Earth and enhanced geomagnetic activity.

Chapter 10 - The lightning from thundercloud toward the ground has been widely studied. It has been found, however, that more elusive lightning forms, *e.g.*, dubbed elves, sprites, and blue jets, flit above the thunderclouds. Recently, giant electric jets, which link the top of thunderclouds to the lower ionosphere, have been observed. These events are a new important factor in the model of the earth's electrical and chemical environment. However, the observations of giant electric jets are not enough to understand the factors causing the formation and propagation of the events because no associated cloud-to-ground (CG) lightning discharges were detected in the underlying thunderstorms. In this article, the authors use the numerical simulation method to study the formation and propagation of giant electric jets. The initiating conditions of streamer formation in laboratory situations are extended to construct a criterion for the formation of giant electric jets. The Monte Carlo technique is used to perform the microscopic estimation of the growth of electrons produced in giant electric jets. A randomly stepped three-dimensional (3D) model is developed to simulate the propagation of giant electric jets. The simulation presents a 3D overall picture for the well-developed giant electric jets, which agrees well with the observations.

Chapter 11 - Based on high resolution observations the authors studied the photometric behavior and morphology of solar fine structures under sunspots influence.

The pictures are white-light photographs taken with a 50-cm aperture and 120-m focal length reflector located in Tadjikistan at the Pamir Mountain Station (spatial resolution 0.3 arcsec). The photographic film characteristics are: 120-mm format, 3 DIN sensibility and a resolution of 120 lines per millimeter. From the film and filter spectra combination a bandwidth of 40 nm centered at λ 650 nm was obtained. The exposure time was 1.5 ms. The photoheliograms shows the sunspot group SD 284/78 (Solnechnye Dannye Classification). The films were scanned with an Arcus 1200 – AGFA scanner at 300 d.p.i. and 256 intensity levels. The authors have obtained a standardized sequence of images to characterize the spatial-temporal behavior of some umbra details of the main sunspot and surrounding granulation. Basic elements of the standardization methodology, and the morphologic and photometric results obtained are presented.

The bright granules spatial distribution and the granules surface density as function of distance from the center of the main sunspot umbra were determined. Bright granules distribute delimiting cells of dimensions in the mesogranular scale. The mean diameter of these cells does not show significant variation with the variation of the magnetic field of the sunspot. Moreover, the granules surface density does not show significant variation with distance to the sunspot umbra. Both results point to a very weak, if any, influence of the sunspot magnetic field at distances greater than 20 arcsec.

Umbral dots and nuclei were detected and their spatial distributions were analyzed. Umbral dot diameter distribution was obtained and discussed.

The authors conduced a threshold process at levels defined by a method based on intensity cumulative histogram to determine the areas occupied by the different photometric elements. The temporal behavior of the measured quantities and their relations were obtained. The authors found a relation between the area of the umbral nucleus and the area of the umbra (K_{nu}) and between the area of the penumbra and area of the umbra (K). Following the statement of some researchers linking the K index to the magnetic strength; the relation between the areas of the umbral nucleus and the area of the umbra means that spots with magnetic field intensity less than a limiting value should not present umbral nucleus.

Chapter 12 - Satellite experiments convincingly showed opportunity of penetration of radiowaves of VLF and LF- signals radiated by ground transmitters and earthquakes above F2-layer of the ionosphere (e.g. French satellite "DEMETR"). Up today the Appleton-Lassen-Hartree formula is used for calculation of dielectric permeability of ionospheric plasma. It is impossible to explain from this formula the propagation of radiowaves of a frequency range below the electron gyrofrequency from the earth surface up to heights above F2-layer of the ionosphere. The purpose of the work is to develop the theory of propagation of radiowaves in the magnetised ionospheric plasma, which explains the penetration VLF and LF- signals above F2-layer of the ionosphere. The problem was resolved in a common case, when the dependence between a vector of an electric displacement and an electric intensity is not linear (in a contrast to the solution of this problem earlier). New formula was developed for dielectric permeability of the ionospheric plasma. The authors also develop the formulae to calculate the critical frequencies of F2-layer of the ionosphere. The result of the frequency calculations with helping of the new formulae, practically, consists with the old ones, when a frequency of a radio wave is higher than the electron gyrofrequency. Thus, it will be difficult to show experimentally advantage of the new or old formulas. At the same time the authors have found conditions, when a radio wave has a frequency below than the electron gyrofrequency and can propagate upward to the altitudes above F2-layer of the ionosphere. The developed formulae can be also new theoretical basis of ray tracing, where the radio wave is used to probe the structure of the ionosphere; and in radio navigational systems and in radio geodesy, as well as in communication-system prediction and management. It is particular useful when detailed knowledge about the interaction of the radio wave and ionosphere is required, such as in over-the-horizon radar and in single location of distant radio transmitters.

EXPERT COMMENTARY

In: Solar Physics Research Trends
Editor: Pingzhi Wang, pp. 3-13

ISBN: 978-1-60021-987-0
© 2008 Nova Science Publishers, Inc.

THE QUASI-BIENNIAL OSCILLATION IN TIME SERIES OF SOLAR ACTIVITY PARAMETERS

Ana G. Elias and Marta Zossi de Artigas

Consejo Nacional de Investigaciones Científicas y Técnicas, CONICET, Argentina
Departamento de Física, Facultad de Ciencias Exactas y Tecnología, Universidad Nacional de Tucumán, Argentina

Abstract

Time series of solar activity parameters, such as the sunspot number Rz and the 10.7cm solar flux F10.7, present an asymmetric quasi-decadal oscillation as their main variation. In fact, the ascending phase of the solar activity cycle is faster than the descending phase. In certain studies, a filtering process is applied to these time series in order to eliminate this main oscillation. As a consequence, a spurius quasi-biennial cycle may appear, which is the result of the filtering process. The quasi-biennial cycle thus obtained may be similar to the well-known quasi-biennial oscillation (QBO) of the stratospheric equatorial zonal winds, but with no physical association. A study is made of the character of this "induced" quasi-biennial cycle. The results shown here do not rule out a physical origin, but point out that a result obtained after a statistical analysis carries, in addition to the physics behind, the spurious byproducts of the method applied.

Introduction

Quasi-biennial oscillations were revealed in solar activity at the beginning of the 20th century by Schuster (1906) who detected a period of 2.69 years in sunspot number but considered it "in all probability only sub-period of the 11 years' variation", and then by Shapiro and Ward (1962). This initially led to a controversy about whether solar forcing could be responsible for the quasi-biennial oscillation (QBO) in stratospheric equatorial zonal winds. Finally, Lindzen and Holton (1968) successfully explained the key features of the QBO by invoking the mean flow effects of gravity waves forced in the troposphere, propagating into the stratosphere.

In the last two decades several authors have studied the interaction between the QBO in stratospheric equatorial winds and the solar activity cycle (Varotsos, 1989; Kodera, 1993;

Labitzke and van Loon, 1993; Soukharev, 1999; Abarca del Rio at al., 2003; Labitzke, 2005). The methods used in these studies consist mainly in grouping data according to the QBO phases or in running correlations. Regarding the latter Kodera (1993) estimated the running correlation of the QBO with the North Polar stratospheric temperature, Soukharev (1997, 1999), with total ozone content over Northeastern European stations, and Salby and Callaghan (2004) with wintertime tendency. They all obtained running correlation coefficients with a clear quasi-decadal modulation statistically related to the 11-year solar activity cycle.

Troshichev and Gabis (1998), Soukharev and Hood (2001) and Gabis and Troshichev (2004) showed that the UV irradiance undergoes quasi-biennial periodicity, which may have an effect on the stratospheric QBO. After this idea, Elias and Zossi de Artigas (2003) and Zossi de Artigas and Elias (2005) estimated the running correlation between the equatorial stratospheric QBO and the quasi-biennial signal in the 10.7 cm solar radio emission, F10.7, and the quasi-biennial signal in the Dst index, finding a quasi-decadal variation in the correlation coefficient series similar to what Kodera (1993), Soukharev (1997, 1999) and Salby and Callaghan (2004) found. But, the assessed quasi-biennial variation in F10.7 and in Dst, as that of the UV irradiance calculated by Troshichev and Gabis (1998), Soukharev and Hood (2001) and Gabis and Troshichev (2004), involves a filtering process. So, care must be taken about the nature of the oscillation we obtain: is it real or spurious? Related to this subject there are statistical studies analyzing the connection between the equatorial stratospheric QBO and the solar activity cycle (Baldwin and Dunkerton, 1989; Hamilton, 1990; Salby and Shea, 1991; Salby et al., 1997, Elias and Zossi, 2006).

This chapter consists of a statistical study regarding the quasi-biennial oscillation that can be obtained after filtering series strongly linked to the solar activity cycle. Running correlations involving this oscillation, which end in results that may be consequence of the statistical tools used, are also analyzed.

Solar Activity Cycle

Solar activity undergoes, among other variations, the well known 11-year cycle. This oscillation has not a constant period over time, nor is it symmetric. Since solar cycle 1, which started in 1755 and ended in 1766, the solar cycle length has varied between 9 and 14 years. In addition, each cycle rises to maximum activity levels faster than it falls to minimum levels.

There are many time series and parameters linked to solar activity that clearly show the 11-year cycle. Among them, the most used are the sunspot number, Rz, and the 10.7 cm solar flux, F10.7, available at the World Data Center-A for Solar-Terrestrial Physics. There are also other indices such as the composite Mg II index that measures ultraviolet solar activity, the Ipe EUV solar flux index, the coronal index that measures the energy emitted by the sun's corona at a wavelength of 530.3 nm, the Calcium plage area, Lyman-α intensity that is the strong emission line of hydrogen at 121.5 nm, the equivalent width of the Helium 1083 nm solar absorption line, the total solar irradiance that describes the radiant energy emitted by the sun over all wavelengths that falls outside the Earth's atmosphere, the solar flare index that quantifies the solar flare activity, cosmic ray fluxes, and many more (all available at http://www.ngdc.noaa.gov). Geomagnetic indices, such as aa, ap, Dst and solar wind aprameters, also present the 11-year cycle.

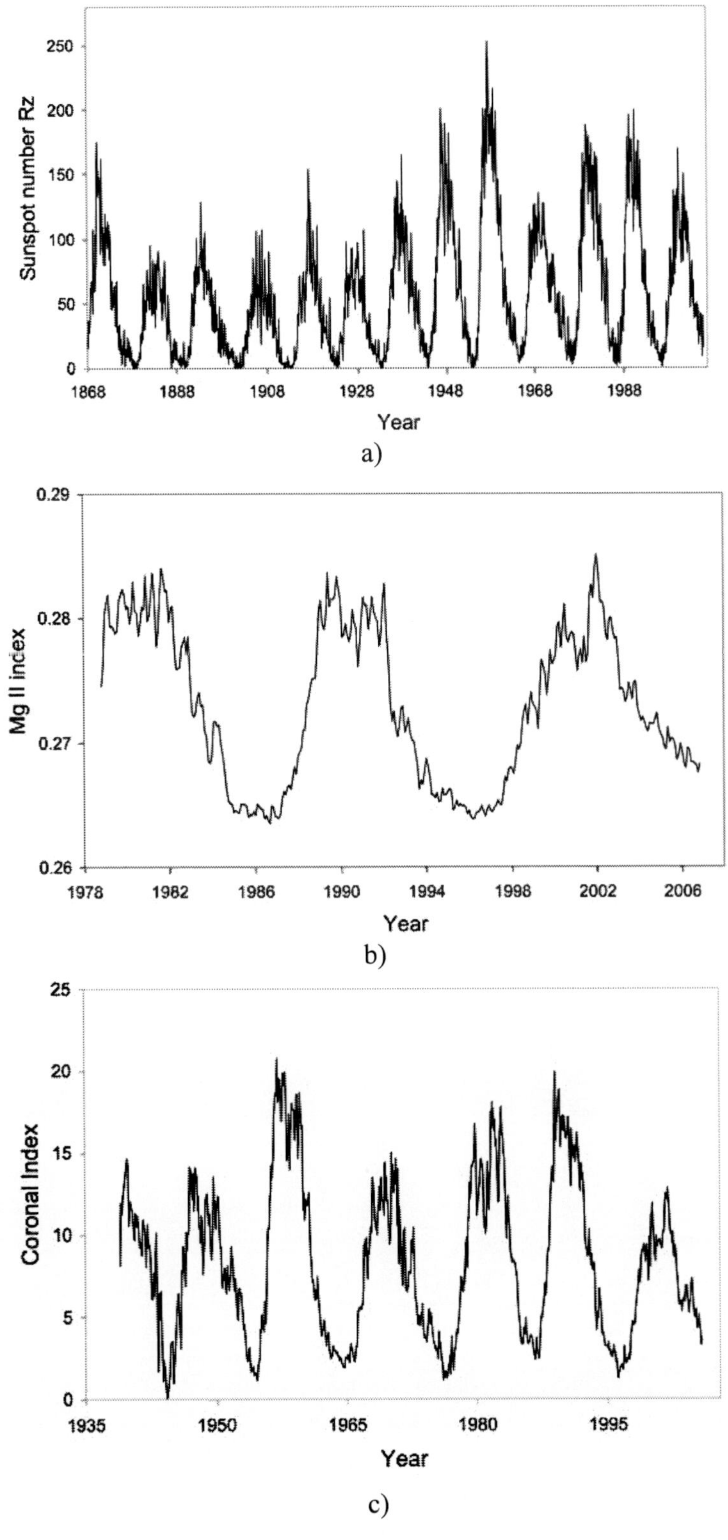

Figure 1. Monthly time series of (a) sunspot number Rz, (b) Mg II index, (d) coronal index.

Figure 1 depicts some of the mentioned time series in order to show the well-defined 11-year cycle that all of them present.

Spectrum of Periodic Assymmetric Time Series

Any periodic function y(t) can be expanded as a discrete series of sine and cosine terms, each term representing one of the harmonics of the fundamental frequency ω_o:

$$y(t) = a_0 + a_1 \cos(\omega_o t) + b_1 \sin(\omega_o t) + a_2 \cos(2\omega_o t) + b_2 \sin(2\omega_o t) + a_3 \cos(3\omega_o t) + b_3 \sin(3\omega_o t) + \ldots\ldots \quad (1)$$

For a truly sinusoidal signal with frequency ω_o, the power spectrum only has one peak at this frequency. But, for a more complex periodic function, harmonics of the fundamental frequency (e.g., $2\omega_o$, $3\omega_o$, $4\omega_o$, $5\omega_o$, ...) will also have non-zero amplitudes.

This is the case of time series linked to the solar activity cycle that present an asymmetric 11-year oscillation and a cycle length that is not constant, as already described. Assuming an "ideal" solar cycle with every cycle lasting 11 years, the asymmetry would produce a spectrum then with peaks not only at the fundamental period T = 11 years, but also at its sub-harmonics corresponding to 5.5 years (11/2), 3.67 years (11/3), 2.75 years (11/4), etc. Figure 2 shows the spectrum of a symmetrical "ideal" 11-year solar cycle and an asymmetrical one. The fundamental frequency corresponds to 0.09 month^{-1}. Figure 2(c) shows an enhanced portion of Figure 2(b) just to show more clearly the presence of peaks at harmonic frequencies of the fundamental one, that is at 0.18, 0.27, 0.36, etc.

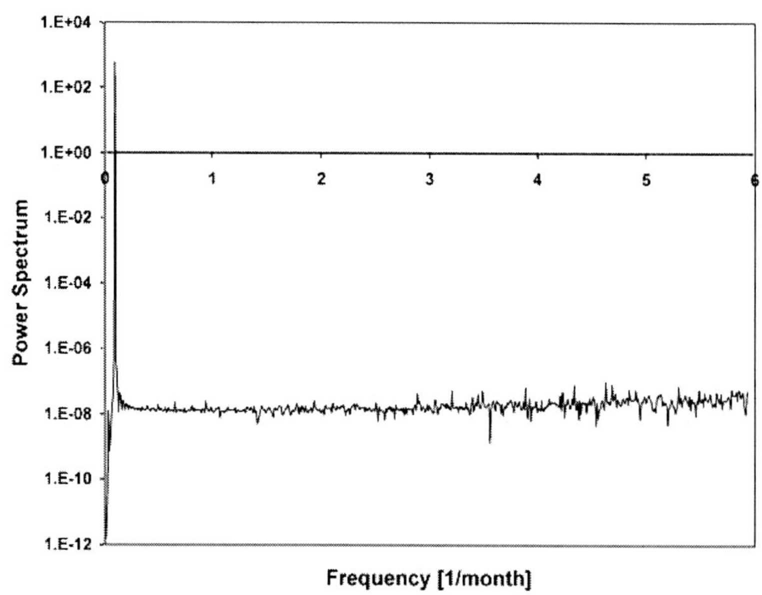

a)

Figure 2. Continued on next page.

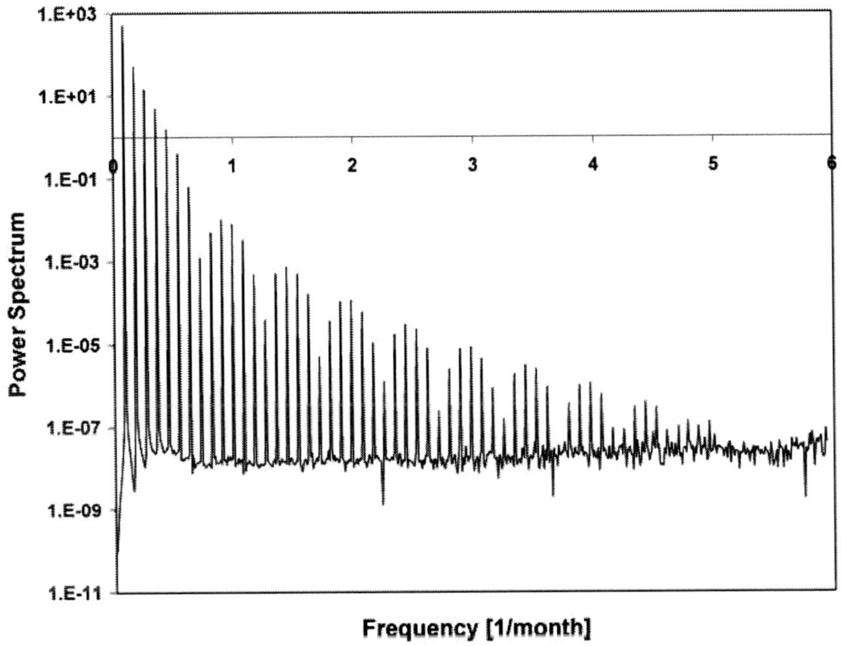

b)

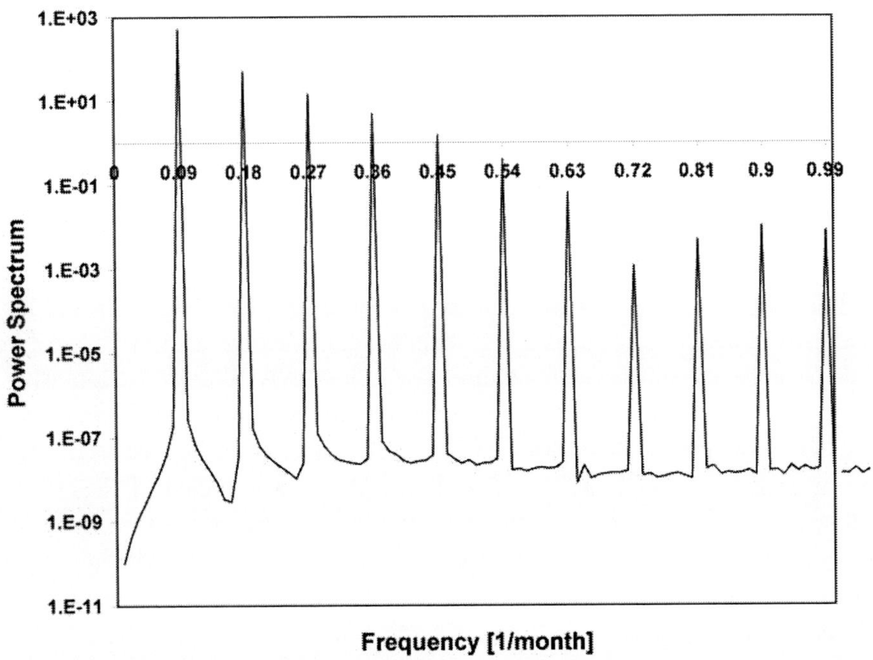

c)

Figure 2. Power spectrum of a series with an 11-year periodicity: (a) symmetric and (b) asymmetric. (c) the same as (b) but for the range 0-1 $month^{-1}$ to highlight the presence of peaks at harmonic frequencies in the asymmetric case.

Filtering Time Series of Solar Activity

In order to eliminate the 11-year cycle of time series strongly linked to solar activity, a filter is applied which usually consists of a moving average. Or, when a certain frequency band is studied, a band-pass filter assessed as a difference of filtered versions of the original time series can be used.

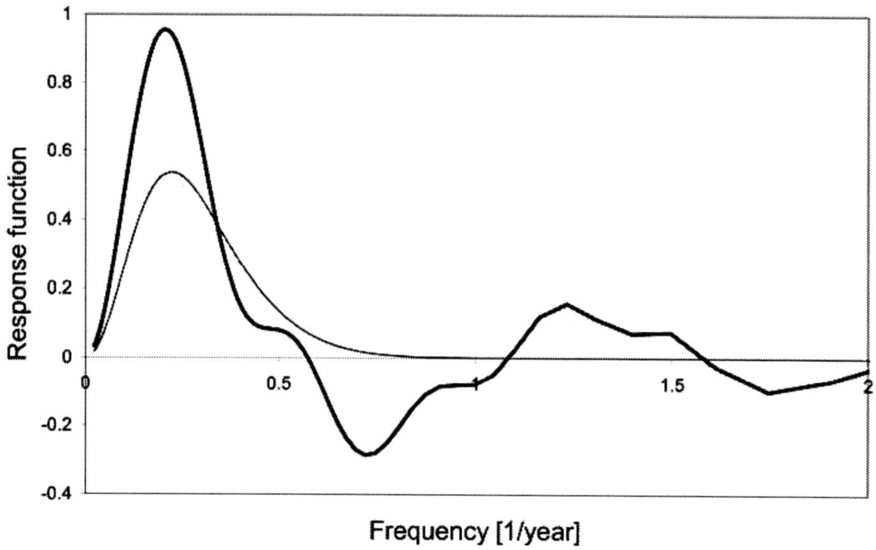

Figure 3. Response function R(ω) of the band-pass filter with frequency bounds of 1/12 month^{-1} and 1/39 month^{-1} using the running mean filter (enhanced line) and the Gaussian bell (solid line). The band-pass filter is obtained as a low-pass filter with cutoff frequency of 1/12 month^{-1} minus a low-pass filter with cutoff frequency of 1/39 month^{-1}.

To analyze a quasi-biennial signal in a time series, frequencies lower and higher than those corresponding to a certain range containing the quasi-biennial cycle are filtered out. This can be done by subtracting the time series low-pass filtered version with a cutoff frequency of 1/12 month^{-1}, for example, from another one with cutoff frequency at 1/39 month^{-1}. The response function R(ω) corresponds to the difference between the response function of each low-pass filter. Figure 3 shows R(ω) for two kind of filters: the running mean filter and the Gaussian bell. Although this function presents a slow drop off (Gibbs' effect) the filter is still useful since the variance for $\omega T>\pi$ is small. Although the Gaussian bell filter presents a more desirable response function, the running mean filter is preferred due to its simplicity and easy application.

The combined effect of the two low-pass filters used to band-pass filter is to emphasize certain of the intermediate frequencies which shows up in the peak of the response function, which in this example is around 0.2 year^{-1}. This effect is known as the Yule-Slutsky effect (Harvey, 1993). In the case of the asymmetrical solar cycle, the filtering process reduce the amplitude of the 11-year peak and enhances those corresponding to 5.5, 3.67 and 2.75 years.

Decadal Variation in Running Correlation Coefficients: A Spurious Consequence of a QBO in Solar Activity Parameters?

We will present now a consequence of using a filtered solar activtiy time series in a running correaltion analysis.

The correlation coefficient, r, between two time series y(t) and u(t) can be estimated as

$$r = \frac{\sum_{t=1}^{N}[y(t)-\bar{y}]\,[u(t)-\bar{u}]}{\sqrt{\sum_{t=1}^{N}[y(t)-\bar{y}]^2 \sum_{t=1}^{N}[u(t)-\bar{u}]^2}} \tag{2}$$

where N is the number of data of each series and, $\bar{y}$ and $\bar{u}$ are the mean of y(t) and u(t) respectively. When a running correlation is assessed, then r becomes time dependent:

$$r(t) = \frac{\sum_{t_i=t-m}^{t+m}[y(t_i)-\bar{y}(t)]\,[u(t_i)-\bar{u}(t)]}{\sqrt{\sum_{t_i=t-m}^{t+m}[y(t_i)-\bar{y}(t)]^2 \sum_{t_i=t-m}^{t+m}[u(t_i)-\bar{u}(t)]^2}} \tag{3}$$

where 2m+1=M is the window width and the mean values of y(t) and u(t), which like r, are also time series now, correspond to:

$$\bar{y}(t) = \frac{1}{M}\sum_{t_i=t-m}^{t+m} y(t_i) \quad \text{and} \quad \bar{u}(t) = \frac{1}{M}\sum_{t_i=t-m}^{t+m} u(t_i)$$

If y(t) and u(t) are waves then the running correlation r(t) will result in a kind of beating between both of them that is a periodic function oscillating with the beat frequency. Figure 4 shows, as an example, the running correlation between series of periodicities ranging between 1.8 and 2.8 years. It can be noticed in this case that a decadal modulation is obtained when the correlated series have periodicities of 2.4 and 2 years.

An analytical expression for r in equation (3) can be obtained by replacing summations by integrals. Then:

$$r = \frac{a_1 - a_2 - 4\dfrac{a_3\, a_4}{m\omega_2\,\omega_1}}{\sqrt{\dfrac{1}{\omega_2\,\omega_1}\left(a_5 - 4\dfrac{a_3^2}{m\omega_2^2}\right)\left(a_6 - 4\dfrac{a_4^2}{m\omega_1^{\,2}}\right)}} \tag{4}$$

where

$$a_1 = \frac{\sin[(\omega_2 - \omega_1)(t + m)] - \sin[(\omega_2 - \omega_1)(t - m)]}{2(\omega_2 - \omega_1)}$$

$$a_2 = \frac{\sin[(\omega_2 + \omega_1)(t + m)] - \sin[(\omega_2 + \omega_1)(t - m)]}{2(\omega_2 + \omega_1)}$$

$a_3 = \sin(\omega_2 t)\sin(\omega_2 m)$
$a_4 = \sin(\omega_1 t)\sin(\omega_1 m)$
$a_5 = \omega_2 m - [\sin(2\omega_2 t)\cos(2\omega_2 m)]/2$
$a_6 = \omega_1 m - [\sin(2\omega_1 t)\cos(2\omega_1 m)]/2$

In this way r is an oscillating function in terms of ω_1, ω_2, t and m. As already mentioned, for a given ω_1 there is a lower and a higher frequency ω_2 for which a quasi-decadal oscillation in r can be obtained.

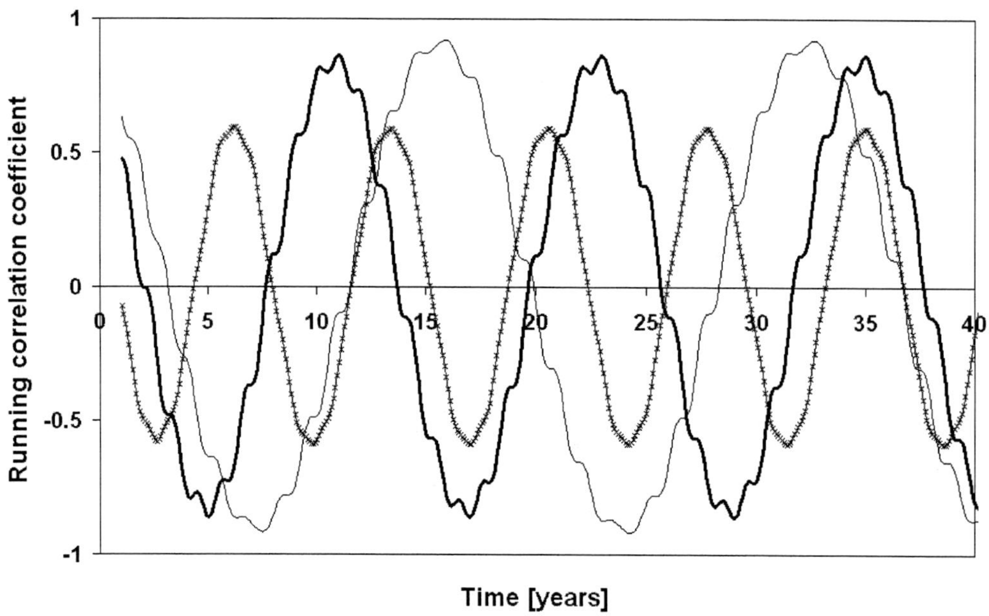

Figure 4. Running correlation with a 4-year window width between monthly sine functions of periods 2.4 and 2.8 years (black thin line), 2.4 and 2 years (black enhanced line) and 2.2 and 1.8 years (black line with crosses).

In the case of time series linked to the solar activity cycle there are sub-harmonics of the 11-year cycle with periods around 2 and 3 years that are enhanced after a filtering process. In a running correlation with the stratosphere QBO, these sub-harmonics lead to a quasi-decadal oscillation in the running correlation coefficient series as deduced from equation (4).

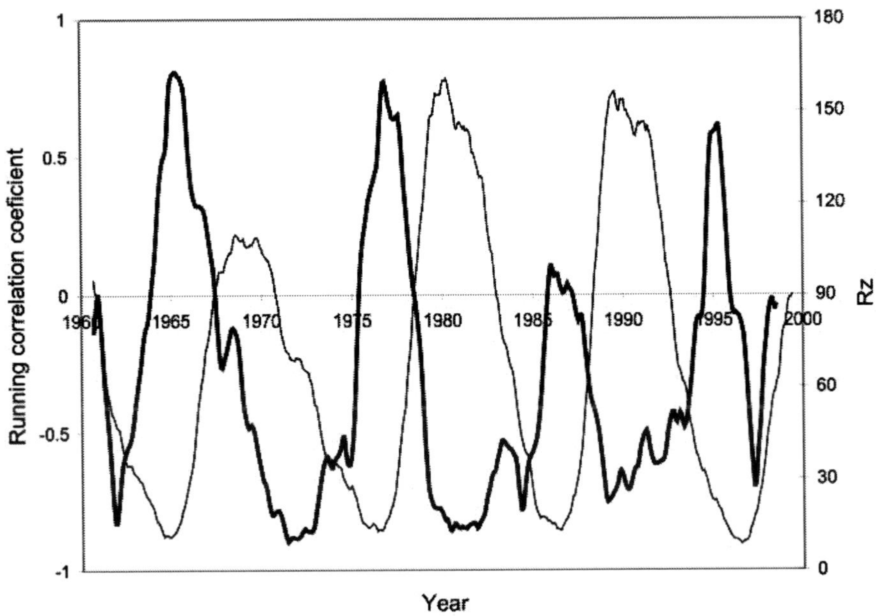

Figure 5. Running correlation coefficient time series with a 3-year window width between F10.7 band pass filtered with frequency bounds corresponding to periods of 12 and 39 months, and the mean zonal equatorial wind at 30 hPa (enhanced line). Solar activity cycle depicted by the sunspot number, Rz, 12-month running mean (thin line).

Figure 5 shows as an example the running correlation between F10.7 band pass filtered with frequency bounds corresponding to periods of 12 and 39 months, and the mean zonal equatorial wind at 30 hPa (taken from a compilation by Naujokat (1986) available at http://dss.ucar.edu/cdroms/karin_labitzke_strat_grids/data). A quasi-decadal oscillation can be noticed with maximum negative correlation around solar maximum and maximum positive correlation at solar minimum. This could mean that during maximum solar activity, higher UV levels occur during the easterly phase of the QBO, and lower UV levels during the westerly phase. During minimum solar activity, this relationship is reversed. Assuming that there is a physical connection between solar variability and equatorial zonal wind on the 25–30 month time-scale, a key point here would be: How can the transition from solar minimum to solar maximum change the sign of the association between them?

One can argue about the statistical significance of these kind of results. In the case of a computed sample correlation coefficient for a sequence of N observations the statistical significance can be computed through the t-statistc with a t-Student distribution with N-2 degrees of freedom:

$$t = r\sqrt{\frac{N-2}{1-r^2}} \tag{5}$$

In our case a significance of 95% and 99% is obtained when r is greater than 0.25 and 0.37 respectively. For this test, the observations are assumed to be independent of one another. But observations in a time series are seldom independent. Autocorrelation essentially reduces the number of independent observations and an "effective" sample size should be used. An equation of the effective sample size for computation, N', of a correlation coefficient between two autocorrelated time series is given by $N' = N (1 - r_{1y} r_{1u}) / (1 + r_{1y} r_{1u})$ (Ezekiel and Fox, 1959), where r_{1y} and r_{1u} are the first-order autocorrelation coefficients of time series y and u. In this case, a significance of 95% and 99% is obtained when r is greater than 0.32 and 0.5 respectively.

Another consideration that should be taken into account is that fluctuations in running correlations between two correlated time series are sometimes over-interpreted. Windowed correlations are expected to wander to some extent purely as a result of stochastic behavior. Gershunov et al. (2001) emphasize this point, and suggest that bootstrap methods be applied to evaluate significance of fluctuations in windowed correlation. In the case of Figure 5, considering that r fluctuates around cero, the bootstrapped confidence limits at a 95% significant level are $\pm$ 0.5, which means that, at a 95% confidence level, the running correlation obtained may be more variable on decadal timescales than should be expected from sampling variability alone. However, this result do not hold on for a 99% significance level.

Conclusion

In solar activity proxies or time series strongly linked to solar activity cycle, quasi-biennial oscillations may appear as a result of the filtering process applied in order to enhance this kind of variation. The nature of this oscillation should be analyzed in detail since it may lead to results of statistical character only. This is probably the case of the running correlation assessed between solar activity filtered series and the stratosphere equatorial zonal wind QBO, where the decadal modulation of the running correlation coefficients may be obtained due to a beating between the frequencies involved.

Our results seem to lead as back to the results obtained by Schuster in 1906. However, it should be taken into account the pure statistical character of the present study and the idealization of some of the series considered.

References

Abarca del Rio, R., Gambis, D., Salstein, D., Nelson, P., Dai, A., Solar activity and earth rotation variability. *Journal of Geodynamics*. 2003, 36, 423-443.

Baldwin, M.P., Dunkerton, T.J., Observations and statistical simulations of a proposed solar cycle/QBO/weather relationship. *Geophys. Res. Let.* 1989, 16, 863-866.

Elias, A.G., Zossi de Artigas, M., A search for an association between the equatorial stratospheric QBO and solar UV irradiance. *Geophys. Res. Let.* 2003, 30, 1841, doi:10.1029/2003GL017771.

Elias, A.G., Zossi de Artigas, M.,The quasi-decadal modulation of running correlations involving the QBO. *J. Atmos. Sol. Terr. Phys.* 2006, 68, 1980-1986.

Ezekiel, M., Fox, K.A., Methods of Correlation and Regression Analysis, John Wiley and Sons, New York, United States, 1959, 548 pp.

Gabis, J., Troshichev, O., Influence of solar UV irradiance on quasi-biennial oscillations in the Earth's atmosphere. *Adv. Space Res.* 2004, 34, 355-360.

Gershunov A., Schneider N., Barnett T., Low-frequency modulation of the ENSO–Indian Monsoon rainfall relationship: signal or noise?. *J. Climate*. 2001, 14, 2486-2492.

Hamilton, K., A look at the recently proposed solar-QBO-weather relationship. *J. Climate*. 1990, 3, 497-503.

Harvey, A.C., *Time Series Models,* Harvester Wheatsheaf, London, Great Britain, 1993, 308 pp.

Kodera, K., Quasi-decadal modulation of the influence of the equatorial quasi-biennial oscillation on the north polar stratosphere temperatures. *J. Geophys. Res.* 1993, 98, 7245-7250.

Labitzke, K., Van Loon, H., Some recent studies of probable connections between solar and atmospheric variability. *Annales Geophysicae*. 1993, 11, 1084-1094.

Labitzke, K., On the solar cycle-QBO relationship: a summary. *J. Atmos. Sol. Terr. Phys.* 2005, 67, 45-54.

Lindzen, R. S., and J. R. Holton, A theory of the quasi-biennial oscillation. *J. Atmos. Sci.* 1968, 25, 1095–1107.

Naujokat, B., An Update of the Observed Quasi-Biennial Oscillation of the Stratospheric Winds over the Tropics. *J. Atmos. Sci.* 1986, 43, 1873 1877.

Salby, M., Shea, D., Correlations between solar activity and the atmosphere: An unphysical explanation. *J. Geophys. Res.* 1991, 96, 22579-22595.

Salby, M., Callaghan, P., Shea, D., Interdependence of the tropical and extratropical QBO: Relationship to the solar cycle versus a biennial oscillation in the stratosphere. *J. Geophys. Res.* 1997, 102, 29789-29798.

Salby, M., Callaghan, P., Evidence of the solar cycle in the general circulation of the stratosphere. *J. Climate*. 2004, 17, 34-46.

Schuster, A., On Sun-spot Periodicities-Preliminary Notice. *Proc. Royal Soc. London*. 1906, 77, 141– 145.

Shapiro, R., and F. Ward, A Neglected Cycle in Sunspot Numbers?. *J. Atmos. Sci.* 1962, 19, 506– 507.

Soukharev, B., The sunspot cycle, the QBO, and the total ozone over Northeastern Europe: a connection through the dynamics of stratospheric circulation. *Annales Geophysicae*. 1997, 15, 1595-1603.

Soukharev, B., On the solar/QBO effect on the interannual variability of total ozone and the stratospheric circulation over Northern Europe. *J. Atmos. Solar Terr. Phys.* 1999, 61, 1093-1109.

Soukharev, B., Hood, L.L., Possible solar modulation of the equatorial quasi-biennial oscillation: Additional statistical evidence. *J. Geophys. Res.* 2001, 106, 14855-14868.

Troshichev, O. A., Gabis, I.P., Variations of solar UV irradiance related to short-term and medium-term changes of solar activity. *J. Geophys. Res.* 1998, 103, 20659-20667.

Varotsos, C., Comment on connections between the 11-year solar cycle, the QBO and total ozone. *J. Atmos. Sol. Terr. Phys.* 1989, 51, 367-370.

Zossi de Artigas, M. and A.G. Elias, The equatorial stratospheric QBO and geomagnetic activity. *J. Atmos. Sol. Terr. Phys.* 2005, 67, 1280-1286.

SHORT COMMUNICATION

In: Solar Physics Research Trends
Editor: Pingzhi Wang, pp. 17-27

ISBN: 978-1-60021-987-0
© 2008 Nova Science Publishers, Inc.

REMOTE RADIO SOUNDING OF PERTURBED PLASMA FORMATIONS AT SMALL DISTANCES FROM THE SUN WITH A SPACECRAFT

***A. N. Afanasiev*[1,a] *and N. T. Afanasiev*[2,b]**

[1]Institute of Solar-Terrestrial Physics, Siberian Division of Russian Academy of Sciences, 126 Lermontov St., Irkutsk 664033, Russia

[2]Irkutsk State University, 20 Gagarin Blvd., Irkutsk 664003, Russia

Abstract

This paper investigates the form of the energy spectrum of a spacecraft monochromatic radio signal propagating through the solar corona at small heliocentric distances. Energy spectra obtained during coronal radio sounding with the VENERA-16 spacecraft at the frequency of 937.5 MHz [Pavel'ev et al., 1988] are analysed. These spectra have a two-component structure, which could appear due to the reflection of radio waves from a traveling plasma formation. Based on the results obtained by Pavel'ev et al. [1988], a mathematical modeling of the energy spectrum is carried out for the case where a large-scale plasma formation with increased electron density occurs at the radio signal path. Calculations performed have shown that the plasma formation reflecting the radio waves can be represented by the leading edge of a coronal mass ejection.

Introduction

There is currently great interest in the study of the generation and propagation in the solar corona of plasma formations such as coronal mass ejections (CMEs). These phenomena can have a pronounced effect on the near-Earth space environment, leading to a deformation of the magnetosphere, causing geomagnetic storms, etc.

Information on CMEs is commonly obtained using white-light coronagraphs onboard spacecraft. Currently, such data are provided by the LASCO coronagraph onboard Solar and

[a] E-mail address: saf@iszf.irk.ru

[b] E-mail address: nta@api.isu.ru

Heliospheric Observatory (SOHO). Coronagraphs provide two-dimensional images of CMEs so that it is possible to determine directly some characteristics of plasma ejections in the plane of sky, such as the angular size of an ejection in latitude, and the velocity of its elements. On the other hand, the electron density in CMEs can be determined from two-dimensional images taken at a single point only by invoking additional assumptions about the three-dimensional structure of a plasma ejection (see, e.g., Thernisien, Howard, and Vourlidas [2006]). Furthermore, it is difficult using observational data from coronagraphs, to derive information on electron density fluctuations in CMEs.

One further opportunity to investigate CMEs can be furnished by observations of radio emission from distant cosmic sources through the solar wind [Lynch, Coles, and Sheely, 2002] as well as by experiments on remote sounding of the corona and solar wind using radio signals from spacecraft [Dolbezhev et al., 1986; Woo, 1988; Efimov et al., 1990; Bird et al., 2006], and signals from planetary radars [Harmon and Coles, 1983]. A CME traversing the radio ray path causes additional changes in the radio wave phase and amplitude with respect to changes introduced by ambient near-solar plasma. Such changes in radio emission characteristics contain information on CME parameters as well as on its fine turbulent structure.

Experiments showed that in the case of sounding coronal plasma using monochromatic radio emission from spacecraft at the passage of CME there arises an additional regular frequency Doppler shift [Dolbezhev et al., 1986; Efimov et al., 1990; Efimov and Rudash, 2002], with an increase of the frequency fluctuation level [Bird et al., 2006]. By analyzing the frequency fluctuations of the radio signal from the GALILEO spacecraft, Bird et al. (2006) the spectral index of turbulent inhomogeneities in a CME as well as the CME's velocity was determined. In this case the passage of the plasma ejection was recorded when the radio ray path proximate point was at about 54 solar radii from the Sun.

On the other hand, in the study of the generation mechanisms of CMEs it is of interest to measure their characteristics at small heliocentric distances. With this aim in view, it is appropriate to use, in addition to the other characteristics of monochromatic radio emission, the energy spectrum, as it contains information not only on electron density fluctuations in a CME but also on its regular properties (mean electron density in particular).

In this paper we investigate the form of the energy spectrum for the case where a coronal plasma ejection traverses the path of signal propagation. We carry out a numerical simulation of the energy spectrum, using the theoretical expressions reported by Afanasiev [2005], and observational data on sounding the corona from the VENERA-16 spacecraft [Pavel'ev et al., 1988]. We discuss the possibility of using the energy spectrum of the sounding signal in probing CMEs.

1. The Technique for Calculating the Energy Spectrum

Let us consider the propagation through the corona of monochromatic radio emission at frequency f. Let the spacecraft that serves as the source of radio emission be in the ecliptic plane at a distance r_s from the Sun's center (Figure 1), and let the receiver be located at Earth.

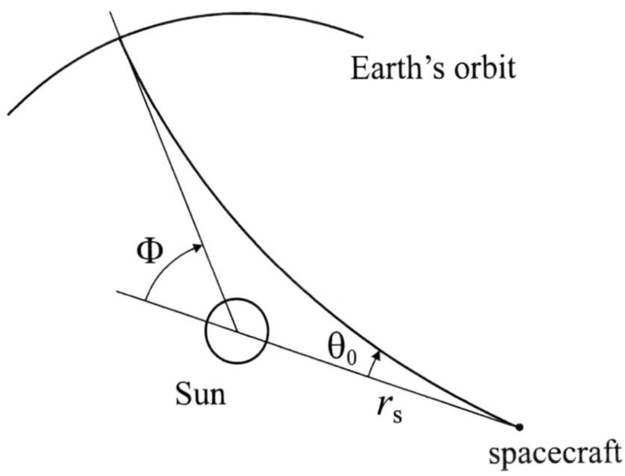

Figure 1. Regular refraction of radio waves in the solar corona.

Nonuniform distribution of mean electron density in the corona causes the radio waves to undergo regular refraction, i.e. the propagation path of radio emission becomes curved (Figure 1). The extent to which the path is curved depends on the radio emission frequency, and on the impact angle θ_0.

As a consequence of regular refraction, in the case of sounding a spherically symmetric corona, the signal arrives at the receiver following two paths: along a direct ray, and along a ray reflected from the Sun [Kucheryavenkov et al., 1985; Afanasiev, 2005]. For a spherically symmetric corona the dependence of the angle Φ on the impact angle θ_0 of the path, referred to as the distance-angular characteristic (DAC), has the qualitative form shown in Figure 2 [Afanasiev, 2005]. The DAC is inferred by solving the system of ray equations [Kravtsov and Orlov, 1980]. The presence of an extremum in the function $\Phi(\theta_0)$ indicates the formation of a regular caustic.

Radio emission propagating through the corona undergoes not only regular refraction but also scattering caused by turbulent electron density inhomogeneities in the corona. Scattering gives rise to phase and amplitude fluctuations and, hence, to a spectral line broadening (or the energy spectrum broadening).

The expression for the energy spectrum of initially monochromatic radio emission propagating through a randomly inhomogeneous solar corona was obtained by Afanasiev [2005] using the interference integral method [Orlov, 1974; Avdeev et al., 1988]. Following this method, a scalar wave field $\widetilde{U}$ in the two-dimensional case, at the point with a radius-vector $\mathbf{r}$, can be represented as the expansion in partial waves:

$$\widetilde{U}(\mathbf{r},t) = \int A(\mathbf{r},t) e^{ik\left[\Psi_0(\mathbf{r},a,t)+\widetilde{\Psi}_1(\mathbf{r},a,t)\right]} da, \tag{1}$$

where the symbol "~" stands for the random character of the quantity, Ψ_0 is a regular eikonal of a partial wave, $\widetilde{\Psi}_1$ is a random addition to the eikonal introduced by the inhomogeneities,

k is the wave number, A is the amplitude of a partial wave, t is time, and $a = r_s \sin\theta_0$ is the impact parameter. Integration in formula (1) is performed over all values of the impact parameter. It is assumed that $\left|\widetilde{\Psi}_1\right| << \Psi_0$.

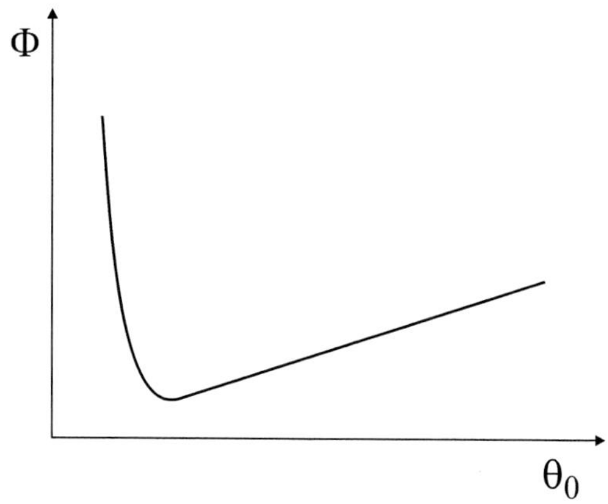

Figure 2. Distance-angular characteristic of a spherically-symmetric corona. The slightly sloping portion of the curve corresponds to the direct rays and the steep portion – to the reflected rays.

On rearrangement, for the energy spectrum

$$R(\Omega) = \frac{1}{2\pi}\int_{-\infty}^{\infty}\left\langle \widetilde{U}(\mathbf{r},t)\widetilde{U}^*(\mathbf{r},t+\tau)\right\rangle e^{i\Omega\tau}d\tau,$$

we can obtain

$$R(\Omega) = \frac{1}{k^2}\int \frac{|A(\mathbf{r},a)|^2}{Q^{1/2}(a)}\exp\left\{-\frac{\sigma_a^2(a)[\Omega - k\Psi_\omega(a)]^2 + k^2\Psi_a^2(a)\sigma_\Omega^2(a)}{2k^2Q(a)} - \right. \quad . \quad (2)$$
$$\left. -\frac{\Psi_a(a)\Gamma_{a\Omega}(a)[\Omega - k\Psi_\omega(a)]}{kQ(a)}\right\} da$$

The following designations are introduced here:

$$Q(a) = \sigma_a^2(a)\sigma_\Omega^2(a) - \Gamma_{a\Omega}^2(a),$$

$$\Psi_a = \frac{\partial\Psi_0}{\partial a},\ \Psi_\omega = \frac{\partial\Psi_0}{\partial t},\ \sigma_\Omega^2 = \left\langle\left(\frac{\partial\widetilde{\Psi}_1}{\partial t}\right)^2\right\rangle,\ \sigma_a^2 = \left\langle\left(\frac{\partial\widetilde{\Psi}_1}{\partial a}\right)^2\right\rangle,\ \Gamma_{a\Omega} = \left\langle\frac{\partial\widetilde{\Psi}_1}{\partial a}\frac{\partial\widetilde{\Psi}_1}{\partial t}\right\rangle,$$

and the angle brackets $\langle\ \rangle$ correspond to a statistical averaging over all realizations. Formula (2) holds true at a distance from the caustic as well as in its neighborhood.

For the model of a spherically symmetric corona, when the receiver is in the illuminated region (where two rays – the direct ray and the ray reflected from the Sun – arrive at each point of the Earth's orbit) at a large distance from the caustic singularity, on some rearrangement, from the expression (2) we can obtain

$$R(\Omega)=\frac{\sqrt{2\pi}}{k^2}\sum_{i=1}^{2}\frac{R_0(a_i)}{\sigma_\Omega(a_i)}\,e^{-\frac{[\Omega-k\Psi_\omega(a_i)]^2}{2k^2\sigma_\Omega^2(a_i)}}\,.$$

According to the resulting expression, the energy spectrum, to a spherically symmetric corona's approximation, consists of two components, one of which $(i=1)$ corresponds to the direct ray, and the other $(i=2)$ corresponds to the ray reflected from the Sun.

Thus, theoretically, in the case of radio sounding of the solar corona the receiver must receive two rays. However, since the energy of the reflected signal is much lower than that of the direct signal (since the reflected ray propagates through deeper regions of near-solar plasma), experiments usually recorded only the signal corresponding to the direct ray. On the other hand, there is experimental evidence [Pavel'ev et al., 1988] that the reflected signal has been recorded.

2. Experimental Results

The experiment on radio sounding of the corona for recording the rays reflected from the Sun was carried out from 6 to 23 June 1984 using the VENERA-16 spacecraft. Sounding was performed at a frequency $f=937.5\ \text{MHz}$. The data processing and results are detailed in a paper of Pavel'ev et al. [1988].

Figure 3 from Pavel'ev et al. [1988] shows the dynamics of the energy spectrum with time according to the data obtained on 19 June 1984 from 13:23 UT to 13:25 UT.

The origin along the abscissa axis in Figure 3 is combined with the frequency of the direct signal. The numbers 1-3 correspond to the observations made in a 60-sec time interval. The narrowest peak at the left corresponds to the direct signal. Its spectrum appears blurred due to radio emission scattering on traveling turbulent inhomogeneities of near-solar plasma. On the right, one can see the more blurred spectrum of the reflected signal which is displaced with the time toward higher frequencies. The arrow marks the middle of this spectrum. It is easy to see that, with the passage of time, the reflected signal is shifted in the frequency, and the width of its spectrum is decreased from 13 to 4 kHz. The power ratio in the maxima of the spectra of the reflected and direct signals is about 10 dB.

An analysis of the aforementioned experimental data for the energy spectrum made by Pavel'ev et al. [1988] showed that the model assuming a spherically symmetric distribution of mean electron density in the corona does not explain the characteristic properties of the spectrum observed. On the other hand, preliminary estimates suggested that the radio emission could be reflected from a perturbed plasma formation with increased electron density traveling in near-solar space at a distance of 3-4 R_S from the Sun (R_S is the solar radius). In this case the frequency of the radio signal that is reflected from the traveling formation, is Doppler shifted. According to the estimate made by Pavel'ev et al. [1988], the radial velocity of the reflecting region at the beginning of the observing run (Figure 3) $V_p = (650 \pm 50)$ km/s. At the end of the observation this velocity increased to $V_p = (1200 \pm 50)$ km/s. For the time of observation the reflecting region moved away from the Sun by a distance of about 10^5 km. The change in the width of the energy spectrum of the reflected signal during the observation was associated with the dynamics of statistical parameters of the plasma formation.

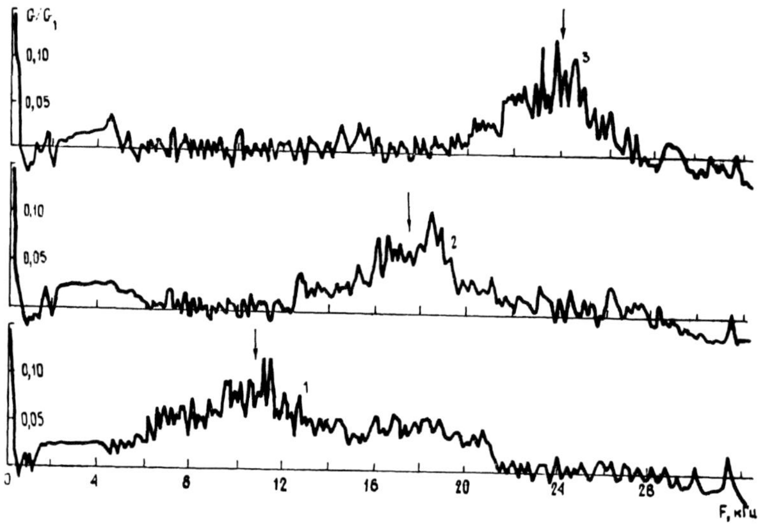

Figure 3. Energy spectra $(f = 937.5\ \text{MHz})$ from data recorded during coronal radio sounding with the VENERA-16 spacecraft [Pavel'ev, et al., 1988]. The horizontal axis indicates the frequency in kilohertz, and the vertical axis indicates the values of spectral power *G*, normalized to the power of a direct signal G_1

3. Numerical Simulation and Discussion

To verify the assumption made by Pavel'ev et al. [1988], we carried out a numerical simulation of the energy spectrum of monochromatic radio emission $(f = 937.5\ \text{MHz})$ for the case where a perturbed plasma formation with increased electron density is present at the sounding signal path. To do this, the electron density distribution in the corona was specified as

$$N_e(r,\varphi) = N_0(r) + \left[N_p(r) - N_0(r)\right]\exp\left[-\left(\frac{r - r_p}{\Delta r}\right)^m - \left(\frac{\varphi - \varphi_p}{\Delta \rho} r_p\right)^m\right], \qquad (4)$$

where (r, φ) are coordinates in a polar heliocentric coordinate system with the polar axis directed toward the point at which the source is located, the parameters r_p, and φ_p specify the coordinates of the "center" of the plasma formation, the parameters Δr, and $\Delta \rho$ specify, respectively, its longitudinal and transverse scales, and the parameter m determines the sharpness of its boundaries.

The function $N_0(r)$ describes a spherically symmetric corona. For this function we took, following Rubtsov, Yakovlev, and Efimov [1987], the relation

$$N_0(r) = A\left(\frac{R_S}{r}\right)^6 + B\left(\frac{R_S}{r}\right)^3 + C\left(\frac{R_S}{r}\right)^2, \tag{5}$$

where $A = 2.3 \times 10^8\ \text{cm}^{-3}$, $B = 8.6 \times 10^6\ \text{cm}^{-3}$, $C = 3.2 \times 10^5\ \text{cm}^{-3}$.

The function $N_p(r)$, characterizing the decrease in electron density in the plasma formation with its distance from the Sun, was specified thus:

$$N_p(r) = N_0(r_0) \cdot \left(\frac{r_0}{r}\right)^2, \tag{6}$$

where $r_0 = 1.3 R_S$. The relation (6) was considered by Jackson and Hildner [1978] in analyzing the variation in excess columnar electron density in a CME with its distance from the Sun. It should be noted, however, that for the time during which the energy spectra in Figure 2 (for 2 minutes) were obtained, the plasma formation studied here was displaced to a short distance (a few fractions of the solar radius). Therefore, of greater importance in this case is not the character of the electron density decrease in the plasma formation but the relative excess of electron density in it over the electron density in the background corona at the same heliocentric distance. In other words, the function N_p could be specified as

$$N_p(r) = \alpha N_0(r),$$

and α could be regarded as one further free parameter in the model of (4). For the relation (6) used in calculations we have $N_p(r_p)/N_0(r_p) \approx 14.5$, with $r_p = 3R_S$.

For different values of the parameters of the model (4) we calculated the DACs. One of them is shown in Figure 4. It is evident that a large-scale plasma formation with increased density gives rise to a specific curvature of the DAC. As a result, the point at Earth's orbit corresponding to the angle Φ_0, receives, in addition to the ray reflected from the Sun, a further three rays: a direct ray, and two rays reflected from the plasma formation. As a consequence of this, the energy spectrum can become multi-component and contain up to 4 components.

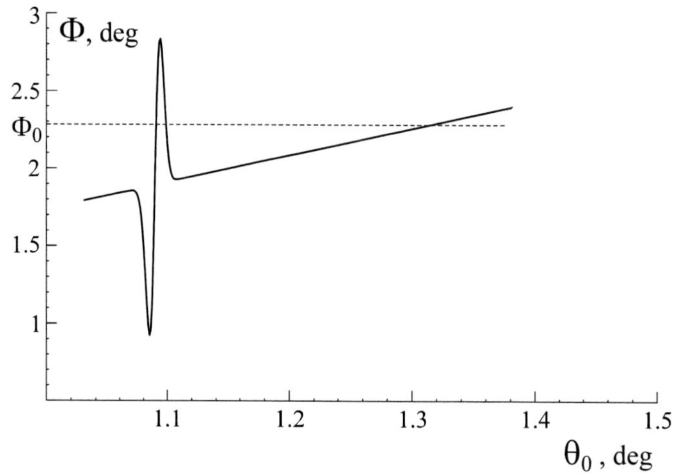

Figure 4. Distance-angular characteristics when there is a large-scale formation with increased electron density in the corona. The portion of the curve, corresponding to the rays reflected from the Sun is not shown here (compare with the Figure 2). The parameters in expression (4) have the following values: $r_p = 3R_S$, $\varphi_p = \pi/2$, $\Delta r = 10^4$ km, $\Delta\rho = 10^6$ km, and $m = 6$

Furthermore, as a result of refraction of radio emission, the plasma formation develops regular caustic singularities of the radiation field. Extrema of the function correspond to them on the DAC plot (Figure 4). The receiver at Earth can be in the neighborhood of a caustic. In this case the reflected signal received at Earth can be anomalously intense.

Calculations of the energy spectrum for the experimental data described above were carried out based on expression (2) (see also Afanasiev [2005]) for different values of random coronal inhomogeneity parameters. To describe turbulent inhomogeneities in the corona, a spatial-time correlation function of dielectric permittivity fluctuations, corresponding to a quasi-homogeneous random field was used:

$$B_0(r_1, r_2, \varphi_1, \varphi_2, t_1 - t_2) = B_m(R_c, \Phi_c) \cdot \left[1 + \left(\frac{\rho}{l}\right)^2\right]^{(n-3)/4} \frac{K_{(n-3)/2}\left(2\pi \frac{\sqrt{\rho^2 + l^2}}{L_0}\right)}{K_{(n-3)/2}\left(2\pi \frac{l}{L_0}\right)}, \quad (7)$$

$$R_c = \tfrac{1}{2}(r_1 + r_2),\ \Phi_c = \tfrac{1}{2}(\varphi_1 + \varphi_2),$$

$$\rho = \rho[r_1, r_2, \varphi_1, \varphi_2, V(t_1 - t_2)],$$

where l and L_0 are, respectively, the inner and outer scales of turbulence, n is the spectral index, $K_\nu(x)$ is the modified Bessel function of the second kind, and V is the velocity of inhomogeneities. The function B_m has the form:

$$B_m(R_c,\Phi_c)=\left(\frac{e^2}{\pi m f^2}\right)^2\sigma_N^2\cdot N_e^2(R_c,\Phi_c),$$

where e end m are the electron charge and mass, respectively, $\sigma_N^2=\left\langle\left(\Delta\widetilde{N}_e/N_e\right)^2\right\rangle$ is the variance of the relative fluctuations of electron density.

The form of the function ρ was chosen by assuming that the hypothesis about the transfer of "frozen" turbulence is valid. In calculations, the velocity V was taken to be V_p within the plasma formation (V_p is the plasma formation velocity) and V_0 in the ambient undisturbed plasma. In the same way, $\sigma_N=\sigma_p$ within the plasma formation and $\sigma_N=\sigma_0$ in the ambient undisturbed corona.

The DAC plotted in Figure 4 corresponds to the initial time of observation. Calculations showed that the DAC changes with the plasma formation crossing the path. However, when the plasma formation moves across the sounding path a small distance, multiplicity of the signal propagation paths remains.

Figure 5 presents the results derived from calculating the energy spectrum for different positions of the plasma formation relative to the sounding signal path. As in Figure 3, the vertical axis indicates the values of the spectral power normalized to the direct signal power, and the horizontal axis indicates the frequency in kilohertz. Noteworthy is the fact that as a result of the scattering of radio emission on the corona's turbulent inhomogeneities, the plasma formation-reflected signal has a one-component spectrum. The two-component structure of the reflected signal spectrum that is suggested by the form of the DAC, is smeared due to scattering effects. Note that the calculated values of the energy spectrum's width for the prescribed model parameters are consistent with the observed spectrum's widths. However, the significant broadening of the reflected signal spectrum could also be associated with effects of diffraction scattering on small-scale inhomogeneities of the plasma formation which are neglected by the method used. Therefore, the values of random inhomogeneity parameters for the plasma formation can be different from those specified in a simulation. Moreover, when compared with Figure 3, the calculated values of power of the signals reflected from the plasma formation are less than those observed. Nevertheless, the energy spectrum maximum of the reflected signal can increase with the Earth's position in orbit due to fact that the point of observation coincides with a caustic. The real electron density distribution in the corona might be different from the accepted model and the intense reflected signal at the bottom panel in Figure 3 might be due to focusing of the radio emission on the caustic.

Calculations of the DACs and of the energy spectra based on the observational data described above generally show that in order for the rays reflected from some plasma formation with increased electron density to appear, this formation must be narrow in a radial variable $(\Delta r << r_p, \Delta r << \Delta\rho)$, and the density gradient in it must be sufficiently large (large values of m are required). It may be suggested that the plasma formation reflecting the radio waves is represented by the leading edge of a CME which is a typical element of many CMEs not infrequently observed with white-light coronagraphs (see, e.g., Krall et al. [2001]).

Unfortunately, the observations with the white-light coronagraph onboard the Solar Maximum Mission (which provided images of the solar corona in 1984) were discontinued at the time when an intense reflected signal from the VENERA-16 spacecraft was recorded (see CME catalogue at http://mlso.hao.ucar.edu/smm).

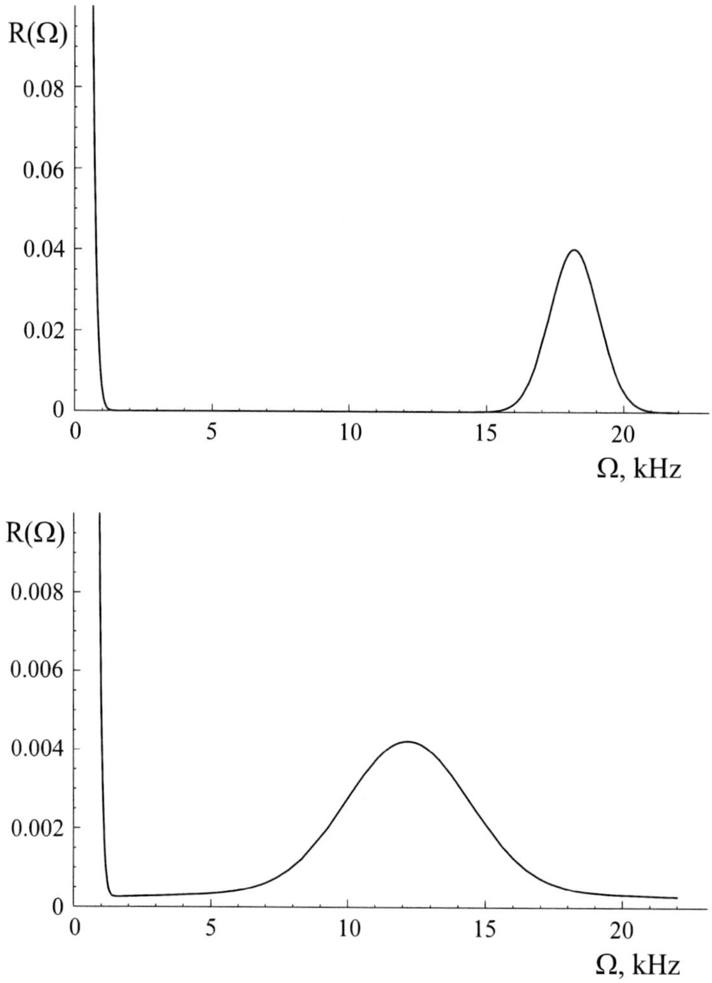

Figure 5. Calculated energy spectra $(f = 937.5\,\text{MHz})$. The parameters of plasma formation have the following values: *bottom panel* – $r_p = 3R_S$, $V_p = 650$ km/s, $\sigma_p = 6\%$; *top panel* – $r_p = 3.15R_S$, $V_p = 1000$ km/s, $\sigma_p = 3.8\%$. The other parameters of model (4) have the same values as in Figure 4. The parameters of the undisturbed corona have the following values: $V_0 = 200$ km/s, $\sigma_0 = 10\%$. The other parameters of model (7) are: $l = 10$ km, $L_0 = 10^4$ km, $n = 11/3$.

Conclusion

We have demonstrated that the energy spectrum of monochromatic radio emission from a spacecraft can be used in probing structural elements of CMEs. If additional components appear in the energy spectrum, then they can be used to estimate a number of characteristics of the reflecting region: the velocity of movement, electron density, as well as the characteristic sizes. Furthermore, when decimetric waves are used in sounding, this method can be used profitably for small heliocentric distances (of about a few solar radii) where electron density in plasma perturbations is high enough for the radio waves to be reflected.

References

Afanasiev, A. N. J. *Atm. Solar-Terrest. Phys.* 2005, 67, 1002-1013.

Avdeev, V. B.; Demin, A. V.; Kravtsov, Yu. A.; Tinin, M. V., Yarygin, A. P. *Radiophys. Quant. Electr.* 1988, 31, 907-921.

Bird, M. K.; Efimov, A. I.; Rudash, V. K.; Samoznaev, L. N.; Chashei, I. V.; Plettemeier, D. *Proc. 36th COSPAR Scientific Assembly*. Held 16-23 July 2006, Beijing, China, 2788-2795.

Dolbezhev, G G.; Efimov, A. I.; Tikhonov, V. F.; Yakovlev, O.I. *Radiotekhnika i electronika*. 1986, 31, 354-360.

Efimov, A. I.; Rudash, V. K. *Zarubezhnaja radioelectronika*. 2002, 9, 28-30.

Efimov, A. I.; Yakovlev, O. I.; Vyshlov, A. C.; Nabatov, A. S.; Laptev, A. S. *Radiophys. Quant. Electr.,* 1990, 33, 729-734.

Harmon, J. K.; Coles, W. A. *Astrophys. J.* 1983, 270, 748-757.

Jackson, B. V.; Hildner, E. *Solar Phys.* 1978, 60, 155-170.

Krall, J.; Chen, J.; Duffin, R. T.; Howard, R. A.; Thompson, B. J. *Astrophys. J.* 2001, 562, 1045-1057.

Kravtsov, Yu. A.; Orlov, Yu. I. Geometrical Optics of Inhomogeneous Media; *Nauka*; Moscow, 1980; 304 p.

Kucheryavenkov, A. I.; Pavel'ev, A. G.; Rubtsov, S. N.; Yakovlev, O. I. *Radiophys. Quant. Electr.* 1985, 28, 549-556

Lynch, B. J.; Coles, W. A.; Sheeley, N. R. *Geophys. Res. Lett.* 2002, 29, 1913, 19-1.

Orlov, Yu. I. *Radiophys. Quant. Electr.* 1974, 17, 788-793.

Pavel'ev, A. G.; Yakovlev, O. I., Kucheryavenkov, A. I.; Rubtsov, S. N.; Andreev, R. A.; Melikhov, A. G. Izvestiya VUZov. *Radiofizika*. 1988. 31, 127-132.

Rubtsov, S. N.; Yakovlev, O. I., Efimov, A. I. *Kosmicheskie Issledovaniya*. 1987, 25, 620.

Thernisien, A. F. R.; Howard, R.A.; Vourlidas, A. *Astrophys. J.* 2006, 652, 763-773.

Woo, R. A. J. *Geophys. Res*. 1988, 93, 3919-3926.

RESEARCH AND REVIEW STUDIES

In: Solar Physics Research Trends
Editor: Pingzhi Wang, pp. 31-94

ISBN: 978-1-60021-987-0
© 2008 Nova Science Publishers, Inc.

Chapter 1

Solar Filaments and Their Eruptions

Boris Filippov
Pushkov Institute of Terrestrial Magnetism, Ionosphere and Radio Wave Propagation, Russian Academy of Sciences (IZMIRAN), Troitsk Moscow Region, 142190, Russia

Abstract

In the hot and rarefied plasma of the solar corona, relatively dense and cold clouds are encountered, with the parameters more typical for the chromosphere, referred to as solar prominences. Some of them are very dynamic and have a short lifetime (loop prominences, surges, sprays) while others are observed over a long period of time and change rather slowly. The latter are usually visible within the chromospheric spectral lines against the background of the solar disk as dark filaments. It is well known that the filaments are elongated along the polarity inversion lines of the radial component of magnetic field in the photosphere. It is the magnetic field that holds the dense substance of filaments high in the corona and prevents it from imminent falling into the chromosphere. Resent research suggests that the magnetic field in the filament environment has a structure of a flux rope, which consists of the helical lines of force. The lower parts of the helixes serve as magnetic traps in the gravitational field, where the dense plasma of prominences can be accumulated. The equilibrium of a filament is determined by the equilibrium of the integral electric current of the flux rope. Its loss of equilibrium or stability leads to the rapid ascent of the filament named eruption. Further development of the process in the corona manifests itself as a coronal mass ejection (CME). Thus, the filaments serve as indicators of the presence of the magnetic flux ropes in the solar atmosphere and on the basis of analysis of their state (the height above the chromosphere, the intensity of internal motions, the upward motion, etc.) it is possible to judge the prospects for sudden disturbances in the solar wind.

1. Introduction

The solar atmosphere shows numerous examples of plasma formations being in the equilibrium in the gravitational and magnetic fields. In contrast to the laboratory conditions, where the walls of a chamber play significant and sometimes determining role in equilibrium and dynamics of plasma, Sun gives the possibility to investigate interaction of plasma with the magnetic field in the pure conditions. Dense photosphere can be considered as the only boundary, which certainly influences the electromagnetic processes in the atmosphere, but it is not difficult to take into account this influence because of simple regular shape of the photospheric surface. Without magnetic field, solar atmosphere would not contain such enormous variety of formations and structures, which amuses the eye of an observer and proposes new puzzles.

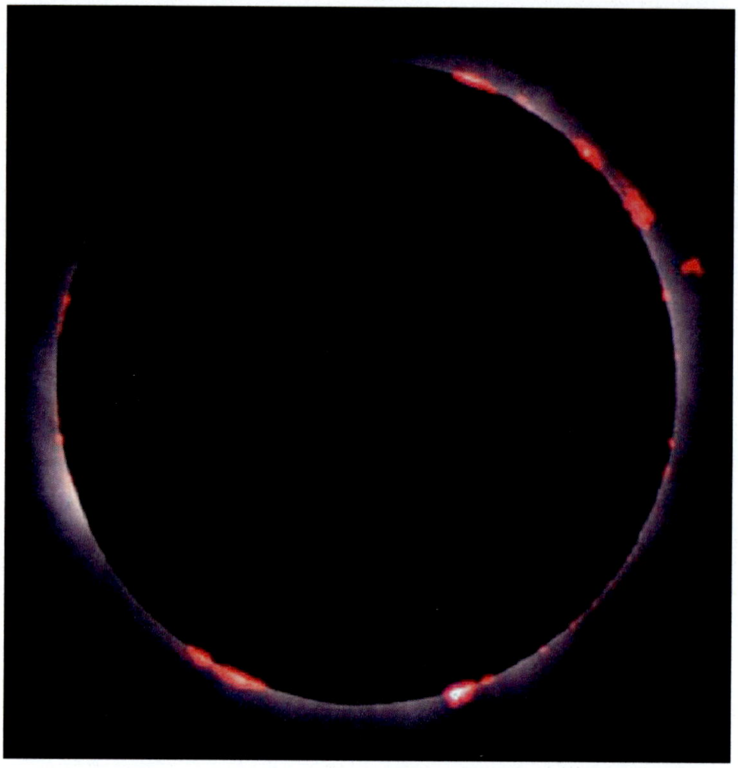

Figure 1. Solar eclipse image on 11 August 1999 showing numerous prominences on the limb (courtesy of S. Koutchmy).

Solar prominences were discovered as bright reddish ridges above the edge of the lunar disk visible during a total solar eclipse. At another ime, prominences can be observed on the limb of the Sun in the light of strong chromospheric spectral lines (for example, Hα). Some prominences are changing rapidly. They can appear and disappear in tens of minutes, while others do not change their appearance in the course of many hours. Long-lived prominences are usually visible on the solar disk as strips of variable width, darker then general chromospheric background, and solar filaments. Some filaments can be traced for several solar rotations (several months). Below we will use the terms 'filament' and 'prominence' as synonyms, keeping in mind that they imply the same physical feature observed in different projections. The history of solar prominence research and their basic properties are described and discussed in great detail in review papers and textbooks (Secchi, 1875; Tandberg-Hanssen 1974; 1995; 1998; Rompolt, 1990; Low, 1996; Martin, 1998). Two IAU meetings were devoted to reviewing and discussion of the new observations, theories and modeling of prominences and their environment in the last two decades (IAU Colloquium 117, 1990; IAU Colloquium 167, 1998).

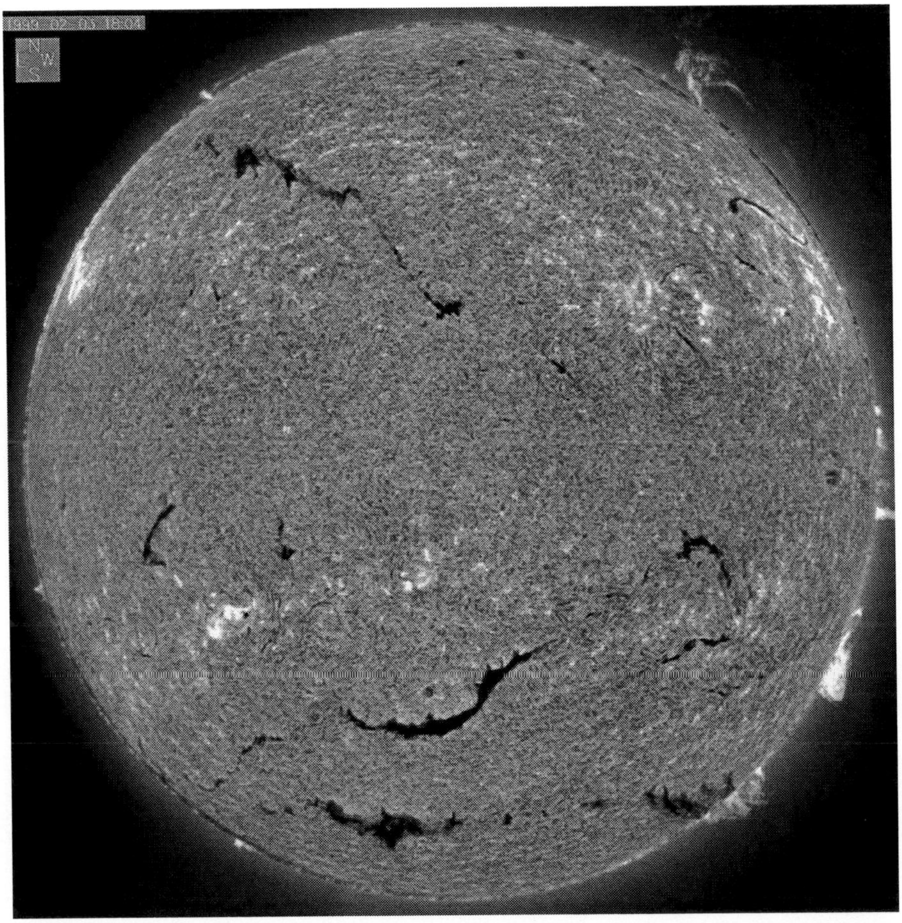

Figure 2. Full disk Hα filtergram on 3 February 1999 (courtesy of Big Bear Solar Observatory/New Jersey Institute of Technology). There are a number of large filaments seen on the disk. Coming to the limb they appear as prominences.

Prominences are the clouds of partly ionized gas hanging in the corona with physical parameters more convenient to the chromosphere than to the corona. Average particle density in a prominence is ~ 10^{11} cm^{-3}, the temperature T is about 7000 K. Thus, the density of prominence plasma is two orders more than that in the surrounding corona, and the temperature is two orders less. These relationships keep the pressure balance on the prominence boundary but cannot retain dense plasma high in the corona from falling into the chromosphere by the action of gravity. Only magnetic forces are capable of supporting the prominence mass. In the absence of the magnetic field in the corona, the substance would drain into the chromosphere under the action of gravity in a few minutes.

Filaments may be separated into two types: active region filaments and quiescent filaments. The active region filaments are located near sunspot groups and other strong magnetic field concentrations. They consist of the bundle of thin threads almost parallel or weakly twisted into a rope. Length of an active region filament is typically 50 Mm, height is near 10 Mm, and width is of a few megameters (Schmieder, 1989). Sometimes one end of a filament enters into a large sunspot (Fig. 3). On the limb active region filaments hardly come forward above the chromosphere.

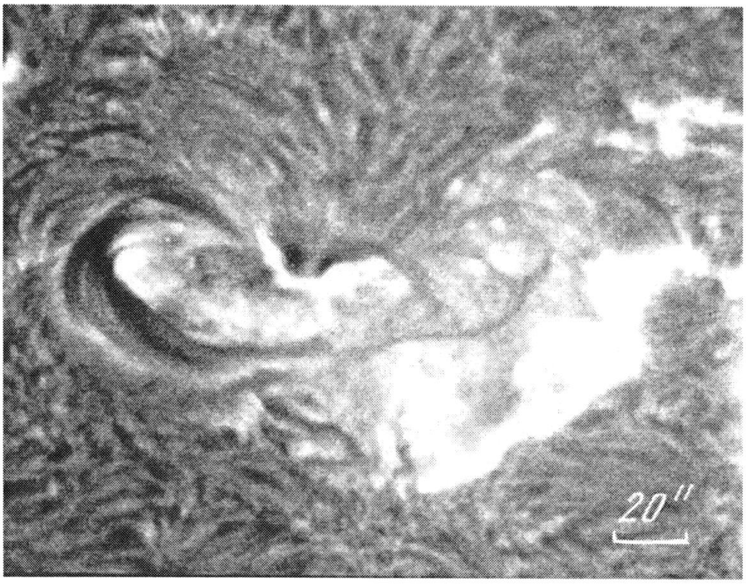

Figure 3. Active region filament within AR McMath 12507 on 7 September 1973 (Kislovodsk Solar Station).

Quiescent prominences have larger sizes and the more friable and diffuse internal structure (Fig. 4). Their length reaches 600 Mm, height can be over 100 Mm, and width is about 15 Mm. Threads, which compose the body of a prominence, are twisted and slightly ragged. Taking into account density variations in the different threads and the presence of dense blobs in separate threads, the filling factor within the body of a filament is estimated as not more than 0,1 (Rompolt, 1990). Sometimes threads are horizontal, sometimes they are predominantly vertical. In certain cases, internal helical structure is clearly recognized with the pitch angle usually greater than that in the active region filaments. Very often observers liken the appearance of a quiescent prominence to a long curtain (Fig 5).

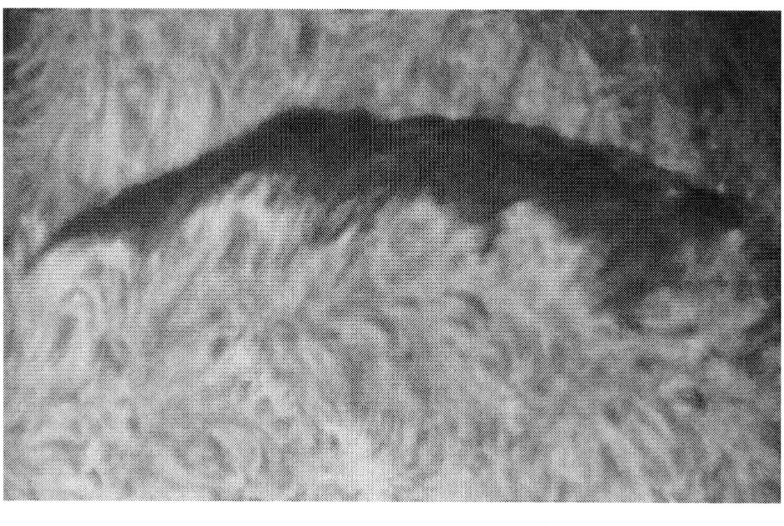

Figure 4. Quiescent filament on 18 June 1989 (Kislovodsk Solar Station).

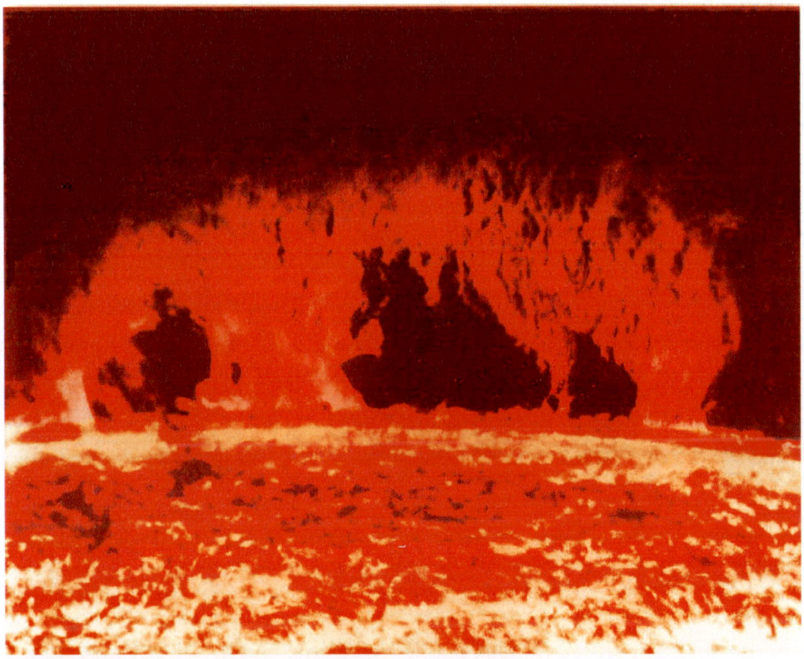

Figure 5. Quiescent prominence observed at Big Bear Solar Observatory in 1974 (courtesy of Big Bear Solar Observatory/New Jersey Institute of Technology).

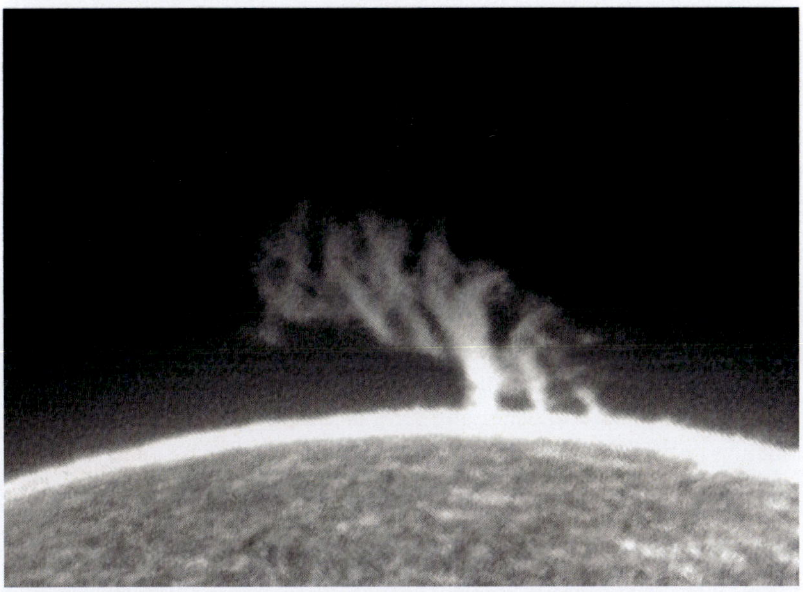

Figure 6. Quiescent prominence on 23 November 1998 with strongly twisted and intertwined internal threads (courtesy of Big Bear Solar Observatory/New Jersey Institute of Technology).

Three different stages can be distinguished in the life of a quiescent prominence. Young, recently formed prominence takes form of one or two arches with the length of one or two supergranule cells and with a height of 15-30 Mm. Arches have pronounced filamentary structure. Threads in the arch are interlaced and frequently twisted into the helical rope. Well developed quiescent prominence of medium age consists of many adjacent arches in the form of a long bridge. Besides the threads extended along the axis of the filament, a system of vertical threads forms a curtain above the arches. During this period, a filament is approximately 200 Mm in length and 50 Mm in height, however some well developed filaments stretch throughout the whole solar disk. An old quiescent prominence has the shape of a huge arch with strongly twisted and intertwined internal threads (Fig. 6). The length of the arch is 150-300 Mm, the height is within 60-200 Mm, and the width (diameter) is 30-80 mm (Rompolt, 1990).

Besides the active region prominences and the quiescent prominences, some observers introduce an intermediate type of prominences (Martin, 1998), which shows to some extent the features of the both types. All these three types very likely reflect the extremes and the middle part in the wide spectrum of the properties of the solar prominences. Differences between them bear quantitative rather than qualitative character.

2. Filaments Position Relative to the Photosphere Magnetic Field

Magnetic nature of prominences manifests itself already in their position on the Sun. Comparison with maps of the photospheric longitudinal (line-of-sight) magnetic field shows that filaments are always located above polarity inversion lines (called also neutral lines) of the field (Babcock, Babcock, 1955; Howard, Harvey, 1964; Smith, Ramsey, 1967; McIntosh, 1972; Snodgrass, Kress, Wilson, 2000; Durrant, 2002). Chains of filaments outline thus large-scale magnetic field cells (Fig. 7). This fact gives grounds to construct so named Hα synoptic maps (Fig. 8), which snow the general topology of the magnetic field sources even for those periods of time when magnetographic measurements were not invented yet. Although it is possible to fix only a sign of the radial magnetic field inside the cells, this information allows calculation in the first approximation of the structure of the global coronal magnetic field.

The distribution of filaments over the solar surface changes with the phase of a solar cycle (Fig. 9). Beginning approximately a year after minimum, more and more filaments appear at middle latitudes of 30°-40° (McIntosh, 1992). These filaments are called the filaments of the first tier (Rust, 2001), and they are forming between active regions. They are more numerous than the active region filaments (Tang, 1987). Slowly displacing toward the poles, the first tier filaments by the time of solar cycle maximum reach the latitudes of 50°-60°, where they stretch along a parallel and form an almost continuous ring named the polar crown. Remaining part of the cycle, filaments of the polar crown spend near the latitude of ~55°. Being located on the polarity inversion line, the polar crowns outline the boundaries of gigantic unipolar cells near the poles. Soon after the beginning of the new cycle, the polar crown filaments start to drift to the poles, while filaments of the first tier of new cycle come to occupy their place. Some time near the sunspot number maximum the polar crown rings around the North and South poles collapse not necessarily simultaneously. An interval between these events showing polarity reversals of the poloidal magnetic field of the Sun in the northern and southern hemispheres can range up to several months.

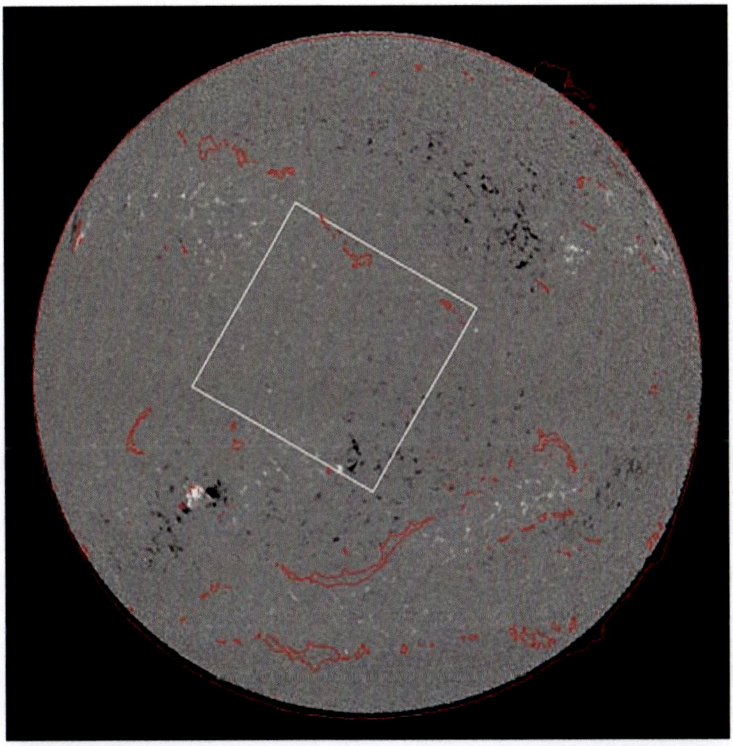

Figure 7. Contours of the filaments (red lines) shown in Fig. 2 overlaid on the SOHO/MDI photospheric magnetogram (courtesy of SOHO/MDI consortium. SOHO is a project of international cooperation between ESA and NASA.).

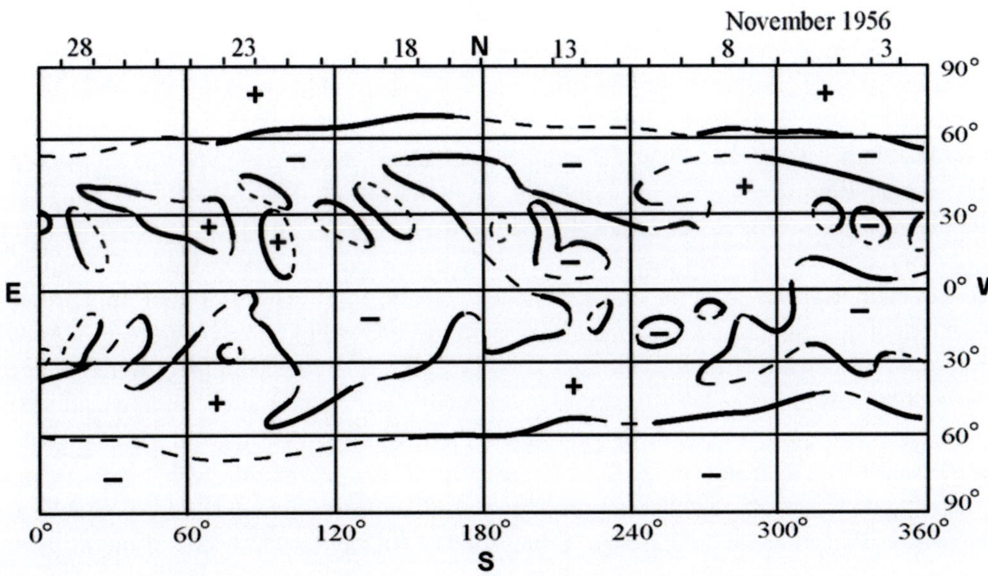

Figure 8. Hα synoptic chart of the large-scale magnetic field of the Sun one year before the polar magnetic field reversal (November 1956, CR 1380). Bold lines show the observed position of filaments and prominences, dashed lines are extrapolation of neutral lines between observed fragments (courtesy of V. I. Makarov).

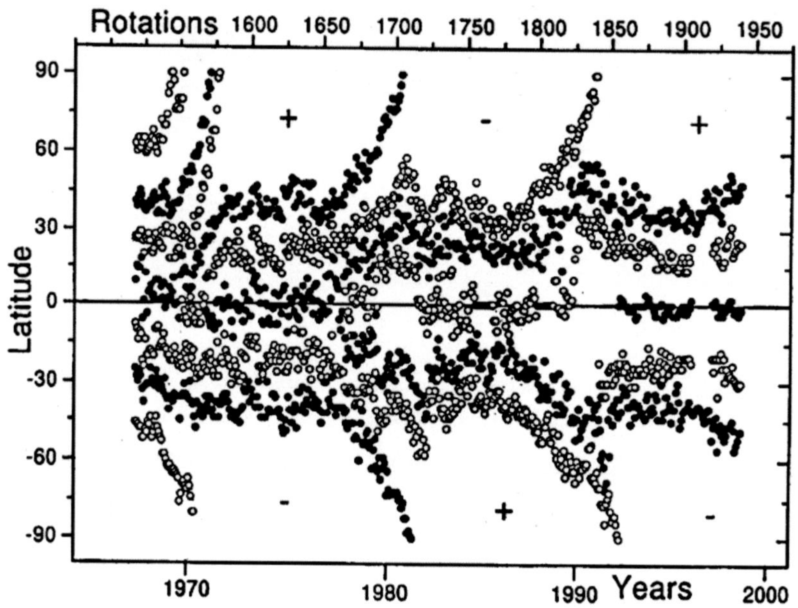

Figure 9. Position of the average latitude of polarity inversion lines of the large-scale magnetic field of the Sun in the period 1968-1998 derived from the prominences and filaments distribution. (Makarov et. al., 1998).

A strip at both sides of a polarity inversion line forms an inversion zone in the chromosphere or a filament channel. Simple potential extrapolation of the photospheric magnetic field gives usually in the inversion zone an arcade of field-line loops straddling the neutral line. Directly above the neutral line, field lines must be horizontal and directed predominantly across it. Actually the magnetic field in the filament channel is considerably more complex. Measurements of the magnetic fields in prominences on the basis of the Zeeman-effect and Hanle-effect showed that the field within them, as a rule, was really horizontal, but it was directed almost along the filament axis and polarity inversion line (Rust, 1967; Ioshpa, 1968; Leroy, 1989; Kim, 1990; Bommier et al, 1994; Demoulin, 1998). The angle between the magnetic vector and the filament axis was 25° on the average. Moreover, the direction of the component perpendicular to the axis was in the majority of cases opposite to the direction expected from the potential field calculation on the basis of photospheric data. Field strength in the quiescent prominences was on the average of approximately 8 G. Active region prominences had stronger field up to 200 G, but measurements in them were hindered because of the low altitude above the limb. In the high quiescent prominences, in which it was possible to measure the field at different heights, the field strength either did not change or a little increased with the height.

Direction of threads, which compose the fine structure of filaments and probably are elongated parallel to the magnetic field, corresponds basically to the results of the magnetic field measurements, except for the vertical threads forming a curtain in large quiescent prominences. Fibril arrangement in a chromospheric filament channel also reveals the direction of the field along the polarity inversion line. Joining of the fibrils into the chains along the channel axis is considered as one of the basic forerunners of the filament formation in the channel (Martin, 1998). The filament channels are more widespread formations than filaments themselves. A channel can contain a filament as well as it can exist without the filament. No more than 60% of channels have detectable filaments. The largest and especially stable filaments appear in channels that have been existed during 4-8 solar rotations. Probability of the appearance of such filaments in younger or older channels is substantially lower (Harvey, 2001).

Bundles of threads, which jut out as barbs from both sides of the main filament body, form lateral feet that connect the filament to the chromosphere. On the limb, they are visible in the form of bases of arches or piers of a "bridge". Feet are arranged along the filament at rater regular way. The distance between the adjacent feet corresponds to the size of supergranules. Some authors believe that the feet connect the filament to field concentrations in nodes of the chromospheric network (Ploceniak, Rompolt, 1973). Other researchers assume that the barbs of the quiescent prominences are ending inside supergranulation cells in regions of weak field (Engvold, 1998), or they are connected to small patches of parasitic polarity or to ephemeral bipolar regions (Martin, Echols, 1994). Difficulties in identification of the barb ends with some features of the photospheric magnetic field arise because of uncertainties in their height above the chromosphere, since a filtergram gives us only one projection.

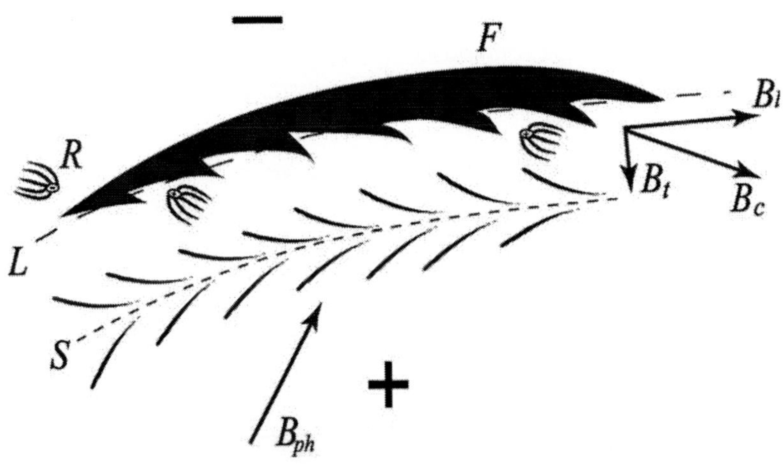

Figure 10. Schematic representation of the structures near a filament. The filament is dextral and rightbearing. L is the polarity inversion line, R is a rosette, S is the axis of the fish-bone structure, B_{ph} is the large-scale photospheric magnetic field, B_l is the field component along the filament axis, B_t is the field component perpendicular to the filament axis.

Polar crown filaments in the northern and southern hemispheres have different orientation of their internal magnetic field relative to the large-scale photosphere field. If one looks at a filament from the side of the positive polarity photospheric field, i.e., along a line of force, the field in the filament that is almost orthogonal to this direction can be directed to the left or to the right (Fig. 10). It was found that in the most of high-latitude filaments of the northern hemisphere, the field is directed to the right and they have been named dextral filaments, while southern polar crown filaments with the field directed to the left have been named sinistral filaments (Martin et al, 1994). This distribution does not change in the following cycles, because both poloidal and toroidal components of the global magnetic field change their direction to opposite in each new cycle. Active region filaments do not follow exactly to this rule, although there is a certain tendency of the predominance of the dextral filaments in the northern hemisphere and of the sinistral filaments in the southern hemisphere.

What is curious, the direction of the axial magnetic field in a filament can be determined on the basis of its general appearance without magnetic measurements. If a filament is viewed from above, the barbs are seen to bear to the right (deflect from the filament axis to the small angle clockwise) or to the left (deflect counterclockwise). As it was found by Martin et al. (1994), all right-bearing filaments were dextral and all left-bearing filaments were sinistral. It is possible also to draw the conclusion about the sign of the magnetic field component transversal to the filament axis. The threads that compose a barb are elongated along the magnetic field. Really, if one looks along the filament axial magnetic field, the clockwise deviation of the threads composing a barb means that there is the transversal component directed to the right (Fig. 10). But the right-bearing filament is sinistral, and there must be positive polarity on the right side of it and negative polarity on the left, which create the field directed from the right to the left quite opposite to what is inferred from the filament fine structure. Thus, there is an evidence of the transversal magnetic field inversion in the filament.

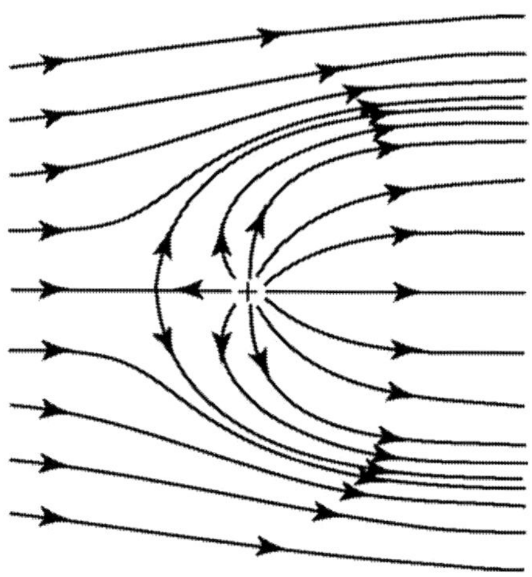

Figure 11. Field lines of the superposition of the homogeneous field and the field of a positive pole. The tail of a rosette shows the direction of the surrounding magnetic field.

This inversion of the transversal component is visible also in the chromospheric filament channel. It manifests itself in the deflection of rosette "tails" located near the filament from the neutral line. A rosette (or a plagette) consists of a bundle of fibrils radially emanating from the central bright area related to a patch of the network magnetic field. Interaction of this field concentration with the surrounding field creates a null point on the side where the fields are antiparallel (Fig. 11). Long radial fibrils remain only on the side where the fields are parallel, and they form a structure resembling a comet tail. Tails of the rosettes emanating from the positive magnetic elements are pointed to the direction of the background field, while tails of the rosettes emanating from negative elements are directed opposite the external field. The tails on opposite sides of the filament are directed predominantly along the neutral line and antiparallel. Since the plagettes on opposite sides of the filament have opposite polarity, the antiparallel streaming of the fibrils implies that the field on both sides is approximately in the same direction (Foukal, 1971). But instead of being slightly inclined toward the neutral line in accordance with the direction of large-scale photosphere field, the rosette tails emanating from the dominant polarity magnetic elements are deflected out of it, which demonstrates the inversion of the transversal field component within the filament channel. The boundary of the zone of inversion frequently is well distinguished due to the specific fish-bone structure (Filippov, 1994) formed by the fibrils (Fig. 10). It can be seen in Fig. 4 and Fig. 12 that the fibrils on one side of the filament are arranged as fish-bones or fir-tree branches. The fibrils are curved in such a way that in the central part of the formation they are directed along the filament axis. This means that the symmetry axis of the fish-bone structure is the line S where the magnetic field component transverse to a filament vanishes. This line is parallel to the axis of a filament closely following the inversion line. The fishbone arrangement of the fibrils also implies that, due to field continuity, it should be assumed that the transversal component changes sign at the symmetry axis. The line S is the singular line of the two-dimensional vector field B_t. A set of integral lines of tangential field passes through each of its point. Thus, the filament channel, which is called the inversion zone keeping in mind a change in the direction of the radial magnetic field component, is also the zone of inversion of the transversal component of the field.

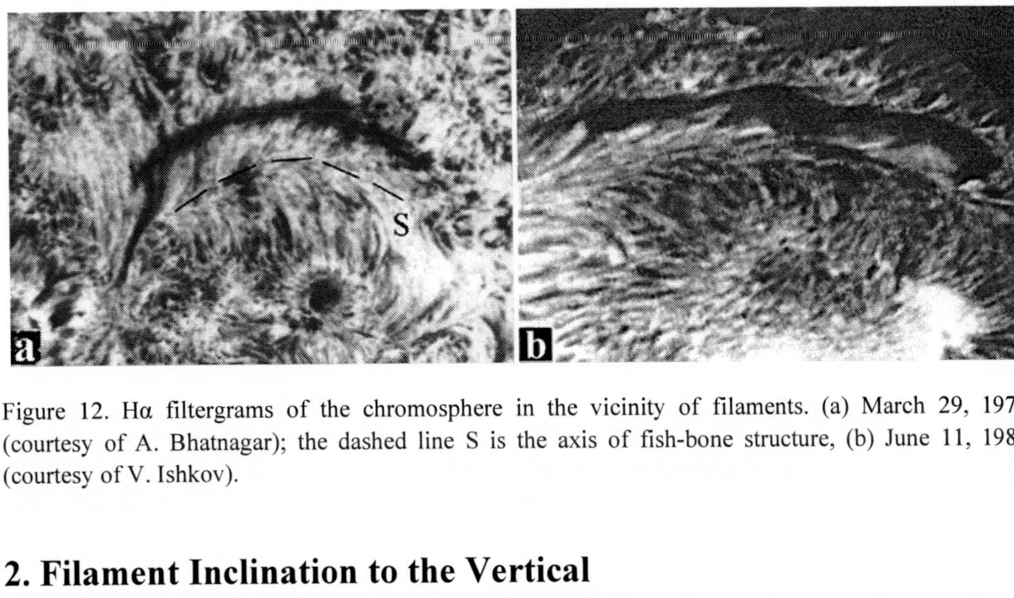

Figure 12. Hα filtergrams of the chromosphere in the vicinity of filaments. (a) March 29, 1971 (courtesy of A. Bhatnagar); the dashed line S is the axis of fish-bone structure, (b) June 11, 1980 (courtesy of V. Ishkov).

2. Filament Inclination to the Vertical

As it was noted, quiescent prominences (filaments) frequently take the form of thin ribbons standing on end, or curtains. If the surface on which the prominence material is concentrated were strictly vertical, filaments extending from north to south would display their narrowest apparent widths at the central meridian when rotating with the Sun. However, the study of d'Azambuja and d'Azambuja (1948) showed that most filaments display their minimum width in the eastern hemisphere, when they are inclined toward the west, in the direction of the solar rotation. d'Azambuja and d'Azambuja (1948) determined that the median inclination of the prominences was about 10°.

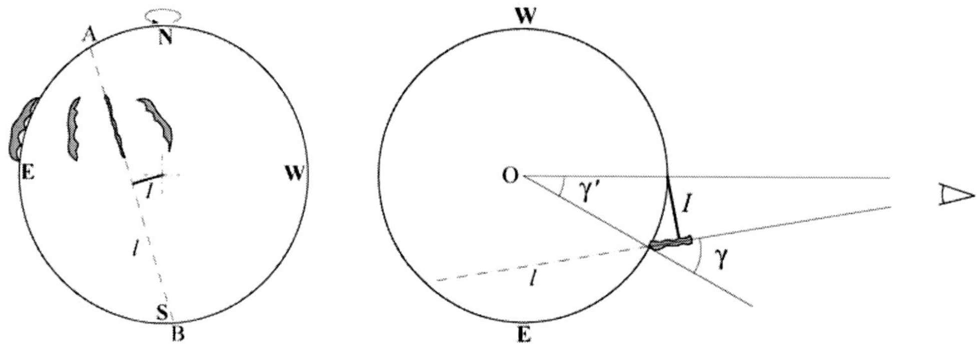

Figure 13. Schematic of the passage of a filament through the solar disk (left) and its position displaying the minimum width (right).

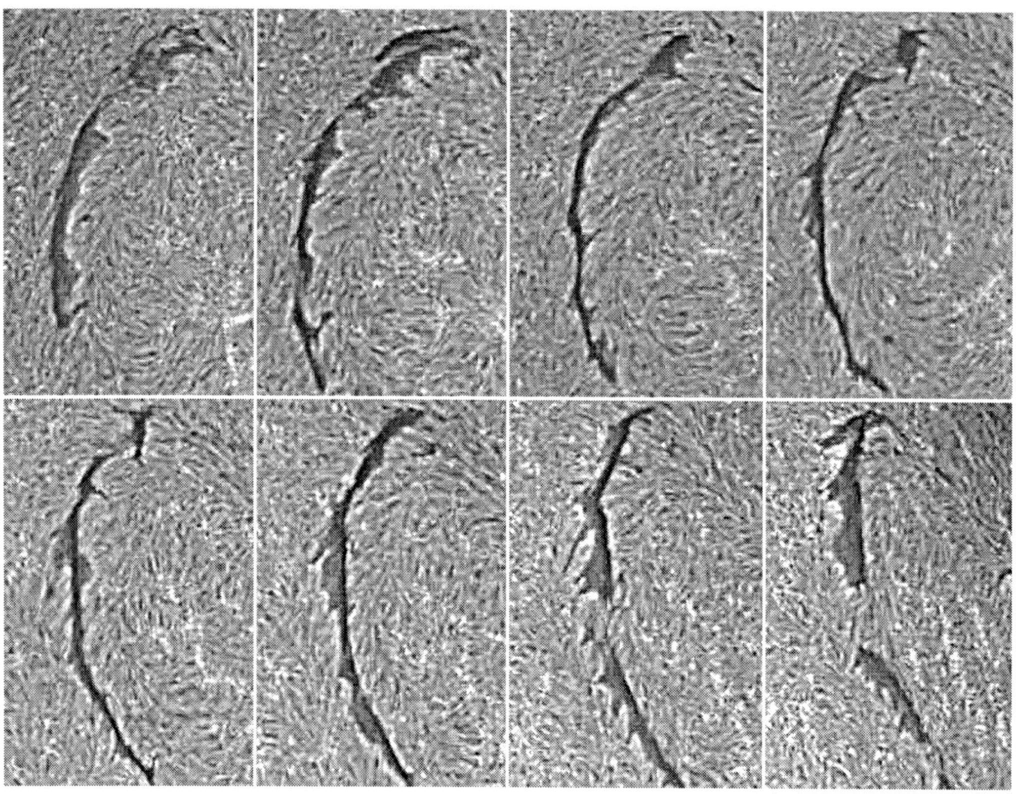

Figure 14. Hα filtergtams for the passage of a filament through the solar disk from 19 to 26 July 2001 (courtesy of Big Bear Solar Observatory/New Jersey Institute of Technology).

We can determine the prominence inclination, if we can measure the distance I between the center of the disk and the line AB passing along the filament axis when the filament's projection onto the disk takes on its minimum width (Fig. 13, left). It is evident that we will observe the minimum width when the prominence plane contains the line of sight (Fig. 13, right). This plane forms a circular cross section l of the spherical solar surface. Since the secant plane makes the same angle to the vertical radius everywhere on the circumference l, it is sufficient to determine the central angle γ', which does not differ from the angle γ by more than 0.25°. Within the required accuracy, we can assume that the distance I is equal to the length of the arc traced on the solar sphere between the circle l and the line of sight, directed to the disk center. Then, $\gamma = I/R$, where R is the solar radius. Figure 14 shows changes in the filament appearance during its passage the solar disk from 19 to 26 July 2001. Since the filament axis is not a straight line, different parts of the filament length demonstrate the minimum width at different times. For the central part of the filament, it was on 21 July (the third frame in the upper row). Additional evidence of seeing along the filament plane is nearly equal length of barbs on both sides of the filament.

The close connection of filaments with the magnetic field suggests that their systematic inclination must reflect some properties of the field. Indeed, Becker (1956) found that prominences lean away from the area of strongest magnetic intensity. On the ground of theoretical model of Kippenhahn and Schluter (1957), which suggested the existence of a dip near the top of magnetic arcade in quadrupolar configuration where stable support of dense plasma is possible, Rust (1966) and Godoli et al. (1974) explained the prominence inclinations using a model with two infinite parallel rows of vertical magnetic dipoles of magnetic moments having non-equal values of the intensity and opposite direction in the two rows. At present time this explanation hardly can be considered satisfactory, since the quadrupolar configuration with the sagging field lines can hardly be typical for the majority of prominences.

Calculations of potential magnetic field in the corona in the filament environments revealed the absence of such dips in most cases (Zagnetko et al., 2005). The potential field determines the general topology of active regions and the locations of flares, prominences, and other active objects. However, the fields in the prominences themselves and in their immediate vicinity are far from potential. Ropes and bundles of thin spiral threads are often observed in the fine structure of filaments and prominences, indicating the presence of strong coronal electric currents, although the potential field of the photospheric sources remains the dominant field governing the overall behavior and morphology of the prominences. This is supported by the fact that filaments follow very accurately polarity inversion lines of the photospheric field.

Comparison of the results of the magnetic field calculations in the corona with the form and arrangement of filaments (Zagnetko et al., 2005) showed that the surface at which the prominence material was concentrated was very close to the surface passing through the apexes of the field line arches of the potential magnetic field (surface $B_r = 0$ or neutral surface}) calculated on the basis of photospheric magnetograms. Neutral surfaces have the same inclination to the vertical (Fig. 15) as it is obtained for filaments using the technique of d'Azambuja. Figure 16 presents superposition of the projections of the calculated neutral surfaces on the corresponding Hα filtergram. Fragments of the neutral surfaces from the chromospheric level to the height of ~ 30 Mm are projected on the figure plane. These projections stand out on the Hα filtergram as light shades. The filaments appear through them as darker formations. Basically, filaments are arranged inside the outlines of the projections of neutral surfaces and to some extent repeat their forms. Thus, the geometric form of prominences and their inclination toward the vertical line reflect the structure the magnetic field created by photospheric sources in the corona.

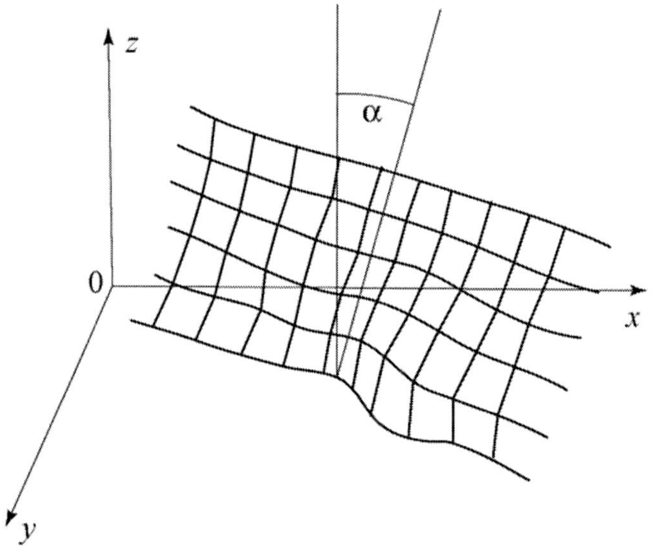

Figure 15. Neutral surface calculated in the potential approximation for the region containing a filament; α is the vertical deviation angle for this surface.

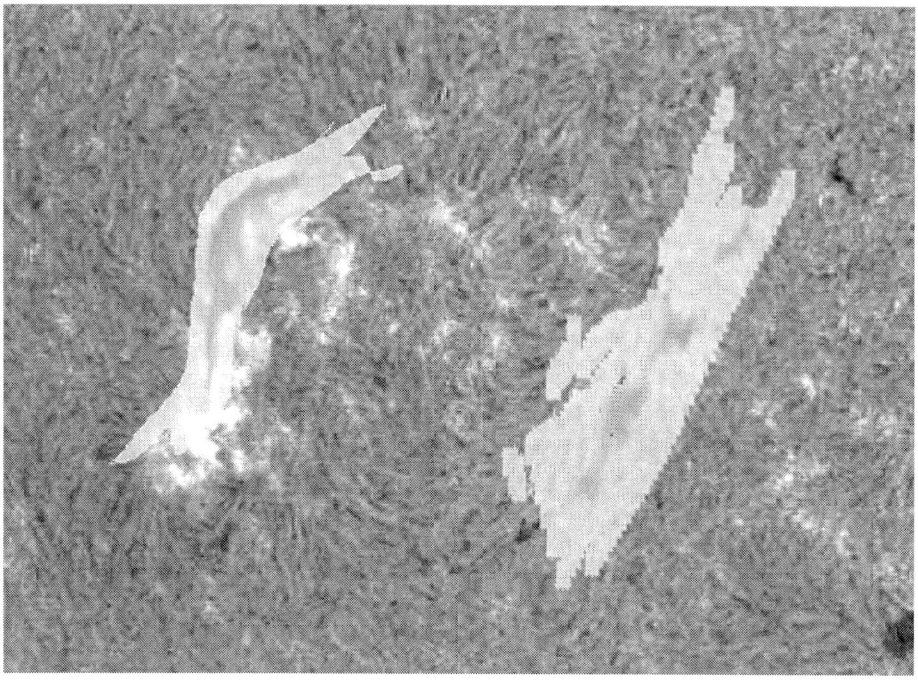

Figure 16. Superposition of the projections of the calculated neutral surfaces on the corresponding Hα filtergram (courtesy of Big Bear Solar Observatory/New Jersey Institute of Technology). The light shade shows the projections of the neutral surfaces onto the sky plane against the background of the chromosphere; the filaments appear through them as darker formations. The neutral surface profile for the right filament is shown in Fig. 15.

3. Motions in Filaments

The term "quiescent" in application to prominences, generally speaking, is not completely precise, since in spite of the external stability and invariability they are dynamic formations. Along individual threads, which compose a filament, intense internal motion of substance is observed with speed of up to several kilometers per second both horizontal and vertical, directed upward and downward (Engvold, 1972; Stellmacher, Wiehr, 1994). Speed even in the vertical threads is much less than that in the free fall, which testifies the action of some retarding mechanism of unclear nature. Frequently the motion has character of oscillations with periods in the range from less than a minute to hours (Yi et al, 1991). Oscillations are noticeably intensified not long before the prominence eruption.

Observations of the characteristics of motion in prominences, filaments, and their environments could help to decide which of two alternative possibilities of the prominence mass loading is more realistic: loading by condensation from the hot and rarefied coronal plasma or by the lift of the cool and dense gas from the chromosphere. The mass of a large prominence (~ 10^{15} g) is a noticeable fraction of the mass of the whole corona; therefore the condensation of coronal substance into the prominence must be accompanied by large-scale motions in the corona, which must be seen in the ultraviolet and soft X-ray ranges. Such motions have not been recorded yet. The lift of substance is determined by the Doppler shift of spectral lines during observations of filaments on the disk (Schmieder et al., 1985). However, usually motions in filaments are recorded both upward and downward (Maltby, 1976; Demoulin et al., 1987), and observations with high resolution reveal counter-streaming flows along closely spaced vertical regions of a prominence (Zirker et al, 1998).

Some prominences demonstrate rotation motions. They are manifested as the displacement of fine structure features in the sky plane and Doppler shifts of the spectral lines (Rusin, Rybansky, 1982; Liggett, Zirin, 1984). Typical velocities of rotation motion are of order of tens km/s. If one assumes that the motion of plasma in the prominences occurs along the magnetic field, the rotation is the consequence the helical shape of field lines. Based on the velocity distributions in the prominences obtained on the basis of the analysis of spectral line profiles, Gigolashvili (1978) found that twisting corresponded to a left screw in the studied prominences of the northern hemisphere and it was in agreement with the shape of a right screw in the prominences of the southern hemisphere. Later, the similar helicity distribution over the hemispheres was discovered on the basis of the magnetic field data and fine structure of the filaments (Martin et al, 1994).

4. Prominence Height above the Chromosphere

The prominences of active regions barely protrude above the average chromospheric level. Their heights usually do not exceed 10 Mm (Schmieder, 1987; 1989). The quiescent prominences reach considerably greater heights, up to 200 Mm (Pettit, 1932; d'Azambuja, 1948). The sizes and heights of quiescent prominences increase during their lifetimes. The height at an early stage is 15–30 Mm, a well-developed filament in the middle phase is at a height of ~50 Mm, and the oldest prominences rise by 60–200 Mm above the limb (Rompolt, 1990). Prominences appear at large heights only in the dynamic phase, during eruption.

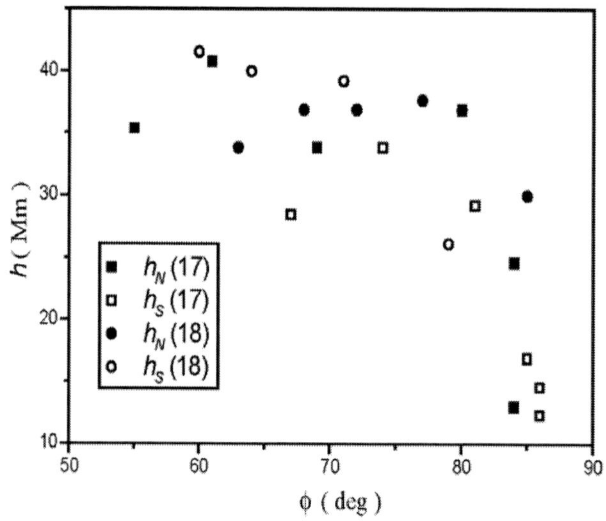

Figure 17. The average height h of prominences vs the latitude of the polar crown filament φ for the northern h_N and southern h_S solar hemispheres in the 17th and 18th cycles (courtesy V.I. Makarov).

The total number of prominences exponentially decreases with increasing height in the range 10–100 Mm (Ananthakrishnen, 1961). Obashev (1963) found the height distribution of all prominences to be similar in shape to Pearson's distribution peaking at a height of 30 Mm. This author found no dependence of the most commonly encountered prominence height on solar-cycle phase from 1955 until 1957, at least in the heliographic latitude range 20°–60°, and on latitude (if the polar regions are excluded). The lack of such dependence was also noted by Dermendzhiev (1977) for 1945–1965. Variation in the yearly mean prominence height with solar cycle was revealed by Cantu et al. (1968) and subsequently confirmed by Makarov (1983) using the 1912–1974 data of the Indian Kodaikanal Observatory. The maximum prominence height is reached near maximum activity or two years earlier. In addition, a systematic increase in the mean prominence height from 25 to 33 Mm for the period under consideration was noted.

Kim et al. (1988a; 1988b) also found a dependence of the mean prominence height on cycle phase for the second half of solar cycle 21, which generally matches the dependence obtained by Makarov. They noted the presence of two peaks in the height distribution of prominences, at 18 and 33 Mm, and surmised that the lack of dependence of the median height on cycle phase could be due its bimodal structure. Having distinguished only high-latitude prominences, which constitute almost continuous polar crowns around the polar caps, Makarov et al. (1992) found that, as the global neutral line drifts poleward, the prominence height above it decreases by more than a factor of 2.

In the cycle 17 (1936-1939), the average height of the polar crown filaments decreased from 33 Mm down to 17 Mm in the northern hemisphere and from 37 Mm down to 16 Mm in the southern hemisphere during their drift from latitude of 40° to the poles (Makarov et al., 1992). In the cycle 18 (1947-1949), a similar pattern was observed (Fig. 17). For 1905-1975, a slight increase of the height of the polar crown filaments (about 10 arc sec, or 7 Mm) was found in the odd cycles contrary to the even ones. This result relates to the different activity of the Sun in the even and odd cycles (Makarov, 1994).

Since the cool dense matter of prominences is kept in the corona from falling into the chromosphere by a magnetic field, the field properties must determine the equilibrium conditions for filaments and, in particular, their heights. Attempts to consider the prominence height as a kind of a magnetic characteristic (Leroy et al., 1984; Kim, Uvakina, 1989) are therefore not accidental, although there is currently no clear understanding of the physical relationship. For instance, it is suggested that the prominence height characterizes the background magnetic field strength (Makarov et al., 1992) or that the height distribution of prominences is related to the distribution of the horizontal gradient in longitudinal magnetic field near the polarity-inversion line (Maksimov, Yermakova, 1985). The latter authors found gradients of 2–10 G Mm^{-1} to correspond to most cases of the existence of filaments above the polarity inversion line in December 1982. No quiescent filaments were observed for gradients exceeding 20 G Mm^{-1}. The existence of a critical transverse gradient in vertical field was also noted in several other studies (Yazev, Khmyrov, 1987; Ioshpa, Kulikova, 1989; Maksimov, Prokop'ev, 1993), although the gradient values differ, being in the range 7–28 G Mm^{-1}. Maksimov and Prokop’ev (1993) partly attribute the discrepancies to the different spatial resolutions of magnetograms.

5. Models of Prominence Equilibrium

Basic forces that act on the plasma of solar atmosphere are the gas pressure gradient ∇p, the electrodynamic force $\mathbf{j} \times \mathbf{B}/c$ named by the different authors as the Lorenz force or the Ampere force, and the gravitation ρg. The gravitational scale height of prominence material (~500 km) is much lower than their typical vertical size and the typical height (10-100 Mm). This means that gas pressure cannot provide their supports. The electrodynamic force remains the only force able to balance the gravity. Although prominences are not quite static formations, the observed internal velocities are much lower than the sound velocity, the Alfven velocity, and the free fall speed. Therefore, at least in the first approximation, they must be in a state of magnetostatic equilibrium.

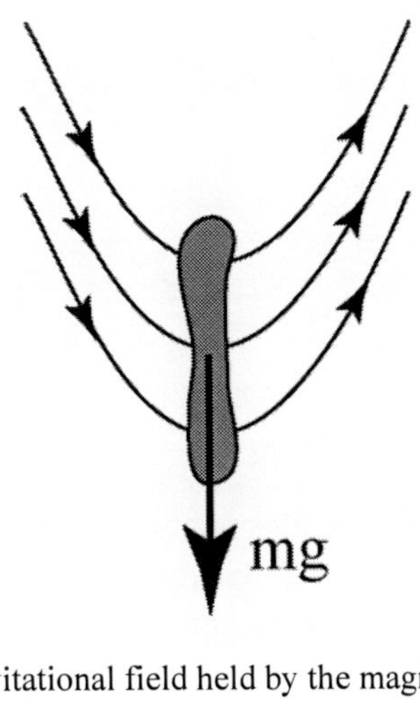

Figure 18. Plasma in the gravitational field held by the magnetic field line tension in a dip.

High-conductivity prominence plasma can rest in a "hammock" of field lines, which protect the plasma from draining into the chromosphere. For the stable equilibrium, the field lines must have the curvature directed upward and create a dip, which is filled up by plasma (Fig. 18). For the first time this type of prominence equilibrium model was proposed by Menzel in 1951 (Menzel, 1951). In 1957 Kippenhahn and Schluter (1957) proposed a model, which for many years became "classic" and generally recognized despite of certain difficulties, which it could not solve. Some of them we will examine below.

Alternative model of prominence equilibrium was proposed in 1974 by Kuperus and Raadu (1974). They assumed that there was strong electric current flowing along the filament axis and took into account that this current should be repulsed from currents induced in the photosphere. The matter is that the time of generation of the coronal current associated with the filament formation is much shorter than the diffusion time of the magnetic field in the photosphere. In view of high inertia, the photosphere does not let the new magnetic field generated in the corona to pass into the interior of the Sun because the magnetic pressure gradient is not enough to drive in motion so dense substance, and diffusion time of the field is too great. Electric currents induced in the photosphere compensate external field in the solar interior, as it occurs on the surface of any high-conductivity body. Above the photosphere, the field of the induced currents is equivalent to a mirror image of the coronal current in the photospheric surface with the opposite direction. As the oppositely directed currents are repulsing, a prominence seemingly hovers over the photosphere exhibiting magnetic levitation.

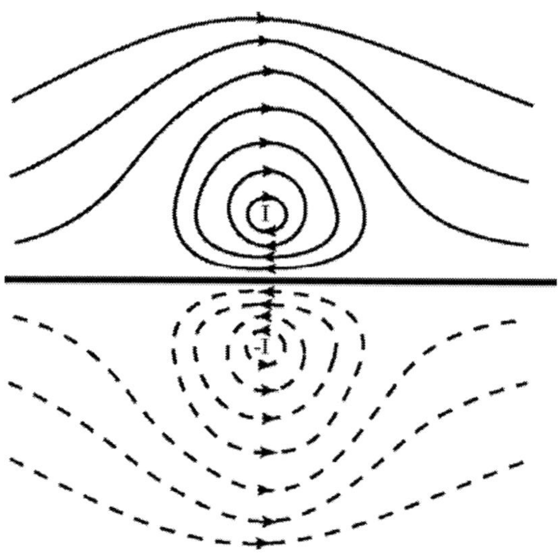

Figure 19. Field lines of a coronal electric current taking into account the field of currents induced in the photosphere (represented by a mirror current).

The field frozen in photospheric plasma, which emerges from the convection zone together with the plasma, effects both on vertical and horizontal equilibrium of the coronal current. The stable horizontal equilibrium is possible only over a polarity inversion line, $B_r = 0$, with the appropriate direction of the current. If a background field forms a simple arcade (over a bipolar region), the stable horizontal equilibrium requires that the vertical component of the Lorenz force acts downward (Van Tend, Kuperus, 1978). This is quite different from the Kippenhahn--Schluter model where the field of subphotospheric sources supports the filament weight.

The two types of the filament equilibrium differ not only by the direction of currents but also by their relative strength. In the Kippenhahn--Schluter model, the current only slightly disturbs the background field, so the field inside a prominence is basically determined by the photospheric sources. In the Kuperus--Raadu model, the field of the current must prevail inside the prominence and at a distance comparable with the prominence height above the chromosphere. Direction of the field inside the prominence might differ significantly from the direction of the background field and even be opposite to it. That is why models of the latter type are referred as the inverse polarity models, while the modifications of the Kippenhahn--Schluter model are named the normal polarity models (Ballester, Priest, 1987; Priest, 1988).

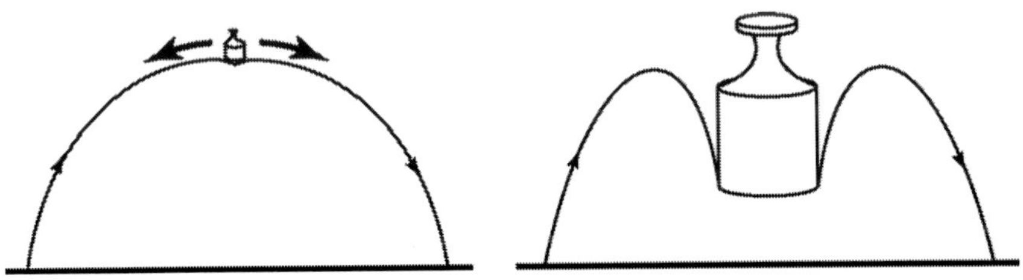

Figure 20. Small weight is unstable at the top of an arch whereas a heavy mass creates large sagging, which ensures stability.

One of the main difficulties, the normal polarity models meet with, is the problem of dip formation at the apex of a magnetic arcade. Simplest and natural way is sagging of field lines under the action of the prominence weight. A small weight, which creates a small dip, stays in unstable equilibrium (Fig. 20) and dense material will rapidly fall down along field lines to one or the other footpoints. A deep dip formed under the action of a large weight can ensure stable equilibrium (Wu et al, 1990b; Choe, Lee, 1992; Schmieder et al, 1991; Fielder, Hood, 1993). However, this mechanism works only in the case of sufficiently high value of the plasma beta $\beta = 8\pi\, nkT/B^2$, whereas in the corona β is usually assumed to be low. Moreover, it is unclear how it is possible to bring enough mass to the top of an arcade. Therefore, most likely one should expect that the dip must be formed in the magnetic field before the accumulation of prominence matter.

As was shown by Anzer (1969), the plasma equilibrium in the Kippenhahn--Schluter model is stable for vertical displacements, provided that the current density decreases with height in a vertical unbounded sheet

$$\frac{dj^2}{dh} < 0, \tag{1}$$

and for horizontal displacements, provided that the field grows with height

$$\frac{dB}{dh} > 0. \tag{2}$$

In a current-free (potential) field, the condition (2) means that the field lines have a geometry with upward curvature.

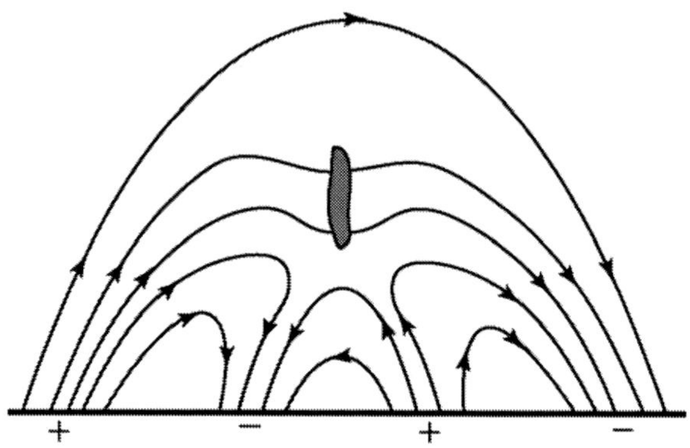

Figure 21. In potential field all sources of which are located under the photosphere, field line sagging in the corona can exist only near a singular saddle point (X-point).

If any of more or less significant sources of the magnetic field is located under the photosphere (the field in the corona is potential), the dip in field lines can exist only near the singular points of the X-type (saddle). For their appearance, it is necessary at least the quadrupolar magnetic configuration (Fig. 21). This geometry is frequently used as the probable magnetic prominence skeleton (Uchida, 1981; Malherbe, Priest, 1983; Demoulin, Priest, 1993). Although quadrupolar configuration was first proposed by Kippenhahn and Schluter (1957) for the stable support of prominences, it can be considered as inverse polarity, since the direction of the field in a prominence is opposite to the direction of the field produced by the sources nearest to the neutral line the filament associated with (Fig. 21). However, in contrast to the vertical sheet extended from the photosphere to infinity, which was studied for stability by Anzer (1969), the region of the sagged field lines has a limited vertical size. It is easy to show that the horizontal and vertical stability in this case require contradictory conditions.

Indeed, for the filament as a whole, the vertical equilibrium implies that

$$F = \frac{I}{c}B(h_0) - mg = 0 , \tag{3}$$

where I is the filament current, h_0 is the height of the filament equilibrium position above the photosphere, B(h) is the magnetic field of photospheric sources, and m is the mass per unit length of the filament. The equilibrium is known to be stable if the second derivative of potential energy is positive,

$$\left.\frac{d^2W}{dh^2}\right|_{h_0} = -\left.\frac{dF}{dh}\right|_{h_0} > 0 . \tag{4}$$

Substituting Equation (3) in (4) and taking into account the fact that g = const on the scale of the vertical prominence extent, we obtain

$$\frac{dB}{dh} < 0\,, \tag{5}$$

which is in conflict with the condition (2) of horizontal stability.

In the simple two-dimensional inverse-polarity model, the force acting in the vertical direction per unit length of the filament is (Van Tend, Kuperus, 1978)

$$F = \frac{I^2}{c^2 h} - \frac{I}{c}B(h) - mg\,, \tag{6}$$

The second term on the right-hand side of (6) represents the part of the total Lorentz force that is due to the potential background field, and the first term represents the remaining part that is due to currents induced in the photosphere (which can be described by a subphotospheric mirror current). These two parts of the Lorentz force are directed downward and upward, respectively. The filament weight mg is sufficient only for equilibrium of very low filaments with weak currents. For high filaments which are ready for eruption, the gravitational term may be ignored. Really, it is follows from Equation (6) that in equilibrium

$$I > chB. \tag{7}$$

Therefore, a lower limit on the upward Lorentz force associated with the photospheric currents is

$$F_B = \frac{I}{c}B > B^2 h\,. \tag{8}$$

For the typical values of B = 5 G and h = $3{\cdot}10^9$ cm, it gives $F_B > 7.5{\cdot}10^{10}$ dyne. The mass m = $2.5{\cdot}10^6$ g per unit length of the filament, needed to counterbalance such a force, is an order of magnitude larger than what is typically inferred from observations. We, therefore, conclude that the gravity term may be left out in Equation (6), so that the total force balance involves only the two Lorentz force terms.

If the filament weight is disregarded, the equilibrium is achieved at height h_0 determined from the equation

$$\frac{I}{ch_0} = B(h_0)\,. \tag{9}$$

The horizontal stability in the inverse-polarity model requires that the relation opposite to (2) be satisfied, which is naturally satisfied for a dipole field. The condition of vertical stability follows from substitution of (6) and (9) in (4):

$$\left.\frac{dB}{dh}\right|_{h_0} > -\frac{B(h_0)}{h_0} . \tag{10}$$

Hence, the filament equilibrium is stable if the vertical gradient in external field lies in the range

$$-\frac{B(h_0)}{h_0} < \left.\frac{dB}{dh}\right|_{h_0} < 0 . \tag{11}$$

If the change of the magnetic field in some range of heights is described by a power-law function

$$B(h) = Ch^{-\kappa}, \tag{12}$$

then we find from (10) that

$$\kappa < 1. \tag{13}$$

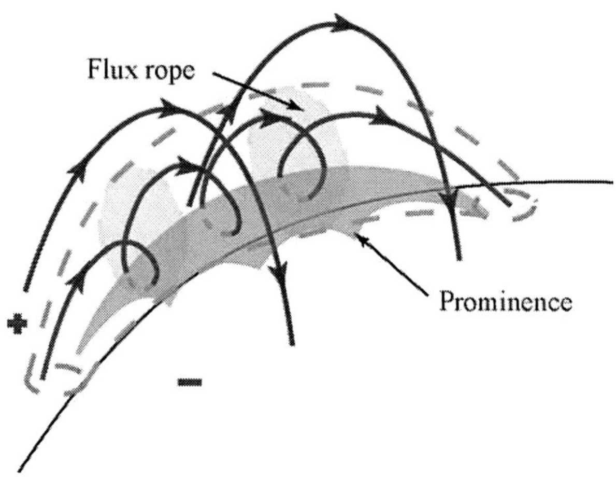

Figure 22. A sketch of the 3D geometry of the magnetic flux rope with a prominence inside.

Thus, for the equilibrium to be stable, the external field in the vicinity of the filament must decrease with height no faster than ~1/ h. This is a result of the height dependence of the magnetic field of the "mirror" current, as suggested by Van Tend and Kuperus (1978). If, beginning from some height h_c, the field falls off faster than ~1/h, then this height is the critical one for quiescent filaments. In models in which the external field is simulated by a linear dipole, the critical height corresponds to the depth at which the dipole lies beneath the photosphere (Molodensky, Filippov, 1987; Martens, Kuin, 1989). For a given background magnetic field distribution, the critical height h_c for stable equilibrium exists defined, as it follows from Equations (12) and (13), by equation (Filippov, Den, 2000; 2001).

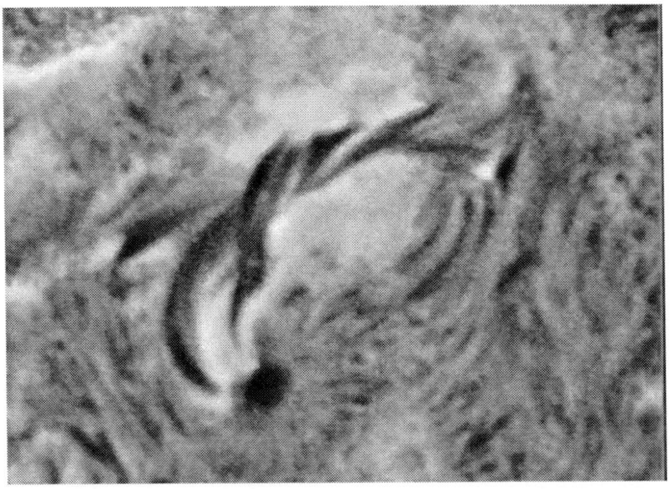

Figure 23. Hα filtergram on 27 June 1980 showing a strongly twisted filament with the internal structure slightly resembling the solution (16) (Crimea Astrophysical Observatory).

$$h_c = \frac{B}{dB/dh\big|_{h_c}} \tag{14}$$

In its pure form, the Kippenhahn--Schluter and Kuperus--Raadu models, of course, are extremely schematic. They illustrate only the principle of equilibrium indicating the main forces critical for the prominence support. In particular, they are two-dimensional and do not consider the magnetic field component directed along the filament axis, whereas in actuality, the filament magnetic field is directed predominantly along the axis. However, this component B_l does not significantly influence the equilibrium conditions, since the filament integral current, in view of its geometry, can be directed only along the axis and, therefore, $I \times B_l = 0$. Surely, it is necessary to take into consideration the axial component in the case of collation of the model's field lines geometry with the structure of threads in a filament. The addition of the axial field creates a shear in the field lines structure in a simple arcade over a polarity inversion line. Closed field lines transform from circles into three-dimensional helixes winding around a cylindrical surface. General 3D structure of the magnetic field in the inverse polarity models is a twisted flux rope inside of a simple arcade (Fig. 22). Since β is small, the field within the flux rope must be close to the force-free field

$$\text{curl } B = \alpha B . \tag{15}$$

One of the simplest solutions of Equation (15) with the cylindrical symmetry and constant α was proposed by Lundquist (1951)

$$\begin{aligned} B_\rho &= 0 \\ B_\varphi &= B_0 J_1(\alpha\rho) \\ B_z &= B_0 J_0(\alpha\rho)\,, \end{aligned} \tag{16}$$

where J_0 and J_1 are the Bessel functions of zero and first order, B_0 is the value of axial component B_z on the symmetry axis, $\rho = 0$. This solution cannon describe precisely the internal structure of a filament because the axial component changes its direction to the opposite in each null of the function J_0. However, if we take only the central part of the rope before the first root of J_0 and sew it with the external field (Molodensky, Filippov, 1988), then equations (16) correspond rather well to observed structure of the twisted filaments: pitch–angle of threads is smallest in the central section of the rope and the threads are elongated almost strictly along the filament axis; on the periphery, the pitch–angle increases (Fig. 23).

A lot of models represent filaments as a part of flux ropes (Van Ballegooijen, Martens, 1989; Amari, Ali, 1989; Isenberg et al, 1993; Forbes, Priest, 1995; Aulanier et al, 1998). Helical field lines, which constitute a flux rope, have dips, capable of supporting the plasma against gravity, at the bottom of each winding. The numerical solution of the system of three-dimensional MHD equations showed the possibility of dip formation in a bipolar arcade deformed by shearing motions near a polarity inversion line (Antiochos et al, 1994; DeVore, Antiochos, 2000; Aulanier et al, 2002). The footpoints of initially bipolar potential field are subjected to linear shearing displacements parallel to the polarity inversion line of the vertical field. These displacements are localized near the inversion line and change sign across it. As the sheared core flux is transported into regions of increasingly weak background field, it begins to inflate upward and eventually to lean over, across the polarity inversion line. The middle sections of these sheared field lines, on the other hand, are restrained from rising by the tension in the overlying, unsheared arcade. When the sheared flux has been displaced sufficiently, dipped field lines are formed in the core. Because the legs at each end of the dipped lines lean across the inversion line, the dipped portions of the field crosses the inversion line in the opposite direction from a simple, current-free arcade. That is, the transverse field within the prominence exhibits inverse polarity. Application of greater shear eventually drives reconnection in the corona, between the highly stressed core flux of the prominence and the overlying, unsheared arcade. The most important consequence of this reconnection is the production of helical field lines threading the prominence and enveloping its core field. The chief disadvantage of this model in the fact that it requires very specific velocity field in the photosphere with the narrow ribbons of antiparallel flows near an inversion line, which is not observed.

It is worth noting that in the three-dimensional case current thread itself can take a form of an arch with a concaved-up central part, which facilitates the accumulation of dense plasma therein (Filippov, 1999). Anchoring of the footpoints in the photosphere requires the current to approach to the photosphere at nearly a right angle. Otherwise, repulsion force from the induced (mirror) currents could not be balanced by the tension of the field lines aligned predominantly along the flux rope axis.

There are morphological models, which attempt to represent the geometry of the filament magnetic field corresponding to the maximum extent to the observed filament fine structure, without worrying about the support of the prominence mass (see, e.g., Martin, Echols, 1994). Such models focus attention on dynamic nature of prominences and imply the action of some unknown mechanism incessantly supplying the magnetic tubes with material and also retarding its free fall. For example, in a dynamic model, support is not necessary if the plasma is continually brought up from the chromosphere and replenished rapidly enough, or if cool plasma threads do not live longer than a few hours but simply represent coronal plasma that is continually condensing and reheating (Priest, 1998).

6. Prominence Height and an Inverse Polarity Model

Within the framework of the existing models prominence equilibrium is possible, in principle, at any height. Question is in stability of the equilibrium. In the classical Kippenhahn--Schluter model, the vertical extent of a prominence is unlimited. As it was shown in Section 5, restriction of the domain with dipped field lines in the vertical size led to the contradictory stability conditions with respect to horizontal (2) and vertical displacements (5). In the model of inverse polarity, stability is determined by the three-dimensional distribution of the magnetic field of photospheric sources. In a given background field, there is a critical height, which can be found from the equation (14). Figure 24 shows distribution of some 80 observed filaments according to their actual height and the critical height calculated for the surrounding magnetic field. The dashed line corresponding to equality of these quantities must be the upper boundary of the distribution. All points are seen to lie actually below this line or close to it.

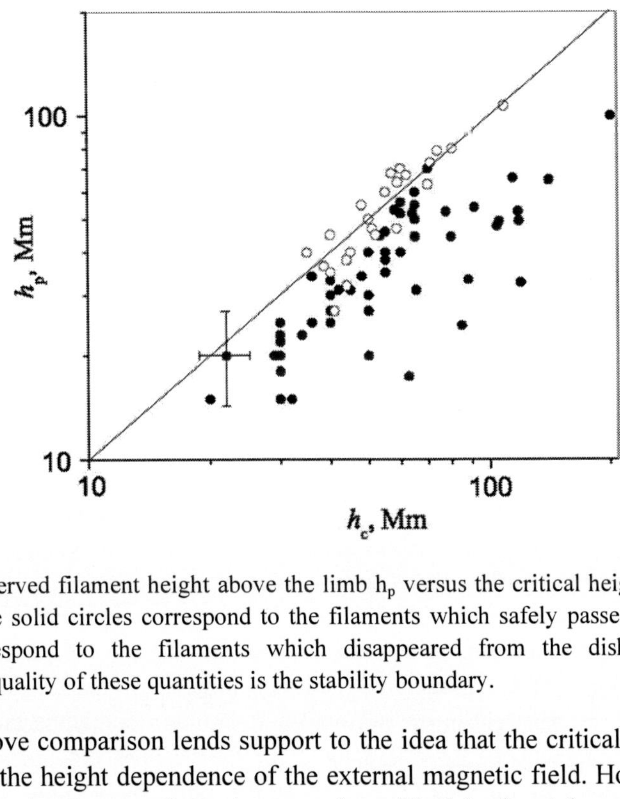

Figure 24. The observed filament height above the limb h_p versus the critical height of stable filament equilibrium h_c. The solid circles correspond to the filaments which safely passed the west limb. The open circles correspond to the filaments which disappeared from the disk. The straight line corresponding to equality of these quantities is the stability boundary.

Thus, the above comparison lends support to the idea that the critical prominence height is determined by the height dependence of the external magnetic field. How can we interpret in this connection the data on the existence of a critical horizontal gradient mentioned in Section 4? We first note that, in a plane current-free field with a translational symmetry along the x axis B = {0, By(y, z), Bz(y, z)},

$$\frac{\partial B_y}{\partial z} = \frac{\partial B_z}{\partial y} . \tag{17}$$

In addition, the fact should be taken into account that the vertical gradient ∂By /∂z, which determines the stability, at the prominence height must be known, whereas the horizontal gradient is measured at the photospheric level. As was already noted above, the critical prominence height for a linear dipole is equal to the depth at which the dipole lies beneath the photosphere. Hence, the z coordinate changes by a factor of 2, while the magnitude of the gradient changes by this factor cubed, i.e., eightfold. Using the inequality (10), for a height of 30 Mm and a field of 25 G typical for quiescent prominences, we obtain

$$\frac{\partial B_z}{\partial y} < 7 \text{ G Mm}^{-1}. \tag{18}$$

This value corresponds to the lower boundary of the measured critical values. Decreasing the height will increase it. For large gradients, the critical height becomes very small, and, probably, a detectable filament cannot form.

In the course of 11-year cycle, magnetic field structure and strength on the Sun change in complex way. It is commonly accepted to assume that the solar global magnetic field is a current-free field between the photosphere and the source at Rs = 2.5R (Schatten, et al., 1969; Altschuler, Newkirk, 1969; Adams, Pneuman, 1976; Altschuler et al., 1977; Hoeksema, et al., 1982). Bravo et al. (1998) studied the behavior of different multipoles throughout the solar magnetic cycle using Hoeksema's source surface current-free model where the photospheric magnetic field is assumed to be radial. They found that at the photosphere level the dipole component (n = 1) is never dominant, not even at sunspot minimum when the octupole component (n = 3) is the strongest.

These two multipoles change in antiphase with the sunspot number having a maximum strength during activity minimum and a minimum strength around activity maximum. Higher-order harmonics roughly follow the sunspot number behavior throughout the cycle. They dominate in photospheric field from the early rising phase of the cycle till the late decay phase, showing the presence of small-scale fields for most of the cycle period. So, there is a phase difference between the lower-order multipoles and higher-order multipoles comprising approximately a quarter of a period of the 22-year magnetic cycle. The higher-order harmonics decrease more rapidly with radial distance than the lower-order ones. Their role in the global field grows smaller when going higher into the corona. Close to the source surface the dipole component dominates at all times.

Polar crown filaments are associated with the most high-latitude large-scale polarity inversion line of the global field in the northern and southern hemispheres. They outline the border of unipolar magnetic cells, which have for their centre the poles. There are three polarity inversion lines in the octupole field at the latitudes 0° and ±51°. In principle, filaments could exist above each of them. The higher the order of the harmonic, the more polarity inversion lines it has on the solar surface. For the ninth harmonic there are nine polarity inversion lines and the highest-latitude lines are located at the latitudes ±76°. So we could interpret the poleward drift of polar crown filaments as the appearance of the new polarity inversion lines at higher and higher latitudes as the result of the dominating of higher and higher harmonics (Makarov, Filippov, 2003).

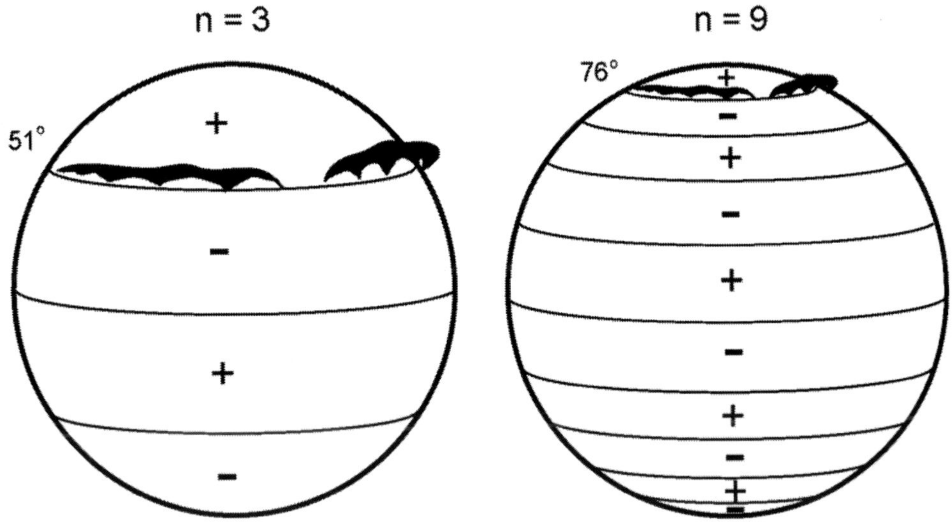

Figure 24. Schematic representation of the poleward migration of the polar crown filaments.

For each harmonic, the limiting height of the prominences is specified by the altitude dependence of the poloidal field component B_θ. We use the standard designations for spherical coordinates r,θ,φ taking into account the relationship between the coordinate θ and the heliographic latitude φ′ in the form θ = 90° - φ′ (for the southern latitudes, the negative sign of φ′ is attributed).

$$B_{\vartheta}^{(n)} = g_n \left(\frac{R_\Theta}{r} \right)^{n+2} \frac{\partial P_n(\cos\theta)}{\partial\theta} \tag{19}$$

where g_n is the coefficient of spherical harmonic n, P_n is a Legendre polynomial. The parameter κ, as it is follows from Equation (12), can be found as

$$\kappa = -\frac{h}{B}\frac{\mathrm{d}B}{\mathrm{d}h}. \tag{20}$$

where h = r - R. Differentiating Equation (19) with respect to h and substituting the result into Equation (20), we obtain

$$\kappa_n = \frac{h(n+2)}{R_\Theta + h}. \tag{21}$$

Finally, from the condition κ = 1, we find

$$h_c^{(n)} = \frac{R_\Theta}{n+1}. \tag{22}$$

As is evident, critical height for the harmonic n is in inverse proportion to its number (degree), increased by one. The position of the inversion lines is determined by the solution of the equations

$$B_r^{\{n\}}(R_\Theta,\theta) = g_n(n+1)P_n(\cos\theta) = 0\,. \tag{23}$$

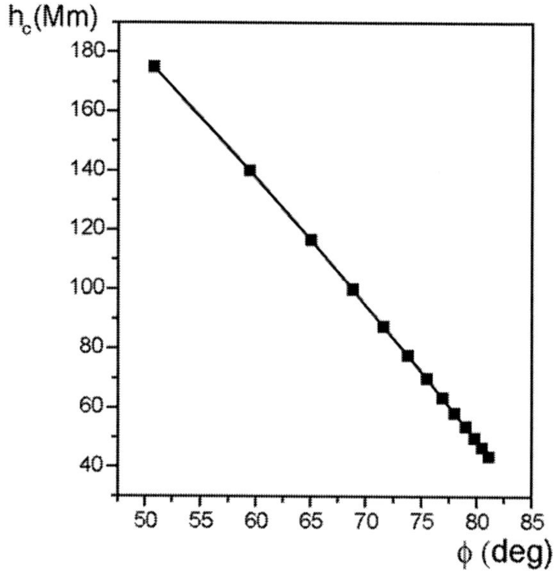

Figure 25. The maximum possible height of a prominence at the given latitude, according to Equation (22).

Taking the smallest root for each n that corresponds to the most high-latitude inversion line we will obtain the dependence the critical height of prominences from the latitude, depicted in Fig. 25. Comparing this plot with the measured average height of prominences shown in Fig. 17, one should bear in mind that even with the uniform distribution of prominences over height; the average value is half of the threshold height. However, it is likely that a prominence spends less time near the threshold level than at the stable heights. Moreover, the measured heights have a tendency to understating because the lower parts of the prominences can be masked by the edge of disk, if they are arranged just beyond the limb. All these factors make intelligible a difference in the scale of y-axes in Fig. 17 and Fig. 25 approximately triply. The general run of the φ-dependence of the height itself corresponds rather well to the experimental data. This allows to state that the factor of two decrease of the height of polar ring filaments during the migration to the poles occurs because of the decrease of the magnetic field scale above the corresponding polarity inversion line.

7. Eruptive Prominences

Despite the fact that filaments are, perhaps, most long-living and stable objects in the solar atmosphere, at times they manifest "sportiveness" and either demonstrate rapid changes and displacements, or disappear from the visibility at all (Fig. 26). Sometimes the activation of a filament is initiated by appreciable external events, for example by a remote flare or fast emergence of new magnetic flux nearby, but more often motion of a filament begins before other manifestations of energetic non-stationary processes in the atmosphere of the Sun. In the majority of cases, there is no any reliable evidence in favor of an external source of filament destabilization.

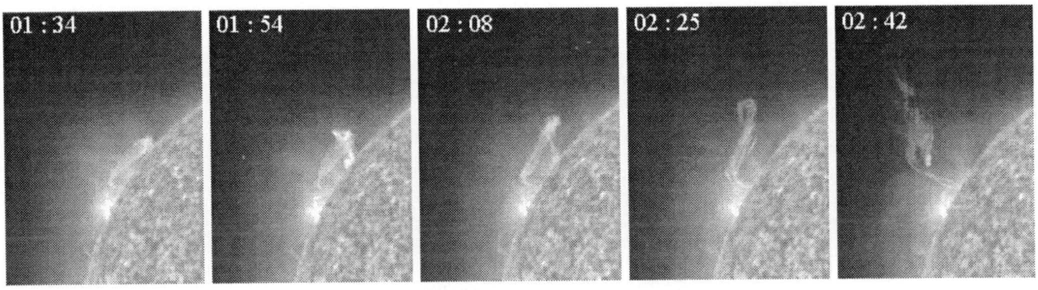

Figure 25. Eruptive prominence in the line He II λ 304 Å on 23 February, 1997 (courtesy of SOHO/EIT consortium. SOHO is a project of international cooperation between ESA and NASA.).

7.1. Precursors of Eruption

A filament before eruption usually increases in sizes and rises above the chromosphere. In the opinion of Zirin (1979), the prominences that have reached a height of 50 Mm are inclined to eruption, which usually occurs within the next couple of days. Vrsnak et al. (1991) found that the relationship between the filament height and the length (the footpoint separation), which reflected the curvature of its axis, was critical for the filament stability. If this value is less than 0.3, the filament remains stable. The existence of the critical height for the filament stability was derived in Section 5 for inverse polarity filaments and we will return to it in Section 11.

Martin and Ramsey (1976) noted the contrast enhancement of filaments in the Hα line before the eruption revealing the density increase. Intense motions are observed along the filament axis, which are gradually reinforced by the beginning of eruption. Frequently oscillations of the whole filament body appear before the eruption with the increasing amplitude. However, all these factors of activation can be found more or less pronounced in the filaments not prone to erupt shortly.

Since prominence existence undoubtedly depends on the magnetic field, it is naturally to search for forerunners of the eruption in field changes. Among the factors destabilizing filaments, there are relative motions of sunspots, which create stress in the field (shear), emergence of new magnetic flux, and the mutual cancellation of flux elements of the opposite polarity (Rust, Kumar, 1994). Bruzek (1952) noted that sudden disappearance of filaments at the distance not more than 25° from new active regions occur twice as frequently as that with the random distribution. Simon et al. (1986) found the relation between the filament activation and generation of new magnetic pores or their motions in immediate proximity of the filament. Feynman and Martin (1995) also noticed that the majority of disappeared filaments were located not far from the place of the emergence of new flux, moreover the probability of eruption grows, if the orientation of the new flux is opposite to existing one and favorable for the reconnection.

Much attention is given to the so-called 'cancellation' of the magnetic flux elements of opposite polarities. This term designates the observed convergence of two small magnetic fragments of opposite polarity, their coalescence, and simultaneous disappearance. There are several alternative explanations of this phenomenon. For example, such field changes in the photosphere will appear during sinking of a small magnetic loop, or lift into the corona of a U-shaped tube (Zwaan, 1985; 1987; Low, Hundhausen, 1995). Corresponding flux tube motion can be a consequence of field lines reconnection, which belong to two isolated loop systems, in a point above the photosphere or in a point under the photosphere (Van Ballegooijen, Martens, 1989; 1990; Martin, Livi, 1992). Convergent motion of the magnetic elements of opposite polarity toward a neutral line and their cancellation are considered as the necessary condition of filament formation and as a contributory factor for eruption (Martin, Livi, Wang, 1985; Hermans, Martin, 1986; Martin, 1990).

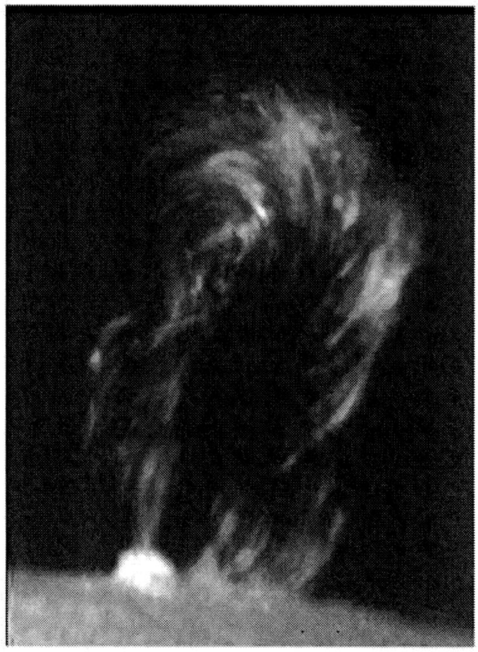

Figure 26. Eruptive prominence in the Hα line at 17:17 UT 18 August 1995 with the pronounced helical structure (courtesy of Big Bear Solar Observatory/New Jersey Institute of Technology).

Helical structure is frequently visible in eruptive prominences (Fig. 26). Some researchers cast doubt on the fact that the helical structure exists in prominences prior to the beginning of eruption. They assume that helical field lines are forming during the eruption as a result of reconnection (Gosling et al., 1995; Antiochos, 1998; Martin, 1998b). The pronounced helical structure is visible, to be sure, not in all prominences and filaments; however, it cannot be assumed as rare in occurrence (see Fig. 3, Fig. 6, Fig. 23). A single image does not allow to reconstruct exactly the three-dimensional structure of a filament. Study of a sequence of images over a long enough time interval, when the projection noticeably changes, does not leave doubts that the helical threads exist, at least in some filaments, long before the eruption The amount of twist is connected with the value of azimuthal (relative to the filament axis) magnetic field and an axial electric current

$$I = \frac{1}{2} c r B_{\varphi} . \tag{24}$$

Assuming that axial magnetic field B_l is approximately of the same magnitude as the potential field of photospheric sources, it is possible to estimate the value of the azimuthal field using the measured pitch-angle δ

$$\tan \delta = \frac{B_{\varphi}}{B_l} \tag{25}$$

and to find the electric current strength. For the filament shown in Fig. 23 the current is estimated as $3 \cdot 10^{11}$ (Kulikova, et al., 1986). Similar estimations of the filament currents were obtained for other twisted filaments (Ballester, 1984; Vrsnak et al, 1988; Srivastova et al., 1991). Usually, all filaments characterized by the pitch-angle $\delta < 40°$ and the total twist along the whole filament length from one footpoint to the other $\Phi < 2\pi$ are stable (Vrsnak et al., 1991). Eruptive prominences have pitch-angle $\delta > 50°$ and the total twist $\Phi > 3\pi$. However, as Vrsnak et al. (1991) pointed out, the large total twist was not sufficient for the destabilization of a filament. In the case of the small ratio of the filament height to its length, it remains stable even with large twist.

Thus, even if there exist some specific signs of approaching filament state to eruption, they are in changes in the position and internal filament fine structure rather than in the processes outside of it. Many factors could play the role of last ‘trigger’, but no one of them could be assumed as the controlling factor and absolutely necessary one. The main thing is that the prominence would be ‘matured’ for the eruption.

7.2. Kinematics of Eruptive Prominences

Initially almost straight and horizontal filament in the beginning of eruption bends in the vertical (or slightly inclined) plane in the form of an arch whose ends remain fixed in the chromosphere (Fig. 25, Fig. 26). Lateral intermediate prominence feet (barbs) detach one after another from the chromosphere and only two feet at the ends of the prominence are remained, which connect prominence body with the chromosphere. At the beginning of eruption, speed grows slowly in contrast to such ejections as surges and sprays, in which acceleration is highest in the initial stage (Nakagawa et al., 1975). Speed can initially be mo more than several km/s. This is typical both for the active region filaments and for the quiescent prominences, but for the former slow phase lasts a minute or a couple of tens of minutes, while for the latter it can have duration of an hour or two. Then rather rapid acceleration follows. Absolute value of acceleration in some cases several times exceeds the acceleration of the free fall on the surface of the Sun ($g = 2{,}7 \cdot 10^4$ cm s^{-2}).

It is necessary to keep in mind that the real beginning of eruption is not always fixed by observers. Filament can be located beyond the limb, and the first, slow phase remains unnoticed. Such type of phenomenon sometimes is difficult to distinguish from flare sprays, in which acceleration occurs very violent and speed rapidly increases to the values frequently exceeding the speed of escape from the solar gravitational field. Kinetic energy density in sprays, apparently, exceeds the magnetic energy density, and material, in contrast to eruptive prominence, is not confined by a magnetic trap and flies away freely as during an explosion.

On the basis of detailed study of about 100 eruptive phenomena Vrsnak (2001) found that the majority of prominences and sprays (87% and 75% respectively) even at the stage of the basic acceleration demonstrate the exponential law of height change versus time. Therefore, the speed and acceleration are directly proportional to the height (distance from the initial position). This stage can be named the linear phase of acceleration. In the late stage of the eruption, the prominence acceleration decreases (25%), speed remains constant (42%) or it decreases (33%). The majority of sprays (67%) are decelerating in the late stage. The peak acceleration of eruptive prominences is on the average equal to 10^4 cm s^{-2} and is reached at the height of approximately 250 Mm, while that of sprays is 10^5 cm s^{-2} at a height of 30 Mm.

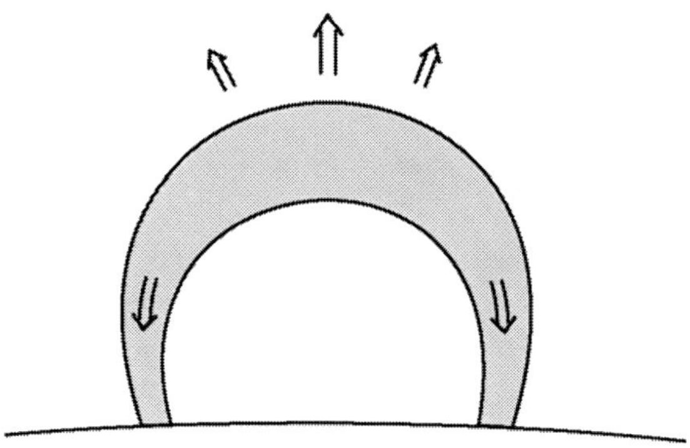

Figure 27. Schematic representation of the material motion in an eruptive prominence.

The arch of eruptive prominence stretches rapidly upward, and the ascending speed reaches value of hundreds km/s. Part of the prominence matter drains along both legs of the arch downward into the chromosphere (Fig. 27), the other part of the matter acquires the speed sufficient to overcome the solar gravity (about 600 km/s) and it is ejected into the interplanetary space. The downward plasma motion is noticeable not only due to the displacement of blobs and inhomogeneities inside the prominence on the limb, but it is also confirmed by the measurements of the Doppler shifts of spectral lines in observations of filament eruptions on the disk. The central section of the eruptive filament, which corresponds to the apex of the arch, is visible in the blue wing of line, whereas the ends adjoining the footpoints are manifested in the red wing. Density lowering near the top of the loop, because of its expansion and partial draining of material, creates the impression of the break of the prominence loop when it is observed in Hα; however, when images are available in ultraviolet lines, it is possible to see that the loop remains continuous with the enormous size too (Fig. 28). Sometimes, only a separate segment of a long filament is destabilized, whereas remaining sections remain in the same place. There are examples when a layer of a filament is seen to separate from the parent filament (Tang, 1986). This layer may erupt immediately, or it may stay in place for tens of minutes before erupting.

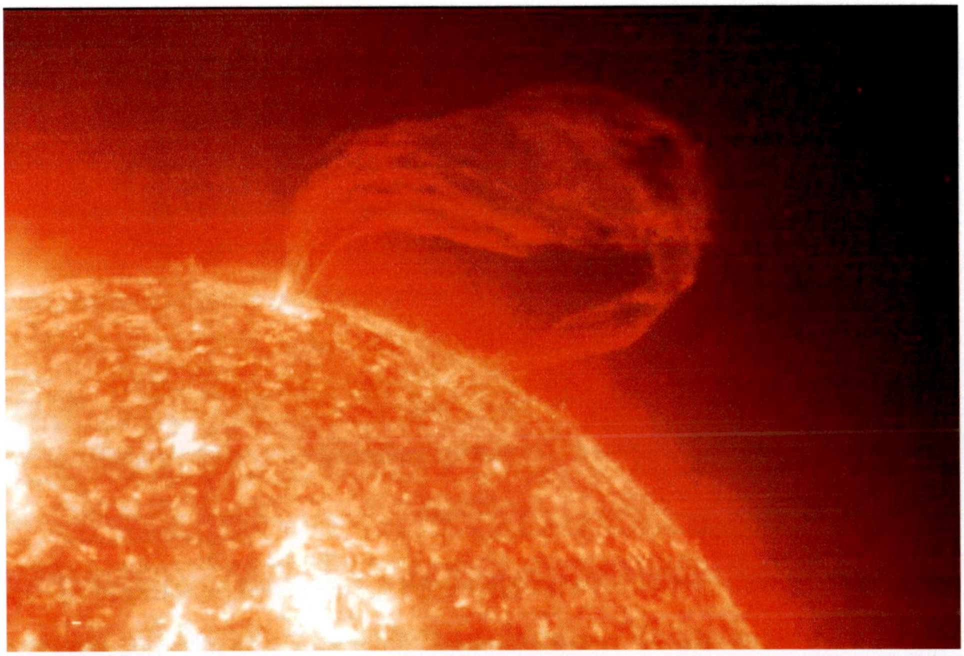

Figure 28. Eruptive prominence in the He II λ 304 Å line (courtesy of SOHO/EIT consortium. SOHO is a project of international cooperation between ESA and NASA.).

8. Catastrophic Loss of Equilibrium

As it has been established Section 5, the most probable magnetic configuration, which supports the plasma of solar prominences high in the corona, is the twisted magnetic flux rope. The flux rope might stay in the stable equilibrium under the action of external and internal forces for a long time (a month or more), but in the case of loss of equilibrium or development of instability, eruption occurs. Since observations do not give clear answer to the question about reasons of the beginnings of eruption, different factors of destabilization of the magnetic configuration are considered in different models.

There are alternative approaches to the problem of location of the reservoir of energy that is transformed into kinetic energy of the ejected mass during the eruption. Energy storage models assume slow accumulation of magnetic energy in the form of coronal currents generated by photospheric motions (rotation the footpoints of magnetic tubes, their rearrangement, shearing of arcades, etc.). Then energy suddenly releases as a result of instability growth or loss of equilibrium. Models with injection of energy suggest sudden rapid penetration of an additional magnetic flux from under-photospheric layers into the corona.

In essence, the problem is analogous to the equilibrium of a toroidal pinch with the longitudinal field studied by Shafranov (1966). The forces acting along the major R and minor a radii of a plasma torus ($R \gg a$) are expressed as

$$F_R = \frac{V}{R}\left(\frac{B_a^2}{4\pi}\left\{\ln\frac{8r}{a} - 1 + \frac{l_i}{2}\right\} + \overline{p} - p_e + \frac{B_e^2 - \overline{B}_i^2}{8\pi} + \frac{B_z B_a R}{2\pi a}\right), \tag{26}$$

$$F_a = \frac{2V}{a}\left(-\frac{B_a^2}{8\pi} + \overline{p} - p_e - \frac{B_e^2 - \overline{B}_i^2}{8\pi}\right), \tag{27}$$

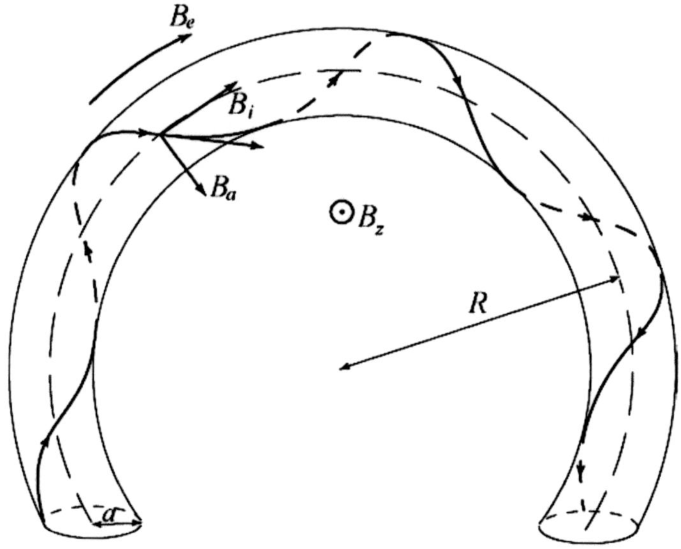

Figure 29. A toroidal pinch with the longitudinal field.

where V is the volume of the torus, B_a is the poloidal component of the magnetic field, $\overline{B}_i$ is the averaged over the cross-section of the radius of a internal longitudinal (toroidal) field, B_e is the external longitudinal field, B_z is the external field perpendicular to the plane containing the torus axis, $\overline{p}$ is the averaged over the torus cross-section internal plasma pressure, p_e is the external pressure, l_i is the internal inductance per unit of the torus length. After adding the gravitational force and the influence of the dense inertial photosphere, is possible to construct different models with one or other geometric and physical simplifications. They represent a filament by the segment of a torus with the ends fixed in the photosphere (Anzer, 1978; Mouschovias, Poland, 1978; Van Tend, 1979; Chen, 1989; Vrsnak, 1990), or by a closed torus over the whole Sun, which hangs above some solar parallel (Lin et al, 1998; Filippov, Gopalswamy, Lozhechkin, 2001), or simply by a straight cylinder (Kuperus, Raadu, 1974; Van Tend, Kuperus, 1978; Molodensky, Filippov, 1987; Martens, Kuin, 1989).

Equilibrium condition $F_a = 0$ along the minor radius a must be fulfilled accurately in the initial stable state and approximately during the eruption, since the filament does not explode with the dispersion in different directions but it rises more or less preserving its form. The main thing for understanding the causes of eruptions is balance of forces along the major radius, thus in some, especially early, models equilibrium inside the torus is assumed to be inevitably retained and is not considered. A flux rope in this case is replaced by a flexible and extensible conductor. Moreover, since the ratio of the major radius to the minor is very great,

R >> a, it is possible to treat the conductor to be rectilinear. Such gross simplification, nevertheless, do not lead to the loss of the most important nonlinear property of the model, namely the possibility the process of the catastrophic loss of equilibrium.

For the first time the idea of existence of critical value for the equilibrium of a straight linear current in the background magnetic field above a conducting surface was expressed by Van Tend and Kuperus (1978). They compared the height dependence of the magnetic field of photospheric sources and the field of a mirror current. Since usually the effective exponent index κ in (12) for the background field changes with height, while for the mirror current field it is equal to 1, there are regions of stable and unstable equilibrium. Molodensky and Filippov (1987) approximated external for the filament current magnetic field in the corona by the field of a vertical two-dimensional dipole M, located at a depth d below the photosphere. Neglecting the filament weight, two equilibrium positions correspond to each value of a current I

$$h_{1,2} = -\frac{Mc}{4I}\left(1 \pm \sqrt{1+\frac{8Id}{Mc}}\right) - d\,. \tag{28}$$

The lower equilibrium point is stable, while upper one is unstable. With an increase in the current I or a decrease of the dipole moment M, the equilibrium points approach each other and merge into one point upon reaching by the current of the critical value I_c determined by vanishing of expressions under the root in (28). The height $h_c = h_1 = h_2$ in this case is equal to the depth d of the bedding of the dipole. Figure 30 shows the schematic dependence of the potential electric current energy from the height. The solid curve corresponds to a current strength lower than the critical value. Current represented by a circle can be in stable equilibrium in the pit and be unstable at the apex of a hump. The dashed curve corresponds to the critical value of current. The pit and the hump have merged, and the current lies on the narrow footstep whence it will fall down to the side of increasing h under any fluctuation.

Immediately after the loss of equilibrium, filament is accelerated very slowly because the electromagnetic forces still practically completely compensate each other (the equilibrium point is a null point created by subphotospheric sources and induced currents). As the filament moves away from the equilibrium point, acceleration grows. Filament mass is now essential, since it defines the inertia term in the equation of motion. In order to obtain the precise kinematical characteristics of motion it is necessary to take into account the equation of induction, which takes the form of the condition of conservation of the total magnetic flux through the contour formed by the filament. The influence of coronal plasma on the filament motion is neglected. According to estimations made for an eruptive filament located near a large spot, the maximum acceleration can many times exceed the value of the acceleration of the free fall (Molodensky, Filippov, 1987).

Similar model was in detail studied by Martens and Kuin (1989) with other direction of the dipole axis (horizontal). Major difference lies in the fact that they assume the development of a current sheet along a neutral line, which appears below the filament during the eruption. Together with its return current induced in the photosphere, the current sheet forms the second current circuit, which is inductively coupled with the filament circuit (the filament and its return current). Appearance of the current sheet makes it possible to include into the model the complex of phenomena described by the theory flaring current sheets

(Parker, 1957; Sweet, 1958; Petschek, 1964; Syrovatskii; 1966; Sonnerup, 1970; Priest, Forbes, 1986; Somov, 1986). Thus, Martens and Kuin attempted to connect more closely eruptive and flare processes, which really were often observed simultaneously.

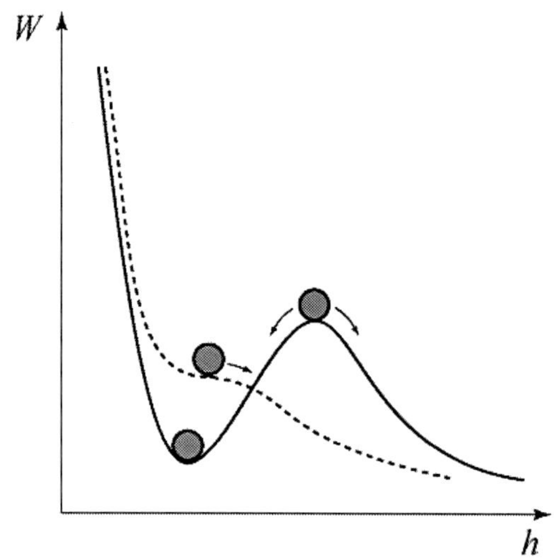

Figure 30. Schematic representation of the dependence of the potential energy of an electric current versus height.

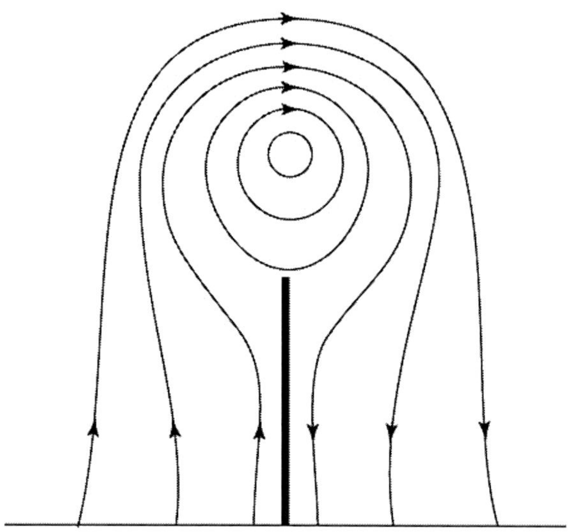

Figure 31. Current sheet below the erupting flux rope without reconnection.

Within the limits of the ideal magnetohydrodynamics, field lines do not reconnect, and the current sheet spreads from the photosphere to the null point under the filament (Fig. 31). Even in this case, the model allows the catastrophic loss of equilibrium, although the flux rope will pass only finite distance to a new equilibrium pint (Forbes, Isenberg, 1991; Forbes et al., 1994; Forbes, Priest, 1995). Depending on the form of boundary conditions in the photosphere, the fraction of initially stored magnetic energy ranged from 1% to 20% is released after the loss of equilibrium. Most of the energy is contained in the current sheet. The following evolution depends on what processes predominate. If reconnection is absent and kinetic energy of the flux rope dissipates, the rope will stop in the upper stable equilibrium point. If there is neither reconnection nor dissipation, the rope will oscillate between the point of loss of equilibrium and some point above the upper equilibrium position. Finally, if reconnection is rapid enough, the rope will rise unlimited, although its motion can slow down after passing the upper equilibrium point. It seems quite probable that reconnection in the corona can be fast, when it is forced by intensive plasma motions (Priest, Forbes, 1990).

Inevitable current sheet formation below an eruptive prominence does not seem so obvious. The idea arose from consideration of ideal conductivity plasma in the two-dimensional (flat) magnetic field with a singular null point. If electric field is applied along the third coordinate, it cannot be compensated by the induction field

$$\mathbf{E} = -\frac{1}{c}\mathbf{v} \times \mathbf{B} \qquad (29)$$

on the line, where B =0, and current would increase unlimitedly. In geometry with such translational symmetry, a neutral line does exist in the center of the hyperbolic magnetic configuration (a saddle singular point or an X- point). The electrodynamic force I × B/c does not act in this case upon the current, so it stays in mechanical equilibrium, although the equilibrium is unstable.

In real situations on the Sun, such high degree of symmetry, naturally, is impossible. Instead of the neutral line there could exist only separate neutral points in the three-dimensional magnetic field, so that the line, where a current sheet could be generated, does not exist. The addition of the third component of the magnetic field changes situation even in the idealized case. The presence of the longitudinal field along the former neutral line seemingly changes nothing, however, if electric field is created only by effects of induction under changes of the boundary conditions of the magnetic field, it is everywhere orthogonal to the magnetic field and no electric field appears along the field line that coincides with the former neutral line.

Generally speaking, it seems that inappropriately high significance is attached to the problem of the field line reconnection in the vicinity of saddle points. The reason is that initially, within the framework of simple two-dimensional models, the hyperbolic magnetic configuration has seemed to be very promising for the accumulation of strong coronal currents and then fast release of magnetic energy providing explanation for the phenomenon of solar flares. Four decades, which had passed from the beginning of intensive theoretical studies in this field, raised, perhaps, more questions than gave answers. Observations did not provide strong proofs of the presence on the Sun of current sheets with those characteristics that follow from the theoretical ideas up to now, although there are numerous attempts to treat separate observations as the manifestation of energy release during the magnetic reconnection (Tsuneta, 1996a,b; Wang, Wang, 1996). In those cases when the saddle shape is clearly visible in the structure of coronal loops and chromospheric fibrils, there are no any manifestations of the presence of currents and additional energy release in the saddles (Molodensky et al., 1990; Filippov, 1995; Filippov, 1999). On the contrary, the nearest environments of a null point seem to have a reduced temperature (Fig. 32) in comparison with the adjacent coronal loops. There are no doubts that reconnection of magnetic field lines in the corona does occur. This is proved, for example, by the transequatorial coronal loops connecting different active regions, which have emerged not simultaneously. The question is does real reconnection in the corona lead to effects that theory expects?

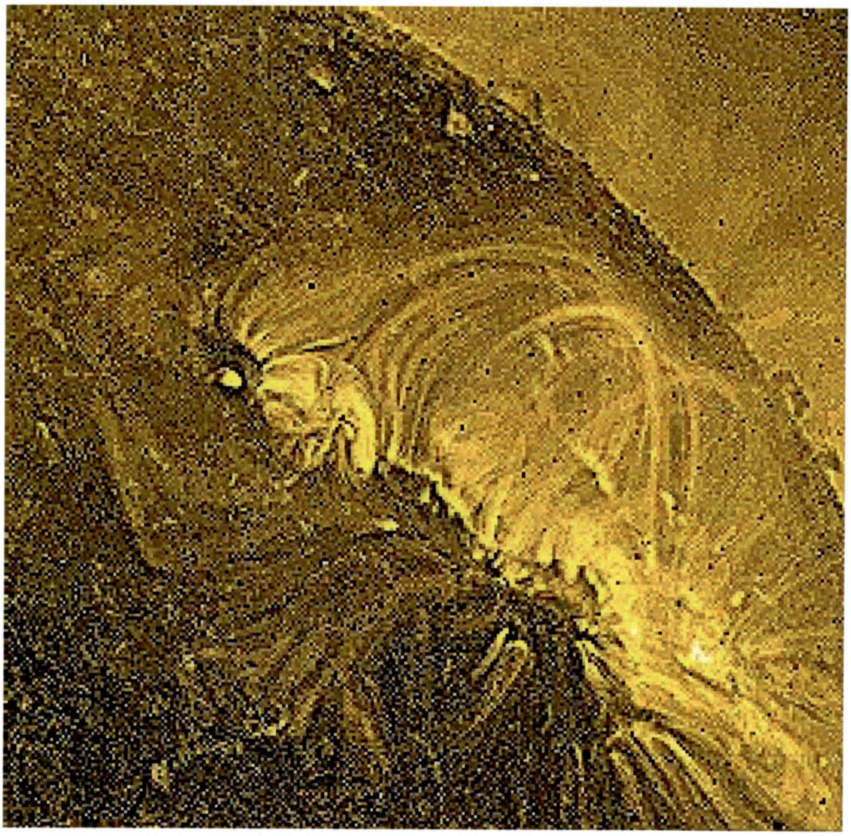

Figure 32. SOHO EIT image of the solar active region NOAA 8113 in Fe XV, 284 Å on 6 December, 1997 at 19:06 UT. The image was filtered using unsharp mask to reveal the structure (courtesy of SOHO EIT Consortium; SOHO is a joint ESA-NASA program). The central part of the saddle does not contain bright loops. The ratio of the image brightness in different EIT channels, which presents the overall temperature distribution of 1 to 2 MK plasma in the corona, shows a relatively low temperature in the environments of the null point (Filippov, 1999).

Possibility to eliminate electric currents in an explicit form from the MHD equations allows speaking about "annihilation" of magnetic field, although physically more clear sense is that we deal with the dissipation of an electric current. In the conditions of the solar corona, where the energy density of magnetic field, as a rule, exceeds the energy density of other forms, it is difficult to expect "the convergence of oppositely directed magnetic fluxes" indicating appearance of a strong local electric current, which is in unstable equilibrium. The possibility of the separate topologically singular points of the magnetic field to influence the general evolution of the field and to permit the reconnection of field lines or not seems no less strange.

Figure 33. Saddle structure containing transequatorial loops is soft X-rays on 7 November 2001 (courtesy of Yohkoh Team).

One way or another, the two-dimensional models of flux rope equilibrium with translational symmetry allow the sudden catastrophic loss of equilibrium with the conversion of the accumulated magnetic energy into the kinetic energy after reaching by parameters of the system some critical values under the action of slow changes in the boundary conditions. Thus, eruption is a manifestation of intrinsic properties of a system, the consequence of its nonlinearity, and does not need strong external triggers. This property remains also in the two-dimensional models with the axial symmetry (Lin et al, 1998; Filippov et al., 2001). Ends of a rope, in contrast to translational symmetry, do not depart to infinity, but they are also not connected to the surface. Extra force, which appears as a result of the curvature of a current, tends to stretch torus along the major radius.

Several models consider the holding action of the ends of a rope anchored in the photosphere. They could be assigned to 3D class if they were not so schematic and analyzed only the equation of motion for the apex of a loop. In this way one more additional force is introduced, which is created by the tension of field lines attached to the surface of the Sun. At the same time, the shape of the eruptive prominence is assumed for simplicity as a segment of a circle in the vertical plane. Models of this type are more frequently used in application to coronal mass ejections. The model of Anzer (1978) probably was the first, in which the motion of an expanding current ring fastened at two footpoints to the photosphere was analyzed. The basic objective of the author was to obtain the speed dependence of the heliocentric distance similar to the observed velocity profiles of CMEs. Initial state in this model was not equilibrium, and the reason for the appearance of this situation was not examined. Van Tend (1979) removed this deficiency, investigating the initial equilibrium of the current ring, which was identified with a prominence. The apex of a loop can stay in the stable equilibrium with the balance of the electrodynamic force stretching the ring and the prominence weight, if the ring radius is less a certain value R_1 or more than R_2. With intermediate values of the radius of the loop, the equilibrium is unstable, therefore, when the prominence reaches the height R_1, as a result of current increase or a drop of the weight, the eruption will occur. Catastrophe can start with anywise slow changes, but as in other models, where the weight is one of the determining factors in the balance of forces, acceleration cannot exceed the certain fraction of the free fall acceleration.

In a number of models not only the stretching action the azimuthal (poloidal) magnetic field of a rope B_a is considered, but also the action of the longitudinal (toroidal) field B_i, which is trying by its tension to shrink the ring. Solov'ev (1985) analyzed the equilibrium of a flux rope in homogeneous medium without magnetic field and with a presence of longitudinal (with respect to the rope axis) potential field or transversal field. He made also some conclusions on stability of the flux rope; for example, in the uniform medium, the rope is stable relative to tension or compression (Solov'ev, 1981). With the balance of the mentioned forces and gravity

$$\frac{B_a^2}{4\pi R_c} - \frac{B_i^2}{4\pi R_c} = -\frac{mGM_\Theta}{r^2}, \tag{30}$$

where R_c is the radius of curvature in the apex of a loop, r is the distance from the center of the Sun, and also the conservation of magnetic flux and the prominence mass m, the motion of the apex of the loop occurs with the constant velocity and it approximately corresponds to the nature of motion of CMEs (Mouschovias, Poland, 1978).

In the model of Vrsnak (1990), the force reaction of the photosphere (or diamagnetic force) is additionally included, expression for which per unit of length coincides with the expression for the straight wire (Kuperus, Raadu, 1974). Analysis of the dependence of equilibrium height of the loop apex on the strength of the longitudinal (toroidal) current showed that with an increase of the twist, i.e., an increase of the ratio B_a/B_i, the profile of equilibrium curve acquires the form letter S, on which the section with the negative inclination corresponds to unstable equilibrium. As in other models, jumping from one steady branch to another has catastrophic nature and it indicates the eruption of the flux rope, which contains prominence.

Combined analysis of the equations of motion of toroidal flux rope with the fixed in the photosphere footpoints along the major radius as well as along the minor one was made in the papers of Chen (1989; 2001). The right sides of the equations of motion are taken in the form (26) and (27), the role of photosphere is limited to the fixation of the current loop footpoints, but its diamagnetic action on coronal currents is not considered. The chain of the coronal current continues into the convection zone, into the area of action of the dynamo mechanism, and the properties of the subphotospheric part of the chain are very important for the model. Total toroidal magnetic flux and poloidal flux are considered as constant. In the absence of external magnetic field, the torus is stable relative to variations of the major radius R only when the ratio of the inductance of the coronal part of the current circuit to the total inductances is more than a certain critical value. So, the rope is stable only when the subphotospheric part of the chain is short and is located close to the photosphere. The model allows transfer from stable state to unstable, but all preparations for this transfer are located under the photosphere and cannot be observed. The latter, however, is in accordance with the absence of visible precursors of eruptions. Numerical calculations of the dynamics of the loop after the loss of equilibrium show that the balance of forces along the minor radius approximately conserves, so that the loop rises without significant increasing in cross-section. A rate of expansion along the minor radius is an order of magnitude less than a rate of expansion along the major radius. This fact can be attributed in support of the models, in which the internal structure of a flux rope is not analyzed.

Under the assumption of the ideal conductivity of the photosphere, when the coronal current is closed through its surface, and in the absence of the external field, Chen's model in the essence closely resembles to the simpler model of Van Tend (1979). The difference between them is in the presence of tension of the toroidal field lines. Probably, because of this difference there are no indications of the existence of the certain interval of unstable equilibrium in Chen's model. Addition of the external field in the corona made it possible to obtain different regimes of equilibrium and the dynamics of the toroidal flux rope (Chen, 1990); however, primary attention in the subsequent papers was paid to the possibility of the "injection" of magnetic energy from subphotospheric layers into the corona along the current chain (Chen, Garren, 1993; Krall et al., 2000). The main application of the model was the description of the dynamics of CMEs and interplanetary magnetic clouds. With sufficiently fast increase of the toroidal current (poloidal magnetic flux), the current loop begins to expand passing consecutively through new equilibrium states. Selecting the law of flux changes, it is possible to obtain the preassigned temporal profiles of position, speed, and the ratio of radii of the torus. It should be noted that for obtaining the observed characteristics it is necessary to increase the poloidal flux 5-10 times within 2-3 hours. Generation of the poloidal flux occurs under the photosphere, possibly, due to the action the mechanism of dynamo. Conceived in the convective zone, the disturbance propagates along the flux tube with the speed of the order of the Alfven speed ensuring the transfer of the poloidal flux into the corona (or equalizing the electric current along the entire chain). Thus, the energy necessary to lifting of the loop and supplying it with kinetic energy is transported from the convection zone into the corona during the process of eruption.

Of course, the question arises, what is the mechanism working in the convection zone, which does not change the poloidal flux in the rope long time before the eruption, then increases it by a factor of ten within several hours, and does again cease its action? Besides, observations do not confirm the presence of the specific rotation motions at a speed of 1-10 km/s during the eruption near the loop footpoints, which should demonstrate the transfer of the poloidal flux. Meanwhile, the authors of the model investigated the case of the catastrophic release of energy stored in the coronal current (Krall et al., 2000). Kinematical characteristics in this case are determined mainly by the structure of background coronal magnetic field. Even with the sufficiently simple and arbitrary dependence of external field on the height, temporary profiles were also in rather good agreement with those observed in CMEs. Basic difference from the case with the injection of the poloidal flux was in the existence of comparatively short initial period of very energetic acceleration, which was not presented, in the opinion of the authors, in CMEs. However, intensive acceleration not long after the loss of equilibrium is rather typical for eruptive prominences, furthermore, by more thorough selection of the external magnetic field distribution, probably, it is possible to attain better agreement with observational data.

Thus, the mechanism of the slow accumulation of magnetic energy in the longitudinal current of magnetic flux rope in the corona and the catastrophic loss of equilibrium upon reaching of the threshold of stability is inherent to a set of models with different geometry and different range of details. The mechanism is particularly attractive because the catastrophe is caused by processes in the corona and the moment of its start can be predicted, in principle, on the basis of observational data.

9. Eruptive Prominences and Flares

Figure 34 shows a series of Hα images with the disappearance of a filament on the disk on 24 June 1999. In viewing a movie, intensive motions along the filament axis in both directions are visible before eruption. Possibly, the motion has an oscillatory nature. Then, after 12:20 UT, the filament begins rapidly to rise. Almost immediately after beginning of the filament ascending, when the displacement of the filament has not yet exceeded its previous height, the

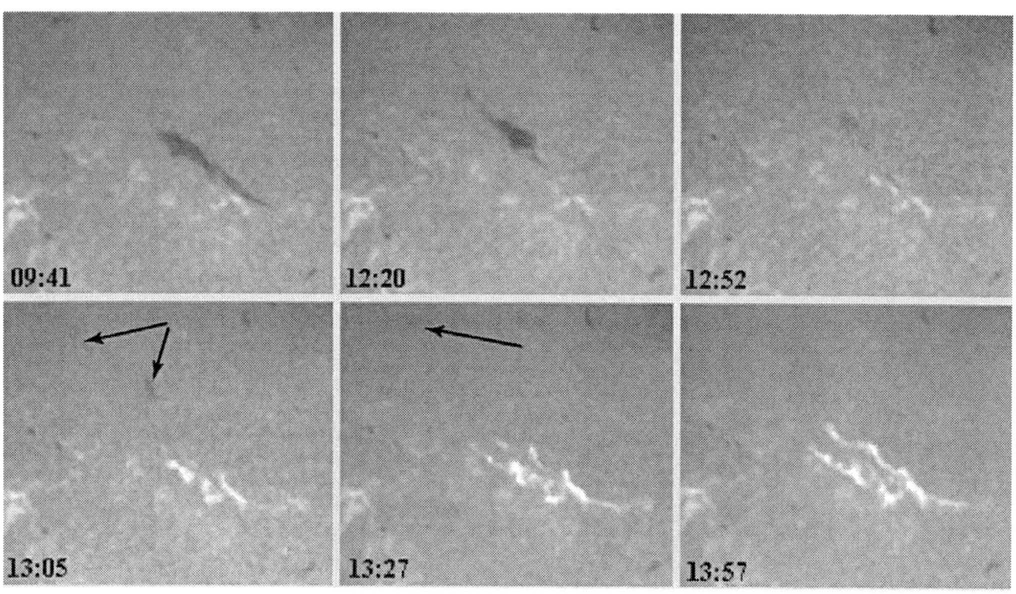

Figure 34. Eruption of a filament in the Hα line on 24 June, 1999 and subsequent development of two-ribbon flare Arrows point to the fragments of eruptive prominence (courtesy of Paris Observatory, Meudon).

brightening appears below it well noticeable already at 12:52 UT. Further flare ribbons spread out along the polarity inversion line, which previously was designated by the filament, and they diverge from it in opposite directions. Several minutes or several tens minutes after the filament disappearance, an arcade of bright loops visible in the X-rays and EUV is formed at the place where the filament was located previously (Fig. 35).

Figure 35. Arcade of postflare loops in the line Fe IX/X 171 Å on 24 June, 1999 at 17:30 UT, which appears after the filament eruption (Fig. 34) (courtesy of SOHO EIT Consortium; SOHO is a joint ESA-NASA program).

This example illustrates the sequence of events in the eruptive phenomenon that occurs out of active regions. Flares far from sunspots proceed only after the filament eruptions. However, in active regions, as a rule, a filament eruption also precedes the beginning of a flare (Zirin, Lackner, 1969; Martin, Ramzey, 1972). Since the active region filaments are thinner and lower than the quiescent filaments, their disappearance or rapid ascent might sometimes be missed.

Eruption and brightening in Hα line happen practically simultaneously, which initiates different opinions on causal relationship between the two phenomena. For example, some authors assume that fast reconnection of field lines at the null point under the filament, which provides flare energy release, decreases the magnetic tension that holds the filament and allows it to rise (Anzer, Pneuman, 1982). The filament in this case passively follows the changes in the magnetic flux under it. Another point of view suggests that the lift of a magnetic flux rope due to the loss of the mechanical equilibrium draws out the field lines of the surrounding magnetic arcade creating a current sheet (Fig. 31). Subsequent reconnection in the current sheet ensures the flare heating, particle acceleration, and the formation of the rising arcade of post-flare loops (Priest, 1981; Moore, 1988; Shibata, 1999). One way or another, unquestionably however that both the filament eruptions and the two-ribbon flares are the result of the process of catastrophic reconstruction of the magnetic field.

10. Prominence Heating during Eruption

Observations in EUV spectral lines with UV telescopes on board SOHO and TRACE satellites show the thermal state of the prominence plasma. Cold and dense material of quiescent prominences absorbs the radiation of coronal structures revealing itself thus as dark features in EUV images (Kucera et al., 1998; Engvold et al., 2001; Heinzel et al., 2001). During eruption some parts of prominences become bright emitting UV radiation in coronal lines. This means that there is an intense heating mechanism that works in the bodies of the prominences. In a very short time, within a minute, the temperature of several parts of a prominence rises to the coronal level. It seems that the heating begins when the prominence as a whole reaches a rather high speed. Heating of the prominence material can be a reason for its disappearance from the field of view of a coronagraph recording just the Hα line. Therefore, in this case a combined type of filament disappearance is possible involving both dynamic and thermal processes (Mouradian et al., 1995).

Figure 36 shows an eruption of a filament on 16 April 2000 observed during the TRACE mission (Filippov, Koutchmy, 2002). In the strict sense, this event did not escape the Sun. The prominence stopped ascending within 30 minutes and the prominence material then drained down back to the low corona. According to the terminology of Tandberg-Hanssen (1995) (see also Gilbert et al., 2000), this is not an eruptive prominence but an active one. The kinematics of the apex of the filament is presented in Fig. 37. It makes no difference from the point of view of the initial stage of phenomenon whether further motion of the filament is finite or infinite (Molodensky et al., 1992). Sudden ascent begins for the same reason of the loss of equilibrium, but in the case of finite motion, the existence of other stable equilibrium point does not allow the filament to slip off from the surrounding magnetic arcade. The eruption began at 03:45 UT with a slow rising of the dark compact filament body. The polarity inversion line seems to be directed along the line-of-sight, thus, the filament arch lies in the plane perpendicular to the line-of-sight and we observe the filament from one end. Then the fast acceleration leads to an ascending filament velocity with a maximum of about 100 km/s. At 03:55 UT, 6 min after the beginning of the sudden acceleration, some parts of the filament body become bright in the EUV images. They look like threads separated by absorbing material and are tilted to the filament axis by an angle of about 45°. A change from absorption to emission in the coronal spectral line means that some parts of the filament are presumably heated up to coronal temperatures of a million degrees. The stage of acceleration lasted less than three minutes, then about five minutes speed practically was not changed, after which the prominence slowed down. By the time of prominence coming to stop at about 04:00 UT, almost all threads in the upper part of the filament arch were bright. Along the separate threads, intensive motion of prominence matter was observed. After gradual cooling of heated prominence plasma, it began to absorb EUV emission of the surrounding corona again. Along the gently sloping trajectories dark blob descended to the chromosphere.

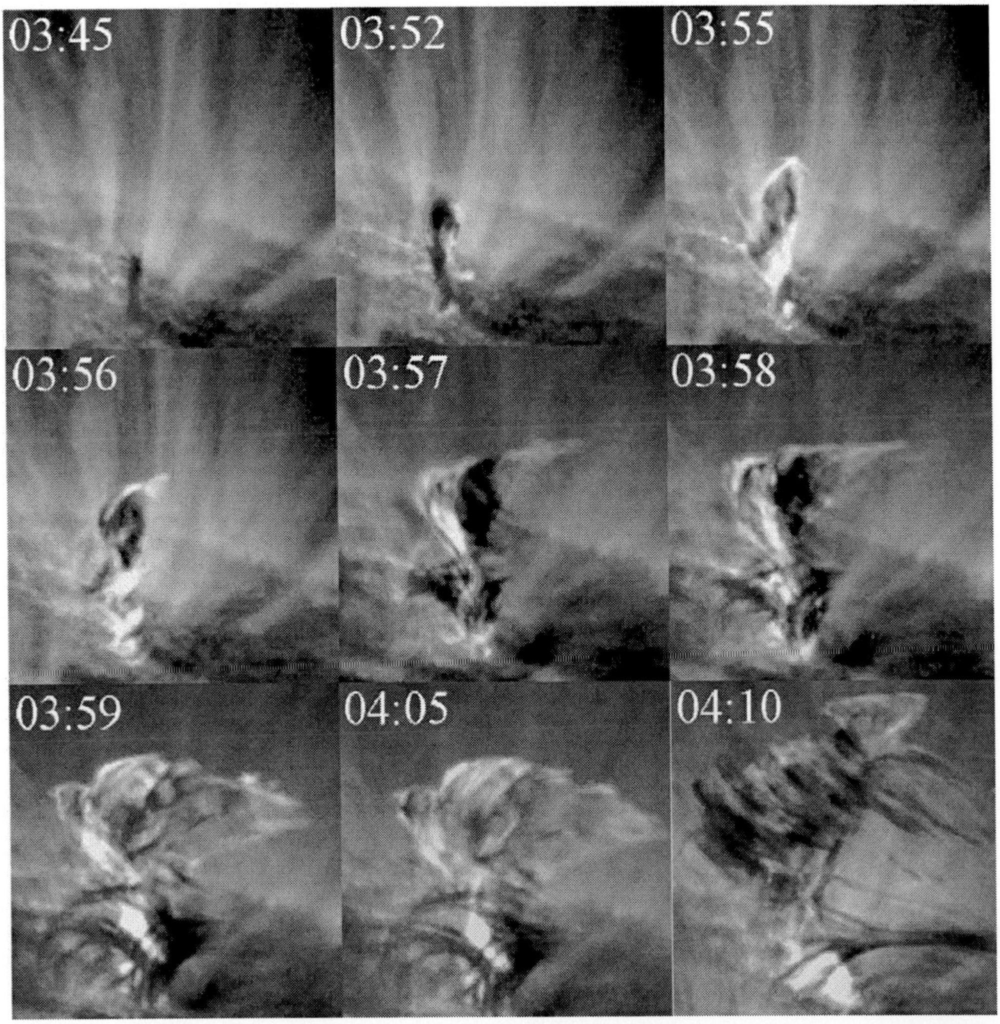

Figure 36. Sequence of TRACE 171 Å images of filament activation on 16 April 2000 (courtesy of TRACE Team. TRACE is a mission of the Stanford-Lockheed Institute for Space Research (a joint program of the Lockheed-Martin Advanced Technology Center's Solar and Astrophysics Laboratory and Stanford's Solar Observatories Group) and part of the NASA Small Explorer program.).

The heating of the moving prominence material seems to be quite different from the heating of post-flare loops, that starts later below the eruptive prominence, at the place where it was located previously, when the prominence is already high in the corona and presumably escaping from the Sun. Loop heating is generally assumed to be due to reconnection of magnetic field lines opened during the eruption (Carmichael, 1964; Sturrock, 1966; Hirayama, 1974; Kopp, Pneuman, 1976; Forbes et al., 1989; Shibata, 1999). Prominence heating can occur within a flux rope that holds its shape and magnetic topology. The heating mechanism should be, therefore, an intrinsic property of the moving prominence which is not at all connected with a possible re-arrangement of the ambient coronal magnetic field. Filippov and Koutchmy, (2002) proposed a mechanism of heating the material of eruptive prominence due to the transformation of a part of prominence macroscopic kinetic energy during evolution of its internal magnetic structure. The primary source of energy is the magnetic energy of the pre-eruptive magnetic configuration stored in the electric current of a flux rope. The prominence eruption occurs due to the loss of the flux rope equilibrium in the external magnetic field. It starts as a catastrophe after a smooth evolution of the system with slow changes of the external magnetic field.

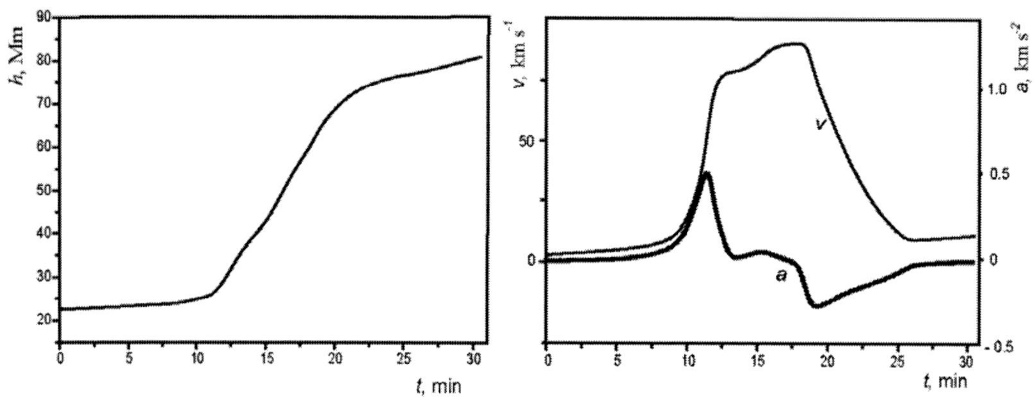

Figure 37. Prominence leading edge height, vertical velocity, and acceleration versus time after 03:39 UT, for the event of April 16, 2000.

The shape of one of the tubes, which model elementary threads forming the filament, can be assumed to be an helix with its axis curved along the half of an ellipse (Fig. 38a)

$$z = h(t)(1 - x^2)^{1/2} + r(t)\sin(2.5\pi(x+1)) \;, \tag{31}$$

$$y = r(t)\cos(2.5\pi(x+1)) \tag{32}$$

where x is the coordinate along a polarity inversion line, z is the vertical coordinate, h(t) is the height of the apex of the filament arch, and r(t) is the radius of the helix. The dimensionless length of the filament along the x-axis is set to 2. Figure 38a shows the shape of the helix at three consecutive moments. In accordance with the frozen-in condition, the plasma is confined within the tube following its evolution ($\beta \ll 1$) and can freely move along the tube axis.

The two-dimensional flows of a plasma in a strong magnetic field ($v_s \ll v \ll v_a$, where v_s is the sound velocity and v_a is the Alfven velocity) were studied by Somov and Syrovatsky (1974). They show that in this case the equations of motion reduce to the equation of freezing-in

$$\frac{d\mathbf{A}}{dt} = \frac{\partial \mathbf{A}}{\partial t} + \mathbf{v}\nabla\mathbf{A} = 0 \;, \tag{33}$$

and the condition of acceleration orthogonality toward the field lines

$$\frac{d\mathbf{v}}{dt} \times \nabla \mathbf{A} = 0 \ , \tag{34}$$

where A is the magnetic field vector potential. Ivanov and Platov (1977) used these equations to model the loop prominence formation. Condition (34) implies that in the case of a curved field line motion occurs towards the curvature center direction, the plasma concentrates near the "bottom" of the curved field line (at a dip).

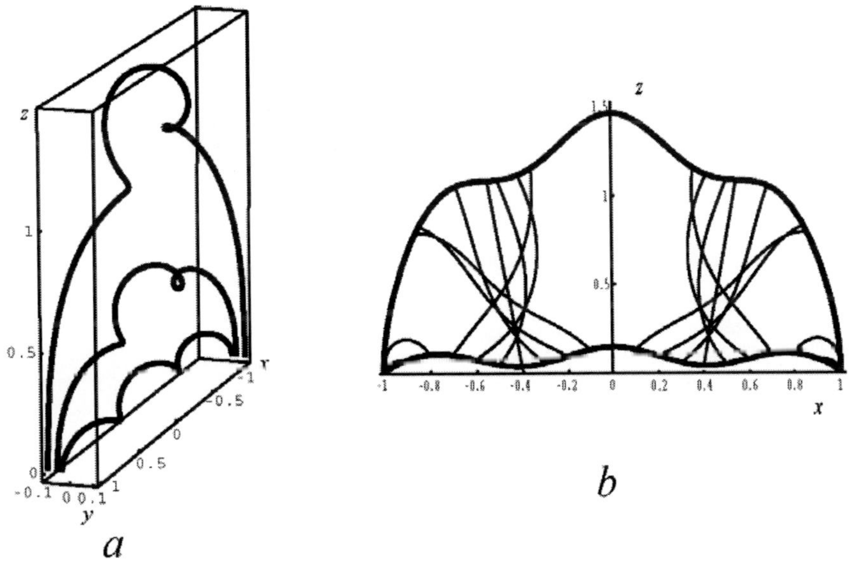

Figure 38. The idealized shape of the helix, at three consecutive moments (a) and trajectories of different fluid particles as the helix apex is rising with a constant acceleration (b).

We can treat (31)-(32) as equations of a field line. For simplicity only Equation (31) will be considered because the eruption occurs mainly in the xz plane. Therefore, we can consider Equation (31) as the first integral of motion, a solution of Equation (33). Differentiating Equation (31) twice with respect to t and using (34), we obtain the following equations of motion of a 'fluid particle':

$$\frac{dv_x}{dt} = -\frac{z'(D-g)}{1+z'^2} \ , \tag{35}$$

$$\frac{dv_z}{dt} = \frac{D-gz'^2}{1+z'^2} \ , \tag{36}$$

where

$$D = -(1-x^2)^{1/2}[\ddot{h} + v_x(2x\dot{h}(1-x^2)^{-1} + hv_x(1-x^2)^{-2}] + (6.25\pi^2 r v_x^2 - \ddot{r})\cos(2.5\pi x) + 5\pi\dot{r}\sin(2.5\pi x) \ , \tag{37}$$

$$z' = \frac{dz}{dx} = -hx(1-x^2)^{-1/2} + 2.5\pi r\cos(2.5\pi(1+x)) \ , \tag{38}$$

The third term in Equations (35)-(36) represents the projection of the gravity acceleration on the tangent direction in the given point.

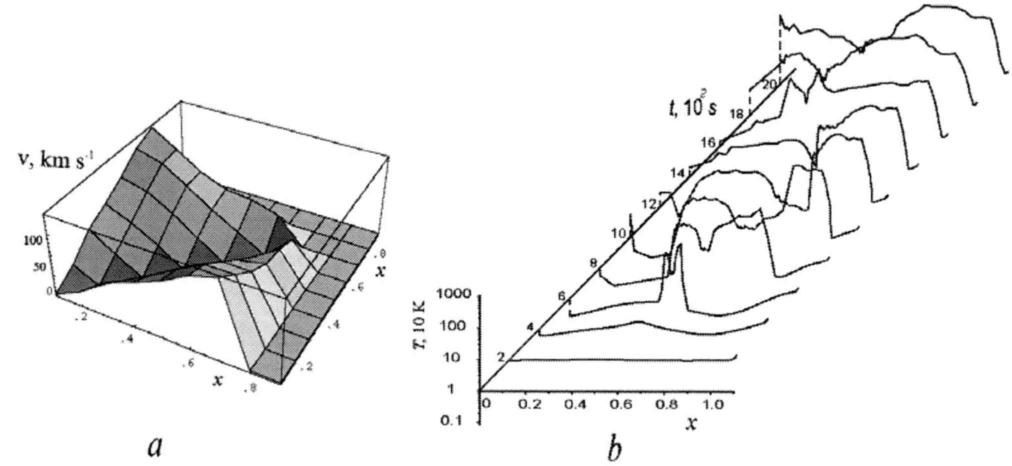

Figure 39. The relative velocities of colliding fluid particles; each of the horizontal axes represent the initial position of each the two colliding particles within the field tube (a) and the distribution of temperature along the tube at different times (b).

We can solve Equations (35)-(36) numerically and find trajectories of motion of each plasma element (a 'fluid particle') in the approximation of negligible gas pressure. Figure 38b shows the trajectories of different fluid particles as the helix apex is rising with the constant acceleration $\ddot{h} = 0.75g$. The radius of the helix also increases simultaneously but with a smaller acceleration $\ddot{r} = 0.05g$. This relationship accounts for the rather moderate increase of the filament cross-section during the onset of the eruption. The initial values of h and r are

$$h_0 = 0.05\ a\,, \qquad\qquad r_0 = 0.025\ a$$

where a = 2 is the dimensionless filament length along the polarity inversion line. In Figure 38b one can see that trajectories of particles that were initially located in the lower parts of the tube with the concavity directed upward, intersect at a certain height.

Actually the motion of gas flows along field lines will produce interactions leading to their braking accompanied by an increase in density and by heating. Since the field tube is inhomogeneously filled with plasma (the filling factor in prominences is small both due to variations in density between different field lines and as a result of plasma concentration into blobs along particular field lines) different plasma volumes in the tube will produce significant relative speed. The surface in Figure 39a represents the relative velocities of colliding fluid particles (only half of the symmetrical filament arch, $0 < x < 1$, is shown in the figure). One horizontal coordinate represents the initial position of the first colliding particle while the other corresponds to a second particle. Therefore, the diagonal corresponds to the collision of the nearest particles which have very small relative velocity. Some particles (mainly near the apex) do not collide at all. The greatest relative velocity corresponds to particles originating from the regions near the neighboring humps of the helix (for instance $x_1 = 0.1$ and $x_2 = 0.7$).

During the collision, the kinetic energy of the macroscopic motion transforms (possibly within shock fronts) into a chaotic kinetic energy or heat. The process is similar to the scenario taking place in a dynamic plasma pinch (see, for instance, Krall, Trivelpiece, 1973). The magnetic field accelerates in that case a cylindrical plasma layer along the radius towards the axis of the cylinder. This envelope collects and picks up all particles on its way as does a snow-plow. When the envelope reaches the axis, the plasma is thermalized. In our curved flux tube, the magnetic field also accelerates the plasma supplying it with some field aligned velocity. Thermalizing with collisions, the plasma could reach the coronal temperature.

We can estimate the final temperature from the equation

$$\frac{\rho v^2}{2} = nkT \,. \tag{39}$$

It gives

$$T = \frac{\mu v^2}{2k} \,, \tag{40}$$

where μ is the average mass particle. A velocity of about 100 km/s is needed to obtain a coronal temperature of $T \sim 10^6$ K.

The possibility of a plasma heating inside a moving flux tube can be illustrated by numerical gas dynamic calculations. Let the tube be filled with an homogeneous plasma of density ρ and temperature T. For the sake of the initial equilibrium we assume the tube to be at first straight and horizontal. This means that the filament axis curvature is large enough while the helix radius is rather small (a very compact filament). The usual Lagrangian form of the one-dimensional hydrodynamics equations for the ideal gas is

$$\frac{d\rho}{dt} = -\rho \frac{\partial v}{\partial l} \,, \tag{41}$$

$$\frac{dp}{dt} = \frac{1}{\rho}(p\gamma + w(\gamma - 1))\frac{\partial \rho}{\partial t} \,, \tag{42}$$

where p is the pressure, γ is the ratio of the specific heats (γ = 5/3 for solar plasma), l is the coordinate along the tube ($dl/dx = \sqrt{1+z'^2}$), w is the artificial viscosity that is added to equations to avoid discontinuities and nonlinear oscillations in numerical calculations (Von Neumann, Richtmyer, 1950).

$$w = \begin{cases} C\rho\Delta l\left(\dfrac{\partial v}{\partial t}\right)^2, & \text{when } \dfrac{\partial v}{\partial t} < 0 \,, \\ 0, & \text{when } \dfrac{\partial v}{\partial t} \geq 0 \,, \end{cases} \tag{43}$$

where Δl is a spatial step of the mesh, and C is a constant ~ 1.

The equation of motion that we obtain from Equation (35), after adding the gradient of the gas pressure, is

$$\frac{dv_x}{dt} = -\frac{z'(D-g)}{1+z'^2} - \frac{1}{\rho\sqrt{1+z'^2}}\frac{\partial}{\partial l}(p+w) \ , \tag{44}$$

The velocity v along the tube that appears in Equation (41) can then be found

$$v = v_x\sqrt{1+z'^2} \ . \tag{45}$$

The boundary conditions were chosen as v = 0 at the ends of the tube. The parameters of the expansion are taken so as to allow the tube to accelerate during the first 500 s with a = 1.5g and then decelerate after during 500 s with a = -1.5g. After 1000 s the shape of the tube did not change. This evolution is very similar to the kinematics of the prominence eruption observed on 16 April 2000, see Figure 37. The initial length of the tube is 160 Mm, the density and the temperature are typical for prominence plasma, $\rho = 1.6 \cdot 10^{-12}$ g cm^{-3}, T = 10^4 K.

Figure 39b shows the distribution of temperature along half of the tube at different times. Narrow peaks of temperature arise at the bottom of the dip. This dense and hot condensation then spreads in both directions, leaving the temperature depression which can be seen at t = 600 s. The density is decreasing and the size of the hot region is increasing. The maximum temperature is about 10^6 K. Only the region near the apex remains cool. The temperature here is actually lower than the initial temperature owing to the evacuation. After the beginning of the deceleration phase the border of the hot region moves to the middle of the tube. Running waves reflect from the ends of the tube and soon all the tube is filled with plasma of coronal temperature.

Numerical calculations presented here have been performed in order to illustrate the possibility of prominence plasma heating during an eruption. We used a very simple adiabatic approximation in the energy equation to demonstrate only the main features of the process. Energy losses due to radiation and thermal conductivity are not taken into account. One of the reasons is the great uncertainty in the parameters of the initial energy balance of a plasma of the solar atmosphere (Mariska, 1992). The mechanism does not include any current sheet formation or dissipation effects which are usually included in models of the heating of coronal structures during flares. However, magnetic field line reconnection should occur freely in the ambient corona to give the way for the flux rope motion in the coronal magnetic field.

The magnetic energy transforms into the macroscopic kinetic energy of the filament motion. Part of this energy turns into the heat through the collisions inside a flux tube. A rather complex structure of the flux tube is necessary for this purpose. Some segments of the tube should have an upward concave form to provide the counter flow acceleration. A typical twisted flux rope easily meets this requirement. The process within the concave tube as it moves upward is in some sense similar to the acceleration of charged particles between approaching magnetic mirrors (Fermi type acceleration). Plasma particles reccive energy from the moving magnetic field.

10. Eruptive Prominences and Coronal Mass Ejections

Hα observations of the onset of the activation of a filament (prominence) frequently end very soon due to the departure of this line from the filter band due to the Doppler effect, the decrease of the emission due to the decrease in the density of expanding material, and the ionization of hydrogen due to heating of the prominence plasma. The subsequent evolution can sometimes be observed in the ultraviolet or radio, but the fate of a prominence usually becomes visible only when it reaches the lower boundary of the field of view of a spaceborne coronagraph (~2R) and becomes visible in white light as a coronal mass ejection (CME). A typical CME consists of three parts: a bright core that is the remnant of an eruptive prominence; a large, dark, lower-density surrounding cavity; and an outer, rather diffuse envelope having the projected shape of a closed loop with its legs fixed on the Sun (Crifo et al, 1983; Sime et al., 1984). Theoretical and observational aspects of problems related to CMEs are discussed in details in reviews of Forbes (2000) and Gopalswamy (2004).

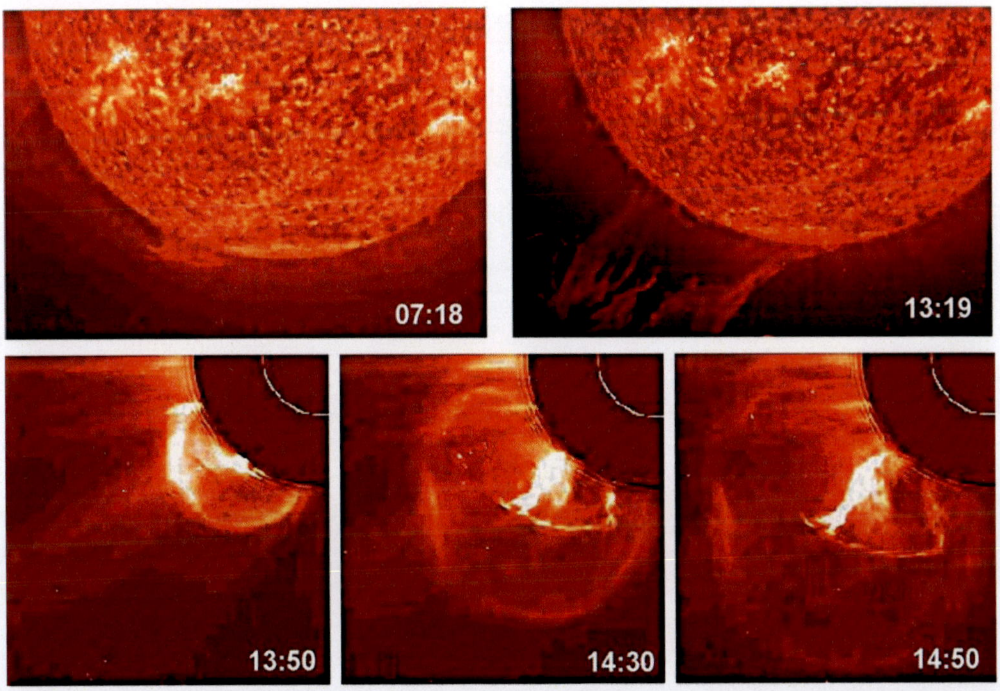

Figure 40. Polar crown filament eruption on 14 June 1999 visible in SOHO EIT He II 304 Å line (top row) and the following CME observed with LASCO C2 (bottom row) (courtesy of SOHO/EIT and SOHO/LASCO consortiums; SOHO is a joint ESA-NASA program).

There is very close association between eruptive prominences and CMEs. There are no doubts, that in the ejections with the bright cores, the material of eruptive prominences is contained in the cores (House et al, 1981). Fig. 40 shows an example of event when an eruptive prominence was recorded in the line of helium He II 304 Å with the ultraviolet telescope EIT on SOHO, and then the remnants of the prominence were well recognised in the structure of a CME up to the distance of several solar radii from the surface. At the same time, there are numerous ejections without bright core, or such events, for which it is impossible to find the conjugated prominence eruption or filament disappearance. Opposite situation, when a CME does not follow after filament eruption, is happened, perhaps, only in the case of "confined" eruption, when the filament does not it fly away too far but stops at some height in a new equilibrium position. Thus, the relationship between these two phenomena is not always obvious, and it is the subject of analysis in many works.

Statistical studies showed the greatest correlation of CMEs with eruptive prominences among all manifestations of activity in the lower layers of the solar atmosphere (Webb et al., 1976; Munro et al, 1979; Webb, Hundhausen, 1987). Important factor is the height, at which the eruptive prominence is observed: the greater the height it reaches, the more probable that a CME will follow. So, all studied by Munro et al. (1979) prominences that reached height of 1.2R and only 60% of prominences that reached the height 1.1R were associated with CMEs. This rises the question of whether the prominences, that at first show sufficiently rapid ascending motion holding its initial shape and then slow down and stop demonstrating finite motion, could be called eruptive ones. Gilbert et al. (2000) proposed to call such prominences as active prominences. On the basis of observations in 1996-1998, they found that CMEs were connected to 46% of such active prominences and to 94% of "real" eruptive prominences, which were capable to overcome solar gravity. Ejections associated with eruptive prominences as a rule had bright cores, while CMEs associated with the active prominences usually did not have them.

Probably, the choice of data for the correlation analysis can strongly influence the result. For example, very low correlation of only 10%-30% between CMEs and prominence eruptions together with sudden filament disappearances obtained by Yang and Wang (2002), possibly, is explained by including into the dataset of the thermal disappearances of filaments not connected with the dynamic events and disregarding of some faint ejections of the halo type. In the same period of time, 82% of eruptive prominences observed with the Nobeyama radioheliograph were associated with CMEs (Gopalswamy et al, 2003). However, only 18% of the prominences that move predominantly along the surface of the Sun ("tangential" phenomena), which, apparently, also can be assigned to active prominences, were possible to associate with CMEs.

Extrapolation of the time dependence of the CME height allows finding the onset time of the CME. Although some uncertainty remains because of the absence of information about the real CME acceleration during a period of time when it was shielded by the occulting disk of a coronagraph, and because of the unknown starting height, nevertheless it is possible to consider that the onset time of a CME coincides rather close (within of half an hour) with the beginning of prominence eruption (Gopalswamy et al, 2003).

The question which of the two phenomena, filament eruption or CME, starts earlier is important for the understanding causal relationship between them. There is a point of view that a rising prominence disturbs the surrounding corona and as a piston pushes out coronal plasma forming a CME; and there is an opposite approach that the prior removal of the coronal structures above a prominence allows it to rise unimpeded. Observations cannot give yet unambiguous answer to this question, but possibly it does not exist. More and more arguments are gathered in favour to the concept that both the prominence eruptions and the CMEs are different observational manifestations of the same common process of the loss of equilibrium of the large-scale coronal current system (magnetic flux rope), and depending on particular initial conditions in the system (density and the mass of the prominence, magnetic field strength, etc.) the pattern of event can be different. So, eruption of very rarefied prominences can remain unnoticed and cannot create sufficiently bright core of a CME.

Strictly speaking, all three eruptive phenomena in the solar atmosphere - flares, eruptive prominences and coronal mass ejections - are closely related and degree of correlation between them depends on special features of particular phenomenon (Lin, 2004). The correlation between CMEs and flares depends on quantity of energy stored in the magnetic configuration prior to eruption. The more free energy is stored, the better correlation. On the other hand, the correlation between CMEs and eruptive prominences depends on amount of matter contained in the initial magnetic structure.

However, a careful comparison of the positions of filaments on the disk that have disappeared suddenly and CMEs reveals some problems in the interpretation of CMEs as continuations of the filament eruptions in the upper corona. The heliolatitudes of the disrupted filaments and of the subsequent coronal ejections sometimes differ by tens of degrees. For example, Cliver and Webb (1998) found that, in a period of maximum activity, the number of CMEs with heliolatitudes > 60° was four times the number of disrupted filaments at latitudes > 45°. If these two phenomena are causally related, it is unclear why their latitudes (position angles) differ so strongly. To solve this problem Filippov et al. (2000; 2001) suggested that the considerable heliographic separation of the observed manifestations of eruption in its initial and final stages may result from nonradial and nonrectilinear motions of flux ropes in the corona due to their interaction with the large-scale magnetic field.

Another objection against the concept of a CME as a further development of the eruption of a prominence in the upper corona is based on the kinematics of these phenomena (Foley et al., 2003). The velocity of an eruptive prominence (bright core of a CME) is usually less than the velocity of the outer loop (envelope of the CME) by a factor as large as 1.5. For this reason, many researchers believe that the prominence cannot "push" the CME and be its driver. The difference in velocities can be also understood in the context of the flux rope model. Formation of the CME structure can be interpreted as a result of drift motion of the coronal material in the presence of an eruptive prominence (Filippov, 1996). Its characteristic is such that the further one gets from the filament, the higher is drift velocity of a plasma. That is why the frontal part of the CME moves faster than the core. This type of motion leads to a rarefaction of plasma and formation of a cavity around the filament. Near the boundary $\beta = 1$ plasma is decelerated, and as a result, coronal density is increasing. Density enhancing near the surface $\beta = 1$ leads to formation of a dense shell which can be correlated with the frontal loop of the CME.

11. Forecast of Prominence Eruptions and CMEs

Modern civilization uses to an increasing extent systems and tools of the global scale which are influenced by not only the local environment but by the conditions in the whole near-earth space. The totality of factors determining the electromagnetic conditions and radiation situation in the vicinity of Earth are named as space weather. Space weather variability is determined mostly by the active processes on the Sun which disturb the interplanetary medium. Investigations of the last decade put coronal mass ejections (CMEs) on the first place as the solar geoeffective phenomena. As it has turned out, geomagnetic storms happen as a rule after the coronal mass ejections (Kahler, 1992; Gosling, 1993) while interplanetary shocks generated by them accelerate charged particles to high energy producing radiation hazard (Kahler, 2003). The most energetic events on the Sun are accompanied by both a CME and a solar flare. Very likely that the two phenomena are interrelated and moreover they are the specific manifestations of the same process revealing itself in different manner depending on the concrete conditions.

Unquestionable significance of the electromagnetic processes in the terrestrial space for the modern civilization makes it actual not only monitoring of space weather and study of the specific mechanisms of its influence on the living on Earth but also the challenge to forecast space weather. There are large centers such as Space Environment Center NOAA (Boulder, Colorado), which analyze a lot of data from ground based and spaceborne observatories and produce a forecast based on statistical and empirical rules and in part on modeling.

Long-time variations of space weather are connected with the 11-year cyclic recurrence of solar activity. Prediction for several years is based on statistical laws and rules governing observed time series. Dynamo theory, the mechanism of magnetic field generation on the Sun, is not able not only to predict the parameters of the following cycle, but even to explain the totality of the known facts. For the prediction of geophysical situation on a time scale of the order of month, the storability the large-scale structure of the corona during a period of solar rotation is of great importance, because it determines the characteristics of the solar wind. Coronal holes, which are sources of the fast solar wind, can exist during several solar rotations and cause recurrent geomagnetic storms upon each entry of the Earth into the zone of the fast flow, if the direction of the interplanetary magnetic field is favorable for reconnection (antiparallel to the terrestrial magnetic field). Onset of such storm can be predicted with a good reliability. Another matter is the disturbances that are caused by sporadic short-term non-stationary phenomena on the Sun, namely flares, eruptive prominences, and CMEs. For their prediction, it is necessary to know, when conditions for the beginning of such phenomena arise on the Sun. This is very complicated problem, because the mechanisms of the non-stationary processes are not clear enough.

Solar flares were considered for a long time as the main and pivotal source of the sudden disturbances of interplanetary medium and geophysical conditions beginning from the date of their discovery in 1859. There were persistent attempts to elaborate a technique of prediction of the solar flare's onset. However, the majority of the proposed techniques was based on statistical and empirical rules and was not very effective (Martin, Ramsey, 1972; Lemmon, 1972; Smith, 1972). Different parameters such as

- magnetic field complexity,
- number of neutral line bends,
- magnetic field sheer,
- magnetic field gradient,
- magnetic flux changes

were used in order to estimate the probability of a flare but no one of them could be assumed as the controlling factor. The reason was probably in the lack of understanding of the physical mechanism of a solar flare.

In the middle of 1970s, a lot of hopes were pinned on MHD models of solar flares that include magnetic field line reconnection (Parker, 1973; Syrovatskii, 1973; Priest, Forbes,

2000). It seemed that the occurrence of the specific hyperbolic magnetic field configuration was the necessary condition for the solar flare and the problem was only to locate such configurations in the solar atmosphere. The source of energy for the flare according to the reconnection models was the stressed coronal magnetic field or, in other words, coronal electric currents concentrated mainly in current sheets. Hyperbolic magnetic structures were the most suitable place for the current sheets. A wealth of theoretical papers relating to the magnetic reconnection was published in last three decades. There were reports of observations of the hyperbolic magnetic structures on the Sun but the problem of the flare forecast was not moved out of the point of the statistical and empirical methods. Moreover, some observations (Fig. 32, Fig. 33) show reduced temperature in the central parts of the hyperbolic (saddle-like) structures in the corona (Filippov, 1999).

The problem of CME prediction is less developed than that of flare prediction. There are some difficulties related not only to the moment of the CME onset, but also to its location. Some researchers believe that CMEs are generated in large-scale closed-field structures, in streamers in most cases (75%) (Hundhausen, 1993; Webb et al., 1994). Others relate CME to open-field regions, namely, coronal holes (Hewish and Bravo, 1986). So-called sigmoid structures, coronal loops bent as the roman letter S, observable in soft X-rays and ultraviolet, are considered to be sources of some CMEs (Canfield et al., 1999; Sterling, 2000). In some cases, sudden disappearance of X-ray transequatorial coronal loops is a close precursor of a CME; this effect is considered as a beginning of the eruption (Akiyama et al., 2000; Khan and Hudson, 2000). However, no real CME precursors, which can be used to predict the possibility of eruption in the corona several hours or days in advance, have been found so far.

Most CMEs are observed near the polarity inversion line of the large-scale magnetic field. This indicates to the electromagnetic mechanism of the acceleration of erupted matter and clears up the statistical relation between CMEs and other events "bound" to the polarity inversion line: streamers, flares, and filaments (prominences). Eruptive prominences are generally connected with CMEs most closely among all events observed in the lower layers of the solar atmosphere, as it was shown in the previous section.

We shall proceed from the assumption that the most probable initial magnetic configuration of a CME is a twisted rope of helical field lines, filling the entire volume of the dark cavity, stretched in the corona along the photospheric polarity inversion line (Low, 1996; Amari et al., 2000; Rust, 2001). Meanwhile, there are other points of view, for example, some researchers assume that a CME is generated in a magnetic arcade subjected to a strong shear deformation (DeVore, Antiochos, 2000). The outer loop is most probably a projection of the envelope formed of coronal plasma at the rope boundary during its ascent (Filippov, 1996), and the dense cold matter of the prominence was accumulated in the lower parts of the helical flux tubes, which serve as magnetic traps for plasma in the gravitational field. The cavity is a rather inconvenient object for observation owing to its reduced emission (Engvold, 2000); therefore, prominences or filaments are the best tracers of the flux-rope position in the corona long before the eruption begins. Many filaments and prominences can be usually simultaneously observed on the Sun even at minimum solar activity. Each filament can be considered as a potential source of a CME. Therefore, the problem of the CME prediction can be reduced to the analysis of the filament equilibrium and estimation of their stability. Some distinctions in treating the problems of flares and eruptions are shown in Table 1. One can state that in flux rope models, the electric currents are concentrating within a simple arcade, very common magnetic configuration on the Sun; magnetic energy transforms into macroscopic kinetic energy and this process can be described in the frame of the ideal MHD.

Table 1. Comparison of the problems of flares and eruptions

The key points	Flare	Eruption
Where does energy store, which is released in an explosive process?	Current sheet	Flux rope
Initial background magnetic configuration:	Hyperbolic (saddle) structure	Simple arcade
Main forms the energy transforms to:	Energy of accelerated particles, thermal energy	Macroscopic kinetic energy
Methodology suitable for modeling of the initial phase of the process:	Resistive MHD	Ideal MHD

We would successful in prediction of an eruption if we were able to estimate the total electric current within the flux rope and to compare it with the critical value. Unfortunately, not only the electric currents but even the magnetic fields are practically impossible to measure in the corona. Some manifestations of the electric currents in the corona are observed as sigmoid coronal structures (Canfield et al., 1999; Sterling et al., 2000) or twisted threads within the filaments (Vrsnak et al., 1991; Rompolt; 1990). They are considered as indicators of a possible eruption, however it is difficult to use them as a quantitative criterion of the possibility of eruption. Vrsnak et al. (1991) analyzed the amount of twist observed in stable quiescent filaments and eruptive prominences. Parameter characterizing the twist is related to the value of the electric current in the filament and could be a good indicator of pre-eruptive state, but, unfortunately, the twist is not always easy to recognize in the filament structure. In any case, filtergrams with high spatial resolution are required.

The model of the flux rope equilibrium discussed in Section 5 shows that the greater is the electric current the higher is the equilibrium point. When current strength achieves a critical value catastrophe happens due to nonlinear properties of the model. The equilibrium is suddenly lost and the flux rope will ascend with increasing acceleration. This means the eruption. Therefore, the height of a filament (prominence) could be in some sense a measure of the electric current strength and, as we will see, a measure of its readiness for eruption. A critical height h_c exists, above which the equilibrium becomes unstable, defined by Equation (14). The parameter h_c can be calculated using magnetograms of the photospheric magnetic field and solving of the boundary-value problem (Cuperman et al., 1990; Rudenko, 2001; Den, 2002). The calculated field is the field generated only by photospheric sources (i.e., a potential field). Although the field of the filament current is not weak compared with the field of the subphotospheric currents, it can be readily separated from the latter. In the flux rope (inverse-polarity filament) model the effect of the induction currents on the photospheric surface neutralizes the normal field component of the filament current on this surface, implying that the field of the coronal currents does not penetrate beneath the photosphere. For this reason, the measured normal field component in the photosphere is entirely determined by the field of the subphotospheric sources and can be used to calculate its potential extension into the corona.

Comparing the observed height of filaments with the parameter h_c it is possible to make conclusion about the prospect of the filament eruption. The problem is that it is necessary to compare two values, one of which is most reliably known at that moment, when the second is practically unknown. The height of a filament can be easily measured when it is observed on the limb as a prominence, while the magnetic field in a photospheric region below the prominence cannot be obtained because the surface of the photosphere nearly parallel to the light of sight. The magnetic field is measured most accurately when a region is located near the disk center. Of course, it is not very easy to measure the height of a filament above the photosphere when it is projected against the solar disk. It should certainly be an ideal case if simultaneous chromospheric images were taken at different angles, e.g., from the Earth and from some distant spacecraft (as in the STEREO project), to determine the filament height from the parallax. Sometimes, when the internal motion is negligible within the filament and the filament shape is invariable, the filament height can be estimated from the difference between the shift of chromospheric details and the shift of the top of the filament due to solar rotation (Vrsnak et al., 1999). One can also use information on the inclination of the symmetry plane of a filament with respect to the vertical in order to estimate the filament's height from its observed width (Zagnetko et al., 2005).

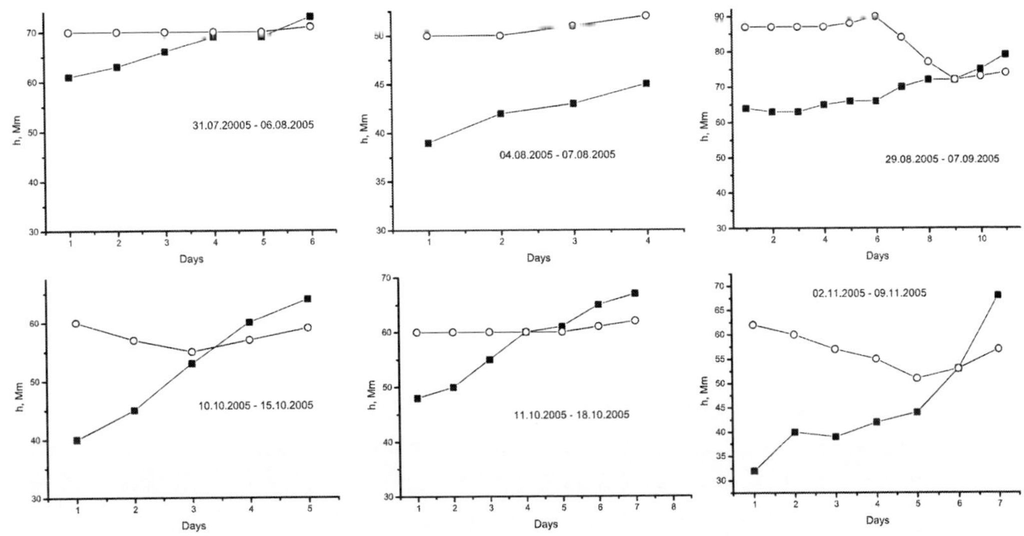

Figure 41. Temporal behavior of the filament height h_p (solid squares) above the photosphere and the critical height h_c (open circles) several days before eruption in the latter part of 2005.

Figure 24 shows distribution of observed some 80 filaments according to their actual height and the critical height calculated for the surrounding magnetic field. The filaments that safely passed the west limb are marked with the solid circles, while the filaments that disappeared from the disk are marked with the open circles. It is seen that all stable quiescent filaments are found to be below the line $h_p = h_c$, which is the limit of stability. The eruptive filaments are clustering about the threshold boundary. This separation of the filaments into two classes shows considerable promise as a means for the prediction of eruptions. Detailed tracing the fate of a particular filament reveals that it prone to erupt within 1-2 days after reaching the critical height (Filippov et al., 2006). On the contrary, the filaments that have the height well below the limiting height are not dangerous for producing an eruption. Figure 41 shows some examples of temporal behavior of both quantities h_p and h_c for several filaments observed in the latter part of 2005. Right ends of the curves correspond to the last moments before a filament erupts or disappears from the disk. It is seen that as a rule the filament erupts soon after it reaches the critical height. However, there are exceptions, possibly due to errors in measurements or a limitation of the model.

12. Conclusion

Detailed correlation of prominence eruptions and coronal mass ejections permit in a number of cases to trace the continuous transfer of one phenomenon into another, which makes it possible to speak about the common eruptive phenomenon. Certainly, a CME is forming in much more rarefied medium and occupies enormous volume. Significant part of the corona and large-scale magnetic field are involved in the CME formation; however, the prominence material is often well distinguished in the general structure of the CME as the bright core. Thus, the problem of sudden ejection of plasma from the corona into the interplanetary space and the problem of prominence existence and death seem to be closely connected one to the other and their solutions should be searched for in the general context.

Theoretical analyses of the phenomenon leads to conception that a CME happens when a large-scale magnetic flux rope in the corona loses its equilibrium and escapes from the coronal magnetic field, which confines it before the ejection. Helical magnetic field tubes of the flux rope serve as magnetic traps in the gravitation field. Dense plasma could collect near the bottom parts of the helical tubes forming prominences or filaments. As in telescopic observations we record a plasma, which emits or absorbs radiation, filaments are the most accessible indicators of the flux ropes presence in the corona. Loss of the flux rope equilibrium appears as a filament eruption. It is the filament eruption that is known is the first manifestation of the solar storm beginning before the onset of a flare or a CME.

The filament height can be a measure of its age and readiness for eruption. The comparison of the real heights of prominences with the calculated critical heights could be a basis for predicting filament eruptions and following CMEs. So, solar storm forecast several days in advance could be made on the basis of the filament monitoring. All necessary data could be obtained, in principle, with groundbased instruments, however spaceborn observations could significantly increase the precision.

References

Adams, J., Pneuman, G. W. *Solar Phys.* 1976, 46, 185.

Akiyama, S., Hori, K., Nakagima, H., Hara, H. In *International Conference on Solar Eruptive Events,* Abstract book; Center for Solar Physics and Space Weather; The Catholic University Of America: Washington, DC, 2000; p. 12.

Altschuler, M. D., Newkirk, G. Jr. *Solar Phys.* 1969, 9, 131.

Altschuler, M. D., Levine, R. H., Stix, M., Harvey, J. W. *Solar Phys.* 1977, 51, 345.

Amari, T., Ali, J. J. *Astron. Astrophys.* 1989, 208, 261.

Amari, T., Luciani, J. F., Mikic, Z., Linker, J. *Astrophys. J.* 2000, 529, L49.

Ananthakrishnen, R. *Astrophys. J.* 1961, 133, 969.
Antiochos, S. K., Dahlburg, R. B., Klimchuk, J. A. *Astrophys. J.* 1994, 420, L41.
Antiochos, S. K., *Astrophys. J.* 1998, 502, L181.
Anzer, U. *Solar Phys.* 1969, 8, 37.
Anzer, U. *Solar Phys.* 1978, 57, 111.
Anzer, U., Pneuman, G. W. *Solar Phys.* 1982, 79, 129.
Aulanier, G., Demoulin, P., van Driel-Gesztelyi, L., Mein, P., DeForest, C. *Astron. Astrophys.* 1998, 335, 309.
Aulanier, G., DeVore, C. R., Antiochos, S. K. *Astrophys. J.* 2002, 567, L97.
Babcock, H. W., Babcock, H. D. *Astrophys. J.* 1955, 121, 349.
Ballester, J.L., Priest, E.R. In *Dynamics and Structure of Solar Prominences*; Ballester J.L., Priest E.R.; Eds.; Universitat de les Illes Balears: Mallorca, 1987; p. 137.
Becker, U. Z. *Astrophys.* 1956, 40, 65.
Bommier, V., Landi degl'Innocenti, E., Leroy, J.-L. *Solar Phys.* 1994, 154, 231.
Bravo, S., Stewart, G.A., Blanco-Cano, X. *Solar Phys.* 1998, 179, 223.
Bruzek, A. Z. *Astrophys.* 1952, 31, 99.
Canfield, R. C., Hudson, H. S., McKenzie, D. E. *Geophys. Res. Lett.* 1999, 26, 627.
Cantu, A. M., Godoli, G., Poletto, G. *Mem. Soc. Astr. Ital.* 1968, 38, 367.
Carmichael, H. In *AAS-NASA Symp. on Solar Flares*; Hess, W. N.; Ed., NASA SP-50, 1964; p. 451.
Chen, J. *Astrophys. J.* 1989, 338, 453.
Chen, J. In *Physics of Magnetic Flux Ropes;* Russell, C. T., Priest, E. R., Lee, L. C.; Eds.; Geophys. Monogr. Ser.; AGU: Washington, D. C., 1990; Vol. 58, p. 269.
Chen, J., Garren D. A. *Geophys. Res. Lett.* 1993, 20, 2319.
Chen, J. *Space Sci. Rev.* 2001, 95, 165.
Choe, G. S., Lee, L. C. *Solar Phys.* 1992, 138, 291.
Cliver, E. W., Webb, D. F. In *New Perspectives on Solar Prominences*; Webb, D.; Rust, D.; Schmieder, B.; Eds.; ASP Conference Series; ASP: San Francisco, CA, 1998; Vol. 150., p. 479.
Crifo, F., Picat, J. P., Cailloux, M. *Solar. Phys.* 1983, 83, 143.
Cuperman, S., Ofman, L., Semel, M. *Astron. Astrophys.* 1990, 227, 583.
d'Azambuja, M., d'Azambuja, L. *Ann. Obs.* Paris, Meudon, 6, 1948, Fasc. VII.
Demoulin, P., Raadu, M. A., Malherbe, J. M. *Astron. Astrophys.* 1987, 183, 142.
Demoulin, P. In *New Perspectives on Solar Prominences*; Webb, D.; Rust, D.; Schmieder, B.; Eds.; ASP Conference Series; ASP: San Francisco, CA, 1998; Vol. 150., p. 78.
Demoulin, P., Priest, E.R. *Solar Phys.* 1993, 144, 283.
Den, O. G. *Astron. Lett.* 2002, 28, 345.
Dermendzhiev, V. S. *Astrofizicheskie Issledovaniya* (Sofia) 1977, Vol. 2, 8.
DeVore, C. R., Antiochos, S. K. *Astrophys. J.* 2000, 539, 954.
Durrant, C. J. *Solar Phys.* 2002, 211, 83.
Engvold, O. *Solar Phys.* 1972, 23, 346.
Engvold, O. In *New Perspectives on Solar Prominences*; Webb, D.; Rust, D.; Schmieder, B.; Eds.; ASP Conference Series; ASP: San Francisco, CA, 1998; Vol. 150., p. 23.
Engvold, O. In *International Conference on Solar Eruptive Events*, Abstract book; Center for Solar Physics and Space Weather; The Catholic University Of America: Washington, DC, 2000; p. 20.

Engvold, O., Jakobsson, H., Tandberg-Hanssen, E., Gurman, J., Moses, S. *Solar Phys*. 2001, 202, 293.

Feynman, J., Martin, S. F. J. *Geophys. Res*. 1995, 100, 3355.

Fielder, R. A. S., Hood, A. W. *Solar Phys*. 1993, 146, 297.

Filippov B. P. *Astron. Letters* 1994,20, 770.

Filippov B. P. *Astron. Astrophys*. 1996, 313, 277.

Filippov, B. P. *Astron. Astrophys*. 1998, 337, 883.

Filippov, B. P. *Solar Phys*. 1999, 185, 297.

Filippov, B. P., Gopalswamy, N., Lozhechkin, A. V. *Solar Phys*. 2001, 203, 119.

Filippov, B., Koutchmy, S. *Solar Phys*. 2002, 208, 283.

Filippov, B.P., Zagnetko, A.M., Ajabshirizadeh, A., Den, O.G. *Solar System Research* 2006, 40, 319.

Foley, C. R., Harra, L. K., Matthews, S. A., Culhane, J. L., Kitai, R. *Astron. Astrophys*. 2003, 399, 749.

Forbes, T. G., Malherbe, J. M., Priest, E. R. *Solar Phys*. 1989, 120, 285.

Forbes, T. G., Isenberg, P. A. *Astrophys. J.* 1991, 373, 294.

Forbes, T. G., Priest, E. R., Isenberg, P. A. *Solar Phys*. 1994, 150, 245.

Forbes, T. G., Priest, E. R. *Astrophys. J.* 1995, 446, 377.

Forbes, T. G. J. *Geophys. Res*. 2000, 105, 23153.

Gigolashvili, M. Sh. *Solar Phys*. 1978, 60, 296.

Gilbert, H. R., Holzer, T. E., Burkepile, J. T., Hunhausen, A. J. *Astrophys. J.* 2000, 537, 503.

Godoli, G. Motta, S., Sciuto, V., Zappala, R. A. In SkylabSolar Workshop; Righini, G.; Ed.; Osservazioni e Memorie Osservatorio di Arcetri: Firenze, Italy, 1974; Vol. 104, 204.

Gopalswamy, N., Shimojo, M., Lu, W., Yashiro, S., Shibasaki, K., Howard, R. A. *Astrophys. J.* 2003, 586, 562.

Gopalswamy, N. In *The Sun and the Heliosphere as an Integrated System*; Poletto, G., Suess, S.; Eds.; ASSL series; Kluwer: Boston, 2004; Chapter 8, p. 201.

Gosling, J. T. J. *Geophys. Res*. 1993, 98, 18,937.

Gosling, J. T., Birn, J., Hesse, M. *Geophys. Res. Lett.* 1995, 22, 869.

Harvey, K. In *Encyclopedia of Astronomy and Astrophysics*; Murdin, P.; Ed.; Coronal Cavities; Inst. of Phys. Publishing: Bristol, 2001.

Heinzel, P., Schmieder, B., Tziotziou, K. *Astrophys. J.* 2001, 561, L223.

Hermans, L., Martin, S. F. In *Coronal and Prominence Plasmas*; Poland, A. I.; Ed.; *NASA Conference Publications* 2442; 1986; p. 369.

Hewish, A., Bravo, S. *Solar Phys*. 1986, 106, 185.

Hirayama, T. *Solar Phys*. 1974, 34, 323.

Hoeksema, J. T., Wilcox, J. M., Scherrer, P. H. J. *Geophys. Res*. 1982, 87, 10331.

House, L. L., Wagner, W. J., Hildner, E., Sawyer, C., Schmidt, H. U. *Astrophys. J.* 1981, 244, L117.

Howard, R. F., Harvey, J. W. *Astrophys. J.* 1964, 139, 1328.

Hundhausen, A. J. J. *Geophys. Res*. 1993, 98, 13177.

IAU Colloquium 117, *Dynamics of Quiescent Prominences*; Ruzdjak, V.; Tandberg-Hanssen, E.; Eds., Lecture Notes in Physics; Springer-Verlag: Berlin, 1990; Vol, 363.

IAU Colloquium 167, *New Perspectives on Solar Prominences*; Webb, D.; Rust, D.; Schmieder, B.; Eds.; ASP Conference Series; ASP: San Francisco, CA, 1998; Vol. 150.

Ioshpa, B. A. In *Structure and Development of Solar Active Regions*; Kiepenheuer, K. O.; Ed.; IAU Symp. 35; 1968; p. 261.

Ioshpa, B. A., Kulikova, E. Kh., In *Solar Magnetic Fields and Corona*; Teplitskaya, R. B.; Ed.; Nauka: Novosibirsk, 1989; Vol. 1, p. 167.

Isenberg, P. A., Forbes, T. G., Demoulin, P. *Astrophys. J.* 1993, 417, 368.

Ivanov, L. N., Platov, Yu. V. *Solar Phys.* 1977, 54, 35.

Kahler, S. W. Annu. Rev. *Astron. Astrophys.* 1992, 30, 113.

Kahler, S. W. Adv. *Space Res.* 2003, 32, 2587.

Khan, J. I., Hudson, H. S. *Geophys. Res. Lett.* 2000, 27, 1083.

Kim, I.S., Klepikov, V. Yu., Koutchmy, S., Stepanov, A. I., Stellmacher, G. *Solnechnye Dannye* 1988a, No. 1, 75.

Kim, I.S., Klepikov, V. Yu., Koutchmy, S., Stepanov, A. I., Stellmacher, G. *Solnechnye Dannye* 1988b, Солнечные данные, No. 5, 77.

Kim, I.S., Uvakina, V. F. In Atmosfera Solntsa, Mezhplanetnaya Sreda, Atmosfera Planet; Gulyaev, R. A.; Ed.; IZMIRAN: Moscow, 1989, p. 125.

Kim, I.S., In *Dynamics of Quiescent Prominences*; Ruzdjak, V.; Tandberg-Hanssen, E.; Eds.; Lecture Notes in Physics, Springer-Verlag: Berlin, 1990; Vol. 363, p. 49.

Kippenhahn, R., Schluter, A. Zitschr. *Astrophys.* 1957, 43, 36.

Kopp, R. A., Pneuman, G. W. *Solar Phys.* 1976, 50, 85.

Krall, N. A., Trivelpiece, A. W. *Principles of Plasma Physics*; McGraw-Hill Book Co.: New York, 1973.

Krall, J., Chen, J., Santoro, R. *Astrophys. J.* 2000, 539, 964.

Kucera, T. A., Andretta, V., Poland, A. I. *Solar Phys.* 1998, 183, 107.

Kulikova, G. N., Molodensky, M. M., Starkova, L. I., Filippov, B. P. In *Solar Magnetic Fields and Corona*; Teplitskaya, R. B.; Ed.;. Nauka: Novosibirsk, 1989; Vol. 2, p. 156.

Kuperus, M., Raadu, M. A. *Astron. Astrophys.* 1974, 31, 189.

Lemmon, J. J. In *Solar Activity Observations and Predictions*; McIntosh, P. S., Dryer, M.; Eds.; Progress in Astronautics and Aeronautics; MIT Press: Cambridge, 1972., Vol. 30, p. 421.

Leroy, J. L., Bommier, V., Sahal-Brechot, S. *Astron. Astrophys.* 1984, 131, 33.

Leroy, J.-L. . In *Dynamics and Structure of Quiescent Solar Prominences*; Priest, E. R.; Ed.; Kluwer Academic Publ.: Dordrecht, Holland, 1989; p. 77.

Liggett, M.; Zirin, H. *Solar Phys.* 1984, 91, 259.

Lin, J., Forbes, T. G., Isenberg, P. A., Demoulin, P. *Astrophys. J.* 1998, 504, 1006.

Lin, J. *Solar Phys.* 2004, 219, 169.

Low, B. C., Hundhausen, A. J. *Astrophys. J.* 1995, 443, 818.

Low, B. C. *Solar Phys.* 1996, 167, 217.

Makarov, V. I. *Solnechnye Dannye* 1983, No. 4, 100.

Makarov, V.I., Tavastsherna, K.S., Davidova E.I., Sivaraman, K.R. *Solnechnye Dannye* 1992, 3, 90.

Makarov, V. I. *Solar Phys.* 1994, 150, 359.

Makarov, V. I., Tavastsherna, K. S., Fatianov, M. I. In *New Cycle of the Solar Activity*; Pulkovo Observatory: Saint Petersburg, Russia, 1998, p. 297.

Makarov, V. I., Filippov, B.P. *Solar Phys.* 2003, 214, 55.

Maksimov, V. P., Ermakova, L. V., *Astron. Zh.* 1985, 62, 558.

Maksimov, V. P., Prokop'ev, A. A. *Astron. Zh.* 1993, 70, 1099.

Malherbe, J. M., Priest, E. R. *Astron. Astrophys.* 1983, 123, 80.

Maltby, P. *Solar Phys.* 1976, 46, 149.

Martens, P. C. H., Kuin. N. P. M., *Solar Phys.* 1989, 122, 263.

Martin, S. F., Ramsey, H. E. In *Solar Activity Observations and Predictions*; McIntosh, P. S., Dryer, M.; Eds.; Progress in Astronautics and Aeronautics; MIT Press: Cambridge, 1972., Vol. 30, p. 371.

Martin, S. F., Livi, S. H. B., Wang, J. *Australian J. Phys.* 1985, 38, 929.

Martin, S. F. In *Dynamics of Quiescent Prominences;* Ruzdjak, V.; Tandberg-Hanssen, E.; Eds.; Lecture Notes in Physics, Springer-Verlag: Berlin, 1990; Vol. 363, p. 1.

Martin, S. F., Livi, S. H. B. In *Eruptive Solar Flares*; Svestka, Z., Jackson, B. V., Machado, M. E.; Eds.; Lecture Notes in Physics; Springer-Verlag: Berlin, 1992, Vol. 399, p. 33.

Martin, S. F., Echols, C. R. In *Solar Surface Magnetism*; Rutten, R. J., Schrijver, C. J.; Eds.; Kluwer Academic Publ.: Dordrecht, Holland, 1994, p. 339.

Martin, S. F. Solar Phys. 1998a, 182, 107.

Martin, S. F., In *New Perspectives on Solar Prominences*; Webb, D.; Rust, D.; Schmieder, B.; Eds.; ASP Conference Series; ASP: San Francisco, CA, 1998b; Vol. 150., p. 419.

Mariska, J. T. *The Solar Transition Region; Astrophys. Series*; Cambridge Univ. Press: Cambridge, 1992; Vol. 23.

McIntosh, P. S. Rev. *Geophys. Space Phys.* 1972, 10, 837.

Menzel, D.H. In *Proc. Conf. of Dynamics of Ionized Media*; University College: London, 1951.

Molodensky, M. M., Filippov, B. P. *Soviet Astron.* 1987, 31, 564.

Molodensky, M. M., Filippov, B. P. *Soviet Astron.* 1988, 32, 202.

Molodensky, M. M., Starkova, L. I., Filippov, B. P. *Soviet Astron.* 1990, 34, 309.

Molodensky, M.M., Filippov, B.P., Shilova, N. S. *Soviet Astron.* 1992, 36, 92.

Moore, R. L. *Astrophys. J.* 1988, 324, 1132.

Mouradian, Z., Soru-Escaut, I., Pojoga, S. *Solar Phys.* 1995, 158, 269.

Mouschovias, T. Ch., Poland, A. I. *Solar Phys.* 1978, 220, 675.

Munro, R. H., Cosling, J. T., Hildner, E., Mac Queen, R. M., Poland, A. I., Ross, C. L. *Solar Phys.* 1979, 61, 201.

Nakagawa, Y., Wu, S. T., Tandberg-Hanssen, E. *Solar Phys.* 1975, 41, 387.

Obashev, S. O. *Izvestia Krymskoi Astrifizicheskoi Observatorii* 1963, 29, 118.

Parker, E. N. J. *Geophys. Res.* 1957, 62, 509.

Parker, E. N. J. *Plasma Phys.* 1973, 9, 49.

Petschek, H. E. In *AAS-NASA Symp. on Solar Flares*; Hess, W. N.; Ed.; NASA SP-50, 1964; p. 425.

Pettit, E. *Astrophys. J.* 1932, 76, 9.

Ploceniak, S., Rompolt, B. *Solar Phys.* 1973, 291973, 399.

Priest, E. R. *Solar Flare Magnetohydrodynamics*; Gordon and Breach: London, 1981.

Priest, E. R., Forbes, T. G. J. *Geophys. Res.* 1986, 91, 5579.

Priest, E. R. In *Dynamics and structure of quiescent solar prominences*; Proceedings of the Workshop, Palma de Mallorca, Spain; Kluwer Academic Publishers: Dordrecht, 1989, p. 1.

Priest, E. R., Forbes, T. G. *Solar Phys.* 1990, 126, 319.

Priest, E. R. In *New Perspectives on Solar Prominences*; Webb, D.; Rust, D.; Schmieder, B.; Eds.; ASP Conference Series; ASP: San Francisco, CA, 1998; Vol. 150., p. 453.

Priest, E. R., Forbes, T. G. Magnetic Reconnection – MHD Theory and Applications; Cambridge Univ. Press: Cambridge, 2000.

Rompolt, B. *Hvar Obs. Bull.* 1990, 14, 37.

Rudenko, G. V. *Solar Phys.* 2001, 198, 5.

Rusin, V., Rybansky, M. *Bull. Astron. Inst. Czechoslovakia* 1982, 33, 219.

Rust, D. M. *NCAR Cooperative Thesis*; No. 6; Univ. of Colorado: Boulder, 1966.

Rust D. M. Astrophys. J. 1967, 150,313.

Rust, D. M., Kumar A. *Solar Phys.* 1994, 155, 69.

Rust, D. M. J. *Geophys. Res.* 2001, 106, 25075.

Schatten, K. H., Wilcox, J. M., Ness, N. F. *Solar Phys.* 1969, 6, 442.

Schmieder, B., Malherbe, J. M., Raadu, M. A. *Astron. Astrophys.* 1985, 142, 249.

Schmieder, B. In *Dynamics and Structure of Solar Prominences*; Ballester J.L., Priest E.R.; Eds.; Universitat de les Illes Balears: Mallorca, 1987; p. 5.

Schmieder, B. In *Dynamics and Structure of Quiescent Solar Prominences*; Priest, E. R.; Ed.; Kluwer Academic Publ.: Dordrecht, Holland, 1989; p. 15.

Schmieder, B., Raadu, M. A., Wiik, J. E. *Astron. Astrophys.* 1991, 252, 353.

Secchi, A. Le Solei; Gauthier-Villars: Paris, 1875.

Shafranov, V. D. In *Review of Plasma Physics*; Consultants Bureau: New York, 1966; Vol. 2, p. 103.

Shibata, K. *Astrophys. Space Sci.* 1999, 264, 129.

Sime, D. G., MacQueen, R. M., Hundhausen, A. J. *J. Geophys. Res.* 1984, 89, 2113.

Simon, G., Gesztelyi, L., Schmieder, B., Main, N. In *Coronal and Prominence Plasmas*; Poland, A. I.; Ed.; NASA Conference Publications 2442; 1986; p. 229.

Smith, J. B., Jr. In *Solar Activity Observations and Predictions*; McIntosh, P. S., Dryer, M.; Eds.; Progress in Astronautics and Aeronautics; MIT Press: Cambridge, 1972., Vol. 30, p. 428.

Smith, S. F., Ramsey, H. E. *Solar Phys.* 1967, 2, 158.

Snodgrass, H. B., Kress, J. M., Wilson, P. R. *Solar Phys.* 2000, 191, 1.

Solov'ev, A. A. *Astron. Zh.* 1981, 58, 597.

Solov'ev, A. A. *Astrofizika* 1985, 23, 393.

Somov, B. V. and Syrovatsky, S. I. *Soviet Phys., Trudy Phys. Inst. Akad. Sci.* 1974, 74, 18.

Somov, B. V. *Astron. Astrophys.* 1986, 163, 210.

Sonnerup, B. U. O. J. *Plasma Phys.* 1970, 4, 161.

Sturrock, P. A. *Nature* 1966, 211, 695.

Stellmacher, G., Weihr, E. *Astron. Astrophys.* 1994, 290, 655.

Sterling, A. C., Hudson, H. S., Thompson, B. J., Zarro, D. M. *Astrophys. J.* 2000, 532, 628.

Sweet, P. A. In *Electromagnetic Phenomena in Cosmic Physics*; Lehnert, B.; Ed.; Cambridge University Press, 1958; p. 124.

Syrovatskii, S. I. *Astron. Zh.* 1966, 43, 340.

Syrovatskii, S. I. *Vestnik Akademii Nauk SSSR* 1973, No. 3, 31.

Tandberg-Hanssen, E. *Solar prominences*; Geophysics and Astrophysics Monographs; D. Reidel Publishing Co.: Dordrecht, Holland, 1974; Vol. 12.

Tandberg-Hanssen, E. In *New Perspectives on Solar Prominences*; Webb, D.; Rust, D.; Schmieder, B.; Eds., ASP Conference Series; ASP: San Francisco, CA, 1998, Vol. 150, p. 11.

Tandberg-Hanssen, E. *The Nature of Prominences*, Astrophysics and Space Science Library; Kluwer Academic Publishers: Dordrecht, Holland, 1995; Vol. 199.

Tang, F. *Solar Phys.* 1987, 107, 233.

Tsuneta, S. *Astrophys. J.* 1996a, 456, L63.

Tsuneta, S. *Astrophys. J.* 1996b, 456, 840.

Uchida, Y. In *Proceedings of the Japan-France Seminar on Solar Physics*; Moriyama, F., Henoux, J. C.; Eds.; 1981, 169.

van Ballegooijen, A. A., Martens, P. C. H. *Astrophys. J.* 1989, 343, 971.

van Ballegooijen, A. A., Martens, P. C. H. *Astrophys. J.* 1990, 361, 283.

Van Tend, W., Kuperus, M. *Solar Phys*. 1978, 59, 115.

Van Tend, W. *Solar Phys*. 1979, 61, 89.

Vrsnak, B. *Solar Phys*. 1990, 129, 295.

Vrsnak, B., Ruzdjak, V., Rompolt, B. *Solar Phys*. 1991, 136, 151.

Vrsnak, B., Rosa, D., Bozic, H. et al. *Solar Phys*. 1999, 185, 207.

Vrsnak, B. J. *Geophys. Res*. 2001, 106, 25249.

Von Neumann, J., Richtmyer, R. D. J. *Appl. Phys*. 1950, 21, 232.

Wang, H., Wang, J. *Astron. Astrophys*. 1996, 313, 285.

Webb, D. F., Krieger, A. S., Rust, D. M. *Solar Phys*. 1976, 48, 159.

Webb, D. F., Hundhausen, A. J. *Solar Phys*. 1987, 108, 383.

Webb, D. F., Forbes, T. G., Aurass, H. et al. *Solar Phys*. 1994, 153, 73.

Wu, S. T., Bao, J. J., An, C. H., Tandberg-Hanssen, E. *Solar Phys*. 1990, 125, 277.

Yang, G., Wang, Y. In *Magnetic Activity and Space Environment*; Wang, H. N., Xu, R. L.; Eds.; Proc. COSPAR Colloq.; Pergamon: Boston, 2002; Ser. 14, p. 113.

Yazev, S. A., Khmyrov, G. M. *Solnechnye Dannye* 1987, No. 12, 75.

Yi, Z., Engvold, O., Keil, S. L. *Solar Phys*. 1991, 134, 275.

Zagnetko, A. M., Filippov, B. P., Den O. G. *Astronomy Reports* 2005, 49, 425.

Zirin, H., Lackner, D. R. *Solar Phys*. 1969, 6, 86.

Zirin, H. In Physics of Solar Prominences; *Proceedings of the IAU Colloquium* **44**; Jensen, E., Maltby, P., Orrall, F. Q.; Eds.; Int. Astron. Union: Paris, 1979; p. 193.

Zirker, J. B., Engvold, O., Martin, S. F. *Nature* 1998, 396, 440.

Zwaan, C. In *High Resolution in Solar Physics*; Muller, R.; Ed.; Lecture Notes in Physics; Springer-Verlag: Berlin, 1985; Vol. 233, p. 263.

Zwaan, C. Ann. Rev. *Astron. Astrophys*. 1987, 25, 83.

In: Solar Physics Research Trends
Editor: Pingzhi Wang, pp. 95-136
ISBN 978-1-60021-987-0
© 2008 Nova Science Publishers, Inc.

Chapter 2

MULTIFRACTAL NATURE OF SOLAR PHENOMENA

Valentyna I. Abramenko
Big Bear Solar Observatory of New Jersey Institute of Technology
40386 North Shore Lane, Big Bear City, CA 92314, USA

Abstract

Last decades, Solar Physics research demonstrated a considerable progress in understanding of macro-scale processes (e.g., the magnetohydrodynamic modeling of the heliosphere, modeling of large-scale coronal magnetic structures, etc.), on one hand and micro-scale phenomena (e.g., turbulence of solar plasma) on the other hand. Further progress seems to be associated with the realization of how various micro-scale processes are involved and manifested in the macro-scale behavior of the entire Sun. A similar problem unavoidably arises in studies of any, other than the Sun, non-linear dynamical dissipative system present in nature. Such systems, that can be placed in between a chaos and a completely determined structure, display a scale-invariant behavior. A broad class of models for scale-invariant structures is now elaborated based on the notion of a fractal, which can be mathematically described as an object with fractional dimensions. The goal of this chapter is to show how the concept of fractality and multifractality is currently elaborated in Solar Physics and how these ideas help us to understand the unpredictable behavior of our closest star. The first section provides introduction to multifractality and outlines physical connections between various aspects of this paradigm. Second section is devoted to reviewing the multifractality approach in the study of the Sun: we will discuss how high-order statistical moments, fractal geometry, multifractality, percolation theory, and the concept of the self-organized criticality are now applied to a broad range of solar phenomena: from explosive events of various scales to geometrical properties of observed filigree structures. The last section outlines new ideas and results based on the concept of multifractality.

1. What Does the *Multifractality* Mean?

When starting to treat any physical system, one have to answer a fundamental question: whether the dissipation of energy can be neglected? In the case of positive answer, the system can be treated as a deterministic system that can be described by a set of differential equations and the uniqueness of the solution is fully determined by the initial conditions. However, for the majority of processes in nature, the energy dissipation cannot be neglected when one tries to obtain a realistic picture for their evolution. Such processes should be treated as a nonlinear dissipative dynamical system (NDDS) and solar plasma is one of them. The universality of such processes (from forest fires and earthquakes to cosmic rays, magnetosphere sub-storms, solar flares, etc.) inspires our efforts to explore them.

The fact that the inner dissipation cannot be neglected anymore results in several important outcomes. First, small fluctuations in the initial state may result in significant deviations of the solution in the future that unavoidably drives the system to chaos. However, chaos can not be set in immediately, it takes some time to destroy a perfectly determined system. A specific state in the evolution of the system appears, when determinacy is lost, but a pure chaos is not reached yet, the state when the system acquires a fractal structure (recall the jagged coast line of Norway) as in spatial, so in temporal domains.

Nevertheless, complete chaos is unreachable due to nonlinear interactions between different scales (again in both spatial and temporal domains), that results in capability of a NDDS to rebuild large- and middle-scale substructures from small-scale chaotic fragments (recall the formation of large clusters in a turbulent interstellar medium, or the inverse magnetic helicity cascade in magnetohydrodynamics). This property of a NDDS results in the existence of an *attractor* in the phase space, i.e., an asymptotic limit of the solutions, which is not necessarily determined by initial conditions (see, e.g., Berge et al. 1988; Henighan 1997).

Figure 1. An example of a self-similar structure. A framed part of the fern leaf is similar to the whole leaf.

Second, a specific state between determinacy and chaos is, in turn, characterized by existence of self-invariance, when parts of an object somehow resemble the whole. To put it in more rigorous way, a self-similar, scale-invariant structure without any characteristic length appears, a *fractal* (Mandelbrot 1982; Feder 1988; Takayasu 1989; Schroeder 2000). A classical example of self-similarity is shown in Figure 1: a whole fern leaf resembles its part (framed), which, in turn, can be considered as a whole leaf and similar smaller part can be cut out, and so on. When we cover a fractal by a mesh, the number of boxes, N, occupied by the fractal structure will be proportional to the scale length (the length of a sample box, r) in a *fractional power*, D_f: $N(r) \propto r^{D_f}$. D_f is called *fractal dimension* and it is an essential characteristic of a fractal. Computer-generated fractals are self-similar over an infinite range of scales, and their geometrical properties can be adequately described by only one parameter, fractal dimension. However, in natural fractals self-similarity does not hold in the strict sense but does hold statistically over a finite range of scales. As soon as we deal with natural fractals, we immediately reveal that the slope of $log(N)$ plotted vs $log(1/r)$ may change inside different scale ranges (Figure 2). Physical reasons for this variable slope are the presence of nonlinear interactions between different scales (see, e.g. Berge et al. 1988; Biskamp 1993) and inhomogeneous character of dissipation (Monin and Yaglom 1975; Frisch 1995). Now the entire range of scales can be divided into an infinite number of subsets and for each subset its own self-similarity law holds with its own fractal dimension. Thus, a natural fractal may be considered as a *multifractal*, i.e., a superposition of an infinite number of monofractals. Local (in sense of scale) character of self-similarity of natural fractals is one of the reasons to make a step from monofractals to multifractals.

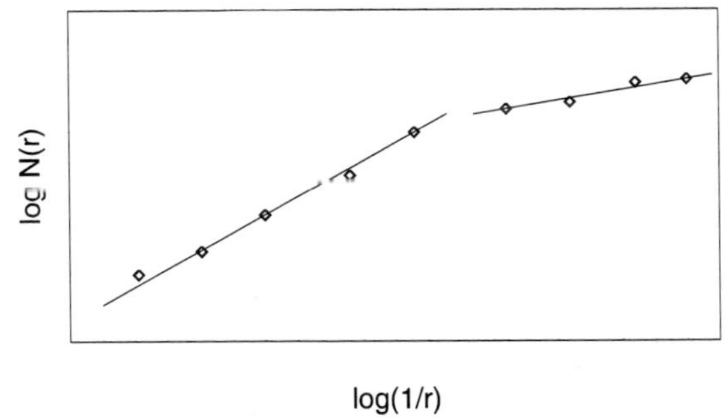

Figure 2. A slope of $log(N)$ versus $log(1/r)$ defines the fractal dimension, D_f. For natural fractals (i.e. for multifractals), the slope can change for different scale ranges.

Another advantage that multifractals present is the possibility to take into account the distribution of density in a fractal. In the case of solar magnetic fields, we can obtain a fractal when we define a magnetic area as a binary mask (Figure 3) equal to unity on the set of all pixels where the absolute value of the vertical component of the magnetic field, B_z, exceeds a given threshold (for example, 50 G, Figure 3, upper left frame), and to zero elsewhere. By using the box-counting method, for example, or some other methods (Takayasu 1989), we can calculate the fractal dimension of the binary mask, which characterizes complexity of its boundary. However, the fractal dimension can tell us nothing about the distribution of the magnetic field density *within* the mask. It is well known that the density distribution may not be homogeneous. Moreover, the density distribution largely determines the complexity of the magnetic structure and, therefore, the underlying physics. Thus, we have to incorporate the information on density. This can be done by introducing an hierarchy of masks defined by increasing threshold. The idea is illustrated in Figure 3: by specifying progressively higher thresholds, say, 500 G (lower left frame), 1000 G (upper right frame), we obtain new fractals and their parameters may vary. By analyzing these new monofractals, we can obtain addition of information on the distribution of increasingly dense pattens in the magnetic structure. The superposition of the infinite number of these monofractals results in a multifractal: a real distribution of magnetic fields on the solar surface (see lower

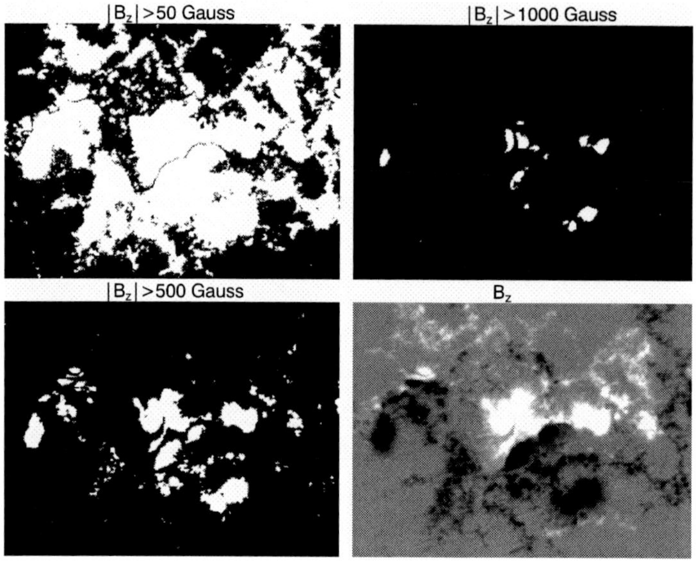

Figure 3. An hierarchy of binary masks of a solar magnetic structure (B_z, lower right frame). Each binary mask (left frames and upper right frame) is equal to unity (*white*) on the set of all pixels where the absolute value of the magnetic field exceeds a given threshold (noted above the frames) and zero (*black*) elsewhere. Superposition of infinite number of such masks/monofractals results in the observed magnetic structure, which is a multifractal. For simplicity sake, the sign of B_z is not considered.

right frame in Figure 3. Please, note that for simplicity sake, the sign of B_z on the masks was disregarded.) To characterize the structure of a multifractal, a continuous spectrum of fractal parameters - a multifractal spectrum - is needed. The broader the multifractal spectrum, the more complex structure it represents.

Structures that arise in the course of evolution of dynamic systems, where dissipative processes may not be neglected, display highly *intermittent* character. Intermittency implies a tendency of a physical entity to concentrate into small-scale features of high intensity surrounded by extended areas of less intense fluctuations (Monin & Yaglom 1975; Biskamp 1993; Frisch 1995). Intermittency manifests itself via burst-like behavior in temporal and spatial domains. Figure 4 demonstrates the phenomenological difference between a Gaussian process (which is non-intermittent) and an intermittent process. Large fluctuations in the intermittent process are not as rare as in the Gaussian process, and they contribute significantly into the statistical moments, which leads to multifractality (see Frisch 1995 for more details). Thus, by analyzing multifractality of a structure, we also explore its intermittency. Generally speaking, intermittency and multifractality are two different terms for the same phenomenon. Historically, the term *Intermittency* is usually applied to time series analysis, whereas the term *multifractality* is used for spatial objects (Takayasu 1989; Frisch 1995). It is worth noting that when a field is intermittent, the structure of its energy dissipation can not be homogeneous but it is intermittent, too (Monin & Yaglom 1975; Frisch 1995). Thus, by exploring intermittency of solar magnetic fields, for example, one may also obtain valuable information on the behavior of magnetic energy dissipation (Lawrence et

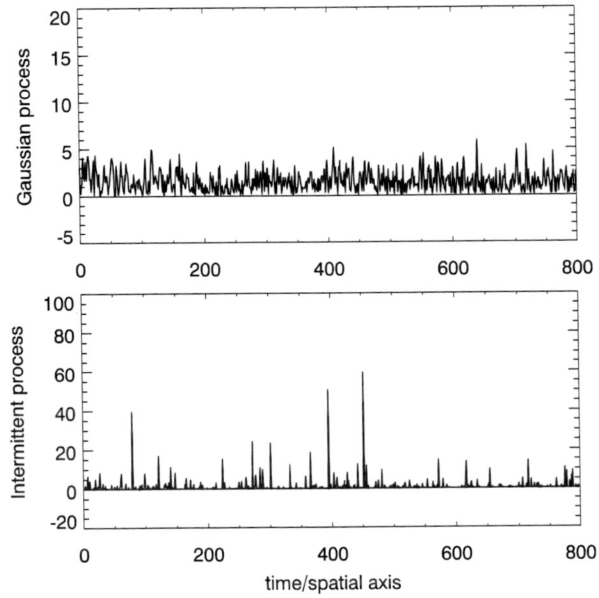

Figure 4. Illustration of phenomenological difference between a Gaussian process (upper frame), which is a non-intermittent process of extremely rare large fluctuations, and an intermittent process (lower frame), which is a multifractal due to the contribution of large fluctuations (which are not so rare now) into the statistical moments. Horizontal axis may represent a time or a spatial coordinate.

al. 1995; Abramenko 2003; Abramenko et al. 2003; Vlahos & Georgoulis 2004).

One of the traits of a multifractal structure (or intermittent process) is a non-Gaussian distribution function. As it follows from the central limit theorem, a Gaussian random process is caused by a *sum* of many independent random variables. The distribution function in this case is completely determined by two non-random parameters: the first and second statistical moments (the mean value and the standard deviation) of the random variable. For an intermittent/multifractal process, the distribution function decreases (as the scale goes to infinity) *slower* than it does for a Gaussian process, which is caused by the increased role of large fluctuations in intermittent process (they are now much stronger and appear more frequently). The growth of the strong peaks population may occur for different reasons, however the most plausible one is that now the fluctuations are not a *sum* of independent (and approximately equal) random variables, but they rather are a *product* of many variables (as it happens in any fragmentation process or random multiplicative cascade). For an intermittent/multifractal variable, the distribution function is no longer determined by the first and second statistical moments. High-order moments play critical role and contribute significantly into the shape of the heavy tail of the distribution function.

To summarize, multifractality is a property of a dynamical system in which energy dissipation cannot be neglected. Multifractality is equivalent to intermittency and manifests itself as a highly jagged, porous structure in spatial domain and via burst-like behavior in time, when extended areas (intervals) of low fluctuations are intermittent with small areas of extremely large fluctuations.

There are several approaches to study multifractality. It can be diagnosed by i) analyzing of high-order statistical moments (e.g., calculation of distribution, structure, and flatness functions); ii) calculating fractal dimensions and multifractality spectra; iii) applying percolation theory; and iv) self-organized critically (SOC) theory. The list is far from being complete and it only includes the methods that are most frequently used in solar physics.

It is worth noting that ideas of multifractality and intermittency are applied also throughout a broad range of scientific studies, from interstellar medium (e.g., Falgarone et al. 1995; Cho et al. 2003; Kritsuk & Norman 2004; Balsara & Kim 2005) to biological and social systems (see numerous examples in, e.g., Schroeder 2000).

2. Studies of the Sun Based on the Multifractality Ideas

2.1. Exploring High Statistical Moments

2.1.1. Distribution Functions

Determination of the distribution function (or probability distribution function, PDF) from observations of a random process in question is usually the first and essential step of any study. Usually, a PDF is determined with one goal in mind: to find the mean value of the data set and its standard deviation. However, the PDF possesses much more information about the process. Strictly speaking, high-order statistical moments can be also evaluated from the PDF (Feder 1988; Takayasu 1989). When the PDF displays a heavy non-Gaussian tail, this is a manifestation of a highly intermittent and multifractal process, prone to eruption-like energy release. Moreover, this also indicates that the system, as a whole, is undergoing a non-linear dissipation with the energy interchange between different scales. Thus, a shape of the PDF is a powerful diagnostics tool. Figure 5 demonstrates a qualitative difference between the distribution functions for a Gaussian and intermittent processes. Distribution functions for a variety of solar processes and structures were derived as of now. First suggestion on a non-Gaussian behavior of our Sun appeared after pioneer space observations of the solar emission in X-rays.

a. Power-laws reign in statistics of X-ray bursts.

First solar X-ray experiments on Orbiting Solar Observatories, OSO-1 (White, 1964), OSO-5 (Frost, 1969), OSO-7 (Daltowe et al. 1974), Explorer 33 and Explorer 35 (Drake, 1971), provided extensive information on hard and soft X-ray bursts. First statistical study of various bursts parameters (Drake, 1971) revealed that distributions of the peak flux of bursts, their time-integrated flux, rise time, and decay time display the frequency distributions that are far from Gaussian. Observed distributions possessed a slowly diminishing at infinity tail and allowed a good fitting with an exponential function and/or a power law function. Later, many authors confirmed this results on the basis of subsequent space missions and provided a variety of power indices for the total energy release, peak count rate, and event duration. Most of them are reviewed by Crosby et al. (1993), Aschwanden et al. (1998) and Charbonneau et al. (2001).

b. Distribution of the magnetic flux content in photospheric magnetic elements.

Bogdan and co-authors (Bogdan et al. 1988) reported on heavy-tailed distribution of umbral areas of solar sunspots that stimulated other efforts to calculate and analyze similar distributions for elements of the magnetic field in the photosphere. Modern observational technique allows us to calculate the distribution function of the magnetic flux content in elements of the magnetic field at the photospheric level. In quiet Sun areas, where the magnetic elements are widely separated, PDFs can be determined relatively more easier than in an active region area, where elements are tightly packed to form clusters of various size and shapes. As a result, studies of the distribution of magnetic flux in quiet sun regions are more frequent. Wang et al. (1995) studied the dynamics and statistics of the quiet-sun magnetic fields using Big Bear Solar Observatory (BBSO) videomagnetograph data. They found the distribution function to be a power law with the power index of -1.68 in areas with the magnetic flux in the range of $(0.2-1) \times 10^{18}$ Mx (intranetwork fields) and -1.27 for flux in the $(2-10) \times 10^{18}$ Mx range (network elements).

Schrijver et al. (1997a) used high resolution data for a quiet sun area from the Michelson Doppler Imager (MDI, Scherrer et al. 1995) on-board of the *Solar and Heliospheric Observatory* (SOHO). They reported that the flux distribution follows an exponential law with a slope of approximately 1×10^{-18} Mx^{-1} in those areas where the flux ranges from 1 to 5 $\times 10^{18}$ Mx. The distribution of magnetic flux in flux concentrations in a plage area (outside of large sunspots) is less steep than that for the quiet sun, with the exponential slope slightly varying as a function of the flux Schrijver et al. (1997b). To explain the shape of the observed distribution, these authors modeled the distribution of flux assuming that three primary processes underlie the distribution of the magnetic flux: merging, cancellation and fragmentation of flux concentrations. Under this assumption, the exponential slope of a model distribution should vary inversely with the square root of the average flux density, and thus a satisfactory agreement between the observed and modeled distributions was reached in the range $(20-150) \times 10^{18}$ Mx.

Later, Abramenko and Longcope (2005) analyzed the distribution of flux content in magnetic elements in the range of flux $\Phi > 10^{19}$ Mx for two well developed active regions. A particular attention was payed to the analytical approximation of the observed distribution, which were calculated from SOHO/MDI high-resolution line-of-sight magnetograms. The first active region NOAA 9077 observed on July 14, 2000 displayed very high flaring activity: it produced the famous Bastille day flare of X5.7 X-ray class, and about 130 lesser flares occurred during the active region passage across the solar disk. A magnetogram of this active region is shown in Figure 3 (*lower right frame*) and in Figure 13. The second active region NOAA 0061, observed on Aug 9, 2002, was a very quiet one: it produced several small C-class flares only. During the time interval under consideration, this active region produced no flares at all. One of its magnetogram is shown in Figure 14. During the MDI observations, both active region were located near the center of the solar disk, therefore, the measured line-of-sight component of the magnetic field very closely represents the B_z component of the magnetic field, which is normal to the solar surface. For each active region, we used a sequence of 248 magnetograms.

To calculate the PDF, we need first to locate and separate each magnetic element in a magnetogram. Our image segmentation code works as follows: we assumed that a given pixel belongs to a magnetic flux concentration when the absolute value of the magnetic field in it exceeds the specified threshold, i.e. $|B_z| > p$. Calculations were performed for a set of thresholds: 25, 50, 75 and 100 G. The local peak of an individual concentration was

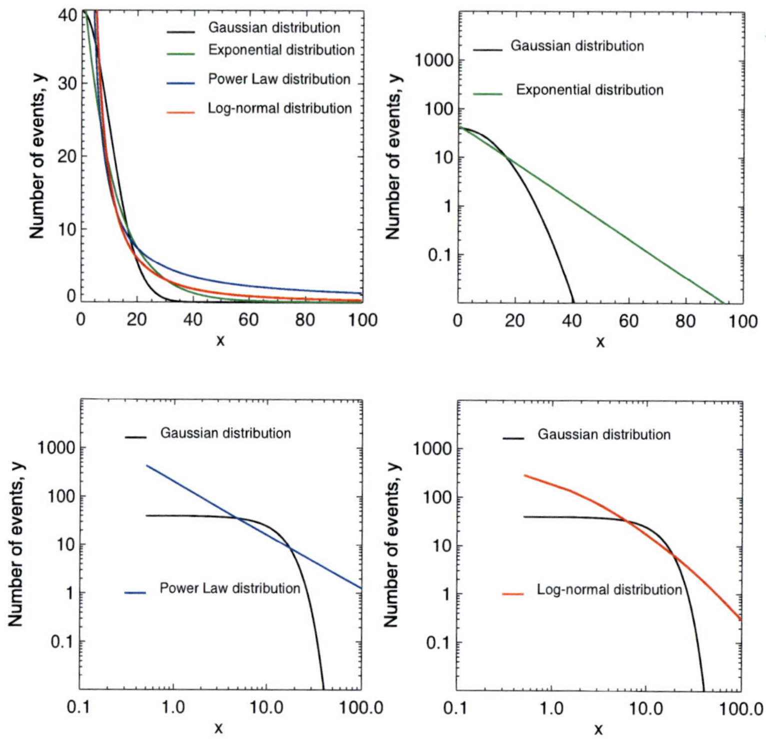

Figure 5. Qualitative comparison of different distribution functions. When plotted in linear scales (upper left frame), it is rather difficult to diagnose the analytical shape of distributions: exponential, power law and log-normal functions look similar, all of them essentially exceed the Gaussian curve at large scales and pass below it at small scales. In the linear-logarithmic plot (upper right frame), the exponential function [Eq. (4)] is a straight line, whereas the power law function [Eq. (5)] is a straight line in the double-logarithmic plot (lower left frame). The log-normal function [lower right frame, Eq. (1)] is slightly curved in the double-logarithmic scales. Thus, the appearance of the observed distribution in different presentations may give a clue for an analytical approximation.

determined as a pixel where both, $d|B_z|/dx$ and $d|B_z|/dy$, change their sign from positive to negative (here x and y are the current coordinate on a magnetogram). The next step was to outline the flux concentrations and to calculate their unsigned flux content. In the case when the unsigned flux density in a given pixel exceeds the threshold, the code determines to which peak a given pixel belongs by moving in a direction of the maximum gradient on the map of $|B_z|$. When a pixel labeled as a peak is encountered, the flux and the area of the pixel are summed to the total flux and total area of a given flux concentration. The result of the application of this image segmentation code to an arbitrary 40×40 pixels fragment of a magnetogram is shown in Figure 6.

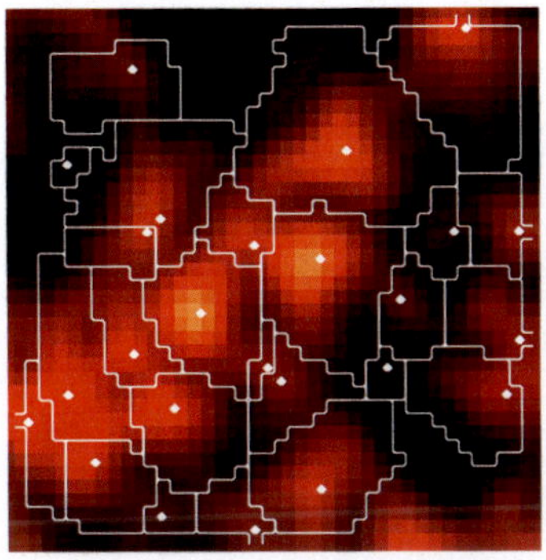

Figure 6. A map of the absolute value of the magnetic field, B_z, for an arbitrary 40×40 pixel area of the AR9077, which illustrates the processing of the image segmentation code. The image is scaled in the range of values $0 - 1200$ G. The white diamonds mark the position of the peaks of flux concentrations. The white lines outline the boundaries between the flux concentrations. Total unsigned flux inside each closed boundary was calculated.

In each magnetogram, approximately 600-800 concentrations were normally localized, depending on the threshold value. For the great majority of the concentrations in both active regions, the absolute value of the magnetic flux did not exceed 8×10^{20} Mx. This value was accepted as the upper limit of the flux range for the histogram. After normalizing each histogram for the total number of concentrations in a magnetogram, they were averaged over the total number of magnetograms, so that an averaged probability distribution function, PDF(Φ), was obtained. These averaged PDFs for both active regions obtained with the threshold $p = 50$ G are shown in Figure 7 (*left panel*).

To apply an analytical approximation to the observed distributions, we first determined the lower value of the magnetic flux, Φ_{cut}, below which the measured flux may be significantly affected by influence of noise, threshold, resolution, etc. Thus, we chose $\Phi_{cut} = 10^{19}$ Mx as flux cut-off above which we believe our sample is reasonably complete.

For an arbitrary magnetogram of each active region, we calculated the cumulative distribution function (CDF) for flux values $\Phi > \Phi_{cut}$. This function, $CDF(\Phi)$, giving the fraction of all concentrations with flux greater than Φ, does not require the bin size be defined. The CDF(Φ) for an arbitrary magnetogram of active region 9077 is shown in Figure 7 (*right panel, red line*).

To quantitatively analyze the obtained distributions, we attempted to fit the observed PDF with three different model distributions: log-normal, exponential and power law.

In a log-normal distribution (Aitchison & Brown 1957; Romeo et al. 2003) the logarithm of the magnetic flux is normally distributed. The corresponding PDF can be written as

$$PDF(\Phi) = \frac{1}{\Phi s(2\pi)^{1/2}} \exp\left[-\frac{1}{2}\left(\frac{\ln(\Phi) - m}{s}\right)^2\right], \quad (1)$$

for $\Phi > 0$. The function's two parameters m and s are the mean and standard deviation of

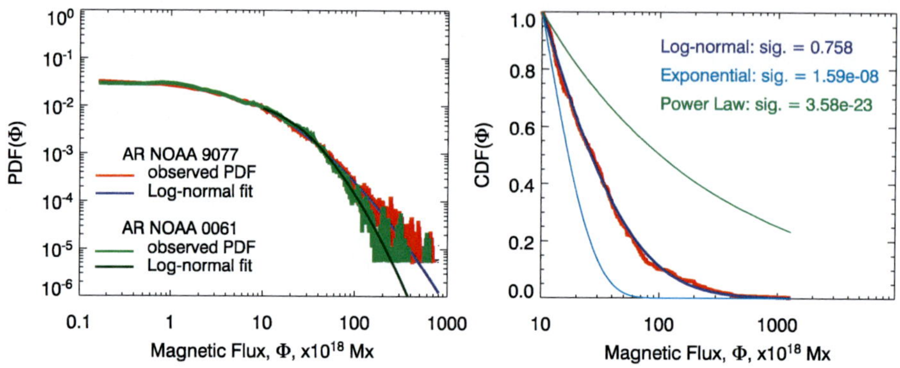

Figure 7. *Left* - Averaged probability distribution functions (PDFs) of the magnetic flux content in flux concentrations for NOAA AR 9077 (*red line*) and for NOAA AR 0061 (*green line*). The best log-normal fits in the range of flux $\Phi > 10^{19}$ Mx to the averaged PDFs are shown by *violet line* for NOAA AR 9077 and by *dark green line* for NOAA AR 0061. *Right* - Observed cumulative distribution function (CDF, the thick red line) calculated in the flux range $\Phi > 10^{19}$ Mx from the magnetogram of NOAA AR 9077 obtained at 08:12 UT on July 14, 2000. The thin violet curve marks the log-normal function, $LN(m, s)$, with parameters $m = 2.20$ and $s = 1.49$, as they were obtained from the averaged PDF. The thin blue curve denotes the the best exponential fit with parameters $C_1 = 0.0135$ and $\beta = 0.0572$, calculated as a best fit to the averaged PDF. The thin green curve marks the power law function with $C_2 = 0.275$ and $\beta = -1.45$, calculated as a best fit to the averaged PDF. All three approximations are calculated over the flux range $\Phi > 10^{19}$ Mx. The level of significance of the Kolmogorov-Smirnov test, "sig", for each type of approximation, is noted. The threshold is 50 G.

$\ln(\Phi)$. The expectation value, μ, and the variance, σ^2, of log-normally distributed flux are defined in terms of these parameters

$$\mu = \exp\left(\frac{2m + s^2}{2}\right), \tag{2}$$

$$\sigma^2 = \exp(2m + 2s^2) - \exp(2m + s^2). \tag{3}$$

The mean flux, μ, exceeds the *median* flux, e^m, by a factor $e^{s^2/2}$, which can be far greater than unity for large values of s. In the opposite limit, $s \ll 1$, the log-normal distribution approaches a Gaussian (normal) distribution and the mean approaches the median.

The averaged PDF was fit with expression (1) by performing a standard, unweighted least-squares Gaussian fit to the function $\Phi \cdot PDF(\Phi)$ versus $\ln(\Phi)$. The mean and standard deviation of this fit yield the parameters m and s. Values for each active region for different threshold values are listed in Table 1. A typical example of the log-normal approximation in the range $\Phi > \Phi_{cut}$ is shown in Figure 7.

The second possibility, an exponential distribution, corresponds to the PDF

$$PDF(\Phi) = C_1 \exp(-\beta\Phi), \tag{4}$$

Table 1. Parameters of the log-normal fit and K-S statistic and significance for the three distributions

	Log-normal fit				Exponential fit		Power Law fit	
Th, G	m	s	d	sig	d	sig	d	sig
NOAA AR 9077								
25	2.113	1.510	0.0389	0.623	0.163	4.71e-09	0.260	1.65e-22
50	2.199	1.491	0.0354	0.758	0.161	1.60e-08	0.270	3.58e-23
75	2.285	1.472	0.0384	0.681	0.153	1.50e-07	0.283	1.25e-24
100	2.378	1.441	0.0344	0.826	0.153	3.60e-07	0.303	5.63e-27
NOAA AR 0061								
25	2.690	0.969	0.0445	0.455	0.1130	1.56e-04	0.275	1.03e-24
50	2.715	0.961	0.0342	0.794	0.0953	2.87e-03	0.288	2.58e-26
75	2.711	0.965	0.0306	0.906	0.0966	3.35e-03	0.290	1.63e-25
100	2.703	0.970	0.0427	0.600	0.1057	1.51e-03	0.295	9.07e-25

where $C_1 = \beta e^{\beta \Phi_{cut}}$ when using the range $\Phi > \Phi_{cut}$. We calculated β as the negative slope when $\ln(PDF)$ is fit, by unweighted least-squares, to a linear function of Φ. Our data for AR NOAA 9077 gave the value $\beta \approx 0.036 \times 10^{-18}$ Mx^{-1} when $\ln(\Phi)$ was fit over a narrow flux range $(20 - 110) \times 10^{18}$. This agrees well with the value found by Schrijver et al. (1997b) for plage area inside an active region over the same flux range (see Figure 2 in Schrijver et al. 1997b). The result of the exponential approximation is shown in Figure 7 (*right panel, blue curve*).

The final possibility, a power law distribution, is given by the PDF

$$PDF(\Phi) = C_2 \Phi^{-\alpha}. \tag{5}$$

In an unweighted least-squares fit of ln(PDF) to a linear function of $\ln(\Phi)$ the slope gives $-\alpha$. An example of the power law fit for the CDF of AR 9077 is shown in Figure 7 (*right panel, green curve*).

The quality of each fit is quantified using the Kolmogorov-Smirnov (K-S) test (Press et al. 1986). The observed CDF (shown by *the thick red line* in Figure 7, *right*) was computed from a single arbitrary magnetogram of NOAA AR 9077. The model CDF (*violet, blue* or *green curve* in Figure 7, *right*) uses parameters found by fitting the average PDF of the same active region. The K-S statistic d, the maximum absolute deviation between the model and the observed CDF, quantifies the likelihood that the data was drawn from the model distribution (Press et al. 1986). The "significance" is the probability of obtaining a greater value of d if data were randomly drawn from the model distribution. For example, when the significance is smaller than 0.01, we may rule out the proposed model distribution with 99% confidence. The K-S statistic and significance are listed in Table 1 for the three distributions. The K-S test eliminates, with very high confidence, all distributions except the log-normal. The log-normal distribution is consistent with the CDFs from both active regions at every threshold value. The results above suggest that the magnetic flux in photospheric concentrations is log-normally distributed, which implies involvement of fragmentation process into evolution of magnetic fields.

In the solar photosphere and convective zone, two essential processes determine the interaction between turbulent plasma flows and magnetic flux tubes. On one hand, a magnetic element tends to be advected by the turbulent diffusion. On the other hand, turbulent motions tend to sweep the field lines together at convergence/sink points of the flow. Fragmentation and concentration processes are in a dynamic equilibrium in homogeneous stationary turbulence. Petrovay and Moreno-Insertis (1997) showed that in inhomogeneous and/or non-stationary situations, turbulent diffusion dominates the concentration causing turbulent erosion of magnetic flux tubes. The idea of the gradual disintegration of sunspots due to erosion of penumbral boundaries was first advanced by Simon & Leighton (1964). Later, Bogdan et al. (1988), on the basis of the observed log-normal distributions of umbral areas, argued that fragmentation of the magnetic elements may be an essential process in the formation of observed magnetic structures. The log-normal distribution of the flux content in magnetic flux elements of an active region suggests that the process of fragmentation dominates the process of flux concentration.

Assuming that the log-normality of flux concentration results from repeated, random fragmentations, we may attribute meaning to the distribution parameter. The variance, s^2, of $\ln(\Phi)$ is proportional to the number of independent fragmentations produced a given concentration from a single initial concentration (Monin & Yaglom 1975, Ch. 25). If the basic fragmentation process is similar in all active regions then s^2 is proportional to the time over which those fragmentations have occurred. Since the magnitude of s^2 for NOAA AR 9077 exceeds that for NOAA AR 0061, by a factor of 2.3, we may further infer that NOAA AR 9077 may be older than NOAA AR 0061 by approximately that factor. Alternatively, NOAA AR 9077 may have undergone more vigorous fragmentation over a comparable lifetime. The latter possibility may also explain the very different level of the flaring activity in those active regions. Note, that a very intensive fragmentation of sunspots during several days before the Bastille day flare in NOAA AR 9077 were reported by Liu & Zhang (2001).

c. Distribution of speeds of coronal mass ejections as measured from LASCO coronographs

Another attempt to apply the distribution function technique to extract the underlying physics was performed by Yurchyshyn and colleagues (Yurchyshyn at al. 2005) in their analysis of initial l speeds of coronal mass ejections (CMEs). CMEs are frequently associated with strong flares on the Sun and expand outward from the Sun, often toward Earth (and beyond) causing strong geomagnetic storms (see, e.g., reviews by Gopalswamy et al. (2006) and Schwenn et al. (2006)). Yurchyshyn and colleagues studied the distribution of plane of sky speeds determined for 4315 CMEs detected by Large Angle and Spectrometric Coronograph Experiment on board Solar and Heliospheric Observatory (SOHO/LASCO). They found that the speed distributions for accelerating and decelerating events are nearly identical and to a good approximation they can be fitted with a single log-normal distribution. This study clearly demonstrates an heuristic idea: when an original distribution function (Figure 8) is very skewed and has a heavy tail, there is a good chance that the *logarithm* of the value in question will be distributed normally (see Figure 9).

The fact that the two groups of CMEs (accelerating and decelerating events) can be modeled by a single distribution may suggest that the same driving mechanism is acting in both slow and fast dynamical types of CMEs. What then can we learn from he fact that CME speeds are distributed log-normally? It is accepted that magnetic reconnection between independent magnetic systems plays a major role in the origin of a coronal ejecta. The log-normal distribution of the speeds of CMEs implies then the presence of nonlinear

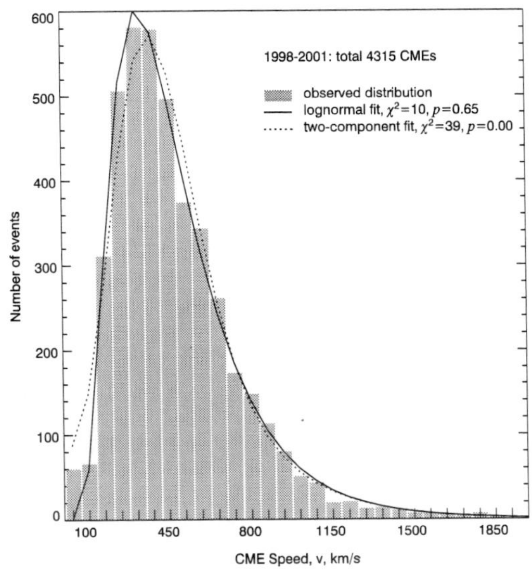

Figure 8. Distribution of the observed speeds of 4315 CMEs (vertical bars). The width of the bar is 70 km s^{-1}. The solid line represents a single log-normal fit to the observed data, while the dashed line is a sum of a Gaussian and a log-normal fits. Courtesy of V. Yurchyshyn.

interactions and multiplicative processes in a system of many magnetic flux loops. When complexity of an evolving magnetic field reaches its peak so the magnetic configuration is no longer stable, a major instability as well as a entire spectrum of smaller scale instabilities may develop in the system of many loops and fluxes, regardless of the assumed initial configuration (Maia et al., 2003). Thus, Kaufmann et al. (2003) interpreted sub-millimeter pulse bursts at the CME onset as signatures of energy release due to small-scale instabilities in the magnetic field. In this sense, the reconnection process in a CME model can be viewed as a multiple reconnection process or as an avalanche of multiple reconnection sites (Lu & Hamilton 1991; Vlahos et al. 1998; Wang et al. 2000; Charbonneau et al. 2001; Abramenko et al. 2003; Kaufmann et al. 2003) in a fragmented magnetic field (Kuklin 1980; Bogdan 1988; Petrovay et al. 1999, Abramenko & Longcope 2005). However, to what extent the multiple reconnection process of fragmented solar magnetic fields is responsible for the log-normal distribution of the CME speeds is yet to be determined.

d. Distribution functions for parameters of the magnetic field in the solar wind

We will discuss the solar wind in section 3 of this chapter. Here we review only a couple of studies that concern the PDFs of the solar wind data.

Burlaga & Lazarus (2000) reported a log-normal distributions of the speed, density and proton temperature measured at the solar wind streams near Earth. Authors emphasized an intermittent, spiky character of measured fluctuations. Recently Vasquez and coauthors (Vasquez et al. 2007) analyzed distributions functions of some parameters of the solar wind observed with ACE spacecraft near Earth. Namely, they determined discontinuities in the structure of the magnetic field that characterize dynamics of processes in solar wind's

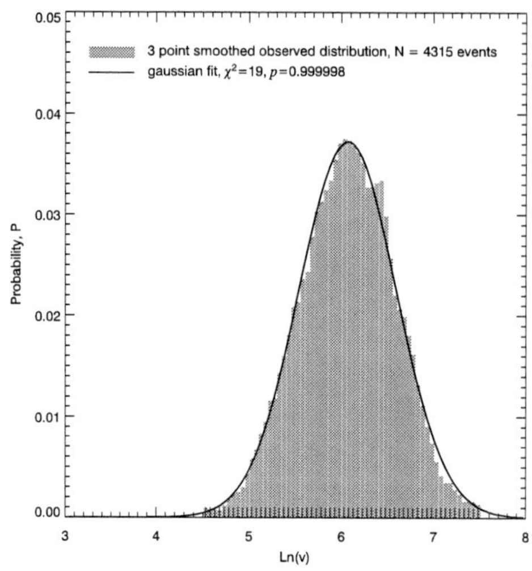

Figure 9. Distribution of the logarithm of velocity, $\ln(v)$, determined for all 4315 CMEs (vertical bars). The solid line represents normal fit to the observed data. Courtesy of V. Yurchyshyn.

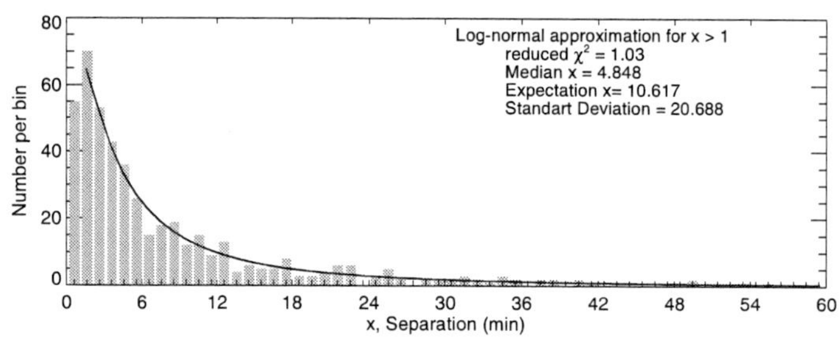

Figure 10. Histogram of the magnetic field discontinuity number as a function of separation in minutes between each discontinuity and the one which follows from January 10 to 13, 2001. The long tail of the distribution is evident. The thin curve denotes the log-normal fit to the data points.

streams on their way from the Sun to the interplanetary space. In particular, the observed distribution function for the separation time intervals between two consecutive discontinuities is shown in Figure 10. The log-normal analytical fit was found to be the best best fit for the observed function. For other analyzed parameters, namely, the width of discontinuity, the spread angle, the increment of the magnetic field density at the discontinuity, log-normal fits were found as best analytical approximations. Analysis of observed distribution functions and comparison with distribution functions derived from numerical experiments led authors to conclude that the Alfvenic turbulence is the most plausible mechanism of the origin of magnetic discontinuities in the solar wind plasma. In particular, the log-normal distribution of separations between discontinuities may indicate that a multiplicative random cascade produces the discontinuities.

2.1.2. Scaling of Structure Functions

Solar plasma, beginning from the base of the corona (the photosphere) up to the outer heliosphere, is in a state of fully developed turbulence. Any system in such a state is an excellent example of a non-linear dynamical dissipative system. Many properties of such systems hold for a turbulent media, as well. At the same time, a turbulent system can be a subject for specific tools elaborated in the framework of the turbulence theory, and those tools eventually provide us with information on their multifractal/intermittent properties. So that multifractality of a turbulent system can be explored by using tools of the turbulence theory. The structure function analysis is one of such tools.

Since Kolmogorov's study (Kolmogorov 1941, hereafter K41), many models have been proposed to describe the statistical behavior of fully developed turbulence. In these studies, the flow is modeled using statistically averaged quantities such as velocities, and structure functions play significant roles. Structure functions were first introduced by Kolmogorov and they are defined as statistical moments of the increments of a turbulent field $\mathbf{u}(\mathbf{x})$ as

$$S_q(r) = \langle |\mathbf{u}(\mathbf{x}+\mathbf{r}) - \mathbf{u}(\mathbf{x})|^q \rangle, \tag{6}$$

where $\mathbf{r}$ is a separation vector, and q is a real number. Structure functions, calculated within the inertial range of scales, r, ($\eta \lesssim r \lesssim L$, where η is a spatial scale where the influence of viscosity becomes significant and L is a scaling factor for the whole system) are described by the power law (K41; Monin and Yaglom 1975; Frisch 1995):

$$S_q(r) \sim (\varepsilon_r(\mathbf{x}) \cdot r)^{q/3} \sim (r)^{\zeta(q)}. \tag{7}$$

where $\varepsilon_r(\mathbf{x})$ is the energy dissipation, averaged over a sphere of size r.

The function $\zeta(q)$ describes one of the most important characteristics of a turbulent field. In order to estimate this function, Kolmogorov assumed, that for fully developed turbulence (turbulence at high Reynolds number, i.e., when the inertial force highly exceeds the viscous force), the probability distribution laws of velocity increments depend only on the first moment (mean value), $\bar{\varepsilon}$, of the function $\varepsilon_r(\mathbf{x})$. Replacing $\varepsilon_r(\mathbf{x})$ in equation (7) by $\bar{\varepsilon}$ we have

$$S_q(r) \sim (\bar{\varepsilon} \cdot r)^{q/3} = C \cdot r^{q/3}, \tag{8}$$

where C is a constant. As a result, function $\zeta(q)$ is defined as a straight line with a slope of $1/3$:

$$\zeta(q) = q/3. \tag{9}$$

Kolmogorov further realized (see also formulation of Landau's objection concerning the original K41 theory in Frisch 1995) that such an assumption is very rigid and turbulent state is not homogeneous across spatial scales. There is a greater spatial concentration of turbulent activity at smaller scales than at larger scales. This indicates that the energy flow and dissipation do not occur everywhere, and that the energy dissipation field should be highly inhomogeneous (intermittent) and also follows a power law,

$$\langle (\varepsilon_r(\mathbf{x})^p \rangle \sim r^{\tau(p)}, \tag{10}$$

where p is a real number. Then equation (7) may be rewritten as

$$S_q(r) \sim (\varepsilon_r(\mathbf{x}) \cdot r)^{q/3} = (\varepsilon_r(\mathbf{x}))^{q/3} \cdot r^{q/3} = r^{\tau(q/3)} \cdot r^{q/3} \tag{11}$$

or

$$\zeta(q) = \tau(q/3) + q/3. \tag{12}$$

Equation (12) is now referred to as the refined Kolmogorov's theory (hereto forward, refined K41) of fully developed turbulence (Kolmogorov 1962a; Kolmogorov 1962b; Monin and Yaglom 1975; Frisch 1995).

One can see from equation (12) that the function $\zeta(q)$ deviates from the straight $q/3$ line. It has been suggested that the deviation is caused by the scaling properties of a field of energy dissipation.

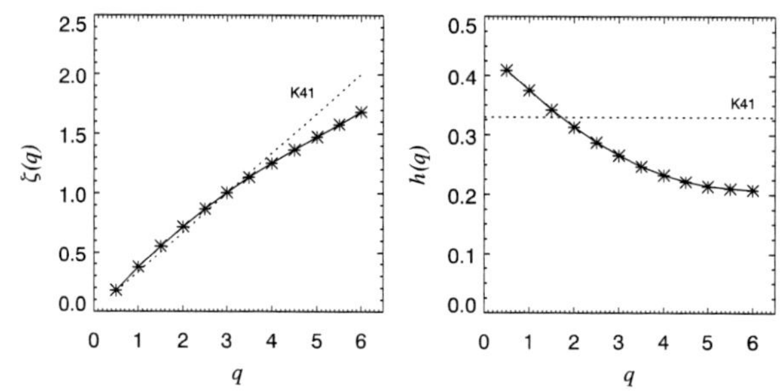

Figure 11. Parameters of fully developed turbulence calculated by using structure functions defined from measurements of wind velocities in the earth's atmosphere. *Left* - exponents $\zeta(q)$ of structure functions are shown as a function of order q (experimental data from Schmitt et al., 1994); *right* - plot of the derivative $h(q)$ of function $\zeta(q)$ calculated by using (15).

Equation (12) for velocity increments has been carefully checked many times over the last four decades and experimental results thus confirmed this suggestion (see Gurvich et al. 1963; Anselmet et al. 1984; Schmitt et al. 1994; Frisch 1995 and references in). In Figure 11 we reproduce a result published in Schmitt et al. (1994). One sees that $\zeta(q)$ is an increasing concave outward function crossing the K41 line at $q = 3$, where $\zeta(3) = 1$, which satisfies both K41 and refined K41 theories as long as $\langle(\varepsilon_r(\mathbf{x})^{3/3}\rangle = \bar{\varepsilon}$.

Important information on a turbulent field can be derived from structure functions $\zeta(q)$, which can be, in principle, obtained from experimental data. For example, the value of the function at $q = 6$ deserves special attention because it defines a power index

$$\beta \equiv 1 - \zeta(6) \tag{13}$$

of a spectrum $E^{(\varepsilon)}(k)$ of energy dissipation $\varepsilon(\mathbf{x})$ that is defined as

$$E^{(\varepsilon)}(k) \sim k^{\beta}, \tag{14}$$

where k is a wave number. Observed values β for a velocity field at fully developed turbulence were found to be in the range of $-0.8 \leq \beta \leq -0.6$ (Gurvich et al. 1963; Anselmet et al. 1984).

Having $\zeta(q)$, as derived from experimental data, and by using equation (12) one can calculate the scaling exponent $\tau(q/3)$ in equation (10) for the energy dissipation field.

The derivative of $\zeta(q)$,

$$h(q) \equiv \frac{d\zeta(q)}{dq}, \tag{15}$$

can also be obtained by using the $\zeta(q)$ function (Figure 11, *right frame*). The deviation of $h(q)$ from a constant value is a direct manifestation of *intermittency* in a turbulence field, which is equivalent to the term *multifractality* in fractal terminology.

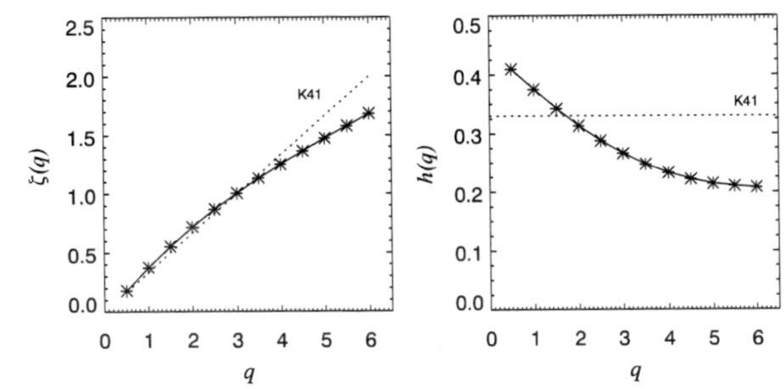

Figure 12. Parameters of fully developed turbulence calculated by using structure functions defined from measurements of wind velocities in the earth's atmosphere. *Left* - exponents $\zeta(q)$ of structure functions are shown as a function of order q (experimental data from Schmitt et al., 1994); *right* - plot of the derivative $h(q)$ of function $\zeta(q)$ calculated by using (15).

According to fractal theory, an intermittent structure consists of subsets, each of which follows its own scaling law. In other words, all values of h, within some range, are permitted and for each value of h there is a fractal set with an h-dependent dimension $D(h)$ near which scaling holds with exponent h. The values of h and $D(h)$ are related by the Legendre transform (Frisch 1995).

Thus, multifractality of a field is manifested via concavity of $\zeta(q)$ function. We accepted the parameter $\Delta h \equiv h_{max} - h_{min}$ to be an estimation of the degree of multifractality of the field (Abramenko 2003). Indeed, larger values of Δh are associated with a broader set of monofractals that together form the resulting multifractal. A typical example of the structure functions for an active region's magnetogram is presented in Figure 12.

The weakest point in the above technique is the determination of the scale range, Δr, where the slope $\zeta(q)$ is to be calculated. To visualize the range of multifractality, Δr, we suggest to use the flatness function (Abramenko 2005b), which is determined as the ratio of the fourth statistical moment to the square of the second statistical moment. Frisch (1995) suggested to use higher statistical moments to calculate the (hyper-)flatness, namely, the ratio of the sixth moment to the cube of the second:

$$F(r) = S_6(r)/(S_2(r))^3. \tag{16}$$

We used Equation (16) to calculate $F(r)$ and, for simplicity, we will refer to $F(r)$ as the flatness function. For monofractal structures, the flatness is not dependent on the scale, r. On the contrary, for a multifractal structure, the flatness grows as a power-law, when the scale r decreases. Usually, for natural fractals, the range, Δr, of the flatness's growth is

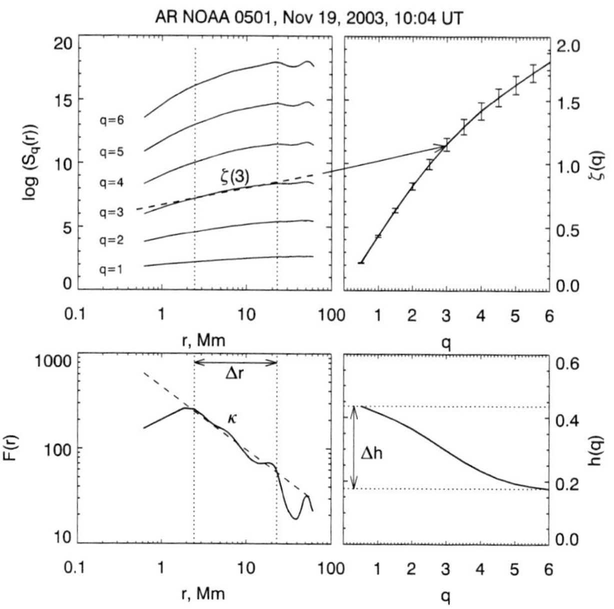

Figure 13. Structure functions $S_q(r)$ (*upper left*) calculated from a magnetogram of active region NOAA 0501 by Equation (6)). *Lower left:* - flatness function $F(r)$ calculated from the structure functions by Equation (16). Vertical dotted lines mark the interval of multifractality, Δr,where flatness grows as power law when r decreases. The interval Δr is also marked in upper left frame. κ is the power index of $F(r)$ determined within Δr. The slope of $S_q(r)$, defined for each q within Δr, is $\zeta(q)$ function (*upper right*), which is a concave for a multifractal and straight line for a monofractal. *Lower right:* - function $h(q)$ is a derivative of $\zeta(q)$. The interval between the maximum and minimum values of $h(q)$ is defined as a degree of multifractality, Δh.

bounded (see lower left frame in Figure 12). Possible causes of the boundness of the interval of multifractality were not comprehensively studied yet. The slope of flatness function, κ, determined within Δr, is related to the degree of multifractality (Figure 12).

a. Structure functions of the magnetic field in active regions

We applied the structure function method to the real magnetograms of different solar active regions (Abramenko et al. 2002; Abramenko 2005b). The result of calculations for two active regions, NOAA 9077 and 0061, are presented in Figures 13 and 14, respectively. We see that scaling behavior of the structure functions is rather different in these active regions. In active region NOAA 9077 there exists a well-defined range of scales, $\Delta r = (4 - 23)$ Mm where flatness $F(r)$ grows with the power index $\kappa = -1.17$ as r decreases. Function $\zeta(q)$ is concave and the corresponding $\Delta h \approx 0.5$. This implies a multifractal structure of the magnetic field in this active region. To the contrary, in NOAA AR 0061 (Figure 14) the flatness function undulates around the horizontal line, which implies a monofractal character of the magnetic field. The function $\zeta(q)$ is nearly a straight line with a vanishing value of $\Delta h \approx 0.05$.

Time profiles of Δh for the two active regions are compared in Figure 15. Two six-hour

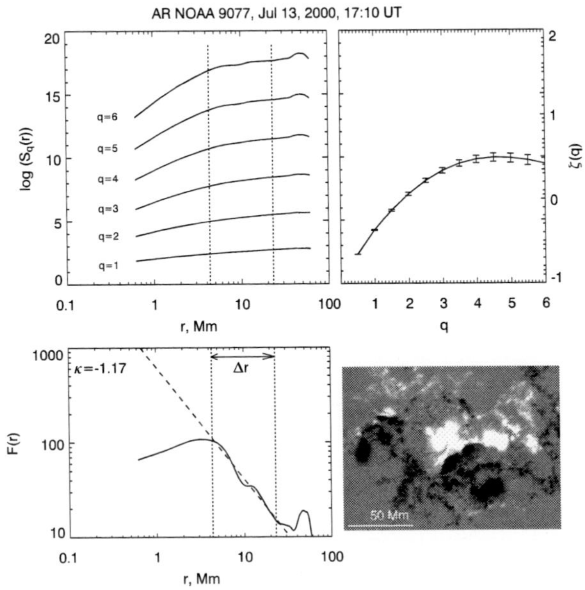

Figure 14. Structure functions $S_q(r)$, flatness function $F(r)$ and $\zeta(q)$ function from one magnetogram of active region NOAA 9077 (*lower right*). κ is a power law index of $F(r)$ calculated inside Δr. The degree of multifractality for this magnetogram $\Delta h = 0.48$. Other notations are the same as in Figure 12.

data sets allowed us to conclude that the non-flaring NOAA AR 0061 persistently displays lower degree of multifractality, as well as lower X-ray flux, than the flaring NOAA AR 9077 does.

It also is worthwhile to note that the third active region NOAA AR 0501 that was analyzed with the structure function method (Abramenko 2005b), showed both intermediate flaring productivity, and intermediate multifractality as compared to AR 0061 and AR 9077 (compare Figures 12, 13, 14). Other examples of the relationship between multifractality and flaring productivity are shown in Figure 16 from Abramenko et al. 2002. One may suggest that higher degree of multifractality of the magnetic field may be associated with stronger flare productivity of an active region. Of cause, a broad statistical study is needed to verify this suggestion.

Day-to-day changes in the flatness function for active region NOAA 0501 are presented in Figure 17. During the 24-hour time interval between magnetograms, the flatness function became steeper ($\kappa = -0.53$ for the earlier moment and -0.73 for the later moment) and the range of multifractality, Δr, extended considerably, especially, at the small-scale side. The degree of multifractality also increased from 0.18 to 0.29. These changes seem to be related to the intensification of the fragmentation process, which, in turn, may lead to a highly intermittent multifractal structure of solar magnetic field (Abramenko & Longcope 2005). Comparison of the magnetograms, presented in the left part of Figure 17, confirms the conclusion on the reinforcement of fragmentation during the discussed time period. Magnetic fields in the 13:57 UT magnetogram are more fragmented, especially inside the

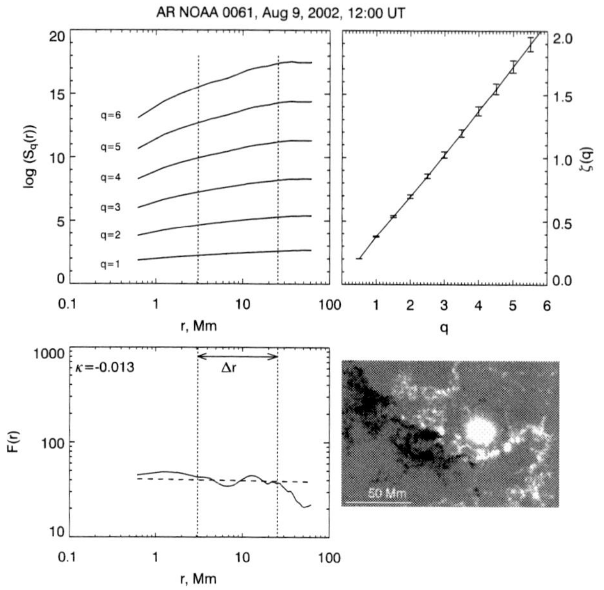

Figure 15. Structure functions $S_q(r)$, flatness function $F(r)$ and $\zeta(q)$ function from one magnetogram of active region NOAA 0061 (*lower right*). Other notations are the same as in Figure 12. Note that for this active region, the function $\zeta(q)$ is a nearly straight line with the vanishing degree of multifractality $\Delta h = 0.058$.

white box, where several large magnetic structures disappeared, while many small-scale features appeared in their places. Reinforcement of the fragmentation seams to be the cause of the extension of the range of multifractality toward smaller scales due to involvement into fragmentation of progressively smaller scales.

b. Structure functions of the of the coronal EUV emission images

Time series of full-disk images of the solar corona obtained by the Extreme Ultraviolet Imaging telescope (EIT, Delaboudiniere et al 1995) on board the SOHO spacecraft in the spectral line Fe XII 195 Å were recently analyzed by Uritsky and co-authors (Uritsky et al. 2007). Two approaches for the spatio-temporal analysis of the coronal brightness above a background EUV threshold were used. Results of the treatment of the coronal brightness structures in the framework of the SOC theory will be discussed below, in Section 2.4 of this chapter. Here we will discuss the results on intermittency of the solar corona explored by the structure functions technique.

EUV emission structures observed near the center of the solar disk (Figure 18, lower panel) were treated as a continuous turbulent field and the corresponding structure functions were calculated for each instant of time with Equation (6). A typical example of such structure functions is shown in Figure 19.

To avoid non-linearity in the structure functions, the extended self similarity (ESS) method was applied under the assumption that $\zeta(3) = 1$ (Kolmogorov scaling). The main idea of the ESS method (Benzi et al. 1993) is to plot $log(S_q(r))$ versus $log(S_3(r))$ and to determine the *relative* scaling exponent, $\alpha_3(q) = \zeta(q)/\zeta(3)$. Thus, inside the linear range

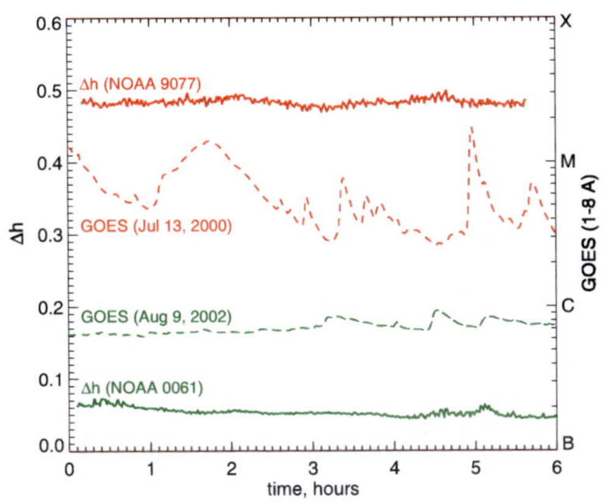

Figure 16. Time variations of the measure of multifractality, Δh (*left axis*), and GOES soft X-ray flux (*right axis, dashed lines*) plotted for six-hour time intervals for the two active regions. Data for NOAA AR 9077 (*red lines*) were obtained between 17:00 and 23:00 UT on July 13, 2000 and data for NOAA AR 0061 (*green lines*) refer to an interval between 11:00 and 17:00 UT on August 9, 2002.

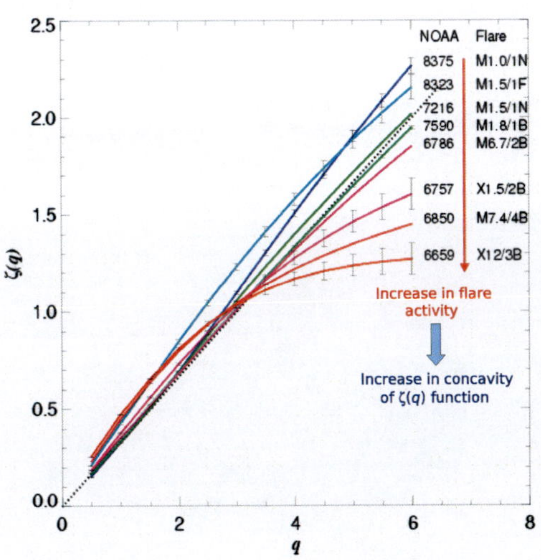

Figure 17. Scaling exponents $\zeta(q)$ of structure functions of order q calculated for eight active regions by Abramenko et al. (2002). The straight dotted line has a slope of $1/3$ and refers to the state of Kolmogorov turbulence. The NOAA number and the strongest flare (X-ray class/optical class) of each active region is shown. Increase of the flaring activity of active regions (from the top down to the bottom) is accompanied by general increase in concavity of $\zeta(q)$ functions.

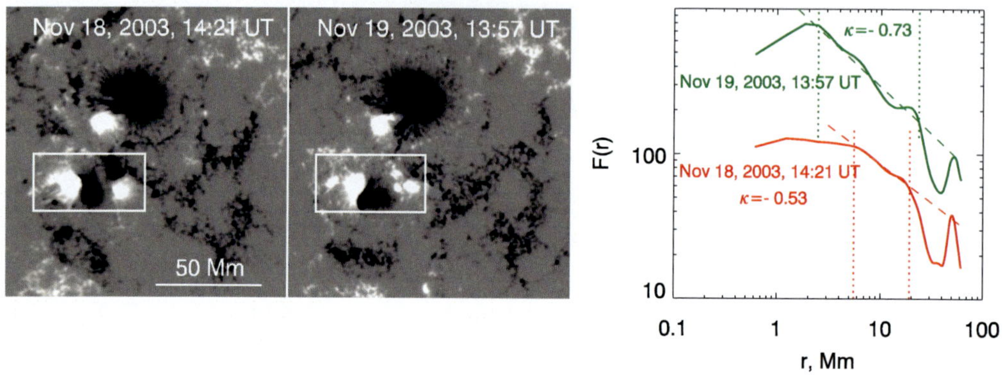

Figure 18. Two magnetograms of NOAA AR 0501 obtained on November 18, 2003 at 14:21 UT and on November 19, 2003 at 13:57 UT. An area where the fragmentation of magnetic elements is the most noticeable, is marked by wight boxes. *Right:* - Flatness functions $F(r)$ for two magnetograms. Green curve, corresponding to the later magnetogram, is shifted up into a "well-readable" position. The ranges of multifractality, Δr, are marked by dotted lines. κ is the power index of $F(r)$ calculated as a linear best fit (*dashed lines*) inside the corresponding intervals Δr.

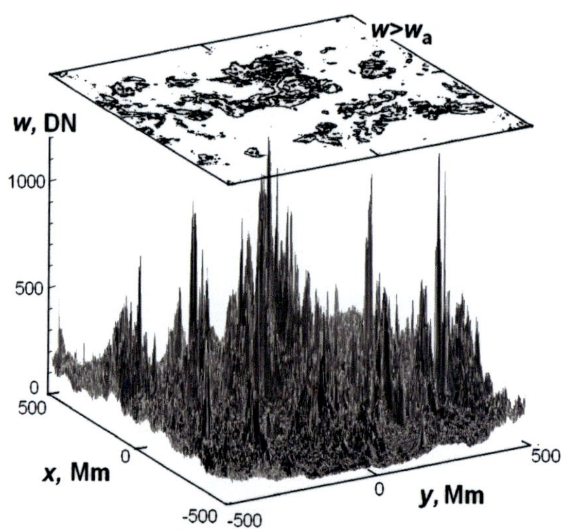

Figure 19. Two views of the coronal EUV structure above an active region on the Sun. *Upper panel:* - a snapshot used to construct avalanches (brightness above a threshold $w_a = 350$). *Lower panel:* - a snapshot of the continuous brightness of the corona at the same time instant. Courtesy of V. Uritsky.

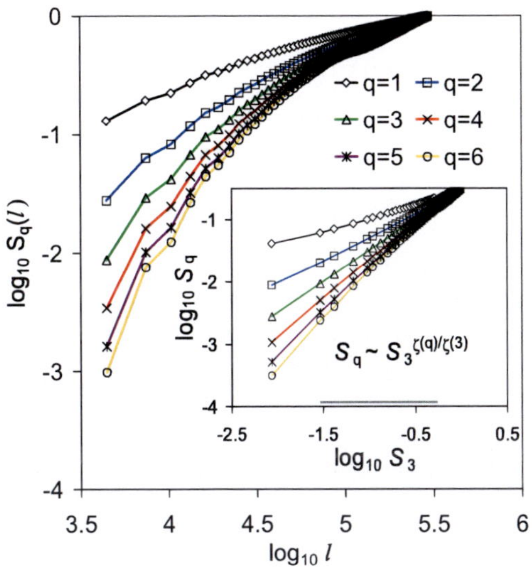

Figure 20. Structure functions of coronal brightness. Each function is normalized by its maximum value. *Inset:* - ESS scaling of the structure functions, The horizontal bar indicated the interval used for calculation of the relative exponents, $\zeta(q)/\zeta(3)$. Courtesy of V. Uritsky.

of $S_q(r)$, we have $\alpha_3(q) = \zeta(q)$. One important advantage appears here: the range of linearity in the $log(S_3(r))$ - $log(S_q(r))$ plot is much more extended than that in the $log(r)$ - $log(S_q(r))$ plot (compare the insert in Figure 19 with the original $S_q(r)$). Therefore, the relative scaling exponent, $\alpha_3(q)$, can be determined with accuracy that is much higher than that for the function $\zeta(q)$. A disadvantage is the unavoidable assumption $\zeta(3) = 1$. All of the above can be repeated for $\zeta(4) = 1$ (Kraichnan scaling).

The relative scaling exponent derived from observational data show the well-pronounced deviation from non-intermittent Kolmogorov-type scaling (Figure 20, dotted straight line), which manifests the highly intermittent nature of the hot solar corona.

c. Structure functions of the photospheric velocity fields

In 1999 Consolini and co-authors (Consolini et al. 1999) reported first attempt to apply the structure functions approach to diagnose the intermittency in the photospheric line-of-sight velocities. High-resolution Dopplergrams obtained in two spectral lines (Fe I 557.61 nm and CI 538.03 nm) with the THEMIS telescope were analyzed by applying the ESS technique. The authors concluded the presence of intermittency in the photosphere. Their results showed that plasma at deeper layer (level of CI 538.03 nm line formation) is more intermittent than that at higher layer (data from the Fe I 557.61 nm line). This finding allowed them to speculate that intermittency is relevant to the photospheric convection.

2.2. Fractal Dimensions and Multifractality Spectrum

For the first time, a fractal dimension appears in the context of solar magnetic fields in the pioneering publication by Tarbell et al. (1990). The authors studied high resolution (0.3 to

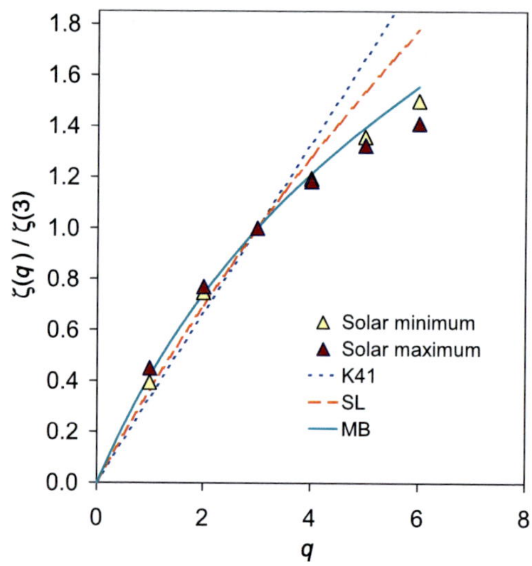

Figure 21. Relative scaling exponents obtained by ESS method. Deviation from the straight K41 line implies intermittency. At solar maximum, the corona is more intermittent than it is at solar minimum. Comparison with the She-Leveque (SL) model and the Müller-Biskamp (MB) model is shown (for details see Uritsky at al. 2007). Courtesy of V. Uritsky.

0.5 arcsec) magnetograms of an active region recorded with the Swedish Solar Observatory on La Palma, Canary Islands. They concluded: "A threshold magnetogram resembles a fractal set, similar to those in Mandelbrot (1982). We can estimate the fractal dimension of this set from a plot of area versus length scale and find self-similar behavior with a dimension of about 1.6 over the entire range considered, from 0.16 to 10 arcsec." This statement was the first experimental proof of fractal nature of magnetic fields on the solar surface, however heuristic ideas about fractality of solar magnetic fields were in air long before, see, for example, Zeldovich et al. (1987).

Later on, more extensive analysis of these excellent data was performed by Schrijver et al. (1992) and Balke et al. (1993). This team analyzed an area on the magnetogram, which included a plage region on the periphery of an active region NOAA AR 5168 so that the fields under study were of moderate intensity. In above studies, the influence of magnetograms noise was studied, as well as a correction of a magnetogram with the modulation transfer function was applied. Authors concluded that "the difference in fractal dimension between the original magnetogram ($D_f = 1.76$) and the enhanced [MTF-corrected- V.A.] magnetogram ($D_f = 1.54$) can be understood as caused by the higher resolution in the enhanced image: with increasing resolution, more details can be observed in clusters, and the fractal dimension will differ more strongly from the Euclidian dimension $D_f = 2$" (Balke et al., 1993). We are inclined to think that such a difference, at least, partially, is caused by the multifractality of the magnetic structure: transition to higher resolution unavoidably involves smaller scales, where D_f is different (see Figure 2). Indeed, the slope in the plot of $log(Area)$ versus $log(Scale)$ [see Figure 3 in Balke et al. (1993) and Figure 1 in Schrijver et al. (1992)] is not constant over the entire range of scales. Thus, even these first studies of fractal dimension in solar magnetograms give a hint that the observed filigree structure is not a monofractal, but rather a multifractal.

Later, Tao et al. (1995) undertook an attempt to numerically simulate spatial distribution of the surface magnetic field. Their simulation was based on the numerical solution of the induction equation with the diffusivity item. This very simple model showed a set of key properties in the behavior of magnetic elements: the crucial role of the transverse velocities in formation of a fractal-like magnetic structure, the loss of memory of the initial conditions during the long-time evolution, as well as the multifractal nature of the resulting flux distribution. During the subsequent decade, numerical simulations of multifractality and intermittency of the solar structures were performed by other groups (e.g., Cadavid et al. 1994; Einaudi et al. 1996; Georgoulis et al. 1998; Dmitruk et al. 1998; Janssen et al. 2003; Nigro et al. 2004).

Fractal analysis was also applied to structures of the solar corona observed in extreme ultra violet (EUV) spectral range by Gallagher et al. (1998). Thresholded images of the corona observed in the range of temperature from 10^4 to 10^6 K above a quiet sun area displayed a fractal dimension in the range of $D_f = (1.3 - 1.7)$, which is in agreement with that reported by Tarbel et al. (1990) and Balke et al. (1993) for the photospheric magnetic structures. We are inclined to think that observed variability of the fractal dimensions D_f is related, at least, partially, to the multifractality of the corona.

Lawrence and co-authors (Lawrence et al. 1993) presented a detailed review [see also Lawrence et al. 1996)] of the multifractal measures and, for the first time, applied the technique of multifractality for magnetograms of quiet and active regions on the Sun. Their results demonstrated multifractality of these both types of magnetic structures. Based on these observational results, Cadavid et al. (1994) suggested a multifractal model of small-scale solar magnetic fields.

Mathematical definitions and calculation formula for the multifractal spectrum can be find, for example, in Feder (1988), Takayasu (1989), Frisch (1995), Schroeder (2000).

Meunier (1999) studied time variations of the fractal dimension of the longitudinal magnetic field on time scales of about 20 months and reported no particular trend with time and no significant correlation with Wolf numbers. Ireland et al. (2003) found the correspondence between the fractal dimension of the photospheric magnetic field and the Hale class of an active region.

Recently, a statistical study of the fractal dimension of the magnetic field in active regions was performed by McAteer et al. (2005). Nearly 10^4 active region images acquired during 7.5 year of MDI observations were analyzed. Authors reviewed the existing methods to calculate the fractal dimension and gathered results of previous calculations of fractal dimension in active regions (see Table 1 in McAteer et al. 2005). They calculated the fractal dimension of the *boundary* of an active region: for a smooth boundary (low complexity), $D_f = 1$, while $D_f \longrightarrow 2$ as the boundary becomes more and more serrated. They found that when $D_f > 1.2$, then M and X class flares are probable within the next 24 hours. Authors concluded that "solar flare productivity exhibits an increase in both the frequency and GOES X-ray magnitude of flares from regions with higher fractal dimension", i. e., with higher complexity of the active region's boundary. Thus, even being treated as a monofractal, the structural organization of the magnetic field exhibits relationship with capability to produce explosive events/flares. This inference seems to be natural, as long as the multifractality in the spatial domain implies highly intermittent (burst-like) behavior in time.

Burst-like behavior in time, in turn, reminds of a percolation process, when a long-lasting period of quietness can abruptly be replaced by an avalanche covering the entire porous array - percolation cluster. This analogy suggests about an ultimate relationship between the multifractality/intermittency and the evolution of a percolation cluster toward an avalanche.

2.3. Percolation Theory Approach

A phenomenon of *percolation*, being a subject of very clear and simple description, brings us to a variety of fractal structures. We will illustrate the basic notion of the percolation theory with an example of a forest fire (Figure 21), i.e. percolation on a two-dimensional lattice.

Let us represent a forest as a square point lattice on witch trees independently occupy nodal points with a probability, p, so that p is equal to the ratio of the total number of trees to the total number of nodal points, $p = n_t/N$. Let the bottom row of trees be ignited (Figure 21 *a*). We will assume that a burning tree will always ignite all of the trees located at adjacent nodal points after one unite of time (i.e., there exists a *bound* between two adjacent nodes when both of them have a tree). If the density of trees, p, is very low, the fire will die out quickly and will not penetrate far from the ignition site.

What will happen when the density of trees increases? Trees, connected with bounds, form clusters of trees. Fire can propagate along bounds until it reaches the boundary of the cluster. On Figure 21 *b*, by igniting the bottom row, we onset five separated energy release events (or five avalanches of energy release) and the largest of them reaches only the 6-th row. When the entire available energy of all burning clusters is exhausted, the fire will die out with no damage for trees located above the 6-th row. In this case, the size of the fire and the amount of released energy are mostly determined by the largest avalanche. In the case illustrated in Figure 21 *b*, $p = 168/(28 \cdot 20) = 0.30$, and the size of the largest avalanche is $\lambda = 5$ pixels. Let us add, say, another 140 trees (Figure 21 *c*). Now, $p = 0.55$ and $\lambda = 17$ pixels, however, the fire still did not reach the upper boundary of the lattice. In this situation, addition of any new tree can be critical and could lead to catastrophic results (assuming that the upper row is, say, a real estate boundary). So that when the trees density increases up to 0.59 (Figure 21 *d*), fire will rich the upper boundary. With addition only the 22 trees to the existing 308 trees, a continuous chain of trees (the global, percolation cluster) is formed, which allows fire to penetrate quickly from the lower to upper boundary of the lattice. Thus, the size of fire can grow up to the size of the entire forest (lattice). This is a *state of percolation*.

First essential property of percolation is the abrupt transition from a state of local coherence to a state of global coherence, in other words, a *threshold character of dissipation*. Numerical modeling on large 2D lattices showed that a cluster, extending over the entire lattice, appears for the first time when the progressively increasing p reaches the value of $p_c = 0.59257 \pm 0.0003$ (Feder 1988). When the size of the lattice, L, goes to infinity, the probability of appearance a global coherent cluster decreases to zero when $p < p_c$ and it goes to 1 when $p > p_c$. The abrupt transition from low to high probability occurs near the percolation threshold, $p = p_c$. If we define the "mass" of a cluster, $M(L)$, as the number

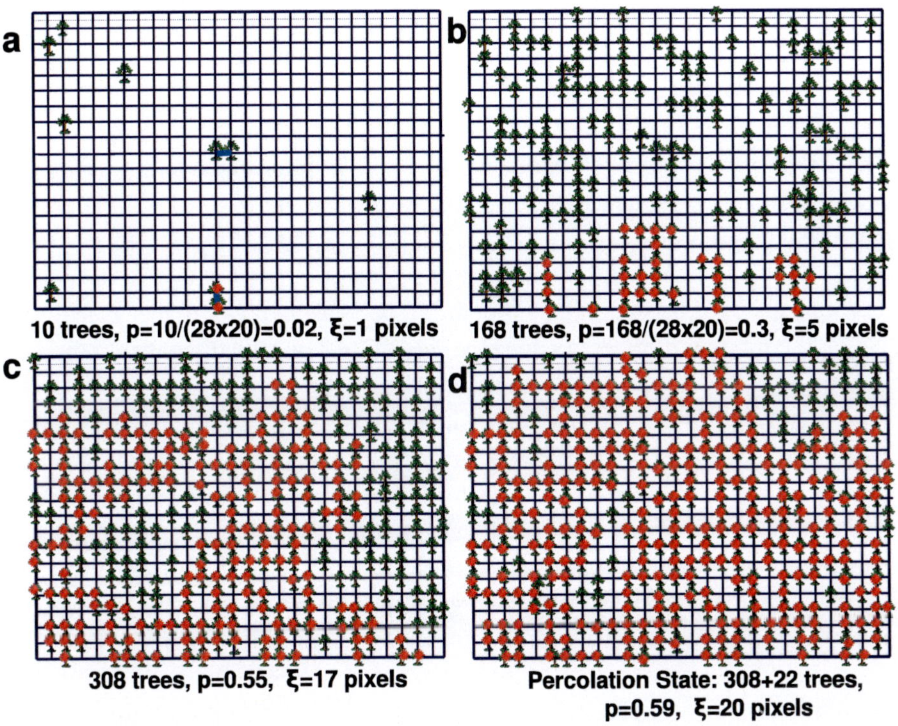

Figure 22. A forest fire illustration of the state of percolation. As the density of trees, p, increases (from *a* through *d*), the the correlation length ζ (a typical size of local fires/avalanches) increases, too. When $p = 0.55$ (frame *c*), addition of only few trees can result in triggering of a global avalanche, when the fire penetrates through the entire forest, so that the state of percolation is reached (frame *d*).

of nodal points belonging to the cluster, then the mass of the largest cluster is related to the lattice size, L, as

$$M(L)_{L\to\infty} \sim \begin{cases} \ln L & \text{if } p < p_c, \\ L^D & \text{if } p = p_c, \\ L^E & \text{if } p > p_c. \end{cases} \tag{17}$$

Therefore, at the percolation threshold, when $p = p_c$, the largest coherent cluster, also called an inner percolation cluster, is a *fractal* with a fractal dimension D_f.

The power-law relation between the scale and the the mass of clusters with a fractional power index at the percolation threshold is the *second essential property of percolation*. At the bound percolation (from a nodal point to another one) on a 2D lattice, $D_f = 91/48 \approx 1.895$ (Feder 1988). The distance, at which the connectedness of nodal points inside a cluster still holds, is determined by a connectedness length, or correlation length, that is usually denoted as ζ, or λ. Phenomenologically, ζ corresponds to the averaged size of clusters on the lattice, or to the correlation length (in turbulence terminology). In our example of the forest fire, it corresponds to the size of the fire. The explicit expression for ζ, which involves the so-called gyration radius of clusters and the probability distribution function for the size of clusters, can be found, for example, in (Feder 1988). An important diagnostic value of the correlation length is that it increases as the system evolves toward the critical state of percolation.

The probability of percolation, $P_N(p)$, defined as $M(L)/L^2$, averaged over a large number of realizations, N, at the percolation threshold is determined as $P_N(p) \sim (p-p_c)^\beta$ when $p > p_c$ and $p \rightarrow p_c$. The index β is equal to $5/36 \approx 0.14$ for the 2D percolation and ≈ 0.4 for the 3D percolation (Feder 1988). In the later 3D case, the percolation corresponds very well to the phenomenon of magnetic phase transition, when the local arrangement of magnetic moments increases with the temperature decreasing down to the critical value, T_c, below which the magnetic moments are arranged similarly over the entire sample and we have got a magnet.

Interpretation of a solar flare phenomenon as a phase transition caused by abrupt percolation of electric currents in a turbulent current sheet was suggested by Pustilnik (1997, 1998). Analyzing plasma instabilities of a turbulent current sheet hosting a flare, author concluded that the equilibrium state of the current sheet is extremely unstable due to various plasma instabilities present: tearing modes, pinch-type instabilities, overheating of turbulent regions in the current sheet, splitting of the current sheet at conductivity discontinuities. This results in an extremely inhomogeneous final state of the flare current sheet. Thus, the current sheet can be treated as a random resistors network consisting of "bad resistors" - turbulent low conductivity domains and "good resistors" - normal plasma domains. Propagation of the electric current through such medium is a percolation process with the fundamental properties: threshold regime of dissipation that explains the explosive nature of flares, and universal power-law distributions of statistical parameters of flares.

Pustilnik has also showed that the percolation in the turbulent current sheet is accompanied by the processes that are, in our opinion, physical realization of the so-called redistribution rules in the SOC theory. These processes are: i) redistribution of currents in the current sheet due to permanent stochastic rebuilding of the resistors network and the current conservation law; ii) redistribution of the heat conductive flux from the heated turbulent domains into surrounding cool plasma which leads to change of the threshold current. This circumstance provides a bridge between the percolation theory and the SOC theory and partially explains why a bulk of interest was shifted towards the SOC theory during the last decade. The latter offers much more broad possibilities for numerical modeling of avalanche processes (that are ultimately the percolation processes) in application to a solar flare phenomenon (see the next section of this chapter).

Successful attempts were undertaken to apply the percolation theory to explain the distribution of magnetic patches over the surface of the Sun. Thus, Wentzel & Seiden (1992, see also Seiden & Wentzel, 1996) modeled appearance (emergence under buoyancy) of sunspots on the solar surface as the percolation through the convective zone. The idea to exploit the percolation approach was stimulated by observations that newly emerging active regions prefer to be located inside the boundary of existing sunspot regions (Zirin 1988, Harvey 1993, Harvey & Zwaan 1993). Under a certain choice of model parameters (such as "the probability that the rise of one flux tube stimulates rise of a neighboring flux tube" and a parameter related to the diffusion of flux tubes), Seiden and Wentzel succeeded to obtain synthetic synoptic maps of "sunspots", which reproduced the intermittent character of the real maps: persistent vast empty regions, corresponding to coronal holes and quiet-sun areas, and compact areas of enhanced magnetism - active regions. Moreover, the size distribution of simulated active regions was in a good agreement with observations reported by Harvey & Zwaan (1993).

Analysis of *real* solar magnetograms in the framework of the percolation theory was presented in the two publications discussed above (see previous section of this chapter), namely, in Schrijver et al. (1992) and Balke et al. (1993). A binary mask (similar to that shown in Figure 3) was calculated with a threshold of 500 G, which produced an ensemble of clusters formed by nearest neighbor pixels with the magnetic flux density exceeding the threshold. Area, A, of each cluster was determined and compared with the size L of the smallest box which can contain the cluster [see Figure 3 in Balke et al. (1993) and Figure 1 in Schrijver et al. (1992)].

The correlation length, $\zeta \approx 3$ arcsec, was calculated by using the gyro-radius technique (Balke et al. 1993). The slope of the $logA$ versus $logL$ plot inside the spatial range $L < \zeta$ allowed to calculate the fractal dimension D_f, that was found to be lower than the fractal dimension $D_f \approx 1.89$ of a cluster at the state of percolation, $p = p_c$. Authors simulated an artificial percolation clusters by randomly filling a 2D lattice of the same size as the magnetogram. The density of filling, $p =$ (number of filled pixels)/(total number of pixels), varied from 0.49 to 0.55. A conclusion was made that "the observed distribution corresponds best with the simulations for which $0.52 \leq p < p_c = 0.59275$." These studies have showed that the magnetic field structure, at least, inside plage areas and below scale of the 3 arcsec, can be treated as a randomly distributed magnetic flux forming a percolation cluster under the percolation threshold. This inference became in future a basis for elaboration of the random walk theory to explore processes of magnetic flux diffusion in the solar photosphere (e.g., Lawrence and Schrijver 1993). "This broad [percolation] theory offers a consistent explanation of geometric properties of both the magnetic flux distribution in instantaneous patterns and of the observed displacements of magnetic flux, without any gouge constants or fudge factors remaining unexplained" (Schrijver et al. 1992).

2.4. Self-Organized Criticality (SOC) Approach

Behavior of highly intermittent/multifractal dynamical systems can be also modeled with the SOC theory. The connection between the multifractality and SOC theory is not as obvious, as it is between the multifractality and the percolation theory. Nevertheless, fractal nature of dissipative non-linear dynamical systems is the basis for the concept of self-organized criticality. Indeed, as we saw in the previous section, an assembly of clusters at the state of percolation possesses a property of fractality, moreover, they form a multifractal system. On the other hand, a process of percolation is an abrupt transition into another state that for many systems implies an explosive event, which, in turn, is an avalanche in the SOC theory.

The SOC theory is mainly focused on analysis of physical conditions and processes that can drive the system to the critical state of catastrophes. In this sense, SOC approach might be more useful for understanding the underlying physics of catastrophes, as compared to the pure geometrical analysis offered by the percolation theory. This also may explain why the SOC approach in solar physics applications is mostly concentrated on statistics of solar *flaring* processes attracting a strong attention of researchers (Lu & Hamilton 1991; Lu at al. 1993; Vlahos et al. 1995; MacKinnon et al. 1996; Georgoulis at al. 2001; Krasnoselskikh et al. 2002; Podladchikova et al. 2002; Podlazov & Osokin 2002; see also a review by Charbonneau et al. 2001 and references in).

It seems to be nearly impossible to avoid the traditional explanation of the SOC ideas by using the sand pile example. Let us consider falling grains of sand on a table surface, so that a pale of sand gradually forms. At the beginning, the height of the pile is increasing and the slopes are steepening. Rare, small-scale avalanches may occur. When the angle of the slope exceeds some critical magnitude ($\theta > \theta_c$), adding one single grain to the pile can provoke a large sand slide (avalanche) that may extend from the top to the very bottom of the pile. The avalanche will redistribute the steepness of the slopes which may result in occurrence of the similar critical situation ($\theta > \theta_c$) at another place, which will cause another avalanche. Continuous falling of grains recurs the system to the previous situation, and so on. The system is now in a state of dynamical equilibrium, when the continuous energy input (slow driving) is balanced by abrupt dissipation events. The state is called a *critical state*. It is important to emphasize that the system reaches such state in a natural way, without any tune factors involved. Because of that, the critical state is called the *self-organized* criticality. Statistical parameters of avalanches obey the power law and the geometry of avalanches displays properties of multifractality (see, for example, Schroeder 2000, chapter 17; McIntosh & Charbonneau 2001).

An excellent review of the SOC theory applications in solar physics was presented by Charbonneau et al. (2001). Here we restrict our self to a brief outline of the problem. Generally, a routine consists of two steps: i) the SOC modeling of a time set of avalanches/flares, and ii) comparison of the statistics of simulated events to that obtained from real observations.

Usually, the modeling starts with specifying of a (uniform) magnetic field on a 2D lattice. After the injection of a randomly placed perturbation, a certain parameter (often, the field gradient or electric current, depending on a model) of the field is recalculated over the lattice. If, at some nodal point, the parameter exceeds a specified threshold, then the point is classified as a *critical* and is attributed to an avalanche. The excess of the parameter at the critical point is then redistributed over adjacent nodal points (the redistribution rule is frequently a delicate issue and it differs from one model to another). After the redistribution, the excess of the threshold may occur at any of the adjacent nodes so that this node is also classified as critical and belonging to the same avalanche. This second nodal point also undergoes the redistribution, and so on, until the avalanche is exhausted. After that, next random disturbance is injected at a random location and the entire routine is repeated. At the very beginning of the model evolution, the total magnetic energy of the system, $\sum B^2/8\pi$, increases. However, after some time period, a dynamical equilibrium is reached and the critical state is settled. Now, the total energy, on average, is constant being superposed by fluctuations due to episodic avalanches. The energy released in avalanches is a small fraction of the total energy of the system. Strictly speaking, an exact prediction of the location, time and size of the oncoming avalanche is impossible.

The above description is a simplified presentation of the general idea for such kind of modeling. Real routines can be much more complicated (some delicate methodological aspects of the SOC concept and related techniques are discussed in Buchlin et al. 2005). A common point of all proposed SOC methods is a presentation of differential equations on a *discrete* lattice. Recently, a new approach to use continuous presentation of equations was suggested by Belanger et al. (2007). Authors argue that this new method may allow some possibilities for short-time forecasting of avalanches, which is extremely important in solar flares application.

Usually, the SOC modeling produces statistical distributions of simulated avalanches that are similar to that obtained from solar observations (see references in Charbonneau et al. 2001). The reason for such a good correspondence might be in the satisfactory accomplishment of the requirements of the SOC theory for the solar atmosphere. According to Parker's conjecture (Parker 1983a, 1983b, 1988), slow driving of the system is performed by random motions of footpoints of magnetic flux tubes caused by convection in the photosphere and below. Various disturbances in the magnetic field propagate up into the corona, which creates braided and intertwined magnetic flux tubes (or, in other words, numerous tangential discontinuities), thus driving the coronal magnetic field to the SOC state. When the density of discontinuities exceeds some critical level, an avalanche of fast small-scale reconnection events (a flare) occurs. Thus, the requirement of slow driving and sudden energy release is met. Furthermore, the amount of energy released in a typical flare is a small fraction of the total magnetic energy stored in a typical active region's magnetic configuration. The medium hosting the flare is an extremely complex, multiscale system of coronal magnetic fields that possesses fractal properties, for which a plausible analog for the redistribution rule can be suggested (for example, redistribution of currents in a resistor network by Pustilnik 1997, see previous section of this chapter). Recently, new ideas were suggested to explain a chain reaction of avalanche-like reconnections under solar conditions, for example, a mechanism of the secondary instability proposed by Dahlburg et al. (2005) and a chain reconnections in flaring process proposed by Kusano et al. (2004).

The ultimate connection between the SOC paradigm and the concept of intermittency for the hot solar corona was demonstrated in a paper by Uritsky et al. (2007) that was discussed above (Section 2.1.2b of this chapter). The shape of the structure functions' scaling exponents (Figure 20) undoubtedly demonstrates the intermittent character of the coronal brightness structures (the 3D structure in Figure 18). For the same continuous brightness structures, authors also calculated thresholded snapshots (top panel in Figure 18) that outline the most heated places in the corona the spectral line can detect. Having the time series, it is possible to determine the lifetime of each cluster of energy release. These clusters were treated as avalanches of energy release. The most important advantage here is that the avalanches were not modeled but rather derived from solar observations. The statistics of avalanches were calculated separately for periods of minimum and maximum of the 23-th solar cycle. The probability distribution functions for the total emission flux, E, the peak area, A, and the lifetime of avalanches, T, are reproduced in Figures 28 - 24, respectively. The distributions were successively fitted with the power law functions which is a necessary condition for the SOC concept. Although the authors' inferences are based on coronal ultraviolet emission data, we are inclined to think that the conclusions are applicable for a broad class of solar phenomena as well. This is because the SOC theory captures an essential property of the solar plasma: multifractality in spatial domain and intermittency in time.

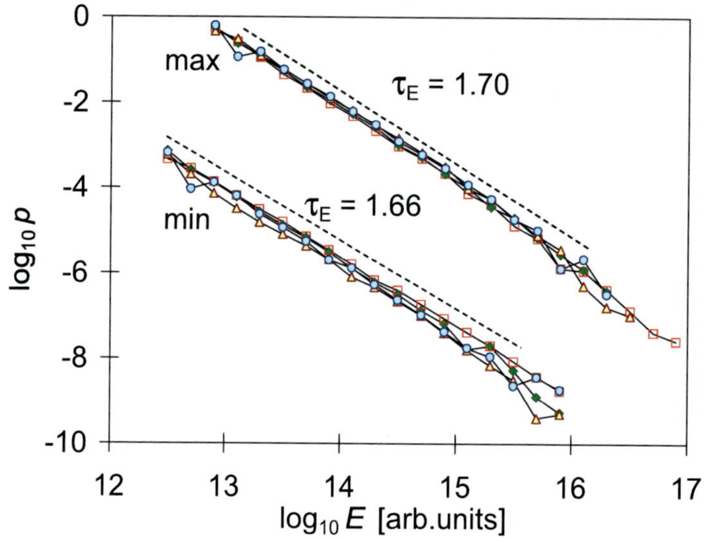

Figure 23. Probability distribution functions of the total emission flux, E, of avalanches shown in the top panel of Figure 17. The data obtained for four different thresholds $w_a = k < w >$ with $k = 0.4$ (squares), $k = 0.8$ (diamonds), $k = 1.2$ (triangles), and $k = 1.6$ (circles). The data points related to the solar minimum (min) are shifted downward for better reading. The calculated indices of the power law approximation, τ_E, are noted. Courtesy of V. Uritsky.

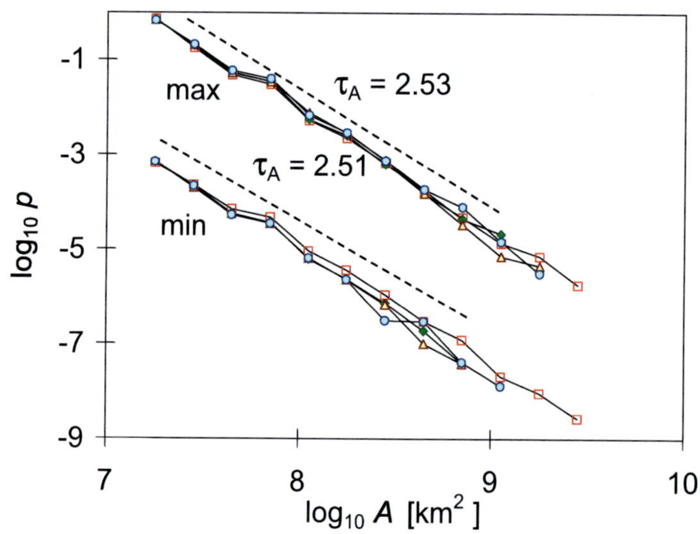

Figure 24. Probability distribution functions of the peak area, A, of avalanches shown in the top panel of Figure 17. Notations are the same as in Figure 21. Courtesy of V. Uritsky.

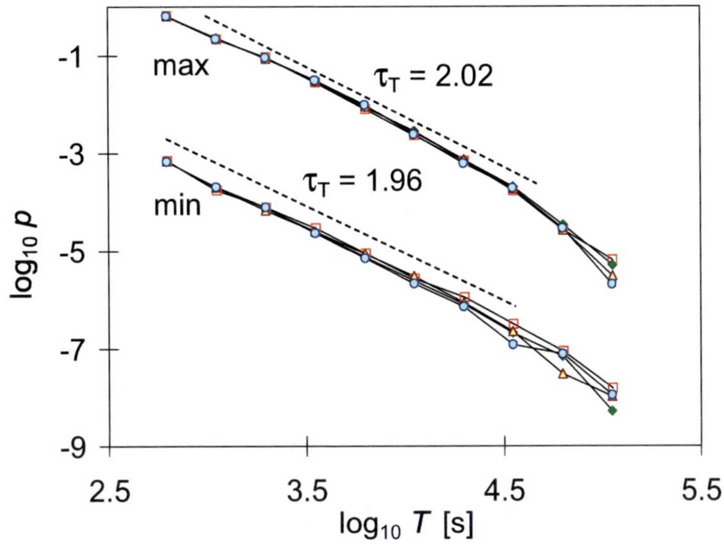

Figure 25. Probability distribution functions of the lifetime, T, of avalanches shown in the top panel of Figure 17. Notations are the same as in Figure 21. Courtesy by Vadim Uritsky.

3. Intermittency in the Heliosphere

This is a very broad topic, that involves a voluminous body of literature and a long history. The whole notion of *fractality* came into solar physics from heliospheric studies, see, for example, pioneering papers by Burlaga & Klein (1986) and Burlaga (1991). Also a great variety of mathematical methods to analyze multifractal/intermittent systems were adopted by solar physicists from numerous studies of the solar wind. We can recommend a recent review by Bruno et al. (2005) and references in.

Even though the solar wind is now observed mostly near Earth, it is an essential part of the Sun's evolution. Solar wind streams are high-speed flows of charged particles propagating in the magnetized medium under the accelerating forces, which largely exceed the gravitational force of the Sun. A possible mechanism of the acceleration is still under debates (see, e.g., Fisk & Gloeckler 2006), however its ultimate energy source is believed to be on account of the Sun.

The solar wind mainly consists of two components: slow solar wind (speed $\sim$ 400 km/s), which originates from areas of closed magnetic fields on the Sun, and fast solar wind (speed $\sim$ $(600 - 800)$ km/s), which originates from areas of open magnetic fields on the Sun, co-called coronal holes. Predominant location of coronal holes varies with the solar cycle: during a solar minimum, coronal holes (of opposite polarity) are concentrated at the poles, whereas during the solar maximum, numerous coronal holes appear at low latitudes, near the solar equator and therefore, Earth becomes a direct target for fast solar wind streams emanated from low-latitude coronal holes.

Because the state of fully developed turbulence is reigning in the heliosphere, various methods of the turbulence theory are frequently used to study intermittency (e.g., Burlaga & Ness 1998; Bruno et al. 2005; Zimbardo et al. 2006; Kaghashvili et al. 2006). Along with this, solar wind is extensively studied with various traditional fractal techniques, such as i) analysis of fractal geometry (Polygiannakis & Moussas 1994; Milovanov & Zelenyi 1999); ii) structure functions (Carbone et al. 2004; Bigazzi et al. 2006); iii) the shell model (Boffetta et al. 1999; Lepreti et al. 2004); iv) the phase coherence method to study non-linear interactions (Koga et al. 2006); v) distribution function technique (Bruno et al. 2001; Lepreti et al. 2004; Leubner & Vörös 2005; Vasquez et al. 2007, the later was discussed in Section 2.1.1d of this chapter). This list is far from being complete and the only aim it pursues is to show a large effort devoted to this kind of research during last decade.

4. Conclusion

In this chapter, we attempted to demonstrate that a broad class of processes and systems on the Sun are neither completely deterministic, nor totally chaotic. They rather are in an intermediate state, which unavoidably possesses a property of multifractality in spatial domain and intermittency in time. We schematically summarize our inferences concerning the place and the role of intermediate processes on diagrams in Figures 25 and 26.

Process:	deterministic	intermediate: multifractal/intermittent	chaos
Energy:	input only	input at large scales ⟶ output at small scales	output only
Dissipation:	no dissipation	dissipation allowed: inhomogeneous in space and highly intermittent (burst-like) in time	homogeneous dissipation
Structurization:	one or several explicitly defined structures	hierarchy of random structures: extended zones of low activity are intermittent with narrow patches of extremely high deviations; possibility to form structures from chaos; large fluctuations are not rare; heavy-tailed distribution functions	homogeneous array of random, predominantly small peaks; extremely rare large peaks. Gaussian distribution function.
Tools:	a set of deterministic equations; a solution is uniquely determined by a boundary/initial conditions	A set of equations includes a dissipation term ↓ A solution (if any) is not determined by a boundary/initial conditions ↙ ↘ To simplify the problem / To find another ways	the central limit theorem; Gaussian statistics; the least mean square technique

Figure 26. The chart to illustrate the intermediate location of multifractal systems

How this knowledge can help us in our efforts to understand the nature in general and the Sun, in particular?

Fist of all, let us consider a methodological aspect. By applying a purely deterministic or purely chaotic approach for any new phenomenon, we may soon realize that the

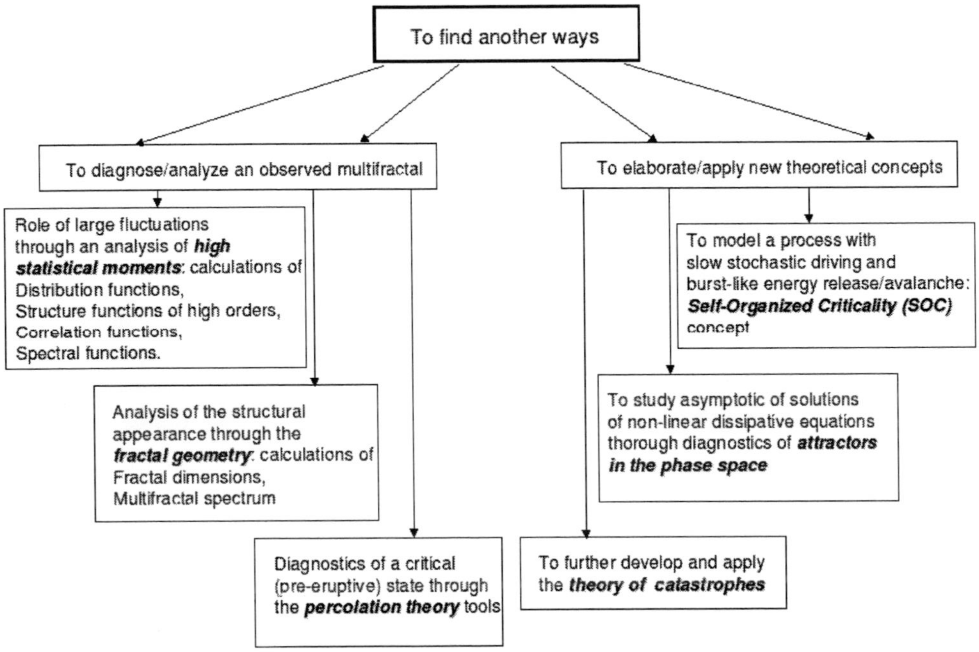

Figure 27. Continuation of the chart presented in Figure 25.

obtained result gives us only one-sided view while some important aspects might escape from our consideration. Besides, processes of integration and differentiation over a fractal measure drastically differs from those undertaken over a Lesbegue (continuous) measure. The same can be said about the averaging over a fractal measure: our traditional tools (such as the central limit theorem, Gaussian statistics, the least mean square technique) might be inapplicable.

As to the usefulness of the multifractality notion for various physical aspects of understanding of solar phenomena, it is worth to mention the long-standing problem of appearance of low plasma conductivity in the corona, especially during a flare. To explain a solar eruption on the scale of an active region that can last of about 100 minutes, it is necessary to imply the presence of super strong electric currents ($\sim 10^{10} A/km^2$) inside a very thin layers ($< 100\ m$, Priest 1982). A fractal concept of coronal magnetic fields can easy meet these requirements. Indeed, a self-similar fractal allows existence of super thin branches (magnetic sheets or tubes) whereas a percolation state implies formation of super strong currents at singular branches of the cluster.

Another intriguing aspect of solar flaring is prediction of flares. A voluminous body of literature (to mention a few, Falconer et al. 2003; Wheatland 2004; Abramenko 2005a; Song et al. 2006; Qahwaji and Colak, 2007; for more references, see Leka and Barnes 2007) and meetings is devoted to the problem, however no dramatic success is achieved as yet. We would like to mention here a conclusion that Leka and Barnes (2007) made from a very thorough and extensive set of four publications devoted to the flare forecast problem: "Our result suggest that the state of photospheric magnetic field at any single time has limited bearing on the occurrence of solar flares". When keeping in mind a fractal nature of the phenomenon, is becomes natural that it is impossible to exactly predict a flare. We can only say whether or not an emerging active region reached the critical state and is able to produce a flare, or it is still in a state of accumulating the energy, when the probability of strong flares is very low. Realization that only a probabilistic statement of the question is physically reasonable, could help much in a treatment of the flare forecasting issue.

Very often studies produce different fittings to the observed distribution functions when analyzing the same solar phenomenon. Which approximation is true? Exponential, power law, or log-normal? What scaling exponents are closer to reality? The multifractal concept tells us that there is no unique answer for this question. The answer depends on the range of scales selected for the fitting: the slope of the distribution may vary for different ranges of scales. For example, for a narrow range, observational data can be successfully approximated by a linear fit in a double-logarithmic plot (a power law), however on a broader range, the observed distribution may be curved. A log-normal fit is capable to capture the curvature of observed PDFs on a broad range of scales, being therefore more successful than the power law fit. Meanwhile, one should keep in mind that all above fittings are only *models* aimed to represent a multifractal system with a *finite number* of scalar parameters. Which is impossible in principle. An adequate description requires, at least, a *continuous spectrum* of parameters.

A widely discussed problem of the heating of the solar corona to multi-million degree temperatures (see, e.g., a recent review by Klimchuk, 2006) strongly relies on the highly inhomogeneous character of coronal magnetic fields and on ultimate interactions of different spatial scales, which both are attributes of multifractality. For example, the so-called direct current (DC) mechanism explains coronal heating as a result of numerous small-scale magnetic reconnections (nanoflares) that occur in places of anomalous resistivity of coronal plasma at numerous magnetic field discontinuities, which, in turn, are generated by random photospheric motions of magnetic flux tubes. This scenario perfectly satisfies both percolation and SOC concepts. Moreover, during last years a progress has been made in explaining why individual large-scale coronal loops do not expend (in their crossection) with height as they should do in force-free field approximation. Klimchuk et al. (2006, 2007) argued that a coronal loop, when treated as "bundles of unresolved, impulsively heated strands" can posses some very specific properties that will differ a real loop from a modeled. This "suggests that the coronal magnetic field is comprised of elemental flux strands that are tangled by turbulent convection. These strands are so small that many tens of them are contained within a single TRACE loop. We suggest that this fine structure is a critical missing ingredient of the extrapolation models and that a combination of footpoint shuffling and coronal reconnection can explain the observed loop symmetry. This has important implications for coronal heating."

Thus, ideas of multifractality, intermittency, the percolation theory, and the SOC theory, all of them share deep common roots, which is caused by natural evolution of non-linear dynamical dissipative systems. Our Sun is one of such systems, and solar physics can gain much by utilizing these concepts, along with a rich set of developed tools, in further understanding of our closest star.

Author would like to thank her co-authors in this field of research: Vincenzo Carbone, Phil Goode, Dana Longcope, Thomas Spirock, Luca Sorriso-Valvo, Haimin Wang, Vasyl Yurchyshyn. I am also grateful to Paul Charbonneau, Lennard Fisk, Marco Velli, and Vadim Uritsky for helpful discussions. The work was supported by NASA NNG05GN34G grant.

References

[1] Abramenko, V.I. *Astronomy Reports* **47**, 151 (2003).

[2] Abramenko, V.I., Yurchyshyn, V.B., Wang, H., Spirock, T.J., Goode, P. R. *Astrophys. J.* **577**, 487 (2002).

[3] Abramenko, V. I.; Yurchyshyn, V. B.; Wang, H.; Spirock, T. J.; Goode, P. R. *Astrophys. J.* **597**, 1135 (2003).

[4] Abramenko, V.I. *Astrophys. J.* **629**, 1141 (2005a).

[5] Abramenko, V.I. *Solar Phys.* **228**, 29 (2005b).

[6] Abramenko, V.I., Longcope, D.W. *Astrophys. J.* **619**, 1160 (2005).

[7] Aitchison,J., Brown, J.A.C. *The Lognormal Distribution* London, Cambridge University Press. (1976).

[8] Anselmet, F., Gagne, Y., Hopfinger, E.J., Antonia, R.A. *J. Fluid Mech.* **140**, 63 (1984).

[9] Aschwanden, M.J., Dennis, B.R., Benz,A.O., *Astrophys. J.* **497**, 972 (1998).

[10] Balke, A.C., Schrijver, C.J., Zwaan, C., Tarbell, T.D. *Solar Phys.* **143**, 215 (1993)

[11] Balsara, D. S., Kim, J. *Astrophys. J.* **634**, 390 (2005).

[12] Belanger, E., Vincent, A., Charbonneau, P. *Solar Phys.* in press (2007).

[13] Benzi, R., Ciliberto, S., Baudet, C., Massaioli, F., Tripiccione, R., Succi, S. *Phys. Rev. E* **48**, R29 (1993).

[14] Berge, P., Pomeau, Y., Vidal, C. *L'Ordre dans le Chaos*, (Hermann, Paris) (1988).

[15] Bigazzi, A., Biferale, L., Gama, S. M. A., Velli, M. *Astrophys. J.* **638**, 499 (2006).

[16] Biskamp, D. *Nonlinear magnetohydrodynamics*, Cambridge University Press (1993).

[17] Boffetta, G., Carbone, V., Guiliani, P., Veltri, P., Vulpiani, A. *Physical Review Letters* **83**, 4662 (1999).

[18] Bogdan, T.J., Gilman, P.A., Lerche, I., Howard, R., *Astrophys. J.* **327**, 1451 (1988).

[19] Bruno, R., Carbone, V., Bavassano, B., Sorriso-Valvo, L. *Advances in Space Research* **35**, 939 (2005).

[20] Bruno, R., Carbone, V., Veltri, P., Pietropaolo, E., Bavassano, B. *Planetary and Space Science* **49**, 1201 (2001).

[21] Buchlin, E., Galtier, S., Velli, M. *Astron. Astrophys.* **436**, 355 (2005).

[22] Burlaga, L. F. *Journal of Geophysical Research* **18**, 69 (1991).

[23] Burlaga, L. F., Klein, L. W. *Journal of Geophysical Research* **91**, 347 (1986).

[24] Burlaga, L. F., Ness, N.F. *Space Science Reviews* **83**, 105 (1998).

[25] Burlaga, L. F., Lazarus, A. J. *Journal of Geophysical Research* **105**, 2357 (2000).

[26] Cadavid, A.C., Lawrence, J.K., Ruzmaikin, A.A., Kayleng-Knight, A. *Astrophys. J.* **429**, 391 (1994).

[27] Carbone, V., Bruno, R., Sorriso-Valvo, L., Lepreti, F. *Planetary and Space Science* **52**, 953 (2004).

[28] Charbonneau, P., McIntodh, S.W., Liu, H.L., Bogdan, T.J., *Solar Physics* **203**, 321 (2001).

[29] Cho, J., Lazarian, A., Vishniac, E.T. *Astrophys. J.* **595**, 812 (2003).

[30] Consolini, G., Berrilli, E., Pietropaolo, E., Bruno, R., Carbone, V., Bavassano, B., Ceppatelli, G. *Magnetic Fields and Solar Processes* (ESA SP-448; Paris: ESA), 209 (1999).

[31] Crosby, N.B., Aschwanden, M.J., and Dennis, B.R., *Solar Physics* **143**, 275 (1993).

[32] Dahkburg, R.B., Klimchuk, J.A., Antiochos, S.K. *Astrophys. J.* **622**, 1191 (2005).

[33] Daltowe, D.W., Hudson, H.S., and Peterson,L.E., *Solar Physics* **35**, 193 (1974).

[34] Delaboudiniere, J.-P.; Artzner, G. E.; Brunaud, J.; Gabriel, A. H.; Hochedez, J. F.; Millier, F.; Song, X. Y.; Au, B.; Dere, K. P.; Howard, R. A.; Kreplin, R.; Michels, D. J.; Moses, J. D.; Defise, J. M.; Jamar, C.; Rochus, P.; Chauvineau, J. P.; Marioge, J. P.; Catura, R. C.; Lemen, J. R.; Shing, L.; Stern, R. A.; Gurman, J. B.; Neupert, W. M.; Maucherat, A.; Clette, F.; Cugnon, P.; van Dessel, E. L. *Solar Physics* **162**, 291 (1995).

[35] Dmitruk, P., Comez, D.O., DeLuca, E.E. *Astrophys. J.* **505**, 974 (1998).

[36] Drake, J.F., *Solar Physics* **16**, 152 (1971).

[37] Einaudi, G., Velli, M., Politano, H., Pouquet, A. *Astrophys. J.* **457**, L113 (1996).

[38] Falconer, D. A., Moore, R. L., Gary, G. A. *Journal of Geophysical Research* **108**, 1380 (2003).

[39] Falgarone, E.; Pineau des Forets, G.; Roueff, E. *Astron. Astrophys.* **300**, 870 (1995).

[40] Feder, J *Fractals*, New York Plenum Press (1988).

[41] Fisk, L. A., Gloeckler, G. *Astrophys. J.* **640**, L79 (2006).

[42] Frisch, U. *Turbulence, The Legacy of A.N. Kolmogorov*, Cambridge University Press, Cambridge (1995).

[43] Frost, K.J., *Astrophys. J.* **158**, L159 (1969).

[44] Gallagher, P. T.; Phillips, K. J. H.; Harra-Murnion, L. K.; Keenan, F. P. *Astron. Astrophys.* **335**, 733 (1998).

[45] Georgoulis, M., Velli, M., Einaudi, G. *Astrophys. J.* **497**, 957 (1998).

[46] Georgoulis, M., Vilmer, N., Crosby, N.B. *Astron. Astrophys.* **367**, 326 (2001).

[47] Gopalswamy, N., Mikic, Z., Maia, D., Alexander, D., Cremades, H., Kaufmann, P., Tripathi, D., Wang, Y. M. *Space Science Reviews* **123**, 303 (2006).

[48] Gurvich, A.S., Zubkovsky, S.L. *Bull. Acad. of Sci. USSR, Ser. Geophys.* **2**, 1856 (1963).

[49] Harvey, K.L. *Ph.D. Thesis*, Univ. Utrecht, Netherlands, (1993).

[50] Harvey, K.L., Zwaan, C *Solar Phys.* **148**, 85 (1993).

[51] Henighan, T. *Strange Attractors* Tesseract Books, (1997).

[52] Ireland, J., Gallagher, P.T., McAteer, R.T.J. in: Star as Suns: Activity, Evolution and Planets, IAU Symp. 219, 21-25 July, 2003 in Sydney, Australia, 255 (2003)

[53] Janssen, K., Vögler, A., Kneer,F. *Astronomy and Astrophysics* **409**, 1127 (2003).

[54] Kaghashvili, E. Kh.; Zank, G. P.; Webb, G. M. *Astrophys. J.* **636**, 1145 (2006).

[55] Kaufmann, P., Gimez de Castro, C. Guillermo, Makhmutov, V. S., Raulin, J.-P.,Schwenn, R., Levato, H., Rovira, M. *J. Geophys. Res.* **108**, 1280 (2003).

[56] Klimchuk, J.A. *Solar Phys.* **234**, 41 (2006).

[57] Klimchuk, J. A., Lopez Fuentes, M. C., Demoulin, P. *American Astronomical Society, SPD meeting N 37*, Abstract 17.06 (2006).

[58] Klimchuk, J. A., Lopez Fuentes, M. C., Demoulin, P. *American Astronomical Society Meeting 210*, Abstract 91.11 (2007).

[59] Koga, D.; Chian, A. C.-L.; Miranda, R. A.; Rempel, E. L. *Physical Review E* **75**, 046401 (2007).

[60] Kolmogorov, A.N. *Dokl. Acad. Nauk SSSR* **30**, 299 (1941).

[61] Kolmogorov, A.N. *Mecanique de la turbulence* Coll. Intern. du CNRS a Marseille. Paris, Ed. CNRS, 447 (1962a).

[62] Kolmogorov, A.N. *J. Fluid Mech.* **13**, 82 (1962b).

[63] Kraichnan,R.H. *Phys. Fluids* **8**, 1385 (1965).

[64] Krasnoselskikh, V., Podladchikova, O., Lefebvre, B., Vilmer, N. *Astron. Astrophys.* **382**, 699 (2002).

[65] Kritsuk, A. G., Norman, M. L. *Astrophys. J.* **601**, L55 (2004).

[66] Kuklin, G.V. *Bull. Astron. Inst. Czechosl.* **31**, 224 (1980).

[67] Kusano, K.; Maeshiro, T.; Yokoyama, T.; Sakurai, T. *Astrophys. J.* **610**, 537 (2004).

[68] Lawrence, J.K., Ruzmaikin, A.A., Cadavid, A.C. *Astrophys. J.* **417**, 805 (1993).

[69] Lawrence, J.K., Schrijver,C.J *Astrophys. J.* **411**, 402L (1993).

[70] Lawrence, J.K., Cadavid, A.C., Ruzmaikin, A.A. *Phys. Rev. E* **51**, 316 (1995).

[71] Lawrence, J.K., Cadavid, A.C., Ruzmaikin, A.A. *Astrophys. J.* **465**, 425 (1996).

[72] Leka, K.D., Barnes, G. *Astrophys. J.* **656**, 1173 (2007).

[73] Lepreti, F., Carbone, V., Guiliani,P., Sorriso-Valvo, L., Veltri, P. *Planetary and Space Science* **52**, 957 (2004).

[74] Leubner, M. P., Vörös, Z. *Astrophys. J.* **618**, 547 (2005).

[75] Liu, Y., Zhang, H. *Astron. Astrophys.* **372**, 1019 (2001).

[76] Lu, E. T., Hamilton, R. J. *Astrophys. J.* **380**, L89 (1991).

[77] Lu, E. T., Hamilton, R. J., McTiernan, J.M., Bromund, K.R. *Astrophys. J.* **412**, 841 (1993).

[78] MacKinnon,A.L., Macpherson,K.P., Vlahos, L. *Astron. Astrophys.* **310**, L9 (1996).

[79] Maia, D., Aulanier, G., Wang, S.J., Pick, M., Malherbe, J.-M., Delaboudinière, J.-P. *Astron. Astrophys.* **405**, 313 (2003).

[80] Mandelbrot, B. *The Fractal Geometry of Nature* W.H.Freeman, San Fransisco, (1982).

[81] McAteer, R.T.J., Gallagher, P.T., Ireland, J. *Astrophys. J.* **631**, 628 (2005).

[82] McIntosh,S.W., Charbonneau, P. *Astrophys. J.* **563**, L165 (2001).

[83] Meunier, N. *Astrophys. J.* **515**, 801 (1999).

[84] Milovanov, A.V., Zelenyi, L.M. *Astrophysics and Space Science* **264**, 317 (1999).

[85] Monin, A.S., Yaglom, A.M. *Statistical Fluid Mechanics*, vol. 2, ed. J.Lumley, MIT Press, Cambridge, MA (1975)

[86] Nigro, G., Malara, F., Carbone, V., Veltry, P. *Physical Review Letters* **92**, 194501 (2004).

[87] Parker, E.N. *Astrophys. J.* **264**, 635 (1983a).

[88] Parker, E.N. *Astrophys. J.* **264**, 642 (1983b).

[89] Parker, E.N. *Astrophys. J.* **330**, 474 (1988).

[90] Petrovay, K., Moreno-Insertis, F. *Astrophys. J.* **485**, 398 (1997).

[91] Petrovay, K., Martez Pillet, V., van Driel-Gesztelyi, L. *Solar Shysics* **188**, 315 (1999).

[92] Podladchikova, O., Dudok de Wit, T., Krasnoselskikh, V., Lefebvre, B. *Astron. Astrophys.* **382**, 713 (2002).

[93] Podlazov, A.V., Osokin, A.R. *Astrophysics and Space Science* **282**, 221 (2002).

[94] Polygiannakis, J.M., Moussas, X. *Astron. Astrophys.* **283**, 990 (1994).

[95] Press, W.H., Teukolsky, S.A., Vetterling, S.A., Flannery, B.P. *Numerical Recipes*, Cambridge Univ. Press (1986).

[96] Priest,E.R. *Solar Magnetohydrodynamics* (Dordrecht: Reidel) (1982).

[97] Pustilnik, L.A. *Astrophysics and Space Science* **264**, 171 (1998).

[98] Pustilnik, L.A. *Astrophysics and Space Science* **252**, 325 (1997).

[99] Qahwaji, R., Colak, T. *Solar Shysics* **241**, 195 (2007).

[100] Romeo, M., Da Costa, V., Bardou, F. *Eur. Phys. J. B* **32**, 513 (2003).

[101] Scherrer, P.H., Bogart, R.S., Bush, R.I., Hoeksema, J.T., Kosovichev, A.G., Schou, J., Rosenberg, W., Springer, L., Tarbell, T.D., Title, A., Wolfson, C.J., Zayer, I. and the MDI engineering team *Solar Shysics* **162**, 129 (1995).

[102] Schmitt, F., Schertzer, D., Lovejoy S., Brunet Y. *Nonlinear Processes in Geophysics* **1**, 95 (1994).

[103] Schrijver, C.J., Zwaan, C., Balke,A.C., Tarbell,T.D., Lawrence, J.K. *Astron. Astrophys.* **253**, L1 (1992).

[104] Schrijver, C.J., Title, A.M., van Ballegooijen, A.A. Hagenaar, H.J., Shine, R.A. *Astrophys. J.* **487**, 423 (1997a).

[105] Schrijver, C.J., Title, A.M., Hagenaar, H.J., Shine, R.A. *Solar Phys.* **175**, 329 (1997b).

[106] Schroeder, M.R. Fractals, Chaos, Power Laws. W.H. Freeman and Company, New York (2000).

[107] Schwenn, R., Raymond, J. C., Alexander, D., Ciaravella, A., Gopalswamy, N., Howard, R., Hudson, H., Kaufmann, P., Klassen, A., Maia, D., Munoz-Martinez, G., Pick, M., Reiner, M., Srivastava, N., Tripathi, D., Vourlidas, A., Wang, Y. M., Zhang, J. *Space Science Reviews* **123**, 127 (2006).

[108] Seiden, P.E., Wentzel,D.G. *Astrophys. J.* **460**, 522 (1996).

[109] Simon, G.W. & Leighton, R.B. *Astrophys. J.* **140**, 1120 (1964).

[110] Song, H., Yurchyshyn, V., Yang, G., Tan, C., Chen, W., Wang, H. *Solar Shysics* **238**, 141 (2006).

[111] Takayasu, H. *Fractals in the physical sciences* Manchester University Press, Manchester and New York (1989)

[112] Tao, L., Du, Y., Rosner, R., Cattaneo, F. *Astrophys. J.* **443**, 434 (1995).

[113] Tarbell, T., Ferguson, S., Frank, Z., Shine, R., Title, A., Topka, K., Scharmer, G. 1990, in J.O.Stenflo (ed.), Solar Photosphere: Structure, Convection, and Magnetic fields, (by the IAU 138), 147 (1990)

[114] Uritsky, V.M., Paczuski, M., Davila, J.M., Jones, S.I. *Phys. Res. Letters, in press*, (2007).

[115] Vasquez, B.J., Abramenko, V.I., Haggerty, D.K., Smith,C.W. *J. Geophys. Res.* in press (2007).

[116] Vlahos, L., Georgoulis, M., Kluiving, R., Paschos, P. *Astron. Astrophys.* **299**, 897 (1995).

[117] Vlahos, L., Georgoulis, M., Kluiving, R., Paschos, P. *Astron. Astrophys.* **299**, 897 (1998).

[118] Vlahos, L., Georgoulis, M.K. *Astrophys. J.* **603**, L61 (2004).

[119] Wang, H., Qiu, J., Denker, C., Spirock, T., Chen, H.J., Goode, P.R. *Astrophys. J.* **542**, 1080 (2000).

[120] Wang, J.X., Wang, H., Tang, F., Lee, J.W., Zirin, H. *Solar Phys.* **160**, 277 (1995).

[121] Wentzel,D.G., Seiden, P.E. *Astrophys. J.* **390**, 280 (1992).

[122] Wheatland, M. S. *Astrophys. J.* **609**, 1134 (2004).

[123] White, W.A. in W.N. Hess (ed), *AAS-NASA Symposium on the Physics of Solar Flares*, Washington (1964)

[124] Yurchyshyn, V., Yashiro, S., Abramenko, V., Wang, H., Gopalswamy, N. *Astrophys. J.* **619**, 599 (2005).

[125] Zeldovich, Ya.B., Molchanov, S.A., Ruzmaikin, A.A., Sokolov, D.D., *Sov. Phys. Usp.* **30**, 353 (1987).

[126] Zimbardo, G., Pommois, P., Veltri, P. *Astrophys. J.* **639**, L91 (2006).

[127] Zirin, H. *Astrophysics of the Sun*, Cambridge: Cambridge Univ. Press. (1988).

In: Solar Physics Research Trends
Editor: Pingzhi Wang, pp. 137-163
ISBN 978-1-60021-987-0
© 2008 Nova Science Publishers, Inc.

Chapter 3

TIME-DISTANCE HELIOSEISMOLOGY OF SUNSPOTS

Sébastien Couvidat
W.W. Hansen Experimental Physics Laboratory, Stanford University

Abstract

Time-distance helioseismology is a recently developed method in solar physics giving us unprecedented access to the local structure and dynamics of subsurface solar layers. Since its inception in 1993, seminal and widely publicized results have been achieved: for instance, the inversion of the sound-speed subsurface structure of a sunspot, of the subsurface flows below and around an active region, the precise measurement of meridional and zonal flows, the mapping of large scale subsurface flows, the measurement of supergranular flows, and even the precise determination of the solar rotation axis. In this chapter, we focus on the results related to sunspots and we describe how recent works on surface effects and the magnetic field interaction with the acoustic wavefield question the relevance of these major breakthroughs.

1. Introduction

The seminal paper by Leighton, Noyes, & Simon (1962) on the observation of periodic oscillations on the solar surface gave birth to one of the most successful field of solar physics, later termed helioseismology. Its recent offspring, asteroseismology, expands the scope of this field to other stars. Until the end of the 80s, helioseismology was restricted to global helioseismology. Solar physicists were using the global oscillation modes —standing waves trapped in a solar cavity— to infer the internal solar structure and rotation profile. Three types of waves have been studied: the acoustic modes (p modes), the surface gravity modes (f modes), and the gravity modes (g modes). G modes are actively looked for but a definite detection is still missing. The 90s saw the emergence of local helioseismology. Waves propagating locally started to be used to analyze localized phenomena, like the solar active regions or some material flows. This chapter focuses on the results of a specific branch of local helioseismology, called time-distance analysis (Duvall et al. 1993). Other methods of local helioseismology, like the ring-diagram analysis and the acoustic holography, are beyond our scope. For more information about all these local helioseismology techniques, the reader is referred to Gizon & Birch (2005). This thorough review describes these techniques, alongside their main results. Local helioseismology is the tool of choice for accessing the internal structure and dynamics of sunspots beneath the solar surface.

The beauty of time-distance analysis resides in its conceptual simplicity. It measures the travel times of wavepackets propagating between different points on the solar surface. By comparing these travel times with the ones derived from a solar model, or from the so-called quiet Sun (quiet meaning absence of strong magnetic activity or organized flow), Tom Duvall, Alexander Kosovichev, and their colleagues demonstrated that it is possible to gain a wealth of information regarding the physics of the solar subsurface layers. The mystery surrounding the regions of strong magnetic field and locally-reduced temperature that are known as sunspots can finally be addressed. In Section 3 we will briefly describe the time-distance formalism. In Section 4 we will focus on some major results obtained by time-distance analysis on active regions. In Section 5 we will review some recent works on the interaction of magnetic field and acoustic waves that cast a doubt on the accuracy of some conclusions mentioned in Section 4. Finally, we will conclude in Section 6.

2. Time-Distance Helioseismology

Were the conditions on the solar surface similar to the ones encountered on Earth, it would be quite easy to follow the temporal evolution of a sound or surface-gravity wave excited by a "sunquake" and then measure its relevant travel times without elaborate mathematical techniques. However, millions of convective cells uniformly distributed excite the solar oscillations below the photosphere. Therefore measuring the travel time of a specific wavepacket traveling between different locations on the solar surface becomes quite challenging and requires the use of cross-covariance calculations. The time-distance method, introduced by Duvall et al. (1993) and formalized by, e.g., D'Silva (1996) and D'Silva et al. (1996), is based on the computation of these cross-covariances $C(\mathbf{r}_1, \mathbf{r}_2, t)$ between the solar oscillation signals at two locations in the photosphere: a source at $\mathbf{r}_1$ and a receiver at $\mathbf{r}_2$. $\mathbf{r} = (x, y)$ is the horizontal vector and t is the time coordinate. In the rest of this chapter, we will work in Cartesian coordinates only. Neglecting the spherical curvature of the Sun is a valid approximation as long as we work locally on small enough spatial regions (typically 200×200 Mm^2 or less). Moreover, to explain the time-distance formalism, we will focus on acoustic waves. The analysis with surface gravity waves is similar, but these waves only sample a shallow 2 Mm below the solar surface and thus cannot be used to obtain 3D information about active regions which are expected to extend deeper inside the Sun. The oscillation signals we consider are Doppler velocities $\phi(\mathbf{r}, t)$ measured by instruments like, e.g., the Michelson Doppler Imager (SOI/MDI; Scherrer et al. 1995) onboard the SoHO spacecraft, or by ground networks like the Global Oscillation Network Group (GONG; Harvey et al. 1996) and the Taiwan Oscillation Network (TON; Chou et al. 1995). These instruments image the solar surface onto CCDs (with 1024×1024 pixels for MDI), producing Doppler velocity and/or intensity maps. Those measuring the Doppler velocity observe the Doppler shifts of specific solar lines. In practice, intensity maps of the solar surface are less used than velocity maps because they produce noisier wave travel times. The raw Doppler velocity cubes are made of velocity maps of the solar surface that are contiguous in time. These cubes are tracked: we compensate for the solar differential rotation so that the feature we follow for, usually, more than 8 hours is always centered in the middle of the (x, y) plane.

2.1. Computation of the Point-to-Point Cross-Covariances

The point-to-point cross-covariance $C(\mathbf{r_1}, \mathbf{r_2}, t)$ can be defined as (Gizon & Birch 2004):

$$C(\mathbf{r_1}, \mathbf{r_2}, t) = \frac{dt}{T - t} \sum_{\tau} \phi(\mathbf{r_1}, \tau)\phi(\mathbf{r_2}, t + \tau) \qquad (1)$$

where dt is the sampling rate, and T is the observation time. Due to the stochastic excitation of acoustic waves by the solar convection zone and also due to the oscillation signal at any location being a superposition of a large number of waves of different travel distances (i.e., of different horizontal phase velocities $v_{ph} = \omega/k$, where k is the horizontal wavenumber and ω is the temporal angular frequency), these point-to-point cross-covariances are very noisy: the datacubes need first to be phase-speed filtered and the cross-covariances need to be averaged (Duvall et al. 1997). The Doppler velocity datacube $\phi(\mathbf{r}, t)$ is phase-speed filtered in the Fourier domain, using a Gaussian filter $F(k, \omega; \Delta)$ for each travel distance $\Delta = |\mathbf{r_2} - \mathbf{r_1}|$. The f-mode ridge is also filtered out. Obviously, when working with f modes, the filtering process is different; we keep the f-mode ridge while discarding other signals. The rationale for using a phase-speed filter instead of, say, a frequency filter, is that different waves with the same phase speed are expected to have the same lower turning points (i.e. they are reflected at the same depth). Therefore, these waves follow roughly the same ray path and form coherent wavepackets.

The standard scheme of averaging depends on the information we try to extract from the data. Indeed, time-distance analysis aims at relating the perturbations in the measured travel times of wavepackets to physical perturbations in the subsurface layers of the Sun. Here perturbations are relative to a solar model that does not include the specific physical phenomenon we try to analyze, or to the quiet Sun for which the phenomenon is absent. The perturbations the travel times are sensitive to are, for instance: a material flow, a sound-speed change, a magnetic field, a frequency-dependent wave damping... If we are concerned by isotropic physical processes (processes that affect the waves independently of their travel direction like a sound-speed perturbation), the standard scheme of averaging is to average the point-to-point cross-covariances over an annulus of radius Δ centered on the source $\mathbf{r_1}$. If we are concerned by anisotropic phenomena, like a horizontal material flow, we usually average the cross-covariances over quadrants. Figure 1 shows the different geometries commonly applied. Two physical phenomena have been investigated so far with time-distance analysis performed on sunspots: sound-speed perturbations and material flows. In the rest of this chapter we will use interchangeably the terms sunspots and active regions.

2.2. Point-to-Annulus Cross-Covariances for Sound-Speed Perturbations

Let us first focus on sound-speed perturbations. To study these perturbations the point-to-point cross-covariances are averaged over annuli of radius Δ. Then, to further increase the signal-to-noise ratio, such point-to-annulus cross-covariances are computed for several distances respectively slightly smaller than, and slightly larger than, Δ. They are averaged to produce a point-to-"broad-annulus" cross-covariance. A detailed explanation of all these

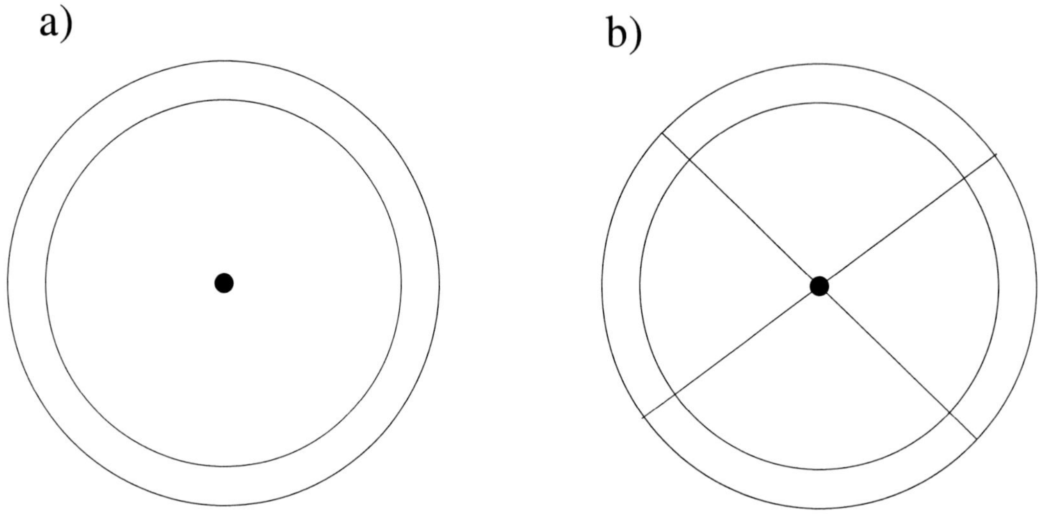

Figure 1. Averaging schemes commonly used in time-distance helioseismology: a) center-to-annulus averaging; b) east-west and north-south quadrant averaging. The central black dots are the source $\mathbf{r}_1$.

steps in the analysis process can be found in, e.g., Zhao (2004) and Couvidat, Birch, & Kosovichev (2006), along with a table of distances Δ and commonly used phase-speed filter characteristics. Different solar physics groups use different parameters for the phase-speed filters. The central phase speed is usually obtained from a solar model: using a ray tracer, we measure the lower turning point of a specific acoustic wave. At this turning point, $k_z = 0$ (where k_z is the vertical wavenumber and z is the vertical coordinate; $z = 0$ being the solar surface) and the horizontal phase speed is equal to the sound speed at that depth. The width of the phase-speed filter is selected empirically. Two travel times of wavepackets are derived from the point-to-annulus cross-covariances: an ingoing (i; from annulus to center) and an outgoing (o; from center to annulus) travel time, $\tau_{\mathrm{i/o}}(\mathbf{r}, \Delta)$. The average of these two travel times, $\tau_{\mathrm{mean}}(\mathbf{r}, \Delta)$, is, in first approximation, sensitive only to the sound-speed $c(\mathbf{r}, z)$ in the region of the Sun traversed by the wavepacket, while the difference $\tau_{\mathrm{diff}}(\mathbf{r}, \Delta)$ in the travel times will be sensitive to flows. It can be argued that τ_{diff} is mainly sensitive to vertical flows: the ray path of the selected wavepackets is almost vertical where these wavepackets are the most sensitive to a physical perturbation. This approximation is more valid for large Δ distances, for which the ray paths are truly vertical near the surface, unlike the ray paths at small Δ distances that have a significant horizontal component. Therefore, $\tau_{\mathrm{diff}}(\mathbf{r}, \Delta)$ for small Δ is also sensitive to the horizontal divergence of the flow (Kosovichev, Duvall, & Scherrer 2000). In the case of f modes, which propagate horizontally, τ_{diff} is a measure of the flow velocity horizontal divergence (Duvall & Gizon 2000). Using a solar model as a reference we can relate the mean travel-time perturbations $\delta\tau_{\mathrm{mean}}(\mathbf{r}, \Delta)$ to the sound-speed perturbations $\delta c(\mathbf{r}, z)$ through an integral relation (e.g., Kosovichev & Duvall 1997):

$$\delta\tau_{\mathrm{mean}}(\mathbf{r}, \Delta) = \iint_S d\mathbf{r}' \int_{-d}^{0} dz\, K_c(\mathbf{r}, \mathbf{r}', z; \Delta)\, \frac{\delta c^2}{c^2}(\mathbf{r}', z) \qquad (2)$$

where S is the area of the region, and d is its depth. The sensitivity kernel for the relative squared sound-speed perturbation is given by K_c. It is worth mentioning that Eq 2 is only approximate because effects on $\delta\tau_{\rm mean}(\mathbf{r}, \Delta)$ other than the sound-speed perturbation are completely ignored. For instance, Brüggen & Spruit (2000) describe the impact of changes in the upper boundary condition in sunspots due to the Wilson depression; Woodard (1997) and Gizon & Birch (2002) demonstrate that increased wave damping in sunspots can introduce shifts in travel times; finally, significant magnetic field effects that will be reviewed in Section 5 also contribute to $\delta\tau_{\rm mean}(\mathbf{r}, \Delta)$. Section 2.5. will deal with the different kind of sensitivity kernels we use and briefly explain how they are obtained. Figure 2 shows an example of a time-distance diagram obtained with SOI/MDI data: the x-axis shows the distance Δ between source and receiver, measured in angular separation at the solar surface, while the y-axis shows the time in minutes. The shape of the cross-covariances is clearly visible, as are the different ridges corresponding to the waves bouncing back at the solar surface in between the source and receiver. This concept of "bounce ridges" is explained on Figure 3, which is a sketch of the acoustic wave propagation inside the Sun in the ray-path approximation: it shows how waves traveling from point A (source) to point B (receiver) at the solar surface can make the journey directly (producing the first bounce ridge on Figure 2), with one bounce at the surface in C (second bounce ridge), or even with additional bounces.

2.3. Point-to-Quadrant Cross-Covariances for Flows

To link the travel-time perturbations to subsurface horizontal flows with velocities $v_x(\mathbf{r}, z)$ and $v_y(\mathbf{r}, z)$, and vertical flows with velocity $v_z(\mathbf{r}, z)$, we average the point-to-point cross covariances over quadrants (for the horizontal components), and over annuli (for the vertical component of the flow, as mentioned in the previous section): over the east-west quadrants in order to be sensitive to east-west flows $v_x(\mathbf{r}, z)$ (with the additional twist that we need to reverse the time t of the data series for one quadrant compared to the other so that we deal with flows traveling in the same direction), and over north-south quadrants to be sensitive to north-south flows $v_y(\mathbf{r}, z)$. From these point-to-quadrant cross-covariances we can derive four different travel times: east-to-west travel time $\tau_{E-W}(\mathbf{r}, \Delta)$, west-to-east travel time $\tau_{W-E}(\mathbf{r}, \Delta)$, north-to-south $\tau_{N-S}(\mathbf{r}, \Delta)$, and south-to-north $\tau_{S-N}(\mathbf{r}, \Delta)$. The difference travel-time perturbations $\delta\tau_{\rm diff}(\mathbf{r}, \Delta) = \tau_{\rm o}(\mathbf{r}, \Delta) - \tau_{\rm i}(\mathbf{r}, \Delta)$ is obtained from the point-to-annulus cross-covariance. We have the following approximate relations:

$$\begin{aligned}\delta\tau_{\rm EW}(\mathbf{r}, \Delta) &= \int\!\!\int_S d\mathbf{r}' \int_{-d}^{0} dz \Big(K_{\rm EW;x}(\mathbf{r}, \mathbf{r}', z; \Delta)\, v_x(\mathbf{r}', z) + \\ &\qquad K_{\rm EW;y}(\mathbf{r}, \mathbf{r}', z; \Delta)\, v_y(\mathbf{r}', z) + K_{\rm EW;z}(\mathbf{r}, \mathbf{r}', z; \Delta)\, v_z(\mathbf{r}', z) \Big) \quad (3)\end{aligned}$$

$$\begin{aligned}\delta\tau_{\rm NS}(\mathbf{r}, \Delta) &= \int\!\!\int_S d\mathbf{r}' \int_{-d}^{0} dz \Big(K_{\rm NS;y}(\mathbf{r}, \mathbf{r}', z; \Delta)\, v_y(\mathbf{r}', z) + \\ &\qquad K_{\rm NS;x}(\mathbf{r}, \mathbf{r}', z; \Delta)\, v_x(\mathbf{r}', z) + K_{\rm NS;z}(\mathbf{r}, \mathbf{r}', z; \Delta)\, v_z(\mathbf{r}', z) \Big) \quad (4)\end{aligned}$$

$$\delta\tau_{\rm diff}(\mathbf{r}, \Delta) = \int\!\!\int_S d\mathbf{r}' \int_{-d}^{0} dz \Big(K_{vz;z}(\mathbf{r}, \mathbf{r}', z; \Delta)\, v_z(\mathbf{r}', z) +$$

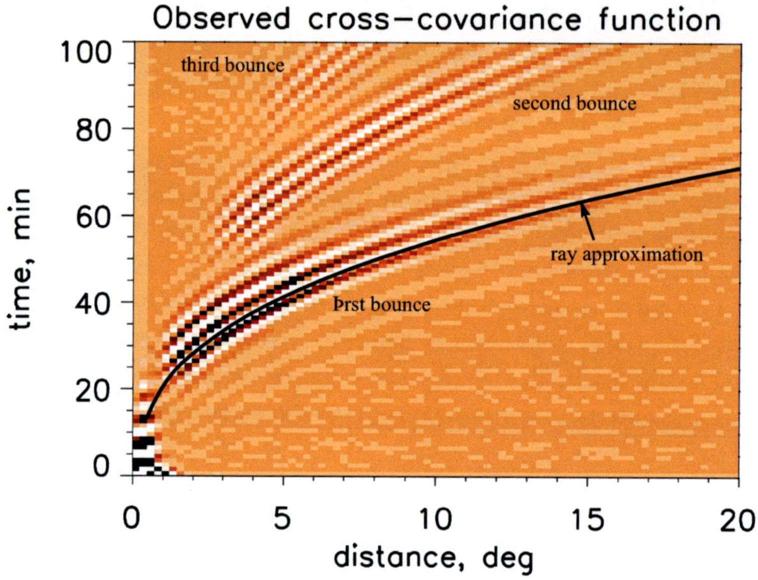

Figure 2. Example of time distance diagram: we clearly see the first-bounce ridge, corresponding to waves traveling directly from the source $\mathbf{r_1}$ to the receiver $\mathbf{r_2}$, and the second-bounce ridge, corresponding to waves that are reflected back at the solar surface in between the source and receiver. From Kosovichev & Duvall (1997).

$$K_{vz;x}(\mathbf{r}, \mathbf{r}', z; \Delta)\, v_x(\mathbf{r}', z) + K_{vz;y}(\mathbf{r}, \mathbf{r}', z; \Delta)\, v_y(\mathbf{r}', z)\Big) \tag{5}$$

where $\delta\tau_{\text{EW}} = \tau_{E-W}(\mathbf{r}, \Delta) - \tau_{W-E}(\mathbf{r}, \Delta)$ is the difference east-west travel-time perturbation, $\delta\tau_{\text{NS}} = \tau_{N-S}(\mathbf{r}, \Delta) - \tau_{S-N}(\mathbf{r}, \Delta)$ is the difference north-south travel-time perturbation, $K_{\text{EW;x}}$ is the sensitivity kernel for velocity in the horizontal x direction for the east-west geometry, $K_{\text{EW;y}}$ is the sensitivity kernel for velocity in the horizontal y direction for the east-west geometry, and $K_{\text{EW;z}}$ is the sensitivity kernel for velocity in the vertical z direction for the east-west geometry. Similarly, K_{NS} are the sensitivity kernels for flow velocity in the north-south geometry, and K_{vz} are the sensitivity kernels in the outgoing-ingoing geometry. It is essential that this set of three equations be solved simultaneously: we cannot infer, say, the vertical velocity independently of the horizontal velocities. Therefore, retrieving the flow velocity, which is a vector, from time-distance analysis is more complex than retrieving the sound speed, which is a scalar.

2.4. Fit for the Travel Times

To obtain the travel times of the wavepackets from the point-to-annulus or point-to-quadrant cross-covariances, two techniques are currently widely used. The first one stems from noticing that by grouping the modes in narrow ranges of angular velocity and applying the

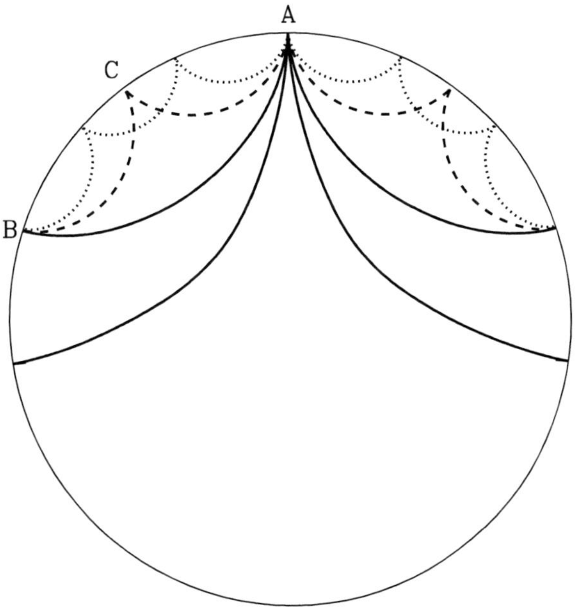

Figure 3. Propagation of acoustic waves inside the Sun in the ray-path approximation. The waves travel from A to B, either directly (solid curve), or by bouncing back at the surface in C (long dashed curve), or by bouncing additional times (dotted curve). Figure kindly provided by A. G. Kosovichev.

method of stationary phase, it appears that the cross-covariances are very similar to Gabor wavelets (Kosovichev, Duvall, & Scherrer 2000). Also, we can fit these cross-covariances with two Gabor wavelets $G(A, \omega_0, \delta\omega, \Delta, \tau_p, \tau_g; t)$, one for the positive times (outgoing waves), and one for the negative times (ingoing waves) (Kosovichev & Duvall 1997):

$$G(A, \omega_0, \delta\omega, \Delta, \tau_p, \tau_g; t) = A\cos(\omega_0(t - \tau_p))\exp\Big(- \frac{\delta\omega^2}{4}(t - \tau_g)\Big) \quad (6)$$

where A is the amplitude of the wavelet, ω_0 is the wavepacket central angular frequency, $\delta\omega$ is the wavepacket frequency width, τ_p is the phase travel time, and τ_g is the group travel time. Even though it is defined modulo $2\pi/\omega_0$, we usually only use the phase travel time because it is less noisy than the group travel time, especially at small Δ distances. Second, Gizon & Birch (2004) developed a new method, very robust with respect to noise, which is linear in the cross-covariance. Their definition of the travel times has some advantages over the Gabor wavelet fit. However, because the sharp drop in the amplitude of the cross-covariance in a sunspot is not accounted for, their definition cannot currently be applied to active regions. Also we will not elaborate on their technique here.

2.5. Derivation of the Sensitivity Kernels

The computation of the sensitivity kernels is a tedious task that can be performed under several assumptions. The ray-path approximation assumes that the wavepackets have an infinitely small wavelength. Within this framework, it is somewhat straightforward to derive the corresponding kernels: this requires a ray-tracer and a solar model (see Giles 1999). Even though the ray-path approximation works for strong perturbations unlike other linearized approximations, such ray-path kernels are somewhat crude for small perturbations. A better way to derive sensitivity kernels is to resort to the Rytov approximation (Jensen & Pijpers 2003): the wave propagation problem is written as a scattering problem that is initially non-linear but can be solved through this Rytov approximation. The resulting sensitivity kernels are broad and extend mainly to the first Fresnel zone (diffraction zone). They have a typical "banana-doughnut" shape (with a zero-sensitivity along the ray path). They take into account the finite wavelength effects of the wavefield. Another approximation can be applied to solve the scattering problem: the Born approximation (Birch, Kosovichev, & Duvall 2004). The resulting kernels are realistic and strongly dependent on the data analysis techniques, especially the phase-speed filtering that determines the frequency content of the acoustic wavepackets (Birch & Kosovichev 2000). Couvidat et al. (2004) compared the ray-path and Fresnel-zone approximations in the context of time-distance helioseismology, showing the superiority of the Fresnel-zone approximation. Similarly, several comparisons between the ray-path and the Born approximations have already been performed in geophysics (e.g. Marquering, Dahlen, & Nolet 1999), and in helioseismology (Birch et al. 2001; Couvidat, Birch, & Kosovichev 2006) confirming that, even though the ray-path kernels give results that are qualitatively valid, a correct retrieval of the location and amplitude of the perturbation requires kernels that take into account the finite wavelength effects. An example of Born-approximation kernel is shown on Figure 4. This figure highlights the dramatic difference between ray-path kernels (which look roughly like an arc of a circle between $\mathbf{r}_1$ and $\mathbf{r}_2$; see the ray paths on Figure 3) and Born-approximation kernels which exhibit a wealth of subtle details. Interestingly, the several averagings (especially over annuli) performed in the course of the time-distance analysis cancel out most of these Born kernels details, explaining why, in the end, azimuthally averaged ray-path kernels are not that different from azimuthally averaged Born-approximation and Fresnel-zone kernels. Recently, Duvall, Birch, & Gizon (2006) developed a very interesting method to measure the sensitivity kernels directly from the Doppler velocity data. This approach, still under construction, will be a powerful way of building highly realistic kernels for a variety of physical perturbations. They used small magnetic elements as approximate point scatterers. They obtained kernels for f modes that look very similar to those obtained theoretically with the Born approximation, confirming the validity of the latter. For the integral relations linking the kernels to the travel times to be applied, the point-to-point sensitivity kernels need to be averaged with the same scheme (annuli, quadrants...) as the one used to obtain the travel times.

2.6. Inversion of Travel-Time Maps

Based on Eqs 2 to 5, the mean, difference, east-west, and north-south travel-time perturbation maps can be inverted for sound-speed perturbation and flow velocity. Two algorithms are commonly used to perform such inversions: LSQR and MCD. LSQR (Paige & Saunders 1982) is an iterative algorithm to solve the linear problem $\mathbf{Ax} = \mathbf{y}$ (where $\mathbf{A}$ is a matrix and $\mathbf{x}$ and $\mathbf{y}$ are vectors) by finding the vector $\mathbf{x}$ that minimizes $||\mathbf{Ax} - \mathbf{y}||_2$. LSQR is close to the conjugate gradients method. The interest of LSQR is that it actually solves

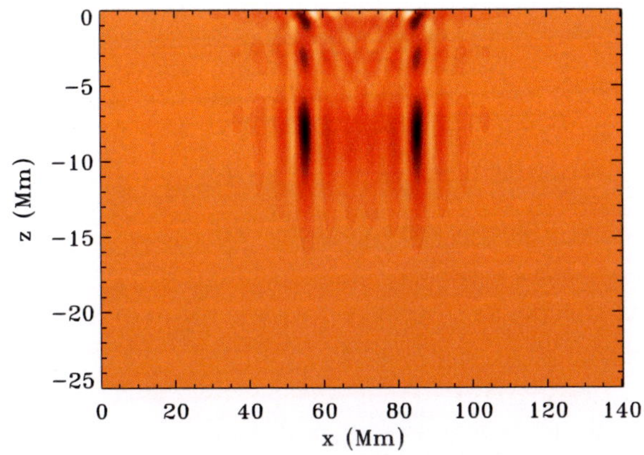

Figure 4. Example of a Born-approximation kernel for relative squared sound-speed perturbation. Vertical cut at $y = 0$ in the point-to-point Born kernel for $\Delta = 30.55$ Mm. The kernel has been multiplied by the sound speed at each depth z. Kernel kindly provided by A. C. Birch. From Couvidat, Birch, & Kosovichev (2006).

$min(||\mathbf{Ax} - \mathbf{y}||_2)$ and $min(||\mathbf{x}||)$ at the same time. Indeed the inverse problem in time-distance helioseismology is ill conditioned and regularization is most needed if we want to obtain a meaningful solution. LSQR includes a regularization by the norm of the solution: it tries to minimize this norm. Any regularization can be seen as a way to bias the solution. Here the physicist performing the inversion assumes that the solution has a small norm, which is taken into account in the inversion procedure. Another regularization scheme (and, it can be argued, more physical) is to minimize the norm of the first derivative of the solution. Such a scheme means that we assume that the solution is "smooth" and does not have steep spatial variations. We have to be careful though when applying this scheme to datacubes with sunspots, because sunspots may have sharp boundaries which can be a problem leading to sidelobes in the inversion results. LSQR is a fast and reliable algorithm, but currently it cannot include the regularization by the first derivative, unlike the MCD algorithm (multi channel deconvolution; Jensen, Jacobsen, & Christensen-Dalsgaard 1998). LSQR and MCD are therefore complementary. However, to apply the MCD algorithm we need an additional assumption: the sensitivity kernels are spatially uniform and isotropic, meaning $K(\mathbf{r}, \mathbf{r}', z; \Delta) = K(\mathbf{r} - \mathbf{r}', z; \Delta)$. Under this assumption, which might not be completely valid in sunspots, Eqs 2 to 5 are convolution products. MCD takes advantage of this property by solving these integral relations in the Fourier domain, where a convolution becomes a mere multiplication. For instance, in the Fourier domain, Eq 2 becomes:

$$\delta\tau_{\text{mean}}(\mathbf{k}, \Delta) = \int_{-d}^{0} dz\, K_c(\mathbf{k}, z; \Delta) \frac{\delta c^2}{c^2}(\mathbf{k}, z) \tag{7}$$

The MCD algorithm transforms a 3D problem in the real space into several 1D problems in the Fourier space. It is then easier and faster to invert each 1D relation, and to finally switch back to the real space. For each vector $\mathbf{k}$ we have, in matrix notation (and after discretization of our problem):

$$K_{ij} = K_c(\mathbf{k}, z_j; \Delta_\text{i}) \quad (8)$$

$$c_j = \frac{\delta c^2}{c^2}(\mathbf{k}, z_j) \quad (9)$$

$$\tau_i = \delta\tau_\text{mean}(\mathbf{k}, \Delta_\text{i}) \quad (10)$$

where K is the kernel matrix, $\mathbf{c}$ is the solution vector, and τ is the data vector. We solve $min(||\tau - K\mathbf{c}||_2^2 + \lambda^2||L\mathbf{c}||_2^2)$, where L is the regularization matrix corresponding to the regularization scheme we choose (for instance, regularization by the first derivative) and λ is the regularization parameter. A possible way to solve this 1D problem is to transform it into standard form (where L is the identity matrix), then compute the singular value decomposition of the transformed matrix K, multiply its singular values by the filter factors corresponding to the regularization scheme, and invert the resulting matrix (Hansen 1998). However, the standard MCD method allows only regularization in the vertical direction, unlike the LSQR algorithm that regularizes in the 3 spatial directions. To fix this issue, R. M. Larsen suggested to use the identity matrix for L (or at least a diagonal matrix) and to replace λ^2 by $\lambda(k)^2 = \lambda_v^2 + \lambda_h^2(k)$, where λ_v is the regularization parameter in the vertical direction and λ_h is the regularization parameter in the horizontal direction (Couvidat et al. 2005). λ_v is a constant while $\lambda_h = \lambda_2\, k$ or $\lambda_h = \lambda_2\, k^2$, where λ_2 is a constant. Indeed, since we invert in the Fourier domain, $k\,\delta c^2/c^2(\mathbf{k}, z)$ is the Fourier transform of the first derivative of $\delta c^2/c^2(\mathbf{r}, z)$, while $k^2\,\delta c^2/c^2(\mathbf{k}, z)$ is the Fourier transform of the second derivative of $\delta c^2/c^2(\mathbf{r}, z)$. Therefore, the k-dependent regularization parameter allows a horizontal regularization by the norm of the first or second derivative of the solution, and a vertical regularization by the norm of the solution. With MCD, one problem is to choose the value of the regularization parameter(s). The so-called L-curve (for one parameter) or L-surface (for two parameters) can be used, but it is often as efficient to follow a pragmatic approach and let the physicist decide empirically which value to select. With LSQR, the number of iterations plays the role of a regularization parameter.

So far, we presented the inverse problem in terms of regularized least-squares (RLS). It must be mentioned that a time-distance inversion can also be performed with an optimally localized averages method (OLA; Backus & Gilbert 1968). We will only outline the main idea behind OLA. A byproduct of the RLS method is the averaging kernels: these matrices $\mathbf{K}_\text{av}$ inform us about the resolution we reach at each point (x, y, z). For instance, with the MCD algorithm, if $\mathbf{K}_\text{av}$ is the averaging kernel at the horizontal wavevector $\mathbf{k}$, then we can write (always in matrix notation): $\mathbf{c} = \mathbf{K}_\text{av}\mathbf{c}_\text{exact}$ where $\mathbf{c}$ is the solution vector at $\mathbf{k}$ (same notation as previously) and $\mathbf{c}_\text{exact}$ is the exact solution. With RLS, the averaging kernels are a result, like the solution vector. With OLA, the averaging kernels are an input: ideally we want averaging kernels as close to a delta function as possible, and the OLA algorithm builds a solution that satisfies this requirement. Therefore, OLA is theoretically a "better" algorithm than RLS. On the other hand, the standard OLA algorithm is also more time consuming than RLS, and it usually gives a similar result.

Another aspect of the inversion process must be mentioned: the noise properties of the data are often ignored. However, if we have access to the covariance matrix C of the data errors, then the least-squares problem we try to solve has to be re-written (Hansen 1998):

$$min(||\Sigma^{-1}(\tau - K\mathbf{c})||_2^2 + \lambda^2(k)||L\mathbf{c}||_2^2) \quad (11)$$

where $C = \Sigma\Sigma^{\mathrm{T}}$ (Cholesky decomposition). A first step in the inclusion of noise properties in time-distance inversion was achieved by Jensen, Duvall, & Jacobsen (2003). Later Couvidat et al. (2005) obtained the covariance matrix C of the noise in the travel-time maps by computing a large enough number of realizations of noise travel-time maps based on the noise model by Gizon & Birch (2004). Taking this noise covariance matrix into account makes it possible to better estimate the noise level on the inversion results, produced by noise propagation through the inverse algorithm.

A more detailed review of inversion techniques in helioseismology (including local helioseismology) can be found in Kosovichev (1999). The reader can also be referred to Christensen-Dalsgaard, Schou, & Thompson (1990) who describe a variety of methods, but apply them only to global helioseismology.

3. Results for the Solar Active Regions

Duvall et al. (1993) predicted that the newly gained ability to measure travel times and distances will be important for helioseismology. They foresaw great benefit to the study of local problems such as inhomogeneities near sunspots. They also already highlighted the possibility to detect subsurface flows, especially in active regions. Indeed, despite having being known for millenia, several specific aspects of sunspots are still not completely understood (Solanki 2003): the subsurface structure of sunspots in terms of magnetic field is still an open question, as are the dynamics of the gas during the sunspot formation, the fraction of the heat flux blocked by the sunspot that is stored in the convection zone, and the whereabouts of the energy absorbed by sunspots from p-mode oscillations. It was solar physicists' hope that time-distance helioseismology could address and help solve these questions.

3.1. Flows Around and Below a Sunspot

One of the first significant result of time-distance helioseismology was the detection of the asymmetry of phase travel times inside a sunspot, by Duvall et al. (1996): using a center-to-annulus geometry, they showed that the travel time of annulus-to-center waves is longer than the one of center-to-annulus waves (see Figure 5). In the ray approximation, near the solar surface, the rays are almost vertical (for large distances Δ). Therefore, as mentioned in Section 3.2, the phase asymmetry can be explained by a vertical flow, and more specifically a downflow. Indeed, a downflow below the sunspot would decrease the travel time of outgoing waves and increase the travel time of ingoing waves. Duvall and his colleagues estimated the velocity of this downflow to be roughly a few km/s. Their observations have strong implications for sunspot models. The model of sunspot developed by Parker (1979) had never been tested before. According to E. Parker, a sunspot is not a monolithic giant flux tube but a collection of small flux tubes held together by material downdrafts. These fluxes are pressed together to form a single flux tube at the visible surface, but are otherwise distinct for layers deeper than a few Mm. In this theory, several vertical magnetic flux tubes emerge at the solar surface and develop downflows around them. Therefore, the detection of downflows by T. L. Duvall and his colleagues was considered a breakthrough in solar

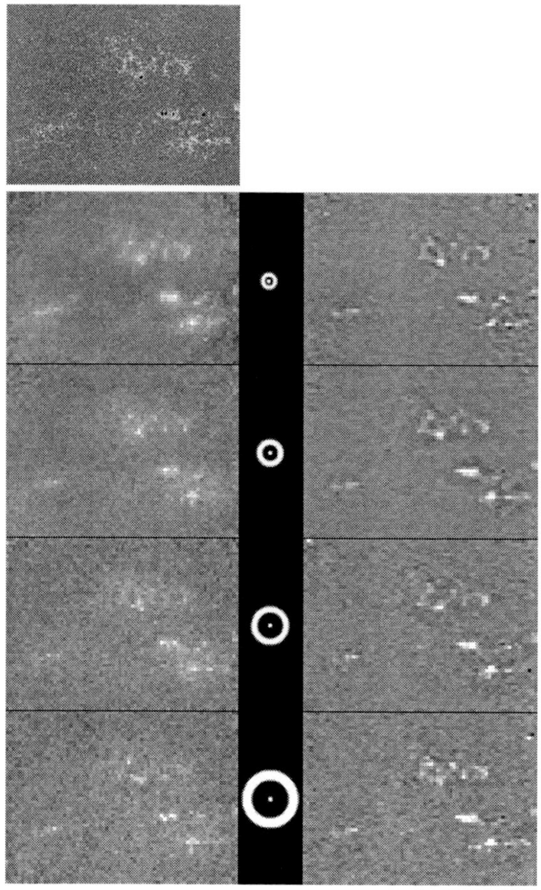

Figure 5. Upper left: image of the Sun in the calcium K-line obtained at the South Pole. The small dark spots are sunspots. Below this image are maps of the mean (left panels) and difference outgoing-minus-ingoing (right panels) travel times. The four sets of travel-time maps correspond to four different annuli, shown in the central panel. For the difference travel-time panels, white color corresponds to outgoing waves traveling faster than ingoing waves. The largest difference is about 1 minute. Figure kindly provided by T. L. Duvall. From Duvall et al. (1996).

physics, because it made it possible to discriminate between the cluster and the monolithic models of sunspots.

A more precise picture of flows in and below a sunspot was later offered by Zhao, Kosovichev, & Duvall (2001) (see Figures 6 and 7). They studied a large sunspot observed by SOI/MDI on June 18, 1998. Using the LSQR algorithm and some ray-path kernels for the inversion of flow velocity, they obtained maps of flow speeds at 10 different layers in depth. Near the surface, at a depth between 0 and 3 Mm, they observed inflows inside the sunspot umbra and penumbra, and a strong downflow immediately below. The amplitude

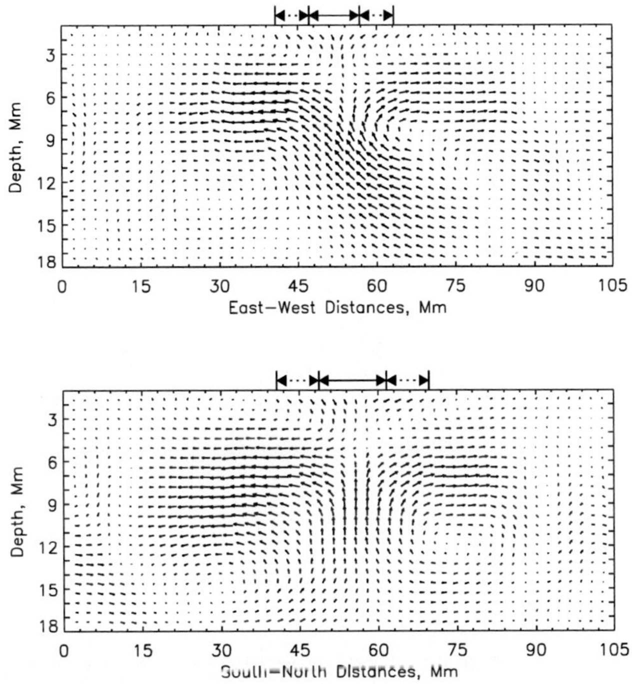

Figure 6. Vertical cuts through a sunspot center of the inverted velocity field. Top panel: cut direction of east-west; bottom panel: cut direction of north-south. The horizontal arrows show the extent of the umbra and penumbra. The longest arrow represents 1.4 km s^{-1}. Figure kindly provided by J. Zhao. From Zhao, Kosovichev, & Duvall (2001).

of the downflow is less than 1 km s^{-1} down to a depth of 5 Mm, while the amplitude of the inflow is less than 1.6 km s^{-1}. Deeper, they observed upflows and outflows inside the sunspot. Their impressive paper was strongly supportive of the cluster model of sunspot. However, Gizon, Duvall, & Larsen (2000) used f modes to detect the horizontal flows (averaged over 2 Mm in depth) in a sunspot, and found outflows slower than 1 km s^{-1} in the umbra and penumbra unlike the inflows seen with p modes. This p- and f-mode discrepancy might be due to shallow Evershed and moat flows, which are known to be concentrated in magnetic filaments; p modes are less sensitive to the surface horizontal flows than f modes, because they propagate mostly vertically near the surface (A. G. Kosovichev, private communication). In a similar fashion, Zhao & Kosovichev (2003) inverted the kinetic helicity in the subsurface layers of a fast-rotating sunspot. The kinetic helicity is defined as $\alpha = \mathbf{v} \cdot (\nabla \times \mathbf{v})/|\mathbf{v}|^2$, where $\mathbf{v}$ is the velocity field. Zhao & Kosovichev used the component of the helicity that corresponds to the vertical components of velocity and vorticity. They linked this helicity to the helicity of the magnetic field, highlighting the fact that twisted magnetic fluxes at the solar surface are also twisted in deeper layers. The authors also showed that rotational motions detected at the photospheric level are present at depths up to 5 Mm (see Figure 8). It might be that magnetic helicity at the solar surface is produced by these helical motions underneath this surface. This paper also linked time-distance helioseismology and solar flares: strong subsurface vortical flows may take part in the magnetic helicity and energy storage that will erupt at the surface, producing a flare.

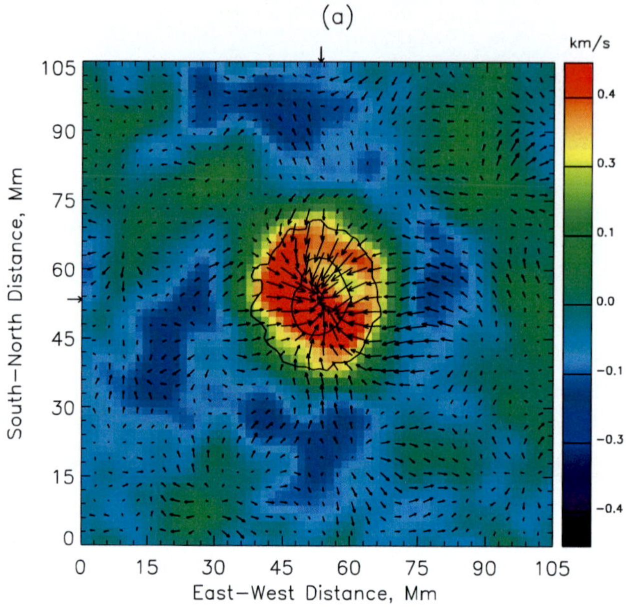

Figure 7. Material flows at a depth of 0-3 Mm in and around a sunspot. Arrows show the direction and magnitude of the horizontal flows, and the background color shows the vertical flows. Positive velocity indicates downflow. The longest arrow represents 1.0 km s^{-1}. Figure kindly provided by J. Zhao. From Zhao, Kosovichev, & Duvall (2001).

Time-distance analysis has since been used to analyze the propagation of acoustic waves excited by solar flares, opening new exciting perspectives for high-resolution helioseismic diagnostics of sunspots (Kosovichev & Zharkova 1998; Kosovichev 2006).

3.2. Subsurface Structure of a Sunspot

In parallel to the inversion of flows, the sound-speed perturbation $\delta c/c$ below a sunspot was inverted for the first time by Kosovichev (1996) and then by Kosovichev, Duvall, & Scherrer (2000). These two seminal papers used high-resolution data from the SOI/MDI instrument and data obtained in 1991 at the South Pole. The well known "coffeecup sunspot" figure (Figure 9) was most publicized and soon became a sort of flagship for time-distance analysis. Ray-path approximation sensitivity kernels were used, and the inversion algorithm was LSQR. A two-region structure was detected below the sunspot. The shallow subsurface layers exhibit a decrease in the sound speed ($\delta c/c < 0$) relative to the quiet Sun, while the deeper layers exhibit an increase in the sound speed ($\delta c/c > 0$). The authors noted that the two-region structure could be due to a thermal or a combination of thermal and magnetic field changes, but they did not favor the sole contribution of the magnetic field. The typical sound-speed perturbation is of the order of $\delta c = 0.3 - 1$ km s^{-1}. Kosovichev et al. (2000) also reported the presence of finger-like structures connecting the deeper region of the sunspot to two pores of the same polarity at the solar surface. This detection was later confirmed by Couvidat et al. (2004) (see Figure 10) and Couvidat, Birch, & Kosovichev (2006) with, respectively, Fresnel-zone and Born-approximation sensitivity kernels and the

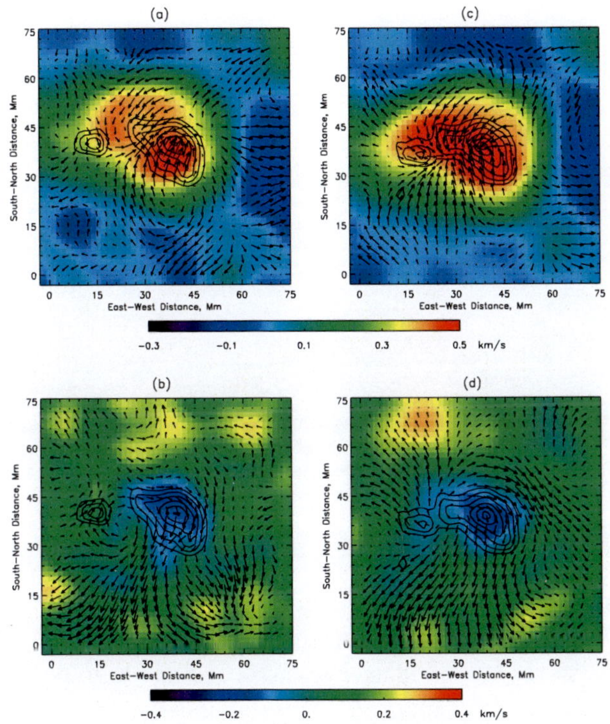

Figure 8. Left panels: background image showing the vertical velocities and the horizontal velocity field (with arrows) obtained for a rotating sunspot at depth intervals 0-3 (top panel) and 9-12 Mm (bottom panel). Right panels: same plots but for data obtained one day later. The longest arrow is 0.5 km s^{-1}. The contour lines are the line-of-sight magnetic field amplitude. Figure kindly provided by J. Zhao. From Zhao & Kosovichev (2003).

MCD algorithm for the inversion. Similarly the two-region structure for the sunspot was confirmed by several groups: a non-exhaustive list includes Jensen et al. (2001); Couvidat et al. (2004); Hughes, Rajaguru, & Thompson (2005) (using GONG data with ray-path kernels and the LSQR algorithm); and Zharkov, Nicholas, & Thompson (2007) (on 6 sunspots observed by SOI/MDI and inverted with the MCD algorithm and the Fresnel-zone kernels of Jensen & Pijpers [2003]). The numerous observations by different groups using different datasets and different inversion methods strongly favored the idea that the two-region structure detected below a sunspot had indeed a physical origin and could not easily be dismissed as an instrumental or algorithm artifact.

4. The Controversy: Significant Impact of the Magnetic Field ?

The results presented in the previous sections were soon open to criticism. The seeds of controversy were even sowed before time-distance helioseismology first results: using the Hankel analysis, Braun, Duvall, & LaBonte (1987) showed that sunspots absorb up to half of the incident acoustic-wave power. This claim has since been disputed by Parchevsky & Kosovichev (2006) who showed that the effect of reduced wave excitation inside a sunspot

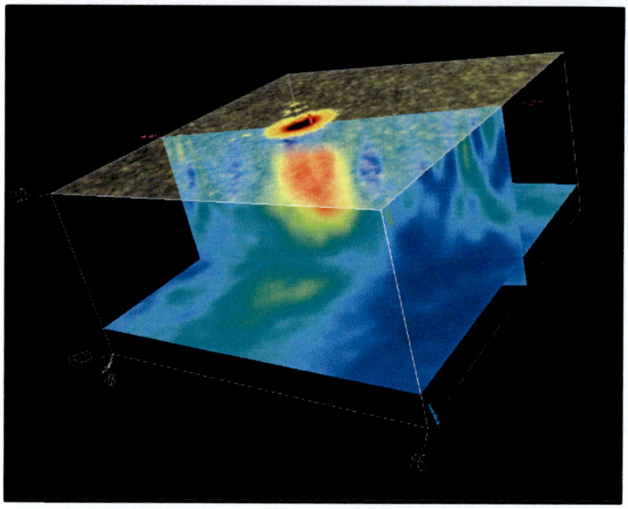

Figure 9. Sound-speed perturbation below a sunspot derived from SOI/MDI data. Three planes are shown: on top the continuum intensity at the surface, showing the sunspot with the dark central umbra surrounded by the somewhat brighter, filamentary penumbra. The second plane is a vertical cut from the surface to a depth of 24 Mm showing areas of faster sound speed as reddish colors and slower sound speed as bluish colors. The third plane (bottom) is a horizontal cut at a depth of 22 Mm showing the horizontal variation of sound speed over a region of 150x150 Mm. Figure kindly provided by A. G. Kosovichev. From Kosovichev, Duvall, & Scherrer (2000).

should also be taken into account. Later, the phase of the incident waves was found to be shifted in sunspot, as was demonstrated by Braun (1995) on South Pole data. Also, at the time Duvall et al. (1996) found that the travel times of ingoing waves (from outside the spot to inside the spot) are larger than the travel times of outgoing waves and they suggested that this travel-time asymmetry is due to downflows below the sunspot, Lindsey et al. (1996) mentioned the issue (raised by M. F. Woodard and D. C. Braun) of acoustic power absorption in sunspot that could mimic the signature of vertical flows by producing this travel-time asymmetry. Moreover, Lindsey (2006), elaborating on previous works with D. C. Braun, argued that mass conservation requires large inflows inside the sunspot to compensate for any downflow below the sunspot photosphere. Despite detection of horizontal inflows in the shallow subsurface sunspot layers by Zhao, Kosovichev, & Duvall (2001), Lindsey (2006) believed that the mass conservation problem remained: the inflows inverted by Zhao et al. (2001) might be too slow to compensate for the rapid mass loss in downflows. The review by Gizon & Birch (2005) emphasized the lack of consistency between f- and p-mode results: as mentioned in section 4.1, f-mode data (Gizon et al. 2000) showed a clear outflow in and around the sunspot, in contradiction to the p-mode data of Zhao et al. (2001). Again, p modes are less sensitive to horizontal flows at the surface than f modes, which might explain the discrepancy. Other authors, and especially Woodard (1997), also demonstrated that acoustic power absorption by sunspots can explain a small part of the travel-time asymmetry (power absorption is not sufficient by itself to explain all the asymmetry though). The issue of whether or not the travel-time asymmetry is due —partially or totally— to down-

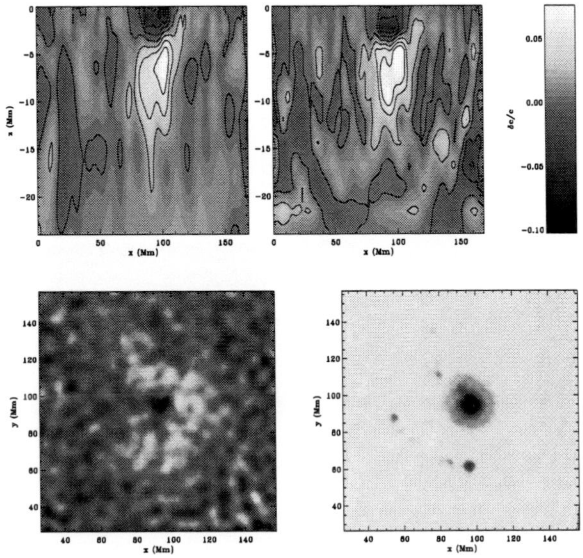

Figure 10. Inversion of the active region NOAA 8243 of 1998 June 20 with the Fresnel-zone (upper left and lower left panels) and the ray-path (upper right panel) kernels and the MCD algorithm. Upper panels: cut in the (x, z) plane of the inversion results. Contour lines have been overplotted to emphasize the structure of the sunspot. Lower right panel: image of the active region taken in the continuum at the solar surface. Lower left panel: cut in the horizontal plane at a depth of 3.3 Mm. Below the main sunspot (located at the center of the image), two finger-like structures are visible, that seem to connect the two pores visible on the lower right panel at 35 Mm from the center of the spot. The greyscale is the same for all the panels but the lower right one. From Couvidat et al. (2004).

flows is currently debated.

In parallel to this problem, Chou, Sun, & Chang (2000) and Chou & Duvall (2000) observed a dependence on frequency of the phase and group travel times in sunspots. They argued that this dependence can be explained by the strong magnetic field of the sunspot. They used time-distance analysis and another method, called acoustic imaging and which is close to acoustic holography, to obtain these results. Recently, Braun & Birch (2006) observed a similar frequency dependence of the travel times within active regions, using acoustic holography (see Figure 11). Couvidat & Rajaguru (2007) also observed this frequency dependence of acoustic wave travel times in active region, using the time-distance analysis (see Figure 12). Any strong frequency dependence could be an indication that the magnetic field inside an active region contaminates the measured travel times. Indeed, there does not seem to be any reason why flows or sound-speed perturbations should be strongly frequency dependent. Even if, in the absence of any perturbation, wavepackets with different frequencies have a slightly different upper turning point that could explain part of a frequency dependence, it is difficult to explain how this effect combined with a flow or a sound-speed perturbation could create the sign reversal in travel-time perturbations that is

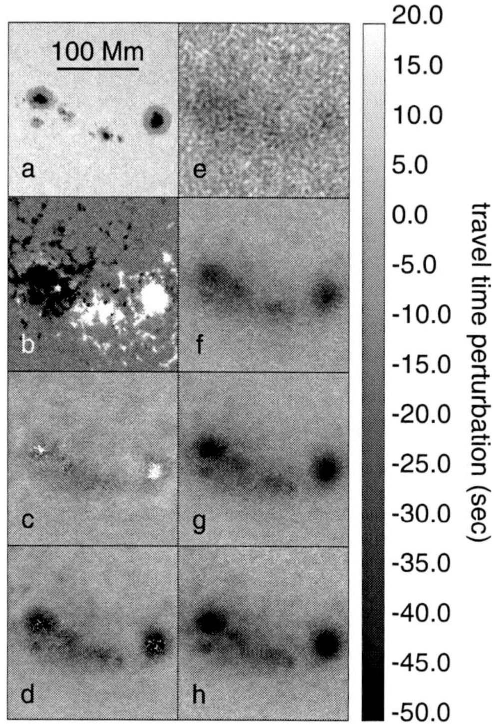

Figure 11. Panels (a): SOI/MDI continuum intensity; (b): line-of-sight magnetogram, (c) to (h): sample maps of the mean travel-time perturbation covering a portion of the region studied and showing sunspot group AR 9885. Panels (c) and (d) show travel-time maps for a specific phase-speed filter and a frequency filter centered at 3 and 4 mHz respectively. Panels (e) to (h) show the travel-time maps for another phase-speed filter and the frequencies 2, 3, 4, and 5 mHz respectively. Figure kindly provided by D. Braun & A. C. Birch. From Braun & Birch (2006).

observed in some parts of an active region: indeed for some specific Δ distances, $\delta\tau_{\mathrm{mean}}$ seems to shift locally from a positive value to a negative one when the frequency content of the wavepacket increases (see Figure 11 —panels c) and d)— and Figure 12). On the other hand, a magnetic field directly alters the dispersion relation of acoustic waves, by converting some of them to slow magneto-acoustic waves (see below). This alteration is clearly frequency dependent. Therefore, such a dependence in the travel times could theoretically be explained by an interaction of the wavefield with a magnetic field.

Similarly, Lindsey & Braun (2003) emphasized the changes in acoustic wave phases that occur when such waves cross strong magnetic field. They named this phenomenon "showerglass" effect (because it behaves like a showerglass scrambling the phases of light rays coming from the shower and preventing any clear observation of what, or who, is inside this shower). They argued that it makes it difficult to properly invert for the sound-speed perturbations and flow velocities below a sunspot. *A priori*, the interaction between the acoustic

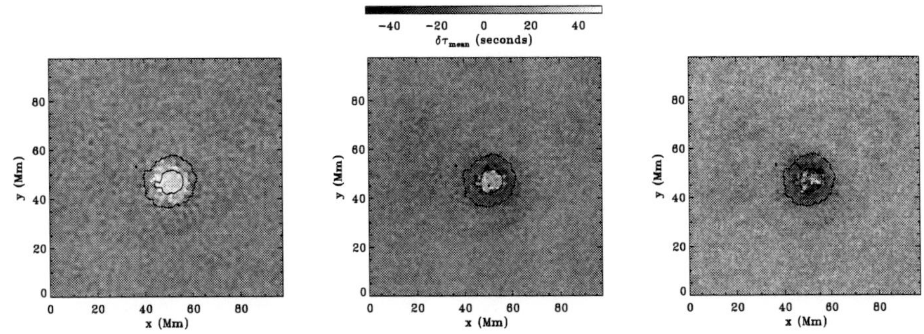

Figure 12. Travel-time maps $\delta\tau_{\text{mean}}(\mathbf{r}, \Delta)$ for the distance $\Delta = 11.6$ Mm and the active Region NOAA 9493. Left panel: with a frequency filter centered on 3 mHz; central panel: filter centered on 4 mHz; right panel: filter centered on 4.5 mHz. The greyscale is the same for all three panels. From Couvidat & Rajaguru (2007).

wavefield and the sunspot magnetic field occurs mainly in a thin layer where the sound and Alfvén speeds coincide (e.g. Cally 2007). Therefore, any inversion of time-distance data that does not account for the surface magnetic field effects could be contaminated mainly in the shallower layers (a few Mm below the surface). Thus the region of decreased sound speed immediately below the sunspot could be partly (or completely) due to a surface effect, instead of a thermal perturbation.

Another puzzling discovery was the anomaly in the seismic signature of sunspot penumbrae mentioned in Lindsey & Braun (2005). A phenomenon they refer to as penumbral acoustic anomaly. This anomaly appears like a phase perturbation for incoming waves: a ring-like structure is conspicuous on the phase maps of the "local ingression control correlation". Similarly, Couvidat & Rajaguru (2007) observed dubious ring-like structures on the travel-time maps of active regions and on the corresponding inversion results. These ring-like structures are mainly located in the penumbra of the sunspot (see Figure 13), and become more and more conspicuous when the frequency content of the acoustic wavepackets shifts to higher frequencies. Cally (2000) highlighted the importance of inclined magnetic fields: the non-vertical field in the penumbra enhances the acoustic power absorption compared to the vertical umbral field. Therefore, the penumbral acoustic anomaly might be due to the interaction of the penumbral field with the acoustic wavefield. Similarly, Jefferies et al. (2006) developed the concept of magnetoacoustic portals: low frequency waves which are evanescent in a non-magnetic atmosphere are able to propagate into the chromosphere through areas of strong and inclined magnetic fields. Schunker et al. (2005), Schunker & Cally (2006), and Schunker et al. (2007) explored the effects of an inclined magnetic field on the acoustic waves. They showed a dependence of the ingression phase of acoustic holography with azimuthal angle within a sunspot penumbra. Schunker et al. (2007) explained the frequency dependence they observed when analyzing their inclined magnetic field effect by reminding that lower frequency acoustic waves have an upper turning point that is located deeper than the one of higher frequency waves, and therefore the lower frequency modes are less affected by the surface magnetic fields.

Zhao & Kosovichev (2006) studied both the showerglass and the inclined magnetic

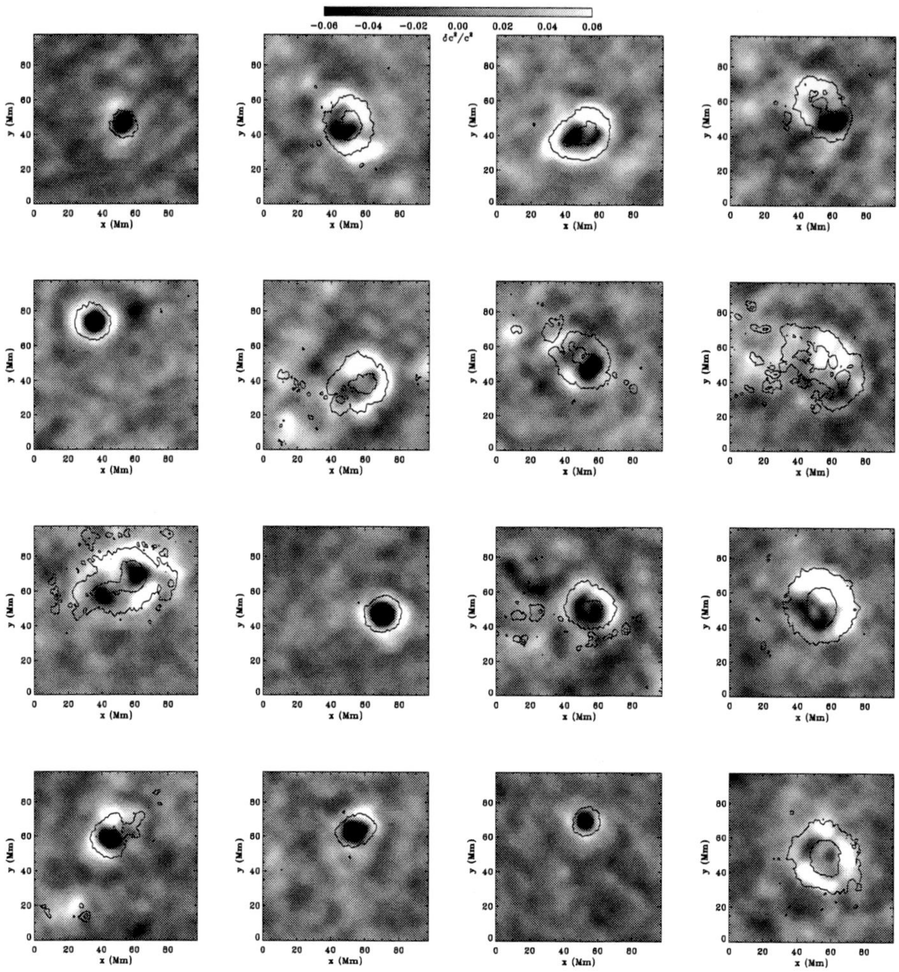

Figure 13. Inversion results $\delta c^2/c^2$: we show a horizontal cut at a depth of $z = -2.38$ to $z = -1.42$ Mm for 16 active regions observed with SOI/MDI. The greyscale is the same for every panel and was truncated to emphasize the bright rings. The boundaries of the umbra and penumbra of each sunspot are overplotted. A white ring of increased sound speed is clearly visible around most sunspot umbrae. From Couvidat & Rajaguru (2007).

field effects with time-distance analysis, to determine whether or not these effects observed with acoustic holography are also present in time-distance helioseismology. Indeed they observed an azimuthal dependence of travel times in a sunspot, but smaller than the results of Schunker et al. (2005). Their inversions of travel-time maps confirmed that the impact of inclined magnetic field seems mainly located in the shallow subsurface layers. They also corroborated the presence of a showerglass effect in time-distance helioseismology: the difference between ingoing and outgoing travel times exhibit a strong dependence on the magnetic field amplitude. Therefore, the travel-time asymmetry varies with the field strength. However, Zhao & Kosovichev (2006) claimed that all travel-time perturbations cannot be accounted for by the showerglass effect, and that some of these perturbations must

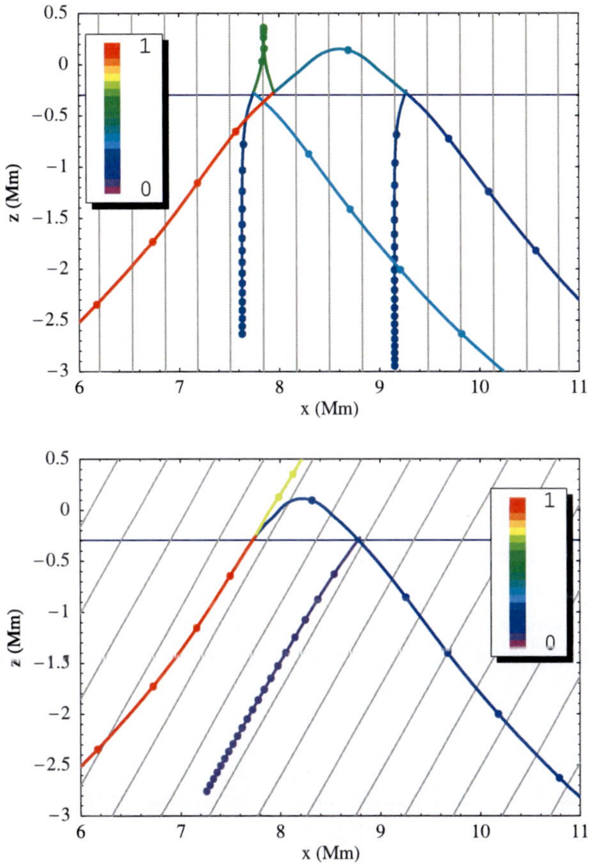

Figure 14. Ray-path diagrams in a vertical plane for 5 mHz acoustic waves launched horizontally from $x = 0$ $z = -5$ Mm in the presence of a 2 kG magnetic field which is vertical (top panel) and inclined at 30° (bottom panel). The colorscale shows the percentage of energy that goes from the incoming acoustic wave (in red) to the slow acoustic mode traveling into the atmosphere and to the fast acoustic wave traveling inside the Sun. The horizontal line indicates the level at which the sound speed is equal to the Alfvén speed. The background grey lines represent the magnetic field. Figure kindly provided by H. Schunker. From Schunker & Cally (2006).

be due to sound-speed perturbation and/or flows underneath the solar surface. Moreover, by using intensity maps instead of Doppler velocity maps, they showed that the inclined magnetic field effect does not exist in intensity observations. Therefore, they believe this effect might actually be due to a projection effect because the horizontal component of the oscillation velocity is neglected. They drew the conclusion that, even though surface effects undeniably contaminate the travel-time measurements, they do not alter the basic conclusions related to the sunspot structure.

So far we only mentioned the possible direct effects of a magnetic field on time-distance measurements, but other effects indirectly caused by the magnetic field might also alter the acoustic wave travel times: for instance, Brüggen & Spruit (2000) suggested that the

Wilson depression might be responsible for some phase shifts, and Gizon & Birch (2002) suggested that wave damping and source suppression in sunspots might affect the travel times. Moreover, an ongoing work by Hanasoge, Couvidat, Rajaguru, & Birch (*submitted to The Astrophysical Journal*) on the effect of inhomogeneously suppressed wave sources, confirms and expands the result that Gizon & Birch (2002) had analytically obtained: mean and difference travel-time shifts are observed in the absence of flows and sound-speed perturbations. This work is based on numerical simulations of the wavefield, mimicking the partial suppression of convection in sunspots because of the magnetic field. The travel-time shifts are due to the resulting wave source suppression, an effect that is not accounted for in traditional time-distance analysis. On top of such physical processes, another problem arises when dealing with Doppler data from sunspots: the Doppler velocity measurements inside active regions are notoriously inaccurate due to contamination by molecular lines, the splitting of the line by Zeeman effect, and the difference in the line formation process in the atmosphere above the sunspot compared to the quiet Sun (Rajaguru et al. 2007). Wachter, Schou, & Sankarasubramanian (2006) and Wachter, Rajaguru, & Bogart (2006) proposed a method to correct for the SOI/MDI Doppler velocity maps: they accounted for the line shape changes that occur in strong magnetic fields. However this method requires all the filtergrams, basic images taken by MDI at slightly different wavelengths, to be available. Therefore, with SOI/MDI the correction can only be applied to data taken during specific campaigns. With the upcoming HMI instrument (Helioseismic and Magnetic Imager) that should be launched in 2008 onboard the SDO spacecraft, we will have all the filtergrams readily available which will allow for a better calibration of the Doppler velocity maps in strong magnetic fields. Finally, Rajaguru et al. (2006) pointed out that the acoustic power reduction in sunspots, combined with the phase-speed filtering procedure, could be a source of systematic errors in the range of $5\% - 40\%$ in the measured travel-time anomalies. They suggested a correction procedure. All the effects mentioned in this section clearly point toward a contamination of travel times in sunspots by phenomena unrelated to flows and sound-speed perturbations.

In parallel to the experimental approaches to the interaction of magnetic field with the acoustic wavefield, a thorough modelling work has progressively emerged. Thus, Cally & Bogdan (1997) simulated in two dimensions the interaction of an incoming acoustic wave with a strong vertical but non-uniform magnetic field. They showed that the p mode is partially converted to a slow magneto-acoustic wave that propagates along the magnetic field lines. This slow magneto-acoustic wave behaves like an Alfv ́en wave away from the conversion zone. This partial conversion may explain the power absorption in sunspots. It occurs in a thin layer where the acoustic sound speed is close to the Alfv ́en speed. Bogdan et al. (2002) also used numerical simulations of nonlinear wave propagation in a 2D magnetic field and observed a similar wave conversion mechanism, although they modeled the solar atmosphere and not the subsurface layers. The simulations of Cally & Bogdan also showed that mode conversion is frequency dependent. However, their model was too simple to quantitatively retrieve the acoustic-wave absorption and phase shifts observed. Cally (2000) improved this model by replacing the vertical magnetic field with a spreading one, more similar to what is found in actual sunspots. This improvement enabled a quantitative comparison of the absorption of p modes by an active region with the helioseismic data. The good match between model and observation confirmed that an inclined magnetic field enhances the acoustic wave conversion to slow modes. Crouch & Cally (2003) found that the peak in acoustic power absorption is reached for a field inclined by 30° from the vertical. Cally, Crouch, & Braun (2003), using the observational data of Braun (1995), suggested that the increase in p-mode phase speed inside a sunspot is mainly due to magneto-acoustic conversion between the solar surface and a depth of about 1 Mm, rather than a thermal perturbation. They also demonstrated that the phase shift and power absorption of wavepackets inside a sunspot depends on the central frequency of this wavepacket. Figure 14 illustrates this acoustic mode conversion mechanism for vertical and inclined magnetic fields.

It seems that a definite answer to the impact of surface magnetic field on travel-time measurements will come from realistic numerical magneto-hydro-dynamic simulations of the solar upper layers. The hydro-dynamic part is already under construction and produces Doppler velocity datacubes that can be used to assess the validity of some time-distance hypotheses (e.g. Hanasoge et al. 2006; Hanasoge & Duvall 2007; Hanasoge, Duvall, & Couvidat 2007; Parchevsky & Kosovichev 2007; and Zhao et al. 2007). Addition of the magnetic field is a work in progress, whose completion will allow an estimation of the degree of contamination of travel times by this field. However, even with a theoretical understanding of the interaction between magnetic field and acoustic wavefield, it might still be difficult to produce sensitivity kernels for time-distance inversions: the Born approximation might still apply in presence of a strong magnetic field, but not very close to the solar surface (Gizon, Hanasoge, & Birch 2006). It seems that time-distance helioseismology has daunting issues to overcome.

5. Conclusion

Time-distance helioseismology produced impressive results and improved our knowledge of the physics of sunspots, with significant impacts both on our knowledge of the Sun and our understanding of the Sun-Earth relationship. It is a difficult field in which the influence of different phenomena are not easily disentangled, raising more questions each time a new result is obtained. The reality of the two-region sound-speed structure and the detected downflows below some active regions is debated. A consensus seems to emerge that travel times of wavepackets are affected by the strong magnetic field of sunspots, which contaminates the inversion results. However, different groups draw different conclusions regarding the degree of contamination. This emerging picture is both encouraging and daunting: the more we understand the physics of sunspots and the different phenomena that impact on the acoustic wave travel times, the more we will need to develop increasingly sophisticated tools to deal with them. With the upcoming launch of the SDO satellite and its HMI instrument, higher resolution Doppler velocity and vector magnetic field maps are expected that will allow to fully take advantage of the time-distance formalism. New discoveries should follow, but in the meantime we need to sort out the different influences that impact on the wave travel times.

Acknowledgments

This work was supported by NASA grants NNG05GH14G (MDI) and NNG05GM85G (LWS). I would like to thank all the physicists who permitted the reproduction of their figures: Aaron Birch, Doug Braun, Thomas Duvall, Alexander Kosovichev, Hannah Schunker, and Junwei Zhao. Moreover, I would like to thank Alexander Kosovichev for the additional comments regarding a first draft of this manuscript, and Hilda Bisangwa for correcting some typographical and grammatical errors.

References

[1] Backus, G. E.; & Gilbert, J. F. *Geophysical Journal of the Royal Astronomical Society* 1968, 16, 169-25

[2] Birch, A. C.; & Kosovichev, A. G. *Solar Physics* 2000, 192, 193-201

[3] Birch, A. C.; Kosovichev, A. G.; Price, G. H.; & Schlottmann, R. B. *Astrophysical Journal* 2001, 561, 229-232

[4] Birch, A. C.; Kosovichev, A. G.; & Duvall, T. L. *Astrophysical Journal* 2004, 608, 580-600

[5] Bogdan, T. J.; et al. *Astronomische Nachrichten* 2002, 323, 196-202

[6] Braun, D. C.; Duvall, T. L.; & LaBonte, B. J. *Astrophysical Journal* 1987, 319, 27-31

[7] Braun, D. C. *Astrophysical Journal* 1995, 451, 859-876

[8] Braun, D. C.; & Birch, A. C. *Astrophysical Journal* 2006, 647, 187-190

[9] Brüggen, M.; & Spruit, H. C. *Solar Physics* 2000, 196, 29-40

[10] Cally, P. S.; & Bogdan, T. J. *Astrophysical Journal* 1997, 486, 67-70

[11] Cally, P. S. *Solar Physics* 2000, 192, 395-401

[12] Cally, P. S.; Crouch, A. D.; & Braun, D. C. *Monthly Notices of the Royal Astronomical Society* 2003, 346, 381-389

[13] Cally, P. S. *Astronomische Nachrichten* 2007, 328, 286-291

[14] Chou, D. Y.; Sun, M. T.; Huang, T. Y. ; et al. *Solar Physics* 1995, 160, 237-243

[15] Chou, D. Y.; & Duvall, T. L. *Astrophysical Journal* 2000, 533, 568-573

[16] Chou, D. Y.; Sun, M. T.; & Chang, H.-K. *Astrophysical Journal* 2000, 532, 622-627

[17] Couvidat, S.; Birch, A. C.; Kosovichev, A. G.; & Zhao, J. *Astrophysical Journal* 2004, 607, 554-563

[18] Couvidat, S.; Gizon, L.; Birch, A. C.; Larsen, R. M.; & Kosovichev, A. G. *Astrophysical Journal Supplement Series* 2005, 158, 217-219

[19] Couvidat, S.; Birch, A. C., & Kosovichev, A. G. *Astrophysical Journal* 2006, 640, 516-524

[20] Couvidat, S.; & Rajaguru, S. P. *Astrophysical Journal* 2007, 661, 558-567

[21] Crouch, A. D.; & Cally, P. S. *Solar Physics* 2003, 214, 201-226

[22] D'Silva, S. *Astrophysical Journal* 1996, 469, 964-975

[23] D'Silva, S.; Duvall, T. L.; Jefferies, S. M.; & Harvey, J. W. *Astrophysical Journal* 1996, 471, 1030-1043

[24] Duvall, T. L.; Jefferies, S. M.; Harvey, J. W.; & Pomerantz, M. A. *Nature* 1993, 362, 430-432

[25] Duvall, T. L.; D'Silva, S.; Jefferies, S. M.; Harvey, J. W.; & Schou, J., *Nature* 1996, 379, 235-237

[26] Duvall, T. L.; & Gizon, L. *Solar Physics* 2000, 192, 177-191

[27] Duvall, T. L.; Birch, A. C.; & Gizon, L. *Astrophysical Journal* 2006, 646, 553-559

[28] Giles, P. In *Time-Distance Measurements of Large-Scale Flows in the Solar Convection Zone*, Ph.D. Dissertation, Stanford University, 1999

[29] Gizon, L.; Duvall, T. L.; & Larsen, R. M. *Journal of Astrophysics and Astronomy* 2000, 21, 339-342

[30] Gizon, L.; & Birch, A. C. *Astrophysical Journal* 2002, 571, 966-986

[31] Gizon, L.; & Birch, A. C. *Astrophysical Journal* 2004, 614, 472-489

[32] Gizon, L.; & Birch, A. C. *Living Reviews in Solar Physics* 2005, 2, http://solarphysics.livingreviews.org/Articles/lrsp-2005-6/

[33] Gizon, L.; Hanasoge, S. M.; & Birch, A. C. *Astrophysical Journal* 2006, 643, 549-555

[34] Hanasoge, S. M.; et al. *Astrophysical Journal* 2006, 648, 1268-1275

[35] Hanasoge, S. M.; & Duvall; T. L. *Astronomische Nachrichten* 2007, 328, 319-323

[36] Hanasoge, S. M.; Duvall, T. L.; & Couvidat, S. *Astrophysical Journal* 2007, *in press*

[37] Hansen, P. C. *Rank-Deficient and Discrete Ill-Posed Problems*; SIAM monograph: Philadelphia, PA, 1998

[38] Harvey, J. W.; et al. *Science* 1996, 272, 1284-1286

[39] Hughes, S. J.; Rajaguru, S. P.; & Thompson, M. J. *Astrophysical Journal* 2007, 627, 1040-1048

[40] Jefferies, S. M.; et al. *Astrophysical Journal* 2006, 648, 151-155

[41] Jensen, J. M.; Jacobsen, B. H.; & Christensen-Dalsgaard, J. In *Structure and Dynamics of the Interior of the Sun and Sun-Like Stars*; Korzennik, S. G. & Wilson, A.; Ed.; ESA, 1998, 635-640

[42] Jensen, J. M.; Duvall, T. L.; Jacobsen, B. H.; & Christensen-Dalsgaard, J. *Astrophysical Journal* 2001, 553, 193-196

[43] Jensen, J. M.; Duvall, T. L.; & Jacobsen, B. H. In *Local and Global Helioseismology: The Present and Future*; Sawaya-Lacoste, H.; Ed.; ESA, 2003, 315-318

[44] Jensen, J. M.; & Pijpers, F. P. *Astronomy & Astrophysics* 2003, 412, 257-265

[45] Kosovichev, A. G. *Astrophysical Journal* 1996, 461, 55-57

[46] Kosovichev, A. G.; & Duvall, T. L. In *SCORe'96 : Solar Convection and Oscillations and their Relationship*; Pijpers, Christensen-Dalsgaard, & Rosenthal; Eds.; Kluwer Academic Publishers; 1997, 225, 241-260

[47] Kosovichev, A. G.; & Zharkova, V. V. *Nature* 1998, 393, 317-318

[48] Kosovichev, A. G. *Journal of Computational and Applied Mathematics* 1999, 109, 1-39

[49] Kosovichev, A. G.; Duvall, T. L.; & Scherrer, P. H. *Solar Physics* 2000, 192, 159-176

[50] Kosovichev, A. G. In *Solar MHD Theory and Observations: A High Spatial Resolution Perspective*; Uitenbroek, Leibacher, & Stein; Eds.; ASP Conference Series; 2006; 354, 154-160

[51] Leighton, R. B.; Noyes, R. W.; & Simon, G. W. *Astrophysical Journal* 1962, 135, 474-499

[52] Lindsey, C.; et al. *Astrophysical Journal* 1996, 470, 636-646

[53] Lindsey, C.; & Braun, D.C. In *Local and Global Helioseismology: The Present and Future*; Sawaya-Lacoste; Ed.; ESA Conference Proceedings, Noordwijk, Netherlands, 2003, SP-517, 23-28

[54] Lindsey, C.; & Braun, D. C. *Astrophysical Journal* 2005, 620, 1107

[55] Lindsey, C. In *SOHO 18/GONG 2006/HELAS I, Beyond the Spherical Sun* ; Fletcher, K.; Ed.; ESA SP-624; 2006; 4L

[56] Marquering, H.; Dahlen, F. A.; & Nolet G. *Geophysical Journal International* 1999, 137, 805-815

[57] Paige, C. C.; & Saunders, M. A. *ACM Transactions on Mathematical Software* 1982, 8, 43-71

[58] Parchevsky, K. V.; & Kosovichev, A. G. *Astrophysical Journal* 2006, *submitted*

[59] Parchevsky, K. V.; & Kosovichev, A. G. *Astrophysical Journal* 2007, *in press*

[60] Parker, E. N. *Astrophysical Journal* 1979, 230, 905-913

[61] Rajaguru, S. P.; Birch, A. C.; Duvall, T. L.; Thompson, M. J.; & Zhao, J. *Astrophysical Journal* 2006, 646, 543-552

[62] Rajaguru, S. P.; Sankarasubramanian, K.; Wachter, R.; & Scherrer, P. H. *Astrophysical Journal* 2007, 654, 175-178

[63] Schunker, H.; Braun, D. C.; Cally, P. S.; & Lindsey, C. *Astrophysical Journal* 2005, 621, 149-152

[64] Schunker, H.; & Cally, P. S. *Monthly Notices of the Royal Astronomical Society* 2006, 372, 551-564

[65] Schunker, H. *Astronomische Nachrichten* 2007, 328, 292-297

[66] Solanki, S. K. *Astronomy & Astrophysics Review* 2003, 11, 153-286

[67] Wachter, R.; Schou, J.; Sankarasubramanian, K. *Astrophysical Journal* 2006, 648, 1256-1267

[68] Wachter, R.; Rajaguru, S. P.; & Bogart, R. S. In *SOHO 18/GONG 2006/HELAS I, Beyond the spherical Sun*; Fletcher; Ed.; 2006, 48 51

[69] Woodard, M. F. *Astrophysical Journal* 1997, 485, 890-894

[70] Zhao, J.; Kosovichev, A. G.; & Duvall, T. L. *Astrophysical Journal* 2001, 557, 384-388

[71] Zhao, J.; & Kosovichev, A. G. *Astrophysical Journal* 2003, 591, 446-453

[72] Zhao J. In *Inference of Solar Subsurface Flows by Time-Distance Helioseismology* , Ph.D. Dissertation, Stanford University, 2004

[73] Zhao, J.; & Kosovichev, A. G. *Astrophysical Journal* 2006, 643, 1317-1324

[74] Zhao, J., Georgobiani, D.; Kosovichev, A. G.; Benson, D.; Stein, R. F.; & Nordlund, A. *Astrophysical Journal* 2007, 659, 848-857

[75] Zharkov, S.; Nicholas, C. J.; & Thompson, M. J. *Astronomische Nachrichten* 2007, 328, 240-244

In: Solar Physics Research Trends
Editor: Pingzhi Wang, pp. 165-194
ISBN: 978-1-60021-987-0
© 2008 Nova Science Publishers, Inc.

Chapter 4

THE QUASI-BIENNIAL OSCILLATIONS IN THE EQUATORIAL STRATOSPHERE: SEASONAL REGULARITIES, DEPENDENCE ON THE SOLAR UV FLUX, AND RELATION TO OZONE DEPLETION IN ANTARCTICA

***I. Gabis*[*] *and O. Troshichev*[**]**
Arctic and Antarctic Research Institute, St. Petersburg, Russia

Abstract

The term Quasi-Biennial Oscillation (QBO) was designated to describe the regular reversals of zonal winds in the equatorial stratosphere occurred with periodicity about 28 months. The phase of QBO cycle, East (E) or West (W) is determined by direction of the zonal winds. Although the wind QBO is mainly a tropical circulation feature, its effects are displayed well beyond the equatorial region. Indeed, the year-to-year fluctuations in the extratropical atmosphere are controlled by QBO, which influence is observed in such parameters as temperature, the total ozone distribution, and others. The most important thing is the well known at present QBO property to modulate the solar activity impact on the stratosphere and troposphere.

Today the dissipation of upward propagating equatorial waves usually examined as a theoretical mechanism for QBO generation. So, it is generally accepted that the proximity of the average QBO period to the double annual one being merely statistical result, whereas the actual period varies from 19 to 36 months. As a result of the high variability and elusive relation to the seasonal cycles, the QBO phase duration seemed to be unpredictable.

A new method for the investigation of the QBO cycle evolution is presented in this chapter. The analysis of the height wind profiles makes it possible to reveal new regularities in the equatorial stratosphere wind reversals. The stagnation stage of descending easterlies always starts in solstice, either in December–January or in June–July, and always completes by the equinox being of quantized duration in different QBO-cycles (about 3, 9, or 15 months). So the length of the complete QBO-cycle, if defined as a period between the

[*] E-mail address: gabis@aari.nw.ru
[**] E-mail address: olegtro@aari.nw.ru

inceptions of the successive stagnation stages, depends on their durations and may be equal to 24, 30, or 36 months. Consequently, the full QBO cycle length can be predicted as soon as end of the stagnation stage is found while observing the zonal wind transformation.

In turn the duration of the stagnation stage depends on the intensity of solar UV flux. The short stalling period (3–4 months) is observed under condition of the high level or steady increase of the UV irradiance during the first equinox in course of the partial QBO cycle. If the UV flux is low or decreases during the proper equinox, the easterly winds do not descend from the upper layer to the lower stratosphere, and the stalling period length at about 20-40 hPa increases to 9–10 or 15–16 months. The intensity of the spring ozone depletion within Antarctic polar vortex is thought to be also dependent on the phase of the equatorial QBO cycle. It is shown the relation of the season regularities of the equatorial QBO-circulation to the interannual fluctuations and the intensity of total ozone (TOZ) destruction over Antarctica during "ozone hole" phenomenon.

Key words: equatorial stratosphere, quasi-biennial oscillations, mean zonal winds, height wind profiles, solar UV irradiance, Antarctic ozone hole.

1. Introduction

Investigation of the solar variability effects on the Earth's atmosphere and, hence, on the weather and climate is one of the key tasks at present. The topic is of long history. Considerable advances in study of the solar activity impact on the troposphere and stratosphere have been made after the finding of K. Labitzke [*Labitzke, K., 1987*], that the character of the solar-atmosphere relations depends to a great extent on the quasi-biennial oscillation (QBO) of zonal winds in equatorial stratosphere. The atmospheric parameter (pressure, temperature, ozone and others) showed a clear correlation/anticorrelation with solar activity when the experimental data were grouped according to the phase of the QBO-cycle [*Labitzke, 1987, 2005; Labitzke and van Loon, 1988, 2000; Chanin et al., 1989; Besprozvannaya et al., 1997; van Loon and Labitzke, 1999, 2000; Georgieva et al., 2005*]. The QBO-phases, East (E) or West (W), were defined by the direction of the zonal wind in the tropical stratosphere.

The QBO is a fundamental property of the equatorial stratosphere circulation. Beginning with first regular measurements of the zonal wind speed in tropical stratosphere in 1953 a long series of observation were accumulated. It made it possible to study the QBO-cycle behavior in detail [*Reed, 1961; Veryard and Edbon, 1961; Naujokat, 1986; Baldwin, et al., 2001 and references in*]. The QBO-phenomenon is observed in the tropical band within a latitude range of about ±12 degrees. The QBO-phase change occurs practically simultaneously in the near-equatorial region irrespective of geographical longitude. It means that the phase identification can be made by the wind observations at any equatorial station. At present the measurements at station Singapore (01°22'N, 103°55'E) are used as a rule. Reversal of the zonal wind occurs in the lower and middle stratosphere from about 70 hPa (~18.5 km) to the ~3 hPa (~40 km) level, the maximum of wind speed (~30 m/s for easterly and ~15 m/s for westerly) being observed at 15-20 hPa.

The full cycle of the wind reversal, from easterly to westerly and back to easterly again, has the average period about 28 months [*Naujokat, 1986; Baldwin, et al., 2001*]. Solely easterly or westerly wind in the entire height region from 70 to 10 hPa occurs rarely: over the 50-year period (1953-2003) the westerly (easterly) at all heights is observed in 3% (0.3%) of time. The oppositely directed winds at higher and lower heights present a common phenomenon. Zonal winds blowing in one direction descend in altitude with time being replaced by winds blowing in the opposite direction. As a result, the time of the wind reversal and, accordingly, the beginning and duration of QBO-phases depends on the altitude. This fact gave rise to the problem of a proper choice of the pressure level for determination of the QBO-phase, because it was unclear a priory which QBO level can be responsible for the extratropical effects. To define the QBO-phase during any given month the different studies choose the various levels in range of 20-50 hPa, or some combination of levels [*Lait, et al., 1989*] or representation of the wind QBO by first two EOFs (empirical orthogonal functions) of monthly wind observations [*Wallace, et al., 1993; Baldwin, 1998*]. It turned out that the correlations involving extratropical parameters depend on level chosen to define the QBO-phase. In this connection it would be emphasized that the phase of QBO-cycle determined by wind direction at a certain altitude level corresponds to quite different height structure of equatorial circulation for different moments of the same QBO-phase. Furthermore, the circulation structures prove to be opposite at the beginning and end of the same QBO-phase.

Altitude of the wind reversal (zero line between the winds of opposite direction) is denoted as a shear zone – easterly or westerly shear zone, when the corresponding wind is higher of the zero line altitude. The time of QBO-phase changes and, consequently, the duration of alternating QBO-phases at given height level is evidently related to rate of the shear zones descending. Analysis of observations shows that westerly shear zones descend more evenly and quickly than easterly one. The later process is frequently slowed in the layer of about 30-40 hPa [*Naujokat, 1986; Dunkerton and Delisi, 1985, 1997; Dunkerton, 1990; Kinnersley and Pawson, 1996; Baldwin, et al., 2001*]. The true rate of descending of the easterly and westerly wind regimes during particular QBO period can significantly differs from the mean rate (~1 km/month). Experimental and model investigations revealed the dependence of the shear zone descending rate and the QBO-phase duration on the 11-year solar cycle [*Soukharev and Hood, 2001; McCormack, 2003; Pascoe 2005*]. Hence the process of wind descending may be influenced by the variability of the solar flux. However, this dependency was vanished during the 1990`s [*Pascoe et al., 2005*]. As was shown in [*Gabis and Troshichev, 2006*], the QBO-phase duration is affected by the solar UV irradiance intensity, but this influence is crucial only at the certain stages in evolution of the wind QBO-cycle.

Up to now the upward propagating tropical waves dissipation is usually examined as a mechanism for QBO generation [*Holton and Lindzen, 1972; Baldwin, et al., 2001 and references in*]. The theoretical QBO models presume that the troposphere wave sources are constant in time. So, it is generally accepted that the QBO period is not controlled by the year or half-year cyclicity, and consistency of the average QBO period to the double annual interval is examined as merely statistical result, the actual period being varied from 19 to 36 months. Nevertheless, there was some experimental evidence that the seasonal cycle might influence on the timing of QBO-phase changes. It was suggested [*Dunkerton and Delisi, 1985, 1997; Dunkerton, 1990*] that the QBO is modulated, but not exactly synchronized, by the seasonal cycle. Despite of the comprehensive information about the main basic morphological features of the equatorial wind QBO the forecast of the moment of the QBO-phase change remains to be unresolved and the problem of the QBO cycle duration remained unpredictable.

Meanwhile, the strong seasonal regularities of the QBO-cycle evolution and relation of the QBO-cycle length to the solar UV flux intensity was shown in the recent publications of authors [*Gabis and Troshichev, 2004, 2005, 2006*]. The revealed regularities can be used to predict the full length of the current QBO cycle and hence the onset of the next one. In the present chapter we summarize and update these earlier results. We also demonstrate relation of the revealed season regularities in the equatorial QBO-circulation to the interannual fluctuations and to the descending trend of total ozone (TOZ) over Antarctica during "ozone hole" phenomenon.

2. Data and Method of the Analysis

Our analysis is based on the wind speed measurements performed at the following equatorial stations: Canton Island (02°46'S, 171°43'W) for January 1953 – August 1967, Gan/Maledive Islands (00°41'S, 73°09'E) for September 1967 – December 1975, and Singapore (01°22'N, 103°55'E) after January 1976. Sets of the height profiles of the monthly mean zonal wind component in layer between 70 and 10 hPa were used to reveal the regularities in the wind changes during QBO-cycle evolution.

The Figure 1 presents, as an example, a transformation of the wind height profiles from August 2004 to April 2007. Positive and negative speeds correspond to the westerly and easterly, respectively. The each plot in Figure 1 contains profiles for two successive months, the previous and current months being shown by curves with solid and open circles, respectively. One can see that the westerly wind is observed from 70 (~18.5 km) to almost 10 hPa (~31.2 km) level in August 2004, at 10 hPa the wind speed U=0.0 m/s. The easterly wind was appeared in September 2004 at altitudes 10-20 hPa and then after four months came down to altitude nearly 30 hPa (~23.5 km). So the mean rate of the easterly shear zone descending during this period was ≈1.9 km/month. From December 2004 to March 2005 the altitude of the shear zone remained approximately invariant (in December, January, February and March at 30 hPa U = 1.4, -1.8, 1.8 and -0.9 m/s, respectively). The westerly and easterly winds were observed at altitudes lower and higher ~30 hPa, respectively. In April 2005 the descent of the easterly shear was resumed and the easterly wind was appeared below 30 hPa, firstly at 35 hPa, then at 40 hPa (in May 2005), and so on. In September 2005 the easterly wind was observed practically in the entire altitude range from 70 to 10 hPa (the wind speeds at 10 and 70 hPa was equal 0.1 and 0.0 m/s, respectively). Consequently, the mean rate of the easterly shear zone descending was ≈0.8 km/month during the period from March 2005 to September 2005. The averaged rate of the easterly shear zone descent for the full period from August 2004 to September 2005 was ≈ 1.0 km/month, but for all that the easterly shear zone during three-four months in December 2004 – March 2005 remained at constant altitude near 30 hPa.

In September 2005 westerly shear was observed at 10 hPa (U=0.1 m/s) and then the westerly wind moved down in time. Consequently W-phase was observed in the altitude range 10-12 hPa in October, at 10-15 hPa in November, at 10-20 hPa in December and so on. In period from October 2005 to August 2006 the westerly came down to the lower layer (70 hPa). It is significant that the westerly shear zone demonstrated the permanent descending without stopping in contrast to the easterly shear zone. The mean rate of the westerly shear zone descent for the period from September 2005 to August 2006 amounted to ≈1.1 km/month.

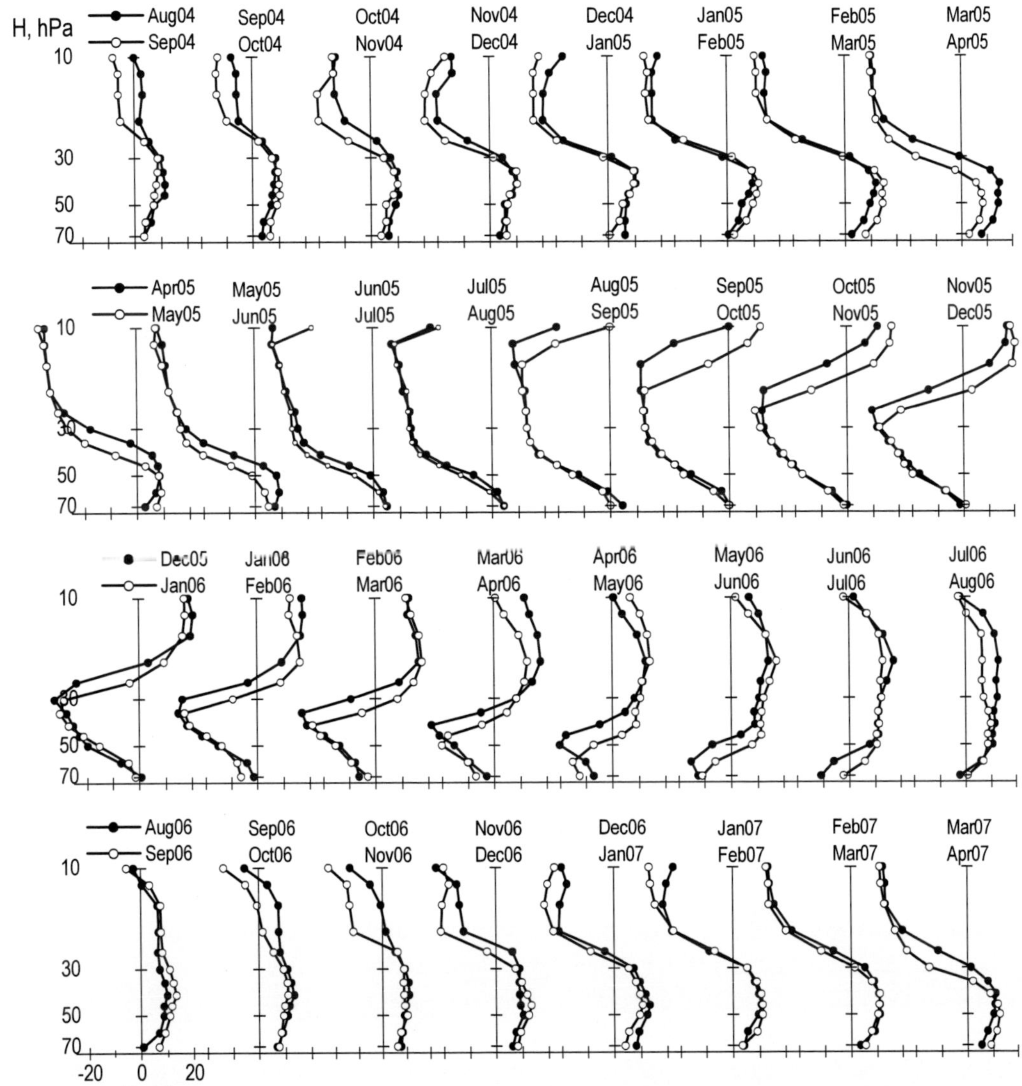

Figure 1. Changes in vertical profiles of the mean zonal wind for the period from August 2004 to April 2007. The each plot contains profiles for two successive months, the current and previous months being marked by open and solid circles, respectively. Positive and negative speeds correspond to the westerly and easterly, respectively.

Commencement of the next easterly shear zone descending occurred in June-July 2006 (the wind speed at 10 hPa level was equal to 1.9 m/s and -2.1 m/s, respectively). Then the easterly wind appeared at the progressively lower levels in the course of event. The easterly shear descending was continued till December 2006. The rate of easterly shear descent during period from June to December 2006 averaged to about ≈1.2 km/month. From December 2006 to March 2007 the changes of height wind profiles near the shear zone altitude were again insignificant, as in previous case, – the differences between wind speed for two adjacent months were below ~5 m/s/month. Hence the altitude of the shear zone remained approximately invariant, but in this case the easterly shear zone halted at slightly higher level than in previous case. The interpolation between measurements at 25 and 30 hPa levels gives the value about 27-29 hPa for the zero wind speed altitude. The wind profiles with easterly in upper layer above the stalling region and westerly lower of the stalling region were observed during three–four months in December 2006 – March 2007. In April 2007 the easterly shear descent was resumed like that in April 2005.

As Figure 1 shows, the QBO phase, determined by wind direction at the certain level, corresponds to the quite different height structure in the equatorial circulation for different periods of the same QBO-phase, and the circulation structures at the beginning and end of the same QBO-phase occurred to be opposite. For example, if the QBO-phase is defined by the wind direction at 40 hPa level, the E-phase starts in May 2005 during descending easterly and terminates in April 2006 during descending westerly. Hence, the E-phase beginning corresponded to westerly in the layer lower ~40 hPa, and to easterly in layer higher ~40 hPa; however, the height wind structure was opposite for the same E-phase end: easterly (westerly) in the layer lower (higher) 40 hPa.

It will be shown in the next section that the above example for 2004-2007 demonstrates the regularly repeating features of the easterly shear behavior. Irrespective of the time for the easterly shear location at 10 hPa level, the three-four months stage of stagnation is observed during the interval from the solstice (December-January) to the first equinox (March-April), and then the easterly shear resumes the descent to the lower stratosphere.

3. Results and Discussion

3.1. Three Types of the QBO-Cycle Evolution SCENARIOS

Analysis of the height profiles for the monthly mean zonal wind during the 55-years period of measurements (1953–2007) makes it possible to conclude that there are three types of the QBO-cycle evolution scenarios. The main peculiarity, which defines this classification, is duration of the stalling period of the easterly shear zone descent.

3.1.1. 24-Month Scenarios

The ten cases with the stalling period, lasting three-four months, took place within the interval from January 1953 to December 2006. It is very likely that just another stalling period of the same duration already started in the beginning of 2007 (see section 3.4). Figure 2 presents the wind profiles occurred from July-August of preceding year to July-August of the subsequent year during ten events. Like Figure 1, each plot in the horizontal row, except the central one, contains profiles for two successive months, the central plot being contained four profiles for interval from December of the preceding year to March of the subsequent year. The December and March profiles are marked by solid and open circles, respectively, January and February profiles are shown by thin lines. One can see that the nearly full cycle of easterly shear zone descent from ~10 hPa to ~70 hPa is completed during every one year period.

During July-December of the preceding years the E-shear moves down from upper layer to ~20-30 hPa, and during March-August of the subsequent years the E-shear descends to the lower stratosphere. In the middle of the examined periods during three-four months (December-March) a spectacular stagnation stages are observed in majority cases. In seven cases of ten the shear height changes not more 5 hPa for December-March. In 1955-1956, 1959-1960 and 1971-1972 the height during December-March changes by 11, 15 and 6-7 hPa, respectively, however before and after this period the shear descent magnitude was higher. Taking into account this repeated regularity, we can say about the short stalling period in process of the easterly shear zone descent.

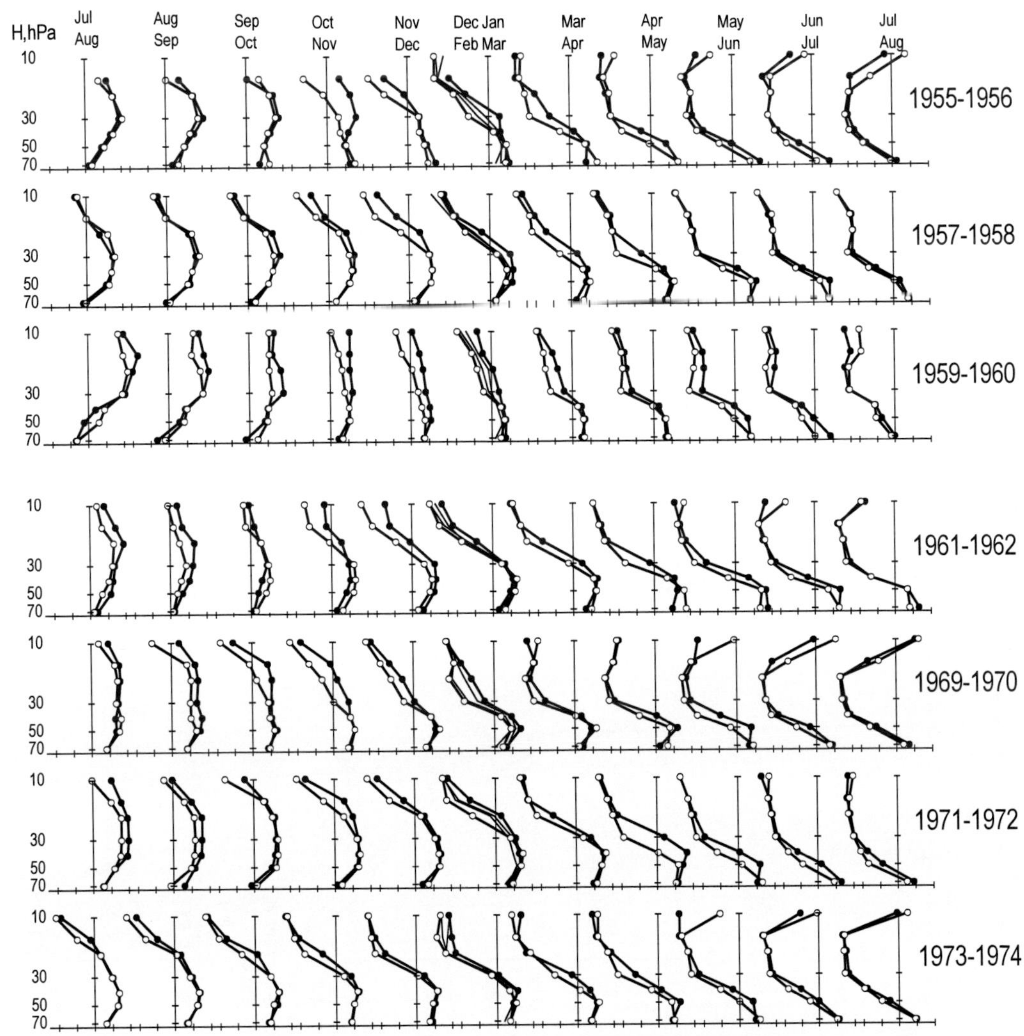

Figure 2. Continued on next page.

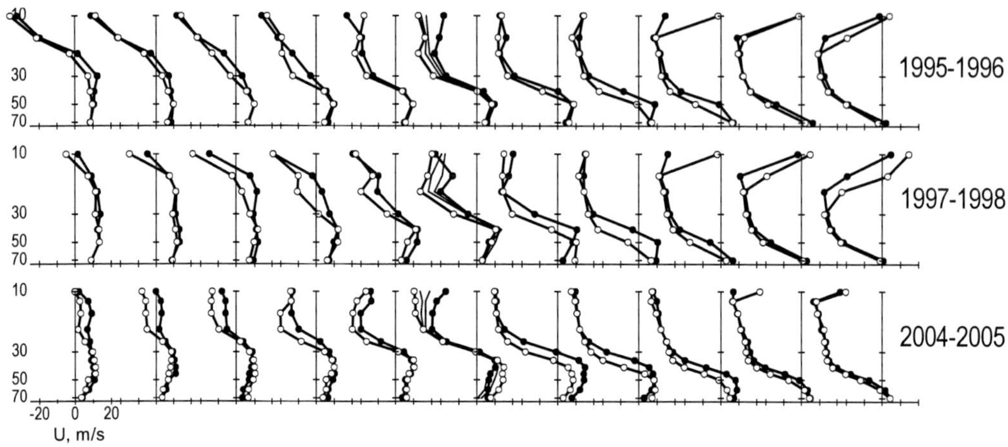

Figure 2. Transformation of height wind profiles nearby the “short” stagnation stage for the 24-months scenarios of QBO-cycle. Shown are ten one year periods from July-August of preceding year to July-August of the subsequent year. The proper months and years are remarked at the upper and right panels of figure, respectively. Each plot in the horizontal row, except the central one, contains profiles for two successive months (the current and previous months being marked by open and solid circles, respectively), the central plot being contained four profiles for interval from December of the preceding year to March of the subsequent year. The December and March profiles are marked by solid and open circles, respectively, January and February profiles are shown by thin lines.

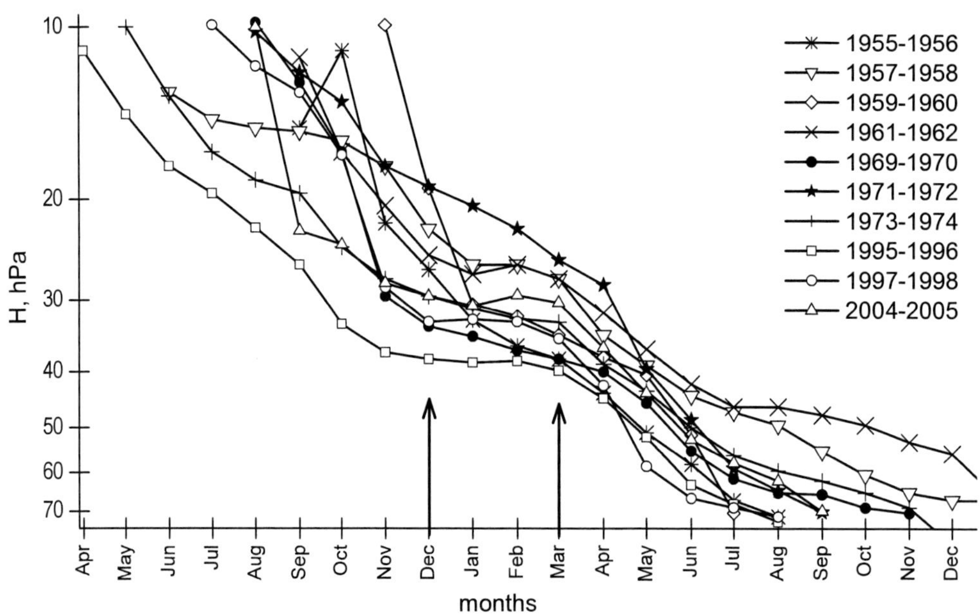

Figure 3. Time-height relationships for ten events of the easterly shear zone descending presented in Figure 2. Each curve illustrates the time position of the zero wind speed altitude for certain E-shear descent, the height interpolation between measurements at the levels, denoted in Figure 2 by solid or open circles, being used. The vertical arrows confine the stalling periods during December-March.

Figure 3 shows the time-height relationship for all ten cases of the easterly shear zone descending presented in Figure 2. Each curve illustrates the time position of the zero wind speed altitude for certain E-shear descent, the height interpolation between adjacent measurements being used. One can see that the shear descent onset, or the 10 hPa level crossing, may be observed within the time interval from April to November. The shear descent end, or the approach to 70 hPa level, may occur within the time interval from July to December of the next year, or later. During the periods from the December of the preceding years to the March of the subsequent years the E-shear is placed at height range 20-40 hPa and demonstrated insignificant altitude variations in each particular case, as mentioned above.

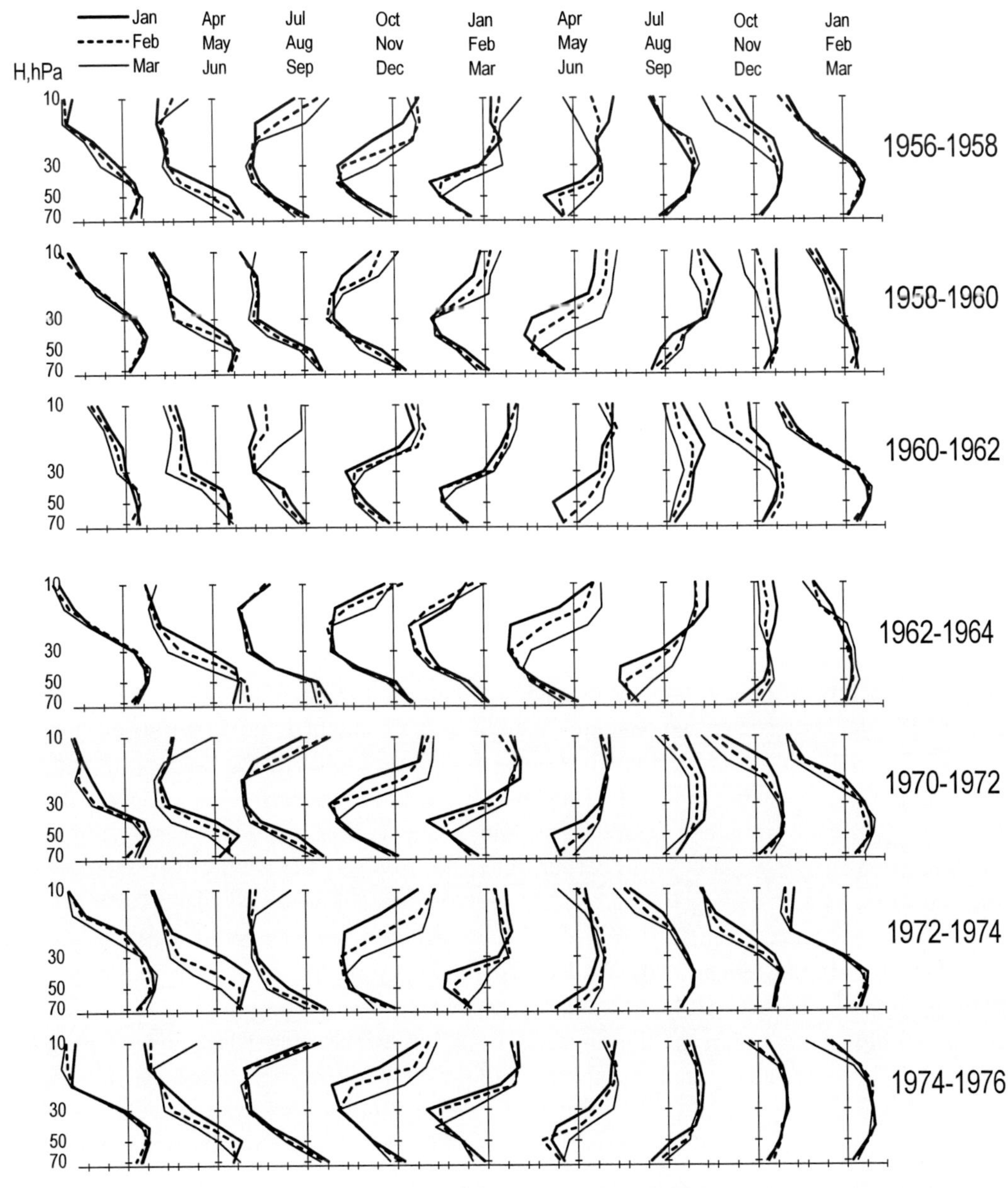

Figure 4. Continued on next page.

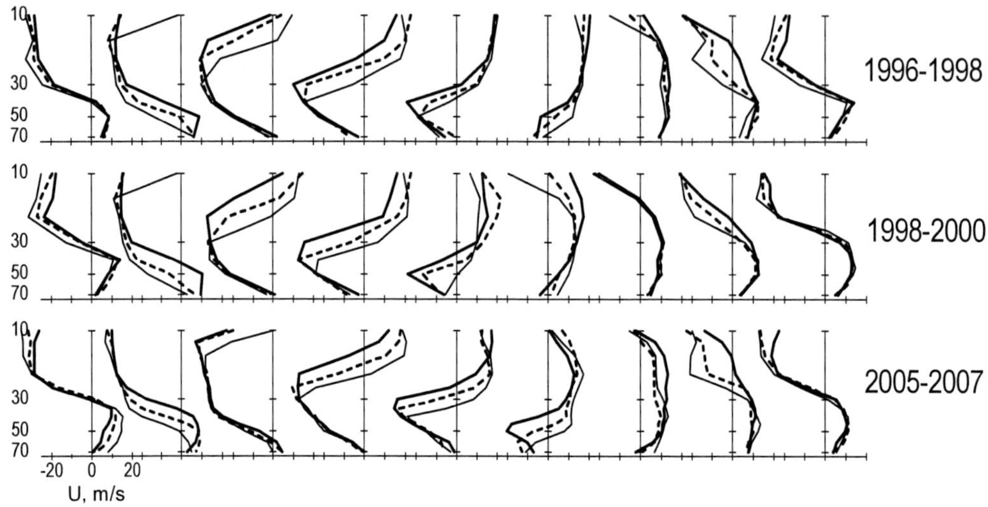

Figure 4. Transformation of height wind profiles for the 24-months scenarios of QBO-cycle. Shown are ten 27-month periods from the stagnation stage onset in January to the stagnation stage of subsequent QBO-cycle. The proper months and years are remarked at the upper and right panels of figure, respectively. Each plot contains three monthly profiles. The first, middle and last profiles on each plot are shown by the hard, dash, and thin line, respectively.

Nevertheless, the zero wind speed altitude in the certain month varies for different E-shear cases, as a result, the zonal wind reversal at the fixed altitude (in range 20-40 hPa) falls on different months for different E-shear cases, depending on the exact height for stagnation stage. It is the reason that the distinct seasonal effect is not revealed in the standard method of the QBO phase derivation, when E-phase or W-phase is determined by the wind direction on fixed level. Figure 3 provides the following very important result: in spite of difference in onsets of E-shear descent at 10 hPa, the stagnation stage starts in solstice (December-January) in all cases, and a new descent resumes in the first succeeding equinox (March-April).

Ten events displayed in Figure 2 are reconstructed in Figure 4 in such a manner that every event starts from the stagnation stage in January-March. Each plot contains three monthly profiles and hence demonstrates the changes of height wind structure during one seasons. The first, middle and last profiles on each plot are shown by the hard, dash, and thin line, respectively. One can see that the stagnation stage in all ten events was observed during one season (three-four months), from the December-January solstice to the nearest equinox in March-April, as it was already noted earlier. After this equinox the wind profile changes are observed lower of the stalling region and the easterly shear descent is finished near 70 hPa within the time interval from July to December in different months for different cases. It is significant that simultaneously with the E-shear arrival to the lower layer the easterly wind speed in higher layer, near 10 hPa, decreases, evidencing the westerly acceleration. The W-shear at 10 hPa starts within the time interval from July to November-December during the first year, in some events before E-shear arrival to the 70 hPa level. The descent of W-shear is completed during four-five seasons, from the onset in June-July of the first year to nearly September-October equinox of the second year. At the same time new E-shear starts to move down from level near 10 hPa in the June-July solstice of the second year and approaches, in all but one case, to nearly the 20-35 hPa level in December-January of the second year. The wind profile changes after the December show insignificant variations near altitude of zero wind speed during some months. Hence the stagnation stage occurred again. Consequently, the stagnation stage was observed during one season, and the full cycle of wind profile changes between the beginnings of two consecutive stagnation stages continues eight seasons. We identify this variant as "24-month scenario" of QBO-cycle evolution.

3.1.2. 30-Month Scenarios

As Figure 4 shows, sometimes the 24-month QBO-scenarios are observed one after other successively. The series of four, three and two repetition of 24-months scenario occurred in 1956-1963, 1970-1975 and 1996-2000, respectively. In other periods (1953-1955, 1964-1969, 1976-1995 and 2000-2004) the wind profile changes demonstrated the larger duration of the stalling period than in 24-month scenario. Figure 5 represents five cases of the QBO-cycles with the stagnation stage which started in December-January solstice. As in the Figure 4 every plot contains three profiles, showing the of height wind structure changes during one season, the first, middle and last profiles being marked on each plot by the hard, dash, and thin line, respectively.

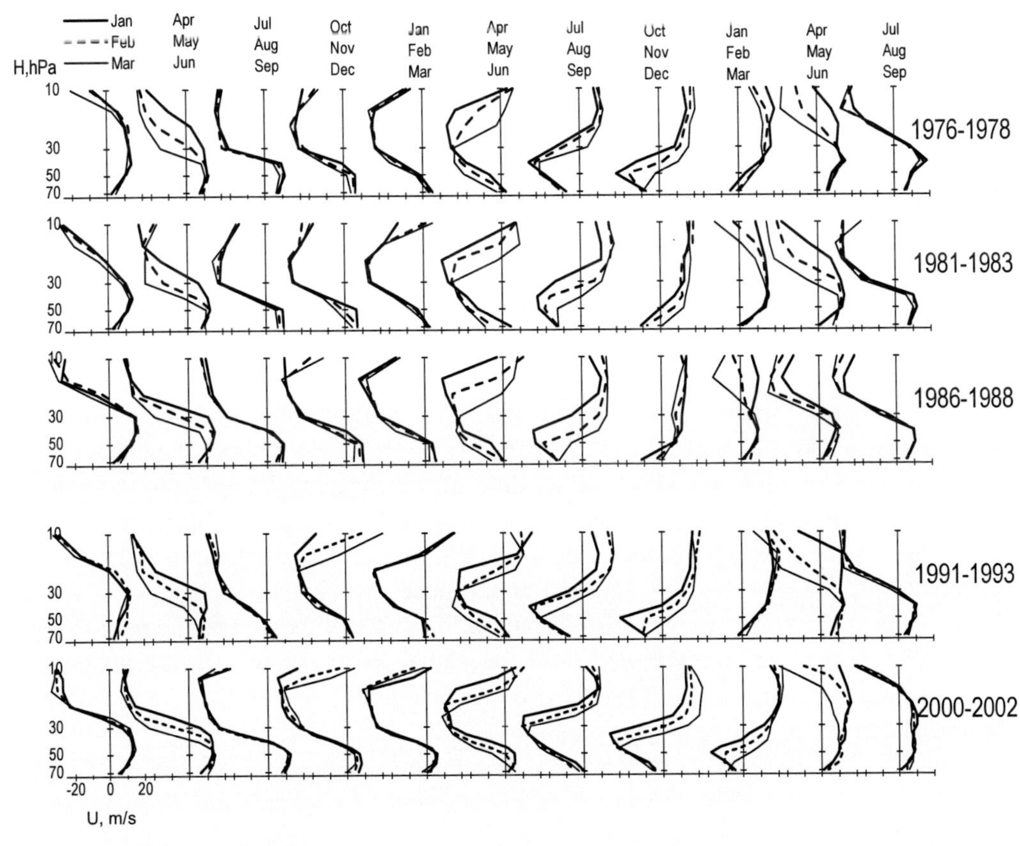

Figure 5. The same as Figure 4, but for January 30-month scenarios. Shown are five 33-month periods from the stagnation stage onset in January to the stagnation stage of subsequent QBO-cycle.

The first plot in each row illustrates insignificant changes in the wind profiles near the easterly shear zone. In the first event (1976-1978) the altitude of zero wind speed was abnormally high (about 13 hPa), but other four cases the zero wind altitude was rather typical (20-28 hPa). The first equinox (March–April) was followed by the slight descent of the easterly shear zone during three-four months. In the next solstice the stopping period occurs again: changes of the wind profile were absent during July-September and the easterly shear zone remained at the constant altitude. However, this altitude was different for different events: 38, 45, 40, 55 and 40 hPa for September 1976, 1981, 1986, 1991 and 2002, respectively. The full period from December-January solstice to the second equinox in September-October turned out to be equal three seasons (or nine – ten months). This period can be regarded as an elongated stagnation stage divided by the intermediate slight easterly shear descent on two steps.

After the second equinox (September-October) the wind profile changes are observed in the lower stratosphere and later the E-shear reaches 70 hPa level. Almost simultaneously the easterly wind starts to decrease in the upper layer, near 10 hPa, and then the westerly shear appears at the 10 hPa level. The subsequent descent of the westerly shear is quite similar to that during 24-month scenarios shown in Figure 4. Throughout four-five seasons, from equinox of the first year (September-October) to the end of the second year, or even later, the W-shear arrives at the 70 hPa level. In the next (third) year the intense wind profile changes occur in the layer higher ~30 hPa. So, it takes about two seasons for the easterly shear descending from 10 to ~30 hPa. As a result in June-July solstice the easterly (westerly) wind are displayed in the middle (lower) stratosphere. This height structures do not change during July-September period, as it seen in Figure 5. So, in each case the beginning of a new stagnation stage is observed in solstice again. The full cycle of the wind profile changes from the initial moment of the first stagnation stage to the initial moment of the stagnation stage in the next QBO-cycle takes ten seasons. The cycle consists of the elongated two-step stagnation stage observed during three seasons and subsequent wind evolution completed within seven seasons. This variant of QBO-cycle evolution we denote as "January 30-month scenario" in conformity with time of the QBO-cycle beginning.

It is obviously that the 30-month January scenario of QBO-cycle with onset in December-January solstice is always followed by the QBO-cycle with beginning in June-July solstice. Figure 6 presents six QBO-cycles, five of which being immediately adjacent to cycles of "January 30-month" scenarios shown in Figure 5. One can see here that in all but one case the wind profile changes were negligible to a great extent during three seasons from June-July solstice to March-April equinox of the next year. In July-September 2002 the altitude of the stagnation stage was higher than in the rest five cases, and the additional E-shear descent took place during period from September to November. However, the shear altitude remained unchanged from December of the first year to the April of the second year. Therefore, in all six events the stagnation stages were observed during three seasons (or nine – ten months), from June-July solstice to March-April equinox of the next year. The subsequent wind profile changes in each case were quite similar to those observed in 24-month and 30-month January scenarios. Seven seasons (or 20-21 months) later the following stagnation stage is observed in December-January solstice. Thus, the full length of QBO-cycles, defined as a period between the successive stagnation stages, turned out to be equal to ten seasons, or 30 months. So, it is a "July 30-month scenario".

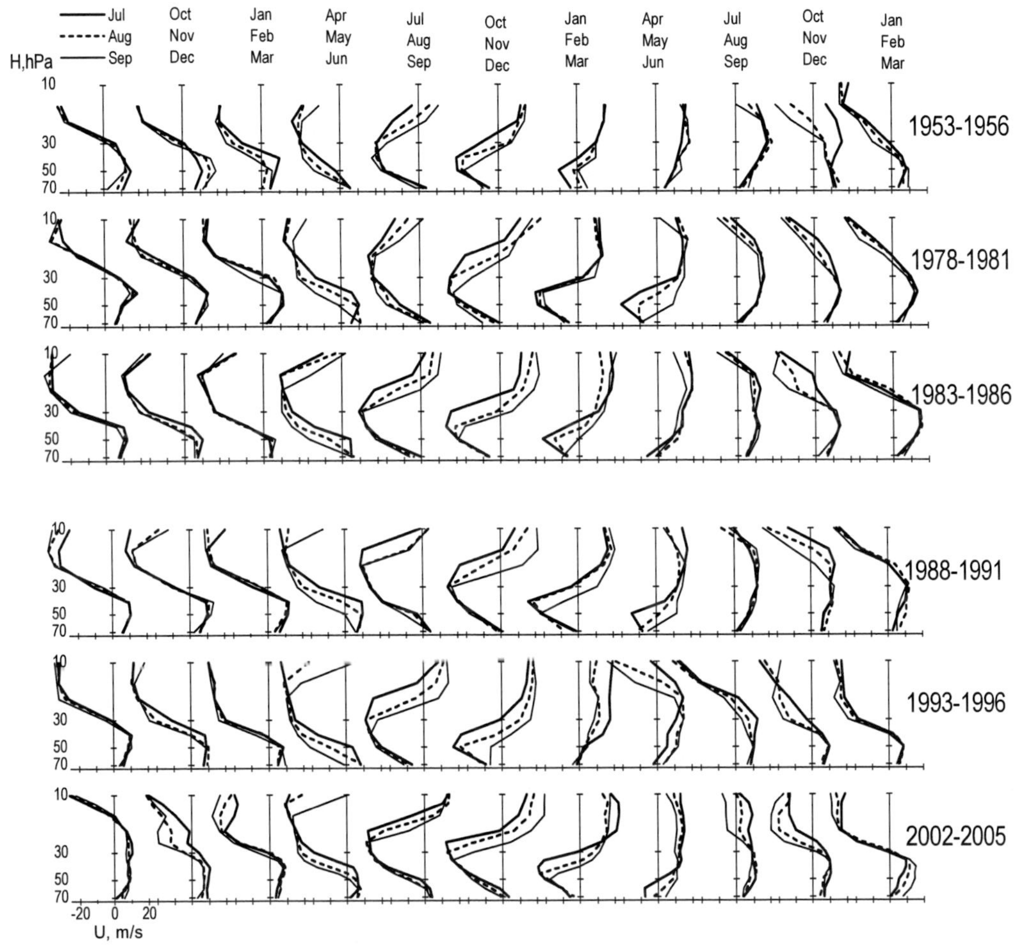

Figure 6. The same as Figure 4, but for July 30-month scenarios. Shown are six 33-month periods from the stagnation stage onset in July to the stagnation stage of subsequent QBO-cycle.

3.1.3. 36-Month Scenarios

Two anomalously long QBO-cycles occurred in 1960s. It was believed [*Dunkerton and Delisi, 1985*] that the unusually slow easterly shear zone descent was a primary cause of lengthened westerly (easterly) QBO-phases at 50 (20) hPa. Figure 7 shows the wind profile changes during two intervals by three years long each: from January 1964 to December 1966 (the upper two rows) and from January 1967 to December 1969 (the lower two rows). Again, like Figures 4, 5 and 6, the plots display the profiles for one season (three months), the first, middle and last profiles on each plot being marked by the hard, dash, and thin line, respectively. As was shown above the series of 24-month QBO-cycles took place in 1956-1963. At the end of the last 24-month QBO-cycles (from January 1962 to December 1963) the decrease of the westerly speed started at the higher layers, however the wind direction remained unchanged. Then westerly was reversed to easterly simultaneously at all altitudes from 24 to 10 hPa in a matter of one month, from December 1963 to January 1964. Later the easterly wind speed in layer higher 29 hPa increases during some months (till July 1964) without essential variations of zero wind speed altitude. From January to June 1964 the altitude changes amount to 5 hPa. Thus, the stagnation stage starts near solstice in December 1963 – January 1964. One can see from Figure 7 that the zero wind speed altitude was very stable near the 30 hPa during the first year. Furthermore, after June 1964 the wind profiles structure at height from 10 to 70 hPa remained practically the same till March-April equinox of the next 1965 year. The easterly shear descent is resumed only 15-16 months, or five seasons, after the stalling period beginning.

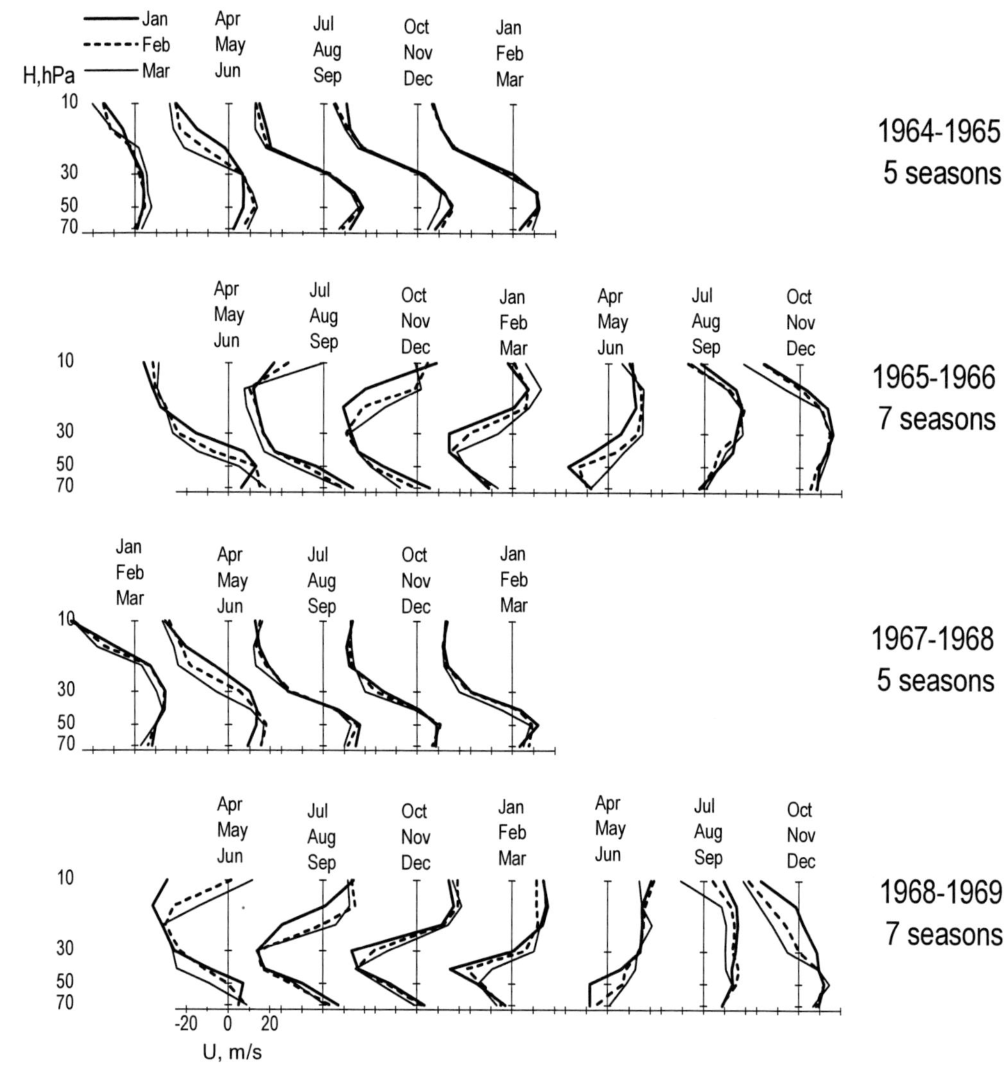

Figure 7. The same as Figure 4, but for 36-month scenarios. Shown are two 39-month periods from the stagnation stage onset in January to the stagnation stage of subsequent QBO-cycle. The first (second) 36-month QBO-cycle is shown in two upper (lower) rows.

As for period till the beginning of a new stagnation stage (marking the following QBO-cycle), the subsequent behavior of easterly and westerly shear descending zones in 1965-1966 was the same as in 24- and 30-month scenarios In such a manner, the cycle was completed in seven seasons (20-21 months) after the end of the stalling period. Thus, the full QBO-cycle length was equal to twelve seasons.

By next solstice (December 1966 – January 1967) the easterly (westerly) were observed again higher (lower) the ~20 hPa level. The altitude of the zero wind speed was kept in the subsequent months, the insignificant changes of the height profiles during December 1966 – March 1967 being indicative of the beginning of subsequent delayed E-shear zone. Behavior of the wind profiles during 1967-1969 was similar to that in 1964-1966 (meaning the durations of the stagnation stage and the full cycle length). The differences between two stagnation stages are concerned only the altitude of halted shear zone but not the moments of the stagnation beginning and end. We denoted the variant of the QBO-cycle evolution, when the total length for each cycle is equal to full three years or twelve seasons, as "the 36-month scenario".

The Table presents the summary of all above examined types of the QBO-cycle evolution observed during period from July 1953 to December 2006.

Table. The QBO-cycles observed during interval from July 1953 to December 2006.

The time intervals	QBO-cycle numbers	24-month scenarios	30-month scenarios		36-month scenarios
			January	July	
2.5 years 1953-1955	1			Jul 53 – Dec 55	
8 years 1956-1963	2	Jan 56 - Dec 57			
	3	Jan 58 - Dec 59			
	4	Jan 60 - Dec 61			
	5	Jan 62 - Dec 63			
6 years 1964-1969	6				Jan 64 – Dec 66
	7				Jan 67 – Dec 69
6 years 1970-1975	8	Jan 70 –Dec 71			
	9	Jan 72 –Dec 73			
	10	Jan 74 –Dec 75			
20 years 1976-1995	11		Jan 76 – Jun 78		
	12			Jul 78 – Dec 80	
	13		Jan 81 – Jun 83		
	14			Jul 83 – Dec 85	
	15		Jan 86 – Jun 88		
	16			Jul 88 – Dec 90	
	17		Jan 91 – Jun 93		
	18			Jul 93 – Dec 95	
4 years 1996-1999	19	Jan 96 –Dec 97			
	20	Jan 98 –Dec 99			
5 years 2000-2004	21		Jan 00 – Jun 02		
	22			Jul 02 – Dec 04	
2 years 2005-2006	23	Jan 05 –Dec 06			

3.2. Seasonal Regularities of the QBO Cycle Evolution

As was noted earlier, the tendency for QBO-phase transitions to be related to annual or seasonal cycles is typical for the lower stratosphere (about 45-50 hPa). Experimental results indicate that the QBO-phase onsets show a preference to occur during northern spring-summer season (April-July) [*Dunkerton and Delisi, 1985; Dunkerton, 1990; Baldwin, et al., 2001; Pascoe et al., 2005*]. However, as was argued in [*Dunkerton, 1990*], the exact seasonal synchronization of the QBO implies that "QBO cycles begin within the limits of data

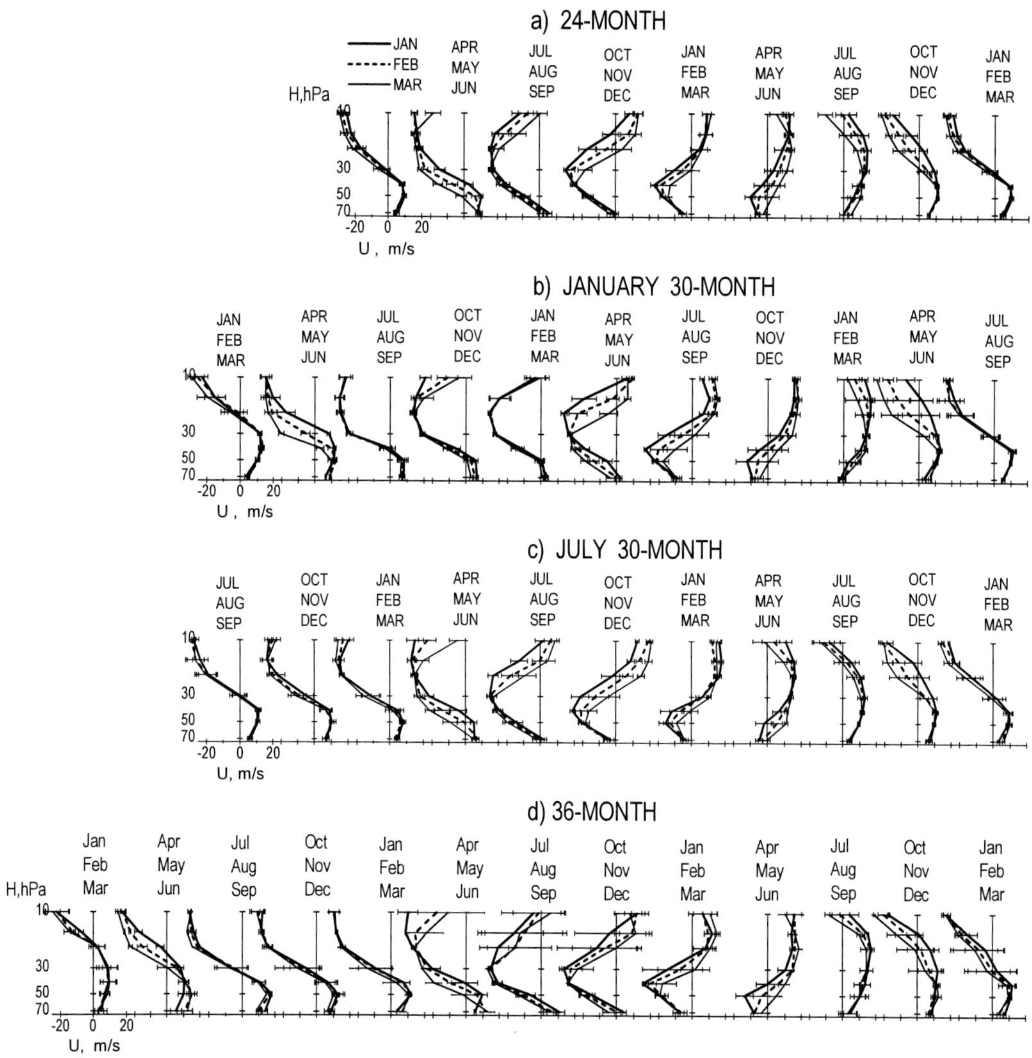

Figure 8. Mean profiles of the wind evolution for QBO cycles with different length: (a) 24-month scenarios, (b) 30-month scenarios, (c) 36-month scenarios. Every plot contains three monthly profiles. Each profile was calculated by averaging the wind speeds at the appropriate heights for the proper months of all QBO-cycles in consistent with dates shown in Table. The first, middle and last profiles on each plot are shown by the hard, dash, and thin line, respectively. The horizontal bars show the standard deviations at each altitude.

accuracy, at exactly the same time or times of the year". But if the QBO-phase (E or W) is determined in the common manner in conformity with the wind direction at a certain fixed pressure level, the phase reversal time can fall at any month of year. In such a case, one can say only about the seasonal modulation in the distribution of the QBO-phase onsets throughout the year.

As it was shown above, analysis of height wind profiles reveals a pronounced seasonal regularity of the QBO-cycle evolution inconspicuous under traditional methods of the QBO examination. The QBO of zonal wind in equatorial stratosphere turned out to be synchronized with the seasonal cycle if the beginning of the easterly shear stagnation stage with westerly (easterly) winds lower (higher) of the stalling region is regarded as the onset for each QBO-cycle. In such a case it is not significant to know the exact altitude of the shear zone during stagnation stage, this altitude varies from one QBO-cycle to other. Figure 8 demonstrates the total scheme of the wind profiles for different scenarios of the QBO evolution. Each profile in Figure 8 was calculated by averaging the wind speeds at the appropriate heights for the proper months of all QBO-cycles in consistent with data shown in the Table. Number of the used QBO cycles at Figures 8a, 8b, 8c and 8d is, respectively, ten, five, six and two. The horizontal bars show the standard deviations for different altitudes. The first plot for each type of scenario shows the profiles observed during first three months after the stagnation stage beginning in solstice. The 24-month, January 30-month and 36-month scenarios begin near the December-January solstice and July 30-month scenarios start near the June-July solstice.

It will be useful to remember that the easterly shear descent from 10 hPa to the stalling region (at about 20-30 hPa)) is observed before the onset of stagnation stage in any type of scenario (see Figures 4-6). Independently from the time of the easterly shear location at 10 hPa level (changing from April to November), the stagnation stage at altitudes 20-30 hPa is formed in solstice (December-January or June-July). The 24-month, both 30-month and 36-month types of QBO scenario are different one season later the stagnation stage beginning. The main feature that defined this difference is the duration (length) of stagnation stage.

One can see from Figure 8a that the stagnation stage in 24-month scenario is finished near the March-April equinox since the easterly shear resumes to descent to the lower stratosphere after this time. It means that the stagnation stage took place during one season. As Figure 8b shows, the two step stagnation stage with slight small intermediate descent during the first equinox is typical of the January 30-month scenario. The following E-shear descent lower the stalling region begins after the second equinox (September-October). Consequently the full duration of stalling period is equals to 9-10 months or three seasons. The July 30-month scenario (Figure 8c) demonstrates the very stable stagnation stage without changes in the zero wind altitude also during 9-10 months or three seasons. And finally the duration of stagnation stage reaches 15-16 month (or five seasons), in 36-month scenario (Figure 8d). Hence the descent of easterly shear zone is stopped at 20-30 hPa layer for over the one, three or five seasons, after that the descent to the lower stratosphere is resumed always in equinox. All cases of the stagnation stage with one season duration were traditionally regarded in research papers as "no stalling events". It was shown in section 3.1 that the easterly shear stalling period is always observed, but, apparently, the one season duration of the stagnation stage is examined as unessential in comparison with three or five seasons, typical of 30 or 36 months scenarios.

The E-shear descent renewal at altitudes lower the stalling region determines the end of the stagnation stage. As Figure 8 shows, the stagnation stage is finished either in March-April

(24-month, July 30-month and 36-month scenarios) or in September-October (January 30-month scenario). The subsequent wind evolution before the next QBO-cycle onset always completed within seven seasons (20-21 month), irrespective of the QBO-scenario. About one season after the stagnation stage end the westerly shear descent starts near the 10 hPa level and reaches the 70 hPa in four-five seasons. Approximately two seasons before a new stagnation stage (identifying the next QBO-cycle) the subsequent easterly shear zone appears in the upper layer and its descent is observed up to solstice, either till December-January (in case of 24-month, July 30-month and 36-month scenarios) or till June-July (in case of January 30-month scenario). Structure of the wind speed changes during these seven seasons varies from cycle to cycle: the westerly and then easterly shear zones appear at 10 hPa level in different time and rate of their succeeding descent to the lower layer varies in large limits. As a result, the wind reversal or QBO-phase transitions times do not reveal the seasonal dependence. At the same time, the tendency for the seasonal modulation of the QBO-phase transitions, mentioned in some investigations, can be easily explained.

The behavior of E-shear descent before the stagnation stage is easily derived from last three-four plots in each row of Figure 8. The westerly speed in layer higher the stalling region starts to decrease about two-three seasons ahead of the stagnation stage beginning that occurs always in solstice (December-January or June-July). Consequently, the transition from W to E QBO-phase at higher levels can be observed either during the second half of year (24-month, July 30-month and 36-month scenarios) or in the first half of year (January 30-month scenario), i.e. throughout the whole year. Nevertheless, there were only five cases of January 30-month scenario over period 1953-2006 in comparison with other eighteens. This fact explains the enlarged probability to observe the W-E transition at higher levels during second half of year.

At 20 hPa (near and slightly higher of the stalling region) the QBO-phase changes from W to E are observed immediately before the December-January solstice that is in last months of the year (October-December). As to the W-E transition at lower layer it is connected with end of the stalling period. Stagnation stage always ends by equinox at altitude about 30-40 hPa. The starts of easterly shear descent from this altitude are observed either in March-April (18 cases) or in September-October (5 cases). Taking into account the mean rate of E-shear descent (~1 km/month) it may be concluded that W-E transition at 45-50 hPa level must be observed more frequently in April-July. The similar explanation is valid also concerning E-W phase transition. The westerly shear appears at 10 hPa about one season after the stagnation stage end. As a result, it occurs at higher layer during first half of year in five cases and during second half of year in eighteen cases. On the contrary, the E-W phase transition in the lower layer takes place during first half of year in eighteen cases and during second half of year in five cases. This circumstance defines the greater probability for E-W transition to be observed at 45-50 hPa in April-June. All above examined regularities in the QBO-phase transitions are in a good consistent with results presented in [*Baldwin, et al., 2001; Pascoe et al., 2005*].

Therefore, it is apparently possible to observe the QBO-phase in any month of year, and the larger or smaller probability of the certain QBO transitions during fixed months or seasons is explained only by the relative number of the realized QBO scenarios. Indeed, the revealed seasonal tendency in distribution of QBO-phase transitions is determined by the larger total number of the QBO events with 24-month, July 30-month and 36-month scenarios in comparison to the events with January 30-month scenario. Since number of July 30-month scenarios is balanced by number of January 30-month scenarios, it is clear that the appointed probability is determined just by group of the 24-month scenarios. Thus, the question of reasons for regular realization of the 24-month QBO scenario is bringing up. It is clear that choice between the 24-month and January 30-month scenarios (i.e. duration of the stalling period) is determined by halt of the E-shear descent during the first equinox after the stagnation stage beginning.

According to [*Pascoe et al., 2005*] whether or not an easterly shear zone becomes stalled depends on the time of year, when the transition reaches the stalling region. If the easterly shear zone arrives at 30 hPa just before the May–July period, the descent to the lower stratosphere occurs without hindrance, because the Brewer-Dobson upwelling at the equator is weakest during this period. It is supposed that the Brewer-Dobson upwelling would act to reinforce the residual upward circulation associated with the easterly QBO shear zone and thereby to stall the descent. This explanation seems us to be unjustified. One can see from Figure 6 that in July 30-month QBO scenarios the shear zone stops in June-July solstice just at ~30 hPa. The monthly wind profiles show that descent in the region higher ~30 hPa proceeded steadily before June-July when the stalling period started. But then in the most of these events only the negligible changes in wind profiles were seen over the subsequent 9-10 months up to the next equinox (March-April), when the easterly shear zone resumed descent to the lower stratosphere. It means that the easterly shear during the first year did not descend in “favored May–July period” (according to [*Pascoe et al., 2005*]), but it moved downward at the same period of next year. Consequently, the enlarged probability (37%) of appearance of the W-E phase transition at 44 hPa in June is determined by occurrence of the easterly shear descent from ~30 hPa altitude to lower stratosphere in March-April in 18 cases from total 23 QBO-cycle observed in 1953-2006.

Thus, the analysis of the height profiles for monthly mean zonal wind has revealed that the process of easterly wind descent to the lower stratosphere always includes the quiet period (not shorter than three months), characterizing by the retardation of the wind speed shear zone in the stalling region at altitudes ~ 20-40 hPa. The stalling period or stagnation stage always starts in the solstice months (December–January or June–July) and ends in one of subsequent equinox, the first, second, or third one. Then seven seasons (20-21 months) are required to complete the current QBO cycle, irrespective of the scenario type. After that a new QBO-cycle starts. Hence, just duration of the stagnation stage is the determining factor for length of the QBO wind rotation cycle. Being dependent on the stagnation stage duration the full QBO cycle can lasts, correspondingly, 24, 30 or 36 months. The minimum length of 24 months is the most likely period for QBO cycle. The longer duration of the cycle (30 months and, sometimes, 36 months) is realized if the beginning of the easterly wind descent to lower layer (or the stagnation stage end) is delayed by no less than two seasons. The aim of the next section is to discover the possible cause of this retardation.

3.3. The Relation of QBO Phases Duration to the Solar UV Irradiance

The distinct seasonal dependence in the zonal wind reversal implies that this process is guided by the internal atmospheric mechanisms. The course of the easterly shear zone descent, being varied for different QBO-cycles, determines the shorter or longer retardation (the stagnation stage) at height near 20-40 hPa. It may be suggested that process of the east wind descent to the lower stratosphere is affected by external conditions, for instance by changes in the solar UV irradiance. As a result, the length of the full QBO cycle may be influenced by the solar UV flux variations through alternation of the of easterly shear zone descent duration.

We examined relations between evolution of the mean height wind profiles and the solar UV irradiance while passing two seasons after the stagnation stage beginning. Intensity of the solar UV irradiance was estimated from the $F_{10.7}$ index, presenting the solar 10.7-sm radio flux. The $F_{10.7}$ index is a standard characteristic of solar activity used in research concerning the solar influence on the Earth's climate. Usage of the $F_{10.7}$ index as the solar activity proxy was established when comparing the time series of the index and UV/EUV measurements [*Donnelly, 1988; Rottman, 2000; Floyd et al., 2005*]. According to [*Soukharev and Hood, 2001; Kane, 2003*] the $F_{10.7}$ index undergoes persistent medium-term variations with quasi-period in the range of 1-3 years, associated with the creation and decay of active regions on the Sun. In our analysis we used the daily $F_{10.7}$ index received from the Space Physics Interactive Data Resource (SPIDR) (http://spidr.ngdc.noaa.gov). To investigate the effects of the solar UV irradiance on quasi-biennial oscillation the medium-term component in $F_{10.7}$ index was derived by subtracting the long- and short-term fluctuations from the index time series. For that the 27-day running mean and the 26-month running mean (the 780 point smoothing) of the index have been evaluated. Then deviation of the 27-day running mean from 780 days smoothed series has been calculated and normalized. The resulting version is denoted as $\Delta F_{10.7}$:

$$\Delta F_{10.7} = \frac{(\langle F_{10.7}\rangle_{27days} - \langle F_{10.7}\rangle_{780days})*100\%}{\langle F_{10.7}\rangle_{780days}}$$

The statistical treatment of the relations between the $F_{10.7}$ index and the wind profiles changes has been performed for period 1953-2005 with use of the epoch superposition method. Results are shown in Figure 9. Three types of the QBO cycles have been examined: the 24-month scenarios, with stagnation stage finished in the nearest equinox and two types of 30-month scenarios ("January" and "July"), with the stagnation stage continued further.

In case of 24-month scenario stagnation stage was always started in December-January, and ended in March-April. Behavior of the $F_{10.7}$ index in these cases was characterized either by the pronounced increase while approaching the stagnation stage end or by the index fluctuations occurring against the high value of $F_{10.7}$. This regularity is demonstrated in Figure 9a, the 1st January within the stagnation stage being taken as "zero date" for the $F_{10.7}$ index. One can see that in case of the 24 months QBO cycle (with minimal duration of stagnation stage) the $F_{10.7}$ index grows, in general, in going from December-January solstice to June-July solstice. The average index $\Delta F_{10.7}$ increases from December to May by 12%, the statistical significance level being better than 0.001. Stagnation stage finishes in March–April against the background of high level of the $F_{10.7}$ index.

In the 30 months scenarios the occurrence of stagnation stage is observed in December-January or in June-July. The appropriate completions of stagnation stage in these scenarios started only during the second equinox (in September-October or in March-April) and stagnation stage lasts about nine months. The low value of the $F_{10.7}$ index in time of the first equinox or pronounced decrease of $F_{10.7}$ was typical of the QBO cycles with 30 months

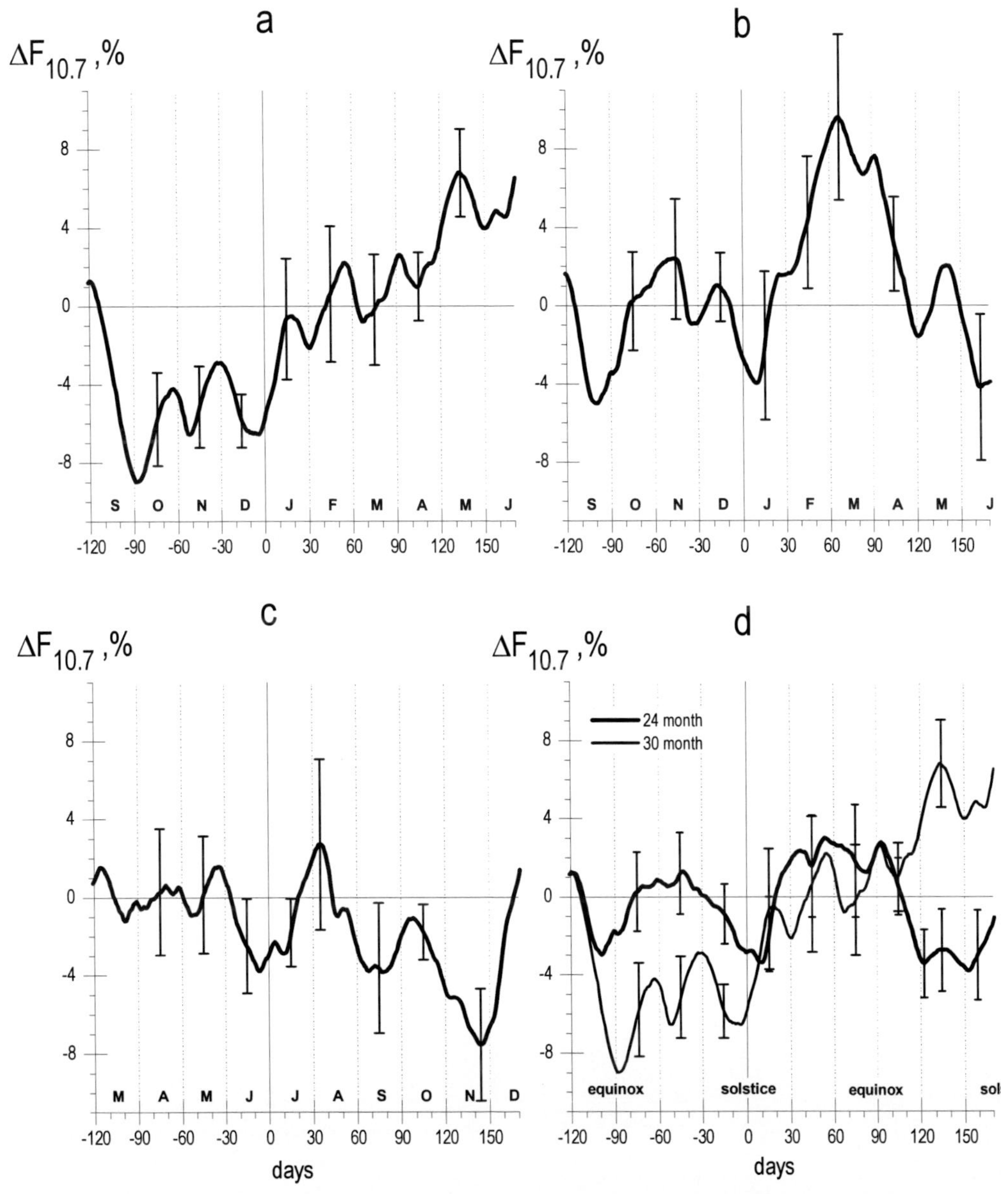

Figure 9. a) Behavior of the mean $\Delta F_{10.7}$ index in case of 24-month scenario of the QBO cycle. The 1st January within the stagnation stage is taken as "zero date" for the $\Delta F_{10.7}$ index. The index $\Delta F_{10.7}$ increases from December to May by 12%, with the statistical significance level better than 0.001.
b) Behavior of the $\Delta F_{10.7}$ index in case of January 30-month scenario of the QBO cycle. The 1st January within the stagnation stage is taken as "zero date" for the $F_{10.7}$ index. The index $\Delta F_{10.7}$ decreases from March to June by 13.8%, the statistical significance level being equal to 0.03.
c) Behavior of the $\Delta F_{10.7}$ index in case of July 30-month scenario of the QBO cycle. The 1st July within the stagnation stage is taken as "zero date" for the $\Delta F_{10.7}$ index. The index $\Delta F_{10.7}$ decreases from August to November by 10%, the statistical significance level being equal to 0.07.
d) Comparison of the mean $\Delta F_{10.7}$ index variations in cases of the 24-month (light curve) and 30-month scenarios (heavy curve), the curve for 30-month scenario being summary of January and July 30-month scenarios shown in Figures 9b and 9c. The difference in the $\Delta F_{10.7}$ index observed at the fifth month after the stagnation stage beginning is 7–9 %, with level of statistical significance equal to 0.008–0.01.

scenarios. The relationship between the wind profile evolution and the $F_{10.7}$ index is shown in Figure 9b for the January 30-month scenario and in Figure 9c for July 30-month scenario, the "zero date" for the $F_{10.7}$ index being taken on the 1st January or the 1st July, respectively. Results of the statistical analysis show that the $\Delta F_{10.7}$ index for January 30-month scenario reduced from March to June by 13.8% (the statistical significance equals to 0.03), the $\Delta F_{10.7}$ index for July 30-month scenario reduced from August to November by 10% (statistical significance equals to 0.07). As a result of the UV decline, the descent of easterly shear zone do not develop below ~ 30 hPa; the finish is delayed for half a year, and total length of the QBO cycle increases up to 30 months.

The quite different character of run of the mean $\Delta F_{10.7}$ index for the 24-month and 30-month QBO scenarios is clearly seen in Figure 9d, where the heavy curve is summary of the both, January and July, types of the 30-month scenario. The difference in the $\Delta F_{10.7}$ index observed at the fifth month after "zero date" in these two scenario types is 7–9 %, with level of statistical significance about 0.01. In case of 36-month scenarios (not shown in the Figure) the $F_{10.7}$ index in course of the first and second equinox either continued to be low ($F_{10.7} \approx$ 70-75 during 1964) or decreased (from mean monthly $F_{10.7} \approx 159$ in March 1967 to ≈ 124 in June 1967 and from ≈ 156 in August 1967 to ≈ 135 in October 1967).

Results of the statistical analysis make it possible to conclude that the easterly winds descend below ~30 hPa only under condition of the high level or steady increase of the UV irradiance occurring just in time of the first equinox in the QBO cycle beginning. If the UV irradiance is low or decreases during the first equinox, the easterly winds typical of the upper layer, do not descend below 30 hPa. The decrease of the UV irradiance seems to be conducive to halt the easterly shear in stalling region (at ~30 hPa). Consequently, whether or not an easterly shear zone is retarded at height ~20-30 hPa depends on level of the UV irradiance in time of the first equinox for the partial QBO cycle.

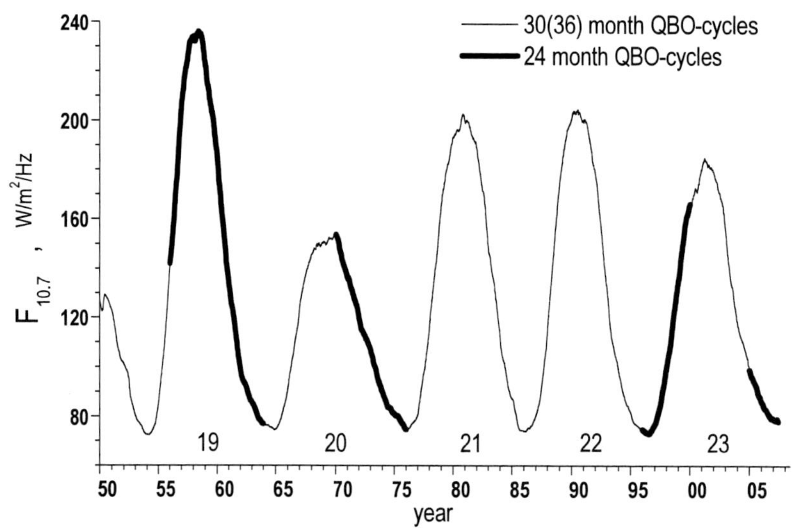

Figure 10. Course of solar activity index $F_{10.7}$ for the period from January 1950 to May 2007. The 26-month running mean (the 780 point smoothing) of the daily $F_{10.7}$ index measurements is presented. The records from 19 to 23 indicate the numbers of solar cycles. Heavy and thing lines correspond to series of short (24 month scenarios) and long (30 or 36 month scenarios) QBO-cycles, respectively.

The QBO-period, duration of the West QBO phase in the lower stratosphere as well as the descent rate of the easterly shear zone demonstrate the periodic variations with a mean period of 10-12 years [*Dunkerton and Delisi, 1985, 1997; Soukharev and Hood, 2001*]. This circumstance explains the attempts to correlate the QBO-period with the 11-year cyclicity of solar activity. In period of solar maxima the rapid descent of easterly shear zone shortens the West QBO phase in the lower stratosphere. During the solar minimum a slow or stalled descent of the easterly shear zone coincides with the lengthy West QBO phase in the lower stratosphere. As suggested in [*Soukharev and Hood, 2001*], this implies more stalling cases in years of solar minimum. Results of our study are not in line with later conclusion. Indeed, the number of lengthy stalling periods resulting in 30 or 36 months scenarios does not depend on phase of the 11-year solar cycle. As Figure 10 shows, the shorter stalling periods (relating to the 24-months scenarios) were dominant in years of solar maximum only during 19th solar cycle (1956-1960). This order broke down in course of the next 20th solar cycle: very long stalling period gave rise to the 36 months QBO scenario observed just in the solar maximum years (1967-1969). Only 30-months scenarios were observed during phases of maximum for the subsequent solar cycles (1978–1983, 1988–1993 and 1999–2003). And vice versa, almost half of the all 24-months scenarios (1962-1963, 1974-1975, 1996-1997 and 2005-2006) occur near solar 11-year minimums. Analysis of all these events provides the basis for our conclusion that delay in the easterly shear zone descending at 30 hPa depends on the intensity of solar UV flux in the proper season. It is well known, that amplitudes of the solar variations at the QBO time-scale (1-3 years) tend to be larger under the solar maximum than under solar minimum conditions [*Donnelly, 1988; Soukharev and Hood, 2001; Kane, 2003; Floyd et al., 2005*]. It makes possible to conclude that observed 11-year variations of the W-QBO phase duration in the lower stratosphere are caused by the 11-year modulation of the shorter-term solar UV oscillations.

According to the QBO numerical models the easterly (westerly) shear zones are related to a cold (warm) temperature anomaly at the equator, and the circulation is characterized by a sinking motion at equator in the westerly shear zones and rising motion in the easterly shear areas [*Reed, 1965; Baldwin et al., 200*]. Based on NCEP reanalysis (1958-2001) *Ribera et al.* [*2004*] reconstructed the spatial and temporal evolution of the several atmospheric variables through a QBO cycle. As they show, the circulation associated with the quasi-biennial oscillation is linked with negative (positive) temperature anomalies caused by the adiabatic rising (sinking) motions over zones of easterly (westerly) wind shear at the equator. Consequently, the solar influence on QBO wind evolution may be related to alterations in distribution and magnitude of the temperature anomalies. The later can be resulted in the additional heating caused by absorption of the increased solar UV radiation at 120-300 nm by ozone and molecular oxygen in the Earth's outer atmosphere.

3.4. Prediction of the QBO Cycle Length

The stagnation stage of the descending easterlies always starts in solstice, either in December-January or in June-July, and having the quantized duration (about 3, 9, or 15 months) always ends by the equinox irrespective of the scenario type. Since the subsequent wind evolution of the particular QBO-cycle is always completed within 21 months, the length of the full QBO-cycle, defined as a period between the beginnings of the successive stagnation stages, will be equal to 24, 30, or 36 months. Consequently, the full length of the current QBO cycle and the appropriate onset of the next one can be predicted while observing the zonal wind transformation during first equinox after the stalling period beginning. As Figure 8 demonstrates, if the westerly winds are observed from 70 to 30-20 hPa and easterly winds occur from 30-20 to 10 hPa in December (or June), this profile will persist without significant changes at least until March of the next year (or until September of the current year). Then, if the easterly shear descent to lower layer has been manifested near the first equinox (in April-May), the length of the current QBO cycle is predictable with certainty.

The previous QBO cycle, starting in December 2004-January 2005, was predicted in 2005 as the cycle developing according to 24-months scenario (see [*Gabis and Troshichev, 2006*]). Prediction was based on experimental data that the easterly shear zone descent resumed in April 2005. It was predicted also that a following QBO-cycle should start in December-January solstice. As a matter of fact, a new QBO-cycle began in December 2006-January 2007 (see Figure 1).

The experimental data available today up to May 2007 make it possible to do the next prediction. Indeed, the analysis of the wind profile changes showed that the easterly shear zone descent from 30 to lower stratosphere is resumed in the first equinox: the zero wind speed altitudes in March, April and May 2007 were equal to 30, 34 and 43 hPa, respectively. It means that stagnation stage starting in December-January solstice lasted only one season and already ended. Therefore, the present QBO cycle (2007-2008) is developed in accordance with 24-months scenario, and the next stagnation stage will start around January 2009.

In the pioneer studies concerning the modulating QBO influence on the solar activity effect on the stratosphere and troposphere [*Holton and Tan, 1980; Labitzke, 1987; Labitzke and van Loon, 1988; Chanin et al., 1989*] the phase of QBO was usually determined by the wind direction at a particular altitude (about 45-50 hPa). To define the QBO-phase during any certain month the different determinations were used in various studying: one of the levels in the range 50-20 hPa, or some combination of levels [*Lait, 1989*], or representation of the wind QBO with use of the first two EOFs (empirical orthogonal functions) of monthly wind observations [*Wallace, et al., 1993; Baldwin and Dunkerton, 1998*]. It is turned out that the correlations involving extratropical parameters depend on the exact level chosen to define the QBO-phase. Furthermore, this level was different in the opposite hemispheres: the strongest QBO-signal was obtained for QBO level of ~40 hPa in the northern hemisphere and ~25 hPa in the southern one [*Baldwin and Dunkerton, 1998*]. According to *Baldwin et al.,* [*2001*] the level for identification of the QBO phase would be chosen in the range from 50 to 25 hPa to maximize the extratropical response in atmospheric processes. Evidently the reason of such uncertainty is the lack of the clear physical mechanisms providing the QBO-influence on the extratropical processes. In addition, the QBO-phase identification by wind direction at a fixed level corresponds to the considerably different height structure of equatorial circulation during various moments of the same QBO-phase.

Our analyses of the height wind profile changes in course the QBO-cycle evolution revealed a clear seasonal regularity. The occurrence of the stagnation stage in a solstice season (December-January or June-July) specifies a quasi-biennial period: 24, 30, or 36 months, depending on the stalling period duration. The entire cycle develops according to certain temporal scales and it is quite insignificant to know the actual altitude of the shear zone during stagnation stage which can vary in different QBO-cycles. As soon as we identify the stagnation stage beginning as the QBO-cycle onset, the QBO transformation of zonal wind in the equatorial stratosphere is synchronized with seasonal cycle.

Firstly QBO cycle was named as "26-months cycle". This length was determined on the basis of data for 4-5 cycles (1953-1962) available in that time [*Reed, et al., 1961; Reed, 1965; Veryard and Ebdon, 1961*]. Accumulation of the observational data led to the increase of mean period to 27.7 months for 1953-1985 [*Naujokat, 1986*], and to 28.2 months for 1953-1995 (Baldwin et al., 2001). The positive trend of the QBO length for last 10 years has been explained by *Mokhov and Eliseev* [*1998*] as due to the influence of the climatic changes. As a matter of fact, the real length of QBO cycle constitutes 24, 30 or 36 months, and "the gradual growth of QBO cycle" is explained by the varying combination of these three lengths in the examined periods. Indeed, as Table shows, there were 3 cycles with 24 months length and only 1 cycle with 30 months length in 1953-1963. The succeeded 20 years included ten cycles and only three of them had 24 months length. As a result, the mean length of QBO cycle essentially increased. After next ten years with four cycles of 30 months, the mean QBO length increased still further. Thus, the 24-month scenario was more common before 1976, and 30-month scenario was more prevalent after 1976. However, the ratio of numbers of the 24- and 30-month cycles level off after 1991, and conclusion about the growth of QBO length can not be regarded as convincing.

3.5. Antarctic Ozone "Hole" and Seasonal Regularities of QBO-Cycle

Earlier it was shown that the level of total ozone (TOZ) over Antarctica during "ozone hole" periods displays significant interannual fluctuations, associated with QBO-phases of zonal wind in the equatorial stratosphere [*Bojkov, 1986; Garcia and Solomon, 1987*]. Later during gathering of experimental data it was revealed the ambiguity concerning relationship of the intensity of TOZ depletion on the QBO-phases [*Lait, et al., 1989; Angell, 1988; 1993; Herman, et al., 1995*]. At present time there are several hypotheses that explain the QBO-modulation at the polar regions by the equatorial circulation influence on meridional exchange and on the strength and isolation of the antarctic winter-spring circumpolar vortex [*Bowman, 1989; O`Sullivan, 1992; O`Sullivan and Salby, 1990; Kinnersley and Tung, 1999; Baldwin, 2001*].

In this section a comparative analysis between monthly mean October TOZ at station Amundsen-Scott (89.98°S; 24.8°W) and direction of zonal circulation in the equatorial stratosphere is presented. TOZ-data are based on ground Dobson-spectrophotometer (1962-2005) [*WOUDC Data Archive: http://www.msc-smc.ec.gc.ca/woudc/index.html*] and satellite TOMS (Total Ozone Mapping Spectrometer) measurements (1978-2005) [*http://toms.gsfc.nasa.gov/*]. The Figure 11 presents the course of TOZ.

It is evidently that for the correct examination of equatorial QBO-circulation influence on "ozone hole" phenomenon it is necessary to take into account the season regularities of the QBO-cycle evolution. During the 24-month and January 30-month scenarios of the QBO-cycles two "ozone holes" are observed. The lasted type of the QBO-cycles is always followed by the July 30-month scenarios starting in June-July solstice. Three "ozone holes" are observed during such QBO-cycles as well as during 36-month scenarios. Here the first "ozone hole" is formed during stagnation stage, i.e. under conditions of the stable circulation in equatorial region with westerly (easterly) in lower (middle) stratosphere. 44 "ozone holes" were observed in Antarctica for the period 1962-2005, and only seven of them occur during stagnation stage. For the most of "ozone holes" either the formation of polar vortex (in May-June) or ozone destruction (in September), or both happen under varying circulation in the equatorial stratosphere.

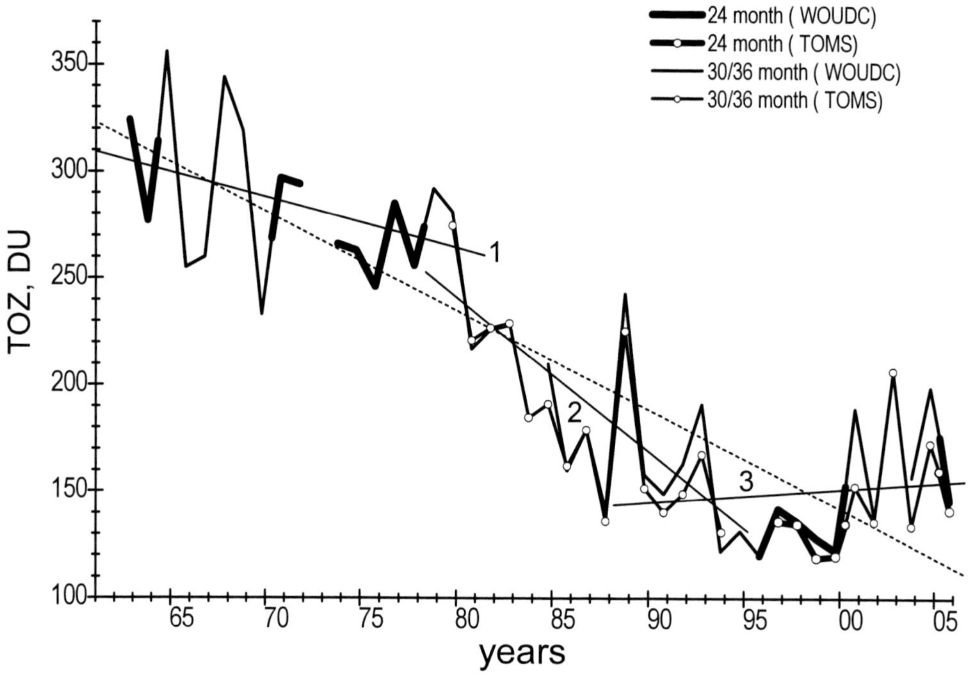

Figure 11. Changes of total ozone (TOZ) during the period 1962-2005. Dash line is trend for period 1962-2005. Thin lines with numbers: 1 – trend for period 1962-1979, 2 – trend for period 1979-1993, 3 – trend for period 1989-2005. Heavy and thing parts of TOZ- curve correspond to series of short (24 month scenarios) and long (30 or 36 month scenarios) QBO-cycles, respectively.

In overwhelming majority cases during the first "ozone hole" in each 24- and January 30-months QBO-cycles TOZ has a relative maximum, while during the second "ozone hole" TOZ has a relative minimum. Nearly always during the third "ozone hole" in July 30-months QBO-cycles TOZ shows a relative minimum. Hence if series of 24-month QBO-cycles is observed in equatorial stratosphere the course of TOZ demonstrates the "true" alternation of relative maxima and minima. Switch from series of 24-month to series of 30-month QBO-cycles results in breakdown of the "true" alternation, because during the July 30-month QBO-cycles the relative TOZ maximum is followed by two minima successively. Figure 11 illustrates that 24-month scenarios were more frequent until the year 1976, and 30-month scenarios more often were observed during 1976-1995. Minor descending trend of TOZ before about 1979 and after 1993 coincides with mixed 24- and 30-months QBO-cycles in the equatorial stratosphere wind. But very large descending trend of TOZ in 1979–1993 was observed when series of 30-months QBO-cycles occurs only. Also one can note the tendency of small amplitude of TOZ-variations during the 24-months QBO-cycles as compared to the amplitude during the 30-month or 36-month QBO-cycles.

4. Conclusion

The changes in equatorial stratospheric wind during QBO-cycle evolution were analyzed with use of the wind speed profiles in height range 70-10 hPa during period from 1953 to 2007. The results showed that the stage of QBO-cycle with easterly above 20-40 hPa and westerly below this height always begins in a solstice and persists for 3-4, 9-10, or 15-16 months without significant changes in the wind structure. This stalling period or stagnation stage is ended by the first, second, or third equinox, respectively, as the descent of the easterly shear to the lower stratosphere is resumed. Seven seasons (20-21 months) after the stagnation stage, the profile with easterly and westerly above and below 20-40 hPa, respectively, is observed again. The full cycle of the wind evolution lasts 24, 30, or 36 months, depending on the stagnation stage duration.

Thus, the reversal in the direction of the equatorial stratospheric zonal mean wind or quasi-biennial oscillations (QBO) have a pronounced seasonal regularity, which is detected when analyzing the height wind speed profiles. The zonal wind transformation turned out to be synchronized with the seasonal cycle if the beginning of the easterly shear stagnation stage is regarded as the onset for each QBO-cycle. In such a case the exact altitude of the shear zone during stagnation stage varies from one QBO-cycle to other, but it is significant that the stalling period always starts in solstice (December-January or June-July), and always ended by the equinox.

The occurrence of the stagnation stage in a solstice season and its different, but quantized duration (about 3, 9, or 15 months) specifies a quasi-biennial period: 24, 30, or 36 months. The minimum length of 24 months is the most likely period for QBO cycle. The longer duration of the cycle (30 months and, sometimes, 36 months) is realized if the easterly wind descending to lower layer (or the stagnation stage end) is delayed by no less than two seasons. The seasonal dependence in the zonal wind reversal implies that this process is guided by the internal atmospheric mechanisms. The start of the easterly shear zone descent lower of the stalling region is varied for different QBO-cycles, suggesting that the process is affected by external conditions, for instance by changes in the solar UV irradiance. As a result, the length of the full QBO cycle may be influenced by the solar UV flux variations through alternation of the start time of the easterly shear zone descending lower stalling region.

If examining the solar UV flux features for different duration of stagnation stage, the following results are obtained. After the short stalling (three–four months) the easterly winds descend below ~30 hPa only under condition of the high level or steady increase of the UV irradiance happening near and after the first equinox in course of the certain QBO cycle. If the UV flux is low or decreases near and after the proper equinox, the easterly winds typical of the upper layer do not descend below 30 hPa, and the stalling period length increases to 9-10 or 15-16 months.

Consequently, the solar UV irradiance influences the QBO cycle length through alternative duration of the easterly shear zone descending from the middle to lower stratosphere. Since the length of the QBO cycle (24, 30 or 36 months) is determined by duration of stagnation stage, the full QBO cycle length can be predicted while observing the zonal wind transformation during the first equinox after the stalling period beginning.

Acknowledgment

The authors are grateful to Barbara Naujokat (Meteorological institute, The Berlin Free University) for kindly providing them the data on monthly mean stratospheric zonal wind components. The work was supported by the Russian Foundation for Basic Research under the grant RFBR-05-05-64633.

References

Angell, J. K. Relation of Antarctic 100 mb temperature and total ozone to equatorial QBO, equatorial SST, and sunspot number, 1958-1987. *Geophys. Res. Lett.* 1988, 15, 915-918.

Angell, J. K. Reexamination of the relation between depth of the Antarctic ozone hole, and equatorial QBO and SST, 1962-1992. *Geophys. Res. Lett.* 1993, 20, 1559-1562.

Baldwin, M. P.; Dunkerton, T. J. Quasi-biennial modulation of the southern hemisphere stratospheric polar vortex. *Geophys. Res. Lett.* 1998, 25(17), 3343-3346.

Baldwin, M. P.; Gray L. J.; Dunkerton, T. J.; Hamilton, K.; Haynes, P. H.; Randel, W. J.; Holton, J. R.; Alexander, M. J.; Hirota, I.; Horinouchi, T.; Jones, D. B. A.; Kinnersley, J. S.; Marquardt, C.; Sato, K.; and Takahashi, M. The Quasi-biennial Oscillation. *Rev. Geophys.* 2001, 39(2), 179-229.

Besprozvannaya, A. S.; Ohl G. I.; Sazonov B. I.; Scherba I. A.; Schuka T. I.; and Troshichev O. A. Influence of short-term changes in solar activity on baric field perturbations in the stratosphere. *J. Atmos. Terr. Phys.* 1997, 59, 1233.

Bojkov, R. D.; The 1979-1985 ozone decline in Antarctic, as reflected in ground level observations. *Geophys. Res. Lett.* 1986, 13, 1236-1239.

Bowman K. P. Global patterns of quasi-biennial oscillation in total ozone. *J. Atmos. Sci.* 1989, 46(21), 3328-3343.

Chanin, M-L.; Keckhut, P.; Hauchecorne, A.; Labitzke, K. The solar activity – QBO effect in the lower thermosphere. *Ann. Geophys.* 1989, 7(5), 463-470.

Donnelly, R. F.. Uniformity in solar UV flux variations important to the stratosphere. *Ann. Geophys.* 1988, 6(4), 417-424.

Dunkerton, T. J. Annual variation of deseasonalized mean flow acceleration in the equatorial lower stratosphere. *J. Meteorol. Soc. Jpn.* 1990, 68(4), 499-508.

Dunkerton, T. J.; Delisi, D. P. Climatology of the equatorial lower stratosphere. *J. Atmosp. Sci.* 1985, 42(4), 376-396.

Dunkerton, T. J.; Delisi, D. P. Interaction of the quasi-biennial oscillation and stratopause semiannual oscillation. *J. Geophys. Res.* 1997, 102, 26107-26116.

Floyd, L.; Newmark, J.; Cook, J.; Herring, L.; McMullin, D. Solar EUV and UV spectral irradiances and solar indices. *J. Atmos. Sol. Terr. Phys.* 2005, 67, 3-15.

Gabis, I. P.; Troshichev, O. A. Influence of solar UV irradiance on quasi-biennial oscillations in the Earth's atmosphere. *Adv. Space Res.* 2004, 34(2), 355-360.

Gabis, I. P.; Troshichev, O. A. QBO cycle identified by changes in height profile of the zonal winds: new regularities. J. *Atmos. Sol. Terr. Phys.* 2005, 67, 33-44.

Gabis, I. P.; Troshichev, O. A. Influence of solar UV irradiance on the quasi-biennial oscillation of zonal winds in the equatorial stratosphere. *J. Atmos. Sol. Terr. Phys.* 2006, 68, 1987-1999.

Garcia, R.; Solomon S. A possible relationship between interannual variability in Antarctic ozone and quasi-biennial oscillation. *Geophys. Res. Lett.* 1987, 14, 848-851.

Georgieva, K.; Kirov, B.; Atanassov, D.; Boneva, A. Impact of magnetic clouds on the middle atmosphere and geomagnetic disturbances. *J. Atmos. Sol. Terr. Phys*. 2005, 67, 163-176.

Herman, J. R.; Newman P. A.; Larko D. Meteor 3/TOMS observations of the 1994 ozone hole. *Geophys. Res. Lett.* 1995, 22(23), 3227-3229.

Holton, J. R.; Lindzen, R. S. An updated theory for the quasi-biennial cycle of the tropical stratosphere. *J. Atmos. Sci.* 1972, 29, 1076-1080.

Holton, J. R.; and Tan, H. C. The influence of the equatorial quasi-biennial oscillation on the global ciculation at 50 mb. *J. Atmosp. Sci.* 1980, 37(10), 2200-2210.

Kinnersley, J. S.; Pawson, S. The descent rates of the shear zone of the equatorial QBO. *J. Atmosp. Sci*. 1996, 53(14), 1937-1949.

Labitzke, K., 1987. Sunspots, the QBO and the stratospheric temperature in the north polar region. *Geophys. Res. Lett.* 14(5), 535-537.

Labitzke, K. On the solar cycle – QBO relationship: a summary. *J. Atmos. Sol. Terr. Phys.* 2005, 67, 45-54.

Labitzke, K.; van Loon, H. Associations Between the 11-year Solar Cycle, the QBO and the Atmosphere, Part I, The Troposphere and Stratosphere in the Northern Hemisphere in winter. *J. Atmos. Terr. Phys*. 1988, 50, 197-206.

Labitzke, K.; van Loon, H. The QBO effect on the solar signal in the global stratosphere in the winter of the Northern Hemisphere. J. *Atmos. Sol. Terr. Phys*. 2000, 62(8), 621-628.

Lait, L. R.; Schoeberl M. R.; Newman P. A. Quasi-biennial modulation of the Antarctic ozone depletion. *J. Geophys. Res*. 1989, 94, 11559-11571.

Kane, R. P. Periodicities of few months in solar indices. *J. Atmos. Sol. Terr. Phys*. 2003, 65, 979-986.

McCormack, J. P. The influence of the 11-year solar cycle on the quasi-biennial oscillation. *Geophys. Res. Lett.* 2003, 30(22), 2162-2165.

Mokhov I. I.; Eliseev A. V. Changes in the characteristics of the quasi-biennial oscillation of zonal wind and temperature in the equatorial lower stratosphere. *Izv. Atmos. Ocean.* Phys. 1998, 34, 291-299.

Naujokat, B. An update of the observed quasi-biennial oscillation of the stratospheric winds over the tropics. *J. Atmos. Sci.* 1986, 43, 1873-1877.

O`Sullivan D.; Salby L. M. Coupling of the quasi-biennial oscillation and extratropical circulation in the stratosphere through planetary wave transport. *J. Atmos. Sci.* 1990, 47(5), 650-673.

O`Sullivan D. Modeling the quasi-biennial oscillation's effect on the winter stratospheric circulation. *J. Atmos. Sci.* 1992, 49(24), 2437-24-48.

Pascoe, C. L.; Gray, L. J.; Crooks, S. A.; Juckes, M. N.; Baldwin, M. P. The quasi-biennial oscillation: analysis using ERA-40 data. *J. Geophys. Res.* 2005, 110, D08105, doi:10.1029/2004JD004941.

Reed, R. J. The present status of the 26-month oscillation. *Bull. Am. Meteorol. Soc.* 1965, 46(7), 374–385.

Reed R. J.; Campbell W. J.; Rasmussen L. A.; Rogers D. G. Evidence of a downward-propagating, annual wind reversal in the equatorial stratosphere. *J. Geophys. Res.* 1961, 66(3), 813-818.

Ribera, P.; Pena-Ortiz, C.; Garcia-Herrera, R.; Gallego, D.; Gimeno, L.; Hernandez, E. Detection of the secondary meridional circulation associated with the quasi-biennial oscillation. *J. Geophys. Res*. 2004. 109, D18112, doi:10.1029/2003JD004363.

Rottman, G. Variations of solar ultraviolet irradiance observed by the UARS SOLSTICE – 1991 to 1999. *Space Sci. Rev*. 2000, 94, 83-91.

Soukharev, B. E.; Hood, L. L. Possible solar modulation of the equatorial quasi-biennial oscillation: additional statistic evidence. J. Geophys. Res. 2001, 106, (D14), 14855-14868.

van Loon, H.; Labitzke, K. The signal of the 11-year solar cycle in the global stratosphere. *J. Atmos. Sol. Terr. Phys*. 1999, 61, 53-61.

van Loon, H.; Labitzke, K. The influence of the 11-year solar cycle on the stratosphere below 30 km: a review. *Space Sci. Rev*. 2000, 94, 259-278.

Veryard, R. G, Ebdon, R. A. Fluctuations in tropical stratospheric winds. *Meteorol. Mag*. 1961, 90, 125-143.

Wallace, J. M.; Panetta R. L.; Estberg J. Representation of the equatorial stratospheric quasi-biennial oscillation in EOF phase space. *J. Atmos. Sci.* 1993, 50(12), 1751-1762.

In: Solar Physics Research Trends
Editor: Pingzhi Wang, pp. 195-232

ISBN: 978-1-60021-987-0
© 2008 Nova Science Publishers, Inc.

Chapter 5

SOME PECULIARITIES OF RESEARCH INTO MULTI-SCALE IONOSPHERIC IRREGULARITIES BY RADIO METHODS

_M. V. Tinin_*
Irkutsk State University, 20 Gagarin
Blvd, Irkutsk 664003, Russia

Abstract

A characteristic feature of ionospheric plasma is the presence of irregularities with different sizes. There is a background irregularity in the form of ionospheric layers with the scales of order 100-400 km along the vertical. Smaller-size irregularities are observed at this background. These ionospheric irregularities have different scales between 100 km and lower. Furthermore, the presence of the geomagnetic field, gravity, wind in the upper atmosphere, and of other factors leads to the fact that ionospheric irregularities are different in size along different directions. Such irregularities are said to be anisomeric. This paper presents some of the latest results derived from developing new methods of describing the propagation of radio waves in a multi-scale inhomogeneous ionosphere and analyzes the possible approaches in perfecting them. In addition to a conventional description of the singly scattered wave from an elongated irregularity smaller than the Fresnel radius in size, we consider the single scattering from a strongly anisomeric irregularity in which the major axis exceeds the corresponding Fresnel radius. Furthermore, we examine the single scattering from a flattened irregularity. This situation is investigated both for the case of a weakly anisomeric irregularity, all scales of which are less than the Fresnel radius, and for the case of a strongly anisomeric irregularity, two axes of which exceed the Fresnel radius. We point to a number of characteristics of the scattering from anisomeric irregularities which, when taken into account, are useful for their diagnostics. With increasing irregularity amplitude, the scattering ceases to be a weak one, and the single scattering approximation becomes invalid. The geometrical optics approximation and its generalizations are used in investigating the scattering in this case. In this paper, we examine some generalizations of the geometrical optics approximation which furnish a means to investigate strong fluctuations in the field, and effects of random irregularities in the neighborhood of the background caustic. We explore the possibility of extending our results obtained in the neighborhood of a regular caustic that is produced in the

* E-mail address: mtinin@api.isu.ru

ionosphere, to similar investigations of the effects of random irregularities under conditions of the solar corona. It is pointed out that mixed integral representations obtained using a double (over the coordinates of the source and observer) weighted Fourier transform hold much promise for the description of the characteristics of radio wave propagation in such strongly regularly and randomly inhomogeneous media as the ionosphere, solar plasma, and fusion plasma.

1. Introduction

Approximate methods are widely used in current theoretical investigations of the wave propagation in inhomogeneous media. Different small parameters are employed in such studied: the geometrical optics (GO) approximation [Kravtsov and Orlov, 1990] and its generalizations [Kravtsov and Orlov, 1999] use, as the small parameter, the ratio of the characteristic size of the field (the size of the Fresnel zone) to the irregularity size, and the theory of the single scattering (the Born approximation) [Rytov et al, 1989; Ishimaru, 1978] takes into account the small contribution from the scattered field caused by the weak deviation of the refractive index. Since the spectrum of ionospheric irregularities usually has a decreasing character [Yeh and Liu, 1982], the single-scattering theory is used when small-scale irregularities are taken into account. Thus most methods are used in investigating the wave propagation either in the presence of large-scale irregularities or in the presence of small-scale irregularities. In the context of the modern theory of wave propagation, these irregularities influence the signal structure in different ways: large-scale irregularities determine the distortion of the phase front and strong fluctuations associated with focusing and interference, whereas small-scale irregularities, generally, make a substantial contribution when the contribution from large-scale irregularities becomes small, in the back-scattering case, for example. As a consequence of this different description of the scattered fields from large-scale or small-scale irregularities, the methods of probing are different for different-scale irregularities of the medium. Thus, computer tomography of large-scale irregularities uses ray tomography that employs the GO model of the signal, while tomography of small-scale irregularities uses diffraction tomography based on employing the single-scattering theory.

However, the presence of different scales is characteristic for ionospheric irregularities. There is a background irregularity in the form of ionospheric layers with the scales of about 100-400 km along the vertical. Beyond-the-horizon communication and radiolocation in the high-frequency range is effected via reflection from ionospheric layers, and these layers make the main contribution to the positioning error in the case of global navigation satellite systems (GNSS).The propagation of the radio wave through the background (regular) ionosphere develops multipathing and focusing at caustics of a different type. Smaller-scale irregularities are observed in this background [Yeh and Liu, 1982]. These ionospheric irregularities have different scales ranging from 100 km to smaller values. They make the main contribution to the residual error in the most accurate dual-frequency GNSS systems, lead to the distortion of spatial-temporal structure of the signal, and to the fact that the wave energy penetrates into the shadow regions that are inaccessible for the propagation in a background ionosphere.

Furthermore, the presence of the geomagnetic field gravity and winds in the upper atmosphere, and some other factors lead to the fact that ionospheric irregularities have a different size along different directions. Such irregularities are referred to as anisomeric

[Kravtsov et al, 1983]. The single scattering from anisomeric irregularities has a number of interesting features, such as strong aspect sensitivity. It is common to do research only for the cases where the irregularities are in the distant zone. The scattering indicatrix is then determined by the three-dimensional spectrum of irregularities [Rytov et al, 1989; Ishimaru, 1978; Kravtsov et al, 1983; Tatarskii, 1971]. However, there rather frequently occur ionospheric irregularities with a sharply pronounced anisomery where, because of the strong elongation of the irregularity, it is in the distant zone with respect to some coordinates, and in the near zone with respect to other coordinates, and vice versa. A conventional approach is inapplicable in this case but requires a modification of the scattering theory even for such, strongly anisomeric irregularities. For instance, the transverse scale of a meteor trail can be about several tens of centimeters, which is much larger than the corresponding Fresnel radius in the VHF range, while the longitudinal size of the meteor trail reaches several tens of kilometers – much larger than the Fresnel radius [McKinley, 1961].

It is very easy to investigate the scattering from large-scale irregularities in the presence of background refraction using the geometrical optics approximation. However, once a background caustic appears, this approximation becomes invalid, so that some modifications of the GO approximation are needed in order to describe the effects of ionospheric irregularities.

This Chapter presents some of the latest results derived from developing new methods for describing the radio wave propagation through a multi-scale inhomogeneous ionosphere and explores the possible avenues for their perfection.

Sections 2 -4 develop the theory of the single scattering of waves from anisomeric irregularities of a different type. Apart from a conventional description of the aspect sensitivity of the scattered way from an elongated irregularity with the size smaller than the Fresnel radius (Section 2), we consider (Section 3) the scattering from a strongly anisomeric irregularity in which the major axis exceeds the corresponding Fresnel radius. Furthermore, Section 4 considers the scattering from a flattened irregularity where two axes of the irregularity far exceed the size of the third axis. This situation is investigated both for the case of weakly anisomeric irregularities, all scales of which are less than the Fresnel radius, and for the case of a strongly anisomeric irregularity, two axes of which exceed the Fresnel radius. Section 5 outlines the GO approach in taking into account background and random refraction at a time, for the propagation through a random inhomogeneous ionosphere. Section 6 is concerned with a generalization of the GO approximation on the basis of the interference integral, making it possible to investigate the effects of random irregularities in the neighborhoods of a background caustic. Also, it considers the limitations of the interference integral method, the Maslov method, and of some other generalizations of the GO approximation, and the possible pathways by which they can be developed further. Section 7 is devoted to the analysis of the possible transfer of results of investigations in the vicinity of a regular caustic that forms in the ionosphere, to similar investigations into the effects of random irregularities under the conditions of the solar corona. Section 8 presents the conclusions regarding further investigation into the radio wave propagation in multi-scale ionospheric plasma.

2. The Single Scattering from an Ionospheric Irregularity

The refractive index $n(\mathbf{r})$ for the ordinary and extraordinary waves of magneto-active ionospheric plasma is defined by the Appelton-Hartrey formula (see, for example, [Budden, 1985; Ginzburg, 1960]) as

$$n^2(\mathbf{r}) \equiv \varepsilon(\mathbf{r}) = 1 + \varepsilon_1(\mathbf{r}), \tag{1}$$

where

$$\varepsilon_1(\mathbf{r}) = -X\left\{1 + iZ - \frac{0.5Y_T^2}{1 - X + iZ} \pm \left[\frac{0.25Y_T^4}{(1 - X + iZ)^2} + Y_L^2\right]^{1/2}\right\}^{-1}. \tag{2}$$

Here plus and minus signs correspond to the ordinary and extraordinary waves, respectively, and following conventional designations are used: $X = \dfrac{\omega_N^2}{\omega^2} = \dfrac{Ne^2}{\varepsilon_0 m_e \omega^2}$, $Z = \dfrac{\nu}{\omega}$, $\mathbf{Y} = \dfrac{e}{m_e \omega}\mathbf{B}$, Y_L and Y_T are the longitudinal (with respect to the propagation of the wave) and transverse components of $\mathbf{Y}$, respectively; $\omega = 2\pi f$ is the emission frequency, ω_N is the plasma frequency; N is electron density; ε_0 is dielectric permittivity of a free space; e and m_e are the electron charge and mass, respectively; ν is the collision frequency of electrons with fixed ions; and $\mathbf{B}$ is the geomagnetic field.

For ionospheric plasma at high frequencies, a perturbation of dielectric permittivity $\varepsilon_1(\mathbf{r})$ is related to electron density $N(\mathbf{r})$ by the relation

$$\varepsilon_1(\mathbf{r}) \approx -X = -80.6\frac{N(\mathbf{r})}{f^2} \tag{3}$$

and the following inequality holds

$$|\varepsilon_1(\mathbf{r})| \ll 1. \tag{4}$$

Let the incident wave be specified in the form of a monochromatic spherical wave

$$U_0(\mathbf{r}) = \frac{A_{sp}}{|\mathbf{r} - \mathbf{r}_0|}\exp\{ik|\mathbf{r} - \mathbf{r}_0|\}, \tag{5}$$

where $k=\omega/c=2\pi/\lambda$, λ is the wavelength in a free space, and the coefficient A_{sp} is determined by the transmitter power, and by the transmitting antenna beam. Then, in view of (4) and using the perturbation method, we can obtain for a single-scattered wave the following expression [Rytov et al, 1989; Ishimaru, 1978]:

$$U_s(\mathbf{r})=\frac{k^2 A_{sp}}{4\pi}\int\limits_V \frac{\exp\{ik\Psi(\mathbf{r}_s)\}}{|\mathbf{r}-\mathbf{r}_s||\mathbf{r}_s-\mathbf{r}_0|}\sin(\gamma(\mathbf{r}_s))\,\varepsilon_1(\mathbf{r}_s)d^3r_s\,, \tag{6}$$

where

$$\Psi(\mathbf{r}_s)=|\mathbf{r}_s-\mathbf{r}_0|+|\mathbf{r}-\mathbf{r}_s|, \tag{7}$$

$\gamma(\mathbf{r}_s)$ is the angle between the electric vector of the incident wave field and the direction from the point of scattering $\mathbf{r}_s$ to the point of reception $\mathbf{r}$.

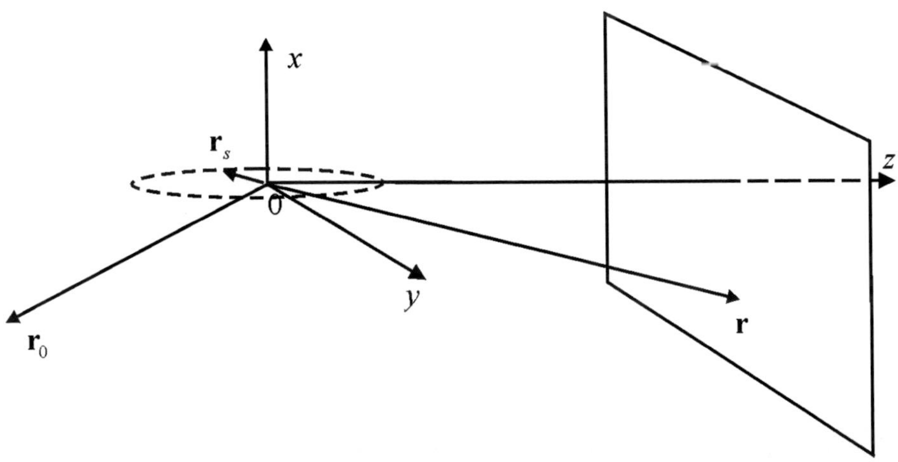

Figure 1. Geometry of the scattering from an anisomeric irregularity.

Let us choose the origin of coordinates $\mathbf{r}=\{x,y,z\}$ at some point inside of the irregularity (see Fig. 1). In the case of a large distance of the transmitter and the receiver from the irregularity the coefficient before $\tilde{\varepsilon}(\mathbf{r}_s)$ in the expression under the integral sign depends weakly on $\mathbf{r}_s$. Therefore, in this coefficient outside the exponential we can assume $\mathbf{r}_s=0$, and only linear terms $O(\mathbf{r}_s)$ can be taken into account in the power series expansion $\Psi(\mathbf{r},\mathbf{r}_s,\mathbf{r}_0)$. This is possible when the irregularity is in the far-field Fraunhofer) zone. As a result of these approximations, the integral in (6) is brought to the form [Rytov et al, 1989; Ishimaru, 1978; Kravtsov et al, 1983; Tatarskii, 1971]:

$$U_s(\mathbf{r}) \approx \frac{k^2 2\pi^2 A_{sp}}{r_0 r} \sin(\gamma(0)) \Phi_{\varepsilon 3}(\mathbf{Q}_3) \exp\{ik\Psi_0\}, \tag{8}$$

where

$$\Psi(0) = |\mathbf{r}_0| + |\mathbf{r}| = r_0 + r \tag{9}$$

is the phase path of the wave arriving at the point of reception after its scattering from the irregularity;

$$\Phi_{\varepsilon 3}(\mathbf{Q}_3) = (2\pi)^{-3} \int \varepsilon_1(\mathbf{r}_s) \exp\{-i\mathbf{r}_s \mathbf{Q}_3\} d^3 r_s \tag{10}$$

is the three-dimensional spatial spectrum of irregularities;

$$\mathbf{Q}_3 = k[\mathbf{n}_s - \mathbf{n}_i] \tag{11}$$

is the scattering vector; $\mathbf{n}_i = -\mathbf{r}_0 / r_0$ and $\mathbf{n}_s = \mathbf{r} / r$ are unit vectors of the incident and the scattered waves.

As the model for a localized plasma irregularity, it is convenient to use a Gaussian irregularity

$$N(\mathbf{r}) = N_m \exp\left\{ -\frac{1}{2}\left[\frac{x^2}{l_x^2} + \frac{y^2}{l_y^2} + \frac{z^2}{l_z^2} \right] \right\}. \tag{12}$$

Substituting into (8) the spectrum (10) for the model of (12) and taking into account the known relationship between the intensity and the received power P_r we obtain

$$P_r = P_t G_t G_r I(\mathbf{r}) \sin^2(\gamma(0)), \tag{13}$$

where P_t is the radiated power; G_t and G_r are the directional diagrams in the power of the transmitting and receiving antennas, respectively; the reduced power $I(\mathbf{r})$ for our model (12) is

$$I(\mathbf{r}) = I_w(\mathbf{r}) = B_1 \left(\frac{2\pi l_x l_y l_z \lambda N_m}{r_0 r} \right)^2 \exp\left\{ -(Q_{3x} l_x)^2 - (Q_{3y} l_y)^2 - (Q_{3z} l_z)^2 \right\}, \tag{14}$$

where $B_1 = 2\pi (20.15 / c^2)^2$, c is the velocity of light in a free space.

For an isomeric irregularity $l_x = l_y = l_z = l$ the scattering is uniformly distributed when $kl \ll 1$, and when $kl \geq 1$ it has a maximum along the forward direction, where $\mathbf{n}_s = \mathbf{n}_i$. The directivity of the forward scattering is enhanced with increasing kl.

3. The Scattering from an Elongated Irregularity

Let us consider how the scattering picture is changed with the increasing size of the irregularities along a single direction, for example, along the axis z (see Fig.1). The geometry of the scattering problem thus gets cylindrical character, and behavior of the reduced power $I_w(\mathbf{r})$ is determined mainly by the projection of the scattering vector Q_{3z}. Hence it follows that when observed in a plane $z = const$, in the transverse plane (see Fig. 1), the scattering will be maximal in some ring in the neighborhood of the intersection of this plane with the side surface of the circular cone

$$n_{sz} = z / r = -z_0 / r_0 = n_{iz} \tag{15}$$

Fig. 2a presents the results from calculating, using formula (14), the scattered power from the irregularity with the sizes $l_x = 0.3m,\ l_y = 0.3m,\ l_z = 10m$, in the plane $z = 100km$ when the source lies at a point with coordinates $x_0 = -50km,\ y_0 = 0km,\ z_0 = -100km$. The wavelength here and in the subsequent discussion is taken equal to 5 m.

A strong dependence of the scattering of anisomeric irregularities on the scattering direction is usually referred to as aspect sensitivity [Rytov et al, 1989; Kravtsov et al, 1983]. With an increase in the longitudinal size l_z the width of the illuminated ring decreases (cf. Fig. 2a with Fig. 2b, where the size l_z is increased to 20 m), and tends to zero in the limit $kl_z \to \infty$. However, with an increase in elongation, the size l_z becomes larger than the corresponding Fresnel size. In this case of a strongly anisomeric irregularity, formula (14) does not hold.

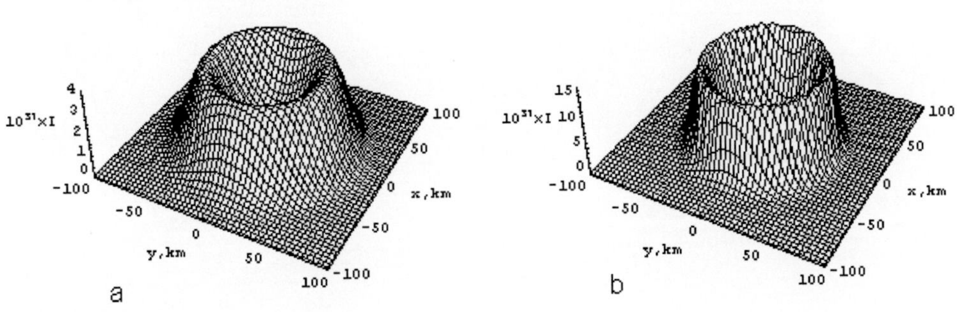

Figure 2. Distribution of the power in the plane $z = 100km$, transverse to the major axis of a weakly anisomeric irregularity with the size $l_x = 0,3m,\ l_y = 0,3m,\ l_z = 10m$ (a) and $l_z = 20m$ (b)

For calculating the scattering from a strongly elongated irregularity in [Tinin and Kim, 2004; 2006; Tinin et al, 2005], the integral over z_s in (6) was evaluated by the method of stationary phase and for the other variable – in the same way as done above. As a result, formula (13) was obtained, in which

$$I(\mathbf{r}) = I_{\ln}(\mathbf{r}) = B_1 \frac{2\pi \left(N_m l_x l_y\right)^2 \lambda^3}{\rho\rho_0\sqrt{\left(\rho+\rho_0\right)^2+\left(z-z_0\right)^2}} \exp\left\{-\frac{z_c^{\,2}}{l_z^2} - Q_{2x}^2\left(z_c\right)l_x^2 - Q_{2y}^2\left(z_c\right)l_y^2\right\}, \quad (16)$$

where

$$z_c = \frac{z\rho_0 + z_0\rho}{\rho+\rho_0}, \quad (17)$$

$$\mathbf{Q}_2(z_c) = k\left[1+\left(\frac{z-z_0}{\rho+\rho_0}\right)^2\right]^{-1/2}\left[\frac{\boldsymbol{\rho}}{\rho}+\frac{\boldsymbol{\rho}_0}{\rho_0}\right], \quad (18)$$

Two-dimensional vectors $\boldsymbol{\rho} = \{x, y\}$ and $\boldsymbol{\rho}_0 = \{x_0, y_0\}$ are the projections of the vectors $\mathbf{r}$ and $\mathbf{r}_0$ onto the plane orthogonal to the major axis of the irregularity, correspondingly.

Since we now are in the near-field zone relative to the longitudinal size of the irregularity l_z increase of l_z leads not to a decrease but to an increase in the width of the illuminated ring on the plane $z = const$. As an illustration of this, Fig. 3 presents the results of calculations by formula (16) for the same parameters as in Fig. 2, but for the longitudinal size of the irregularity which is taken equal to 6 km (Fig. 3a), and to 16 km (Fig. 3b).

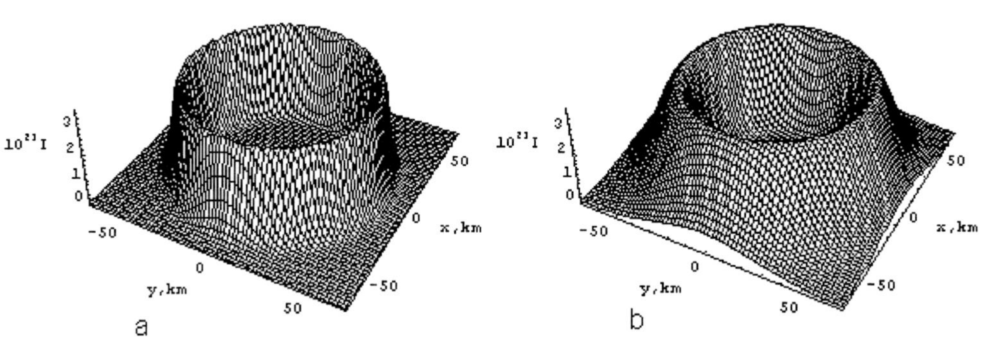

Figure 3. Distribution of the power on the plane $z = 100km$, transverse to the major axis of a strongly anisomeric irregularity with the sizes $l_x = 0.3m,\ l_y = 0.3m,\ l_z = 6km$ (a) and $l_z = 16km$ (b)

4. The Scattering from Flattened Irregularity

We now consider the case where one of the transverse sizes of the elongated irregularity, l_y, for example, increases. In this case the irregularity acquires a flattened, disk-shaped character. For the scattering from a weakly anisomeric irregularity we can use formula (14).

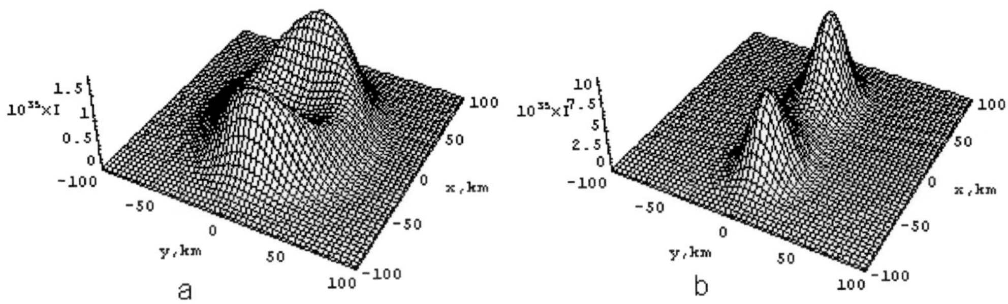

Figure 4. Distribution of the reduced power in the plane $z = 100km$ for the scattering from a weakly anisomeric flattened irregularity with the sizes $l_x = 0.3m,\ l_z = 10m,\ l_y = 2m,$ (a) and $l_y = 5m$ (b)

Fig. 4a presents the results from calculating the scattering from the irregularity in which, unlike the results presented in Fig. 2a, the size l_y is increased from 0.3 m to 2 m. As is evident from the modeling results, and from the analysis of formula (14), the intensity for the ring of aspect sensitivity becomes nonuniformly distributed, with maxima determined by the conditions of (15), and by

$$n_{sy} = y/r = -y_0/r_0 = n_{iy} \tag{19}$$

In particular, when $y_0 = 0$ from this it follows that $y(z) = 0$, and from (15) we obtain

$$x(z) = \pm z\frac{x_0}{z_0}. \tag{20}$$

Thus, with an increase of l_y and l_z the scattered field is concentrated in the infinitesimal vicinities of two scattered rays: the forward scattered ray, and the specularly scattered ray (see Fig. 4b, where a further increase in the size l_y to 5 m occurs). As a matter of fact, however, when $l_y \to \infty$ and $l_z \to \infty$ the size of the region illuminated by the scattered field, does not go to zero; instead, it remains a finite one. This is because here, as in the case of a strongly elongated irregularity above, formula (14) does not hold.

To obtain the formula describing the scattering from a strongly flattened irregularity for the case where its size along the axes z and y exceed the corresponding Fresnel radius, we evaluate in (6) by the method of stationary phase not only the integral over z_s, but also the integral over y_s. As a result, we obtain the expression (13) in which

$$I(\mathbf{r}) = I_{pl}(\mathbf{r}) = B_1 \frac{(N_m l_x)^2 \lambda^4}{\left[|x_0|+|x|\right]^2} \exp\left\{-\frac{z_{cp}^{\ 2}}{l_z^2} - \frac{y_{cp}^{\ 2}}{l_y^2} - Q_1^2 l_x^2\right\} \tag{21}$$

where

$$z_{cp} = \frac{z|x_0| + z_0|x|}{|x_0|+|x|}, \qquad y_{cp} = \frac{y|x_0| + y_0|x|}{|x_0|+|x|} \tag{22}$$

$$Q_1^2 l_x^2 = 2\left[\frac{kl_x}{D_r}(|x_0|+|x|)\right]^2 \left(1 + \frac{x_0 x}{|x_0||x|}\right) \tag{23}$$

$$D_r = \sqrt{(|x_0|+|x|)^2 + (y-y_0)^2 + (z-z_0)^2} \tag{24}$$

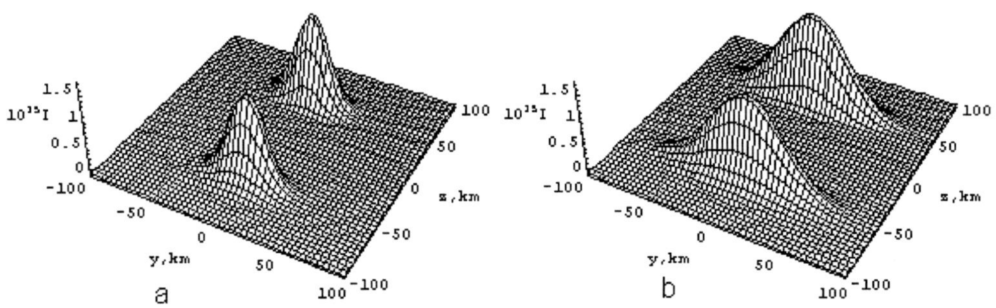

Figure 5. Distribution of the reduced power in the plane $z = 100km$ for the scattering from a strongly anisomeric flattened irregularity with the sizes $l_x = 0.3m,\ l_z = 10km,\ l_y = 10km$ (a) and $l_y = 20km$ (b)

Now, with an increase in the size l_y and l_z, the width of the illuminated region does not decreases; on the contrary, it increase. This is apparent from Fig. 5, showing the results of calculations for $l_x = 0.3m,\ l_z = 10km,\ l_y = 10km$ (Fig. 5a) and $l_y = 20km$ (Fig. 5b).

Noteworthy in the analysis of (21) is the weak influence of other parameters on the scale $I_{pl}(\mathbf{r})$ of the distribution, with the exception of the scales of the irregularity and the distances from the irregularity plane of the source $|x_0|$ and the receiver $|x|$, respectively. If the plane $x = 0$ of a disk-shaped irregularity is parallel to the plane of the Earth on which the transmitter and the receiver are located (which is typical of ground-based ionospheric investigations), then (21) is simplified to

$$I_{pl}(\mathbf{r}) = B_1 \frac{(N_m l_x)^2 \lambda^4}{4x_0^2} \exp\left\{-\frac{(z+z_0)^2}{l_z^2} - \frac{(y+y_0)^2}{l_y^2} - \frac{16k^2 l_x^2}{D_r^2}\right\} \quad (25)$$

From (25) we have a weak dependence of the scale of intensity distribution across the terrestrial surface on the separation between the receiver and transmitter for $kl_x < 1$.

This is also seen by comparing the computational results presented in Fig. 6a for the radar case where the transmitter is at the point $x_0 = -100km$, $y_0 = 0km$, $z_0 = 0km$ below the irregularity with the sizes $l_x = 0.3m$, $l_y = 10km$, $l_z = 10km$ with the results shown in Fig. 6b for spaced-receiver reception where the transmitter is at the point $x_0 = -100km$, $y_0 = 0km$, $z_0 = -500km$.

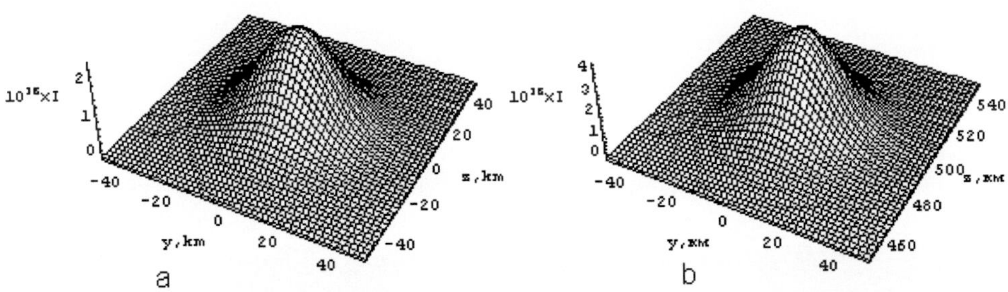

Figure 6. Distribution of the reduced power across a plane Earth $x = -100km$ for the scattering from a strongly anisomeric, flattened irregularity with the sizes $l_x = 0.3m$, $l_z = 10km$, $l_y = 10km$ for the cases of the transmitter on the ground below the irregularity (a), and at a distance $500km$ from it (b).

An entirely different picture is observed for the intensity distribution of a scattered field across the terrestrial surface for an elongated irregularity along the axis z (meteor trail). In this case, using formula (16) we perform calculations for the irregularity with the size $l_x = 0.3m$, $l_z = 0.3m$, $l_y = 10km$. Fig. 7a presents the modeling results for the case where the transmitter is at the point $x_0 = -100km$, $y_0 = 0km$, $z_0 = 0km$, and Fig. 7b – when $x_0 = -100km$, $y_0 = 0km$, $z_0 = -500km$.

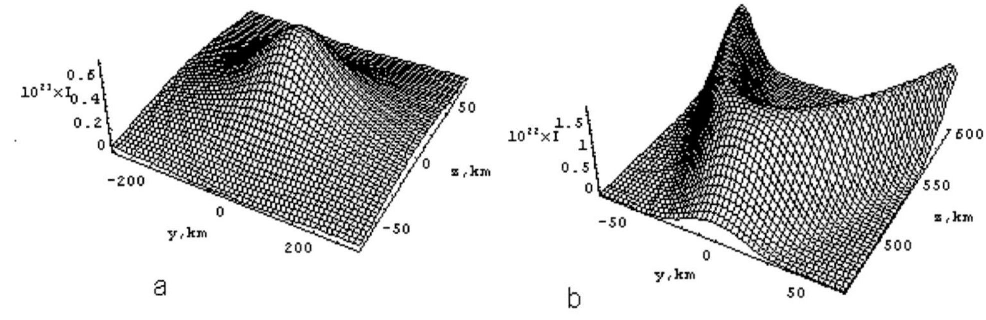

Figure 7. Distribution of the power across a flat Earth $x = -100km$ for the scattering from a strongly anisomeric, elongated irregularity with the sizes $l_x = 0.3m,\ l_z = 0.3m,\ l_y = 10km$ for the cases of the transmitter on the ground below the irregularity (a), and at a distance $500km$ from it (b).

It is evident from the modeling results that, unlike the scattering from a flattened irregularity, in the case of the scattering from an elongated irregularity the picture of intensity distribution for the quasi-radar case (Fig. 7a) differs substantially from the picture of spaced-receiver reception (Fig. 7b).

5. The Geometrical Optics Approximation in Random Inhomogeneous Ionosphere

5.1 The Geometrical Optics Approximation in Inhomogeneous Media

Ionospheric irregularities have different amplitudes, and different scales. With an increase in the irregularity amplitude, the scattering ceases to be a weak one, and the single-scattering approximation used above becomes invalid. Yet, by virtue of the characteristics of the spectrum of ionospheric irregularities, a larger size corresponds to larger amplitude. However, when the characteristic size of the medium is much larger than the wavelength and the other spatial characteristics of the field, the geometrical optics approximation (GO) and its generalizations can be used.

Neglecting the depolarization of the electromagnetic wave in a medium with large-scale irregularities (see, for example, [Tatarskii, 1971]), for the field U (that is, for any component $\mathbf{E}$ or $\mathbf{H}$) outside of the source we can obtain the Helmholtz equation.

$$\Delta U + k^2\varepsilon(\mathbf{r})U = 0\,. \tag{26}$$

The boundary conditions for $U(\mathbf{r})$ can be specified, for example, in the form $\lim_{r\to\infty} U(\mathbf{r}) = 0$ and

$$U(\mathbf{r})\big|_S = U_0(\xi) \tag{27}$$

where ξ are two-dimensional coordinates on the surface S. Very often the spherical surface is used as S, which corresponds to specifying the incident spherical wave. However, in the case of a large distance of the source from the observer and of the irregularity as the initial surface S it is more convenient to use the plane, which implies the incidence of the plane wave on an inhomogeneous medium.

In the GO approximation the field has the form [Kravtsov and Orlov, 1990]:

$$U(\mathbf{r}) = A(\mathbf{r})\exp\{ik\psi(\mathbf{r})\}, \tag{28}$$

where the eikonal $\psi(\mathbf{r})$ and the amplitude $A(\mathbf{r})$ are the solutions to the equations

$$\left(\nabla\psi(\mathbf{r})\right)^2 = \varepsilon\left(\mathbf{r}\right), \tag{29}$$

$$2\left(\nabla A\nabla\psi\right) + A\Delta\psi = 0 \tag{30}$$

with the specified boundary conditions

$$\psi(\mathbf{r})\big|_S = \psi_0(\xi), \tag{31}$$

$$A(\mathbf{r})\big|_S = A_0(\xi) \tag{32}$$

determined when specifying the initial field on the surface S in a form similar to (28).

One of the methods to solve the eikonal equation and the transfer equation involves solving the ray equations

$$\frac{d\mathbf{r}}{d\tau} = \mathbf{P} \tag{33}$$

$$\frac{d\mathbf{P}}{d\tau} = \frac{1}{2}\nabla\varepsilon. \tag{34}$$

Here

$$\mathbf{P} = \nabla\psi \tag{35}$$

and we introduced an auxiliary parameter τ which, for high-frequency plasma, corresponds to the signal delay in plasma.

The initial conditions, at $\tau = \tau_0$ for the rays are represented by the conditions of the escape from the surface S:

$$\mathbf{r}\left(\tau_0\right) = \mathbf{r}_0\left(\xi\right). \tag{36}$$

The initial conditions for the tangent pulse components $\mathbf{P}_\xi$ are derived from the determination (35), and the condition for the orthogonal component $P_\perp$ is inferred from the eikonal equation

$$P_\perp = \pm\sqrt{\varepsilon(\mathbf{r}_0) - P_\xi^2} = \pm\sqrt{\varepsilon(\mathbf{r}_0) - \left(\frac{\partial\psi}{\partial\xi}\right)^2}, \tag{37}$$

where the sign is determined in accordance with the wave propagation direction (relative to the surface S).

By solving the equations with the initial conditions, we obtain a system of rays $\mathbf{R}(\xi,\tau)$, along which the eikonal and transfer equations can be integrated (see [Kravtsov and Orlov, 1990] for more detail) to obtain

$$\psi(\xi,\tau_k) = \psi_0(\xi) + \int_{\tau_0}^{\tau_k} \varepsilon[\mathbf{R}(\xi,\tau)]\,d\tau, \tag{38}$$

$$A(\xi,\tau_k) = A_0\sqrt{\sqrt{\frac{\varepsilon(\mathbf{r}_0)}{\varepsilon(\mathbf{r})}}\left|\frac{d\mathbf{a}_0}{d\mathbf{a}}\right|} = A_0\sqrt{\sqrt{\frac{\varepsilon(\mathbf{r}_0)}{\varepsilon(\mathbf{r})}}\left|\frac{D(\tau_0)}{D(\tau_k)}\right|}, \tag{39}$$

where $d\mathbf{a}$ is the cross-section of an infinitely narrow ray tube formed by a bundle of nearby propagating rays, and $D(\tau)$ is the Jacobian of transformation from Cartesian $\mathbf{r}$ to ray coordinates ξ,τ_k. For the description of the field (28) not in ray coordinates ξ,τ_k, but in Cartesian coordinates $\mathbf{r}$ we can use the solution of the ray equations

$$\mathbf{r} = \mathbf{R}(\xi,\tau_k). \tag{40}$$

Solving (40) we can obtain

$$\begin{cases} \xi = \xi(\mathbf{r}) \\ \tau_k = \tau_k(\mathbf{r}) \end{cases}. \tag{41}$$

It can be said that the value of the parameters (41) adjust (shoot) the ray trajectory to a given point $\mathbf{r}$. Substituting (38)-(39) into (28) and using (41) to switch from ray coordinates to Cartesian coordinates we obtain the GO approximation of our problem.

If there are several systems of parameters $\{\xi^i,\tau_k{}^i\}$, satisfying the adjusting equation (40), this implies that several rays arrive at the specified point. Assuming that the propagation along each ray is proceedings independently, the solution is represented as a superposition of corresponding GO waves.

The GO solution can be somewhat simplified by decreasing the number of ray equations (33)-(34) from six to four. To do this, we replace the parameter τ by one of the coordinate, z for example, and use the eikonal equation (29) in order to exclude the z component of the pulse $\mathbf{P}$. As a result, in the Cartesian system of coordinates we obtain the following system of ray equations

$$\frac{d\boldsymbol{\rho}}{dz}=\mathbf{p}, \tag{42}$$

$$\frac{d\mathbf{p}}{dz}=\frac{1}{2\sqrt{\varepsilon(\boldsymbol{\rho},z)-p^2}}\frac{\partial}{\partial\boldsymbol{\rho}}\varepsilon(\boldsymbol{\rho},z), \tag{43}$$

$$\boldsymbol{\rho}\big|_{z_0}=\boldsymbol{\rho}_0,\ \mathbf{p}\big|_{z_0}=\mathbf{p}_0.$$

Here $\boldsymbol{\rho}=\{x,y\}$, $\mathbf{p}=\{P_x,P_y\}$, and the initial conditions are specified on the plane $z=z_0$.

The solution of the eikonal equation (38) can now be recast as

$$\psi(\mathbf{r})=\psi_0(\boldsymbol{\rho}_0)+\Phi(\mathbf{r}), \tag{44}$$

where

$$\Phi(\mathbf{r})=\int_{z_0}^{z}\sqrt{\varepsilon(\boldsymbol{\rho}(z'),z')}\sqrt{1+\left(\frac{d\mathbf{p}}{dz'}\right)^2}dz'=\int_{z_0}^{z}\frac{\varepsilon(\boldsymbol{\rho}(z'),z')dz'}{\sqrt{\varepsilon(\boldsymbol{\rho}(z'),z')-p^2}} \tag{45}$$

along the ray trajectory before the turning point and

$$\Phi(\mathbf{r})=\left(\int_{z_0}^{z_t}+\int_{z}^{z_t}\right)\sqrt{\varepsilon(\boldsymbol{\rho}(z'),z')}\sqrt{1+\left(\frac{d\mathbf{p}}{dz'}\right)^2}dz'=\left(\int_{z_0}^{z_t}+\int_{z}^{z_t}\right)\frac{\varepsilon(\boldsymbol{\rho}(z'),z')dz'}{\sqrt{\varepsilon(\boldsymbol{\rho}(z'),z')-p^2}} \tag{46}$$

along the ray trajectory after the turning point z_t, which is determined from the equation

$$\varepsilon(\boldsymbol{\rho}(z_t),z_t)=p^2(z_t) \tag{47}$$

The initial values for $\boldsymbol{\rho}(z)$ and $\mathbf{p}(z)$ are determined both by the conditions for the passage of the ray through the point of observation and by the specification of the incident wave phase. In the case of a point source the incident wave is a spherical one, and the initial values are given at $\mathbf{r}\to\mathbf{r}_0$, where $\mathbf{r}_0\equiv\{\boldsymbol{\rho}_0,z_0\}$ is the point of emission. In this case

$\psi_0 = const = 0$, and the arrival of the ray at the point of observation is warranted by the selection of $\mathbf{p}_0$.

When the source is at a large distance from the irregularity and from the observer (that is, when $z_0 \to -\infty$) the incident wave in the plane $z = z_s$ assumes the form of a plane wave. In the initial plane $z = z_s$ the "pulse" $\mathbf{p}_s = \mathbf{p}_0$ then becomes constant, and the arrival of the ray at the specified point of observation is ensured by a corresponding choice of its coordinate $\boldsymbol{\rho}_s = \boldsymbol{\rho}(z_s)$ in the initial plane $z = z_s$.

5.2. Caustics in Inhomogeneous Media

When multipathing occurs in an inhomogeneous medium, there appear caustic surfaces separating regions with a different number of rays. The aforementioned scheme of construction of the GO approximation can not be used in the vicinities of these surfaces, because the solution of the system (40) is absent here, and the cross-section of the ray tube $|d\mathbf{a}|$ goes to zero. Hence, the amplitude (39) of the GO approximation becomes infinite, whereas the solution of the wave equation (26) must be finite. From this follows a need to use, in the neighborhood of the caustics, generalizations of the GO approximation.

Let us consider some very simple examples of ray patterns with caustics.

1 *Reflection of the plane wave from a plane layer*

Let the plane wave be incident at an angle θ_0 to an inhomogeneous semi-space $z > 0$, whose dielectric permittivity $\varepsilon(z)$ does not depend on coordinates x, y and decreases monotonically with increasing z (see Fig. 8).

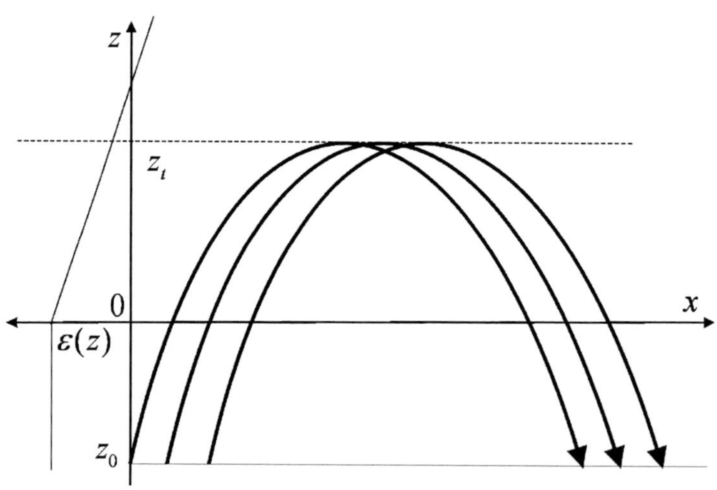

Fig. 8. Reflection of the plane wave from a plane-layered smoothly inhomogeneous medium for $p_y = 0$

Taking into consideration that $\varepsilon(z)$ is independent of x, y, the Snell law follows from the ray equations (42)-(43)

$$\mathbf{p} = const = \mathbf{p}_0 \tag{48}$$

In view of (48) the expressions for the eikonal (44) and the amplitude (39) for the reflected wave assume the form

$$\psi(\mathbf{r}) = \psi(\boldsymbol{\rho}_0, z_0) + \mathbf{p}\bar{\boldsymbol{\rho}}(\mathbf{p}) + \left(\int_{z_0}^{z_t} + \int_{z}^{z_t}\right)\sqrt{\varepsilon(\boldsymbol{\rho}(z'), z') - p^2}\, dz' \tag{49}$$

$$A = A(z_0)\sqrt{\frac{\sqrt{\varepsilon(z_0)}\cos\theta_0}{\sqrt{\varepsilon(z)}\cos\theta}} = A(z_0)\sqrt[4]{\frac{\varepsilon(z_0) - p_0^2}{\varepsilon(z) - p^2}}, \tag{50}$$

where

$$\bar{\boldsymbol{\rho}}(\mathbf{p}) = \left(\int_{z_0}^{z_t} + \int_{z}^{z_t}\right)\frac{\mathbf{p}\, dz'}{\sqrt{\varepsilon(\boldsymbol{\rho}(z'), z') - p^2}}, \tag{51}$$

and

$$\boldsymbol{\rho}_0 = \boldsymbol{\rho} - \bar{\boldsymbol{\rho}}(\mathbf{p}). \tag{52}$$

In this case the caustic coincides with the plane of the turning points

$$\varepsilon(z_t) = p^2, \tag{53}$$

where the amplitude (50) tends to infinity

2 *The incidence of the plane wave on a plane layer.*

Let a point source be at the point with coordinates $\mathbf{r} = \mathbf{r}_0 \equiv \{\boldsymbol{\rho}_0, z_0\}$ (see Fig. 9). In this case

$$\psi(\mathbf{r}) = \mathbf{p}(\boldsymbol{\rho} - \boldsymbol{\rho}_0) + \Phi(\mathbf{p}), \tag{54}$$

where $\mathbf{p}(\boldsymbol{\rho}, z)$ is determined from the adjusting equation

$$\boldsymbol{\rho} - \boldsymbol{\rho}_0 = \bar{\boldsymbol{\rho}}(\mathbf{p}, z). \tag{55}$$

For the amplitude we have (see [Kravtsov and Orlov, 1990] for more detail):

$$|A(\mathbf{r})| = B^{(0)} \sqrt{\frac{\sqrt{\varepsilon(z_0)}}{\sqrt{\varepsilon(z) - p^2}} \left| \sin\theta_0 \middle/ \left(\bar{\rho} \frac{\partial \bar{\rho}}{\partial \theta_0} \right) \right|} \tag{56}$$

If the axis x passes through the source and the observer, then $\bar{\rho}(\beta_0) = \bar{x}(\beta_0)$ and $p_y = 0$.

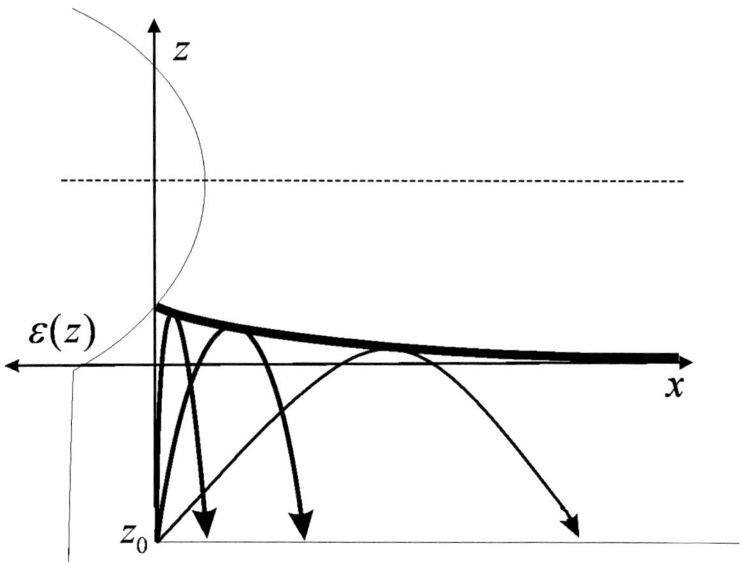

Figure 9. Reflection of the spherical wave from a plane-layered opaque smoothly inhomogeneous medium for $p_y = 0$

Unlike the plane wave, the initial coordinate of the ray $\boldsymbol{\rho}_0$ in this case is determined by the position of the point source, and the condition of ray passage through the point of observation is warranted by the adjusting equation (55).

It is evident from (56) that the caustic of a spherical wave lies not on the surface of the turning points as in the case with the plane wave but on the surface where

$$\frac{\partial \bar{\rho}}{\partial \beta_0} = 0. \tag{57}$$

Thus the caustic depends not only on the inhomogeneity of the medium but also on the kind of incident wave.

The ray pattern for the point source depends on the degree of transparency of the layer.

If in the layer, with an increase of z, the function $\varepsilon(z)$ decreases monotonically down to a negative value, then, as in the case of the plane wave, there is one branch of the caustic (see Fig. 9). Below the caustic there are two rays: the incident ray, and the reflected ray, and there are no rays above the caustic. This caustic, when $\boldsymbol{\rho} = 0$, coincides with the turning point and tends to the lower boundary of the layer when $\rho \to \infty$.

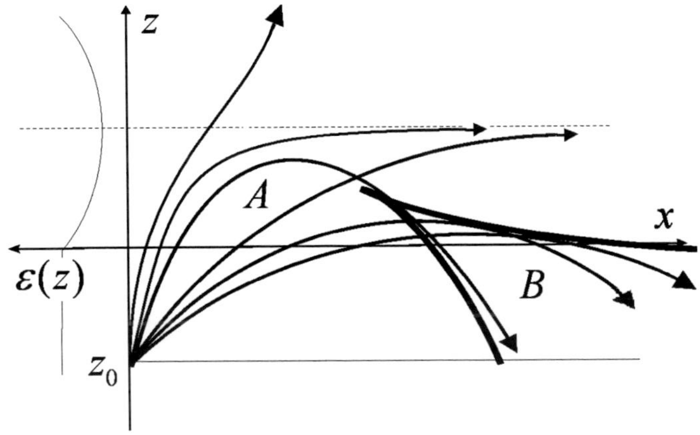

Fig. 10. Rays and caustic of the spherical wave in a semi-opaque smoothly inhomogeneous layer for $p_y = 0$

When the function $\varepsilon(z)$ is not a monotonic one and does not reach a negative value even in its minimum, some of the rays (see Fig. 10) penetrate the layer. In this case the caustic has two branches separating the space into two regions (see Fig. 10). In region A there is one ray which corresponds below the layer to the incident wave, and above the layer to the refracted wave. In region B, two layer-reflected rays are added. In the theory of ionospheric propagation of radio waves, these rays are called the upper (Pedersen) ray, and the lower ray. In the case of the lower ray, with distance away from the observed along the layer, the turning point tends to the lower boundary of the layer. The turning point of the upper ray comes closer to the height of the minimum of $\varepsilon(z)$ (or, in accordance with (1), (3) to the height of the maximum of electron density $N(\mathbf{r})$), and the ray becomes grazing, approaching a critical one. The characteristics of the quasi-critical propagation are investigated in [Tinin, 1973].

If the source is emitting short radio pulses, then two reflected pulses will be observed. As the boundary of the dead zone (the intersection of the terrestrial surface with the lower branch of the caustic) is approached, the amplitude of these pulses increases, and the interference of two waves is enhanced; after that, the pulses merge together on the boundary of the dead zone. Taking into consideration the dependence (1), (3) of dielectric permittivity on the frequency, a similar picture can be observed at a fixed location of the observer but with a changing radiation frequency. With an increase in frequency, the boundary of the dead zone comes closer to the observer, and at the frequency referred to as the maximum usable frequency (MUF), the observer is on the boundary of the dead zone (at the caustic). Above the MUF, no ionosphere-reflected signals are observed within the GO approximation.

However, it is not uncommon to observe, at the background of a large-scale ionospheric irregularity, random ionospheric irregularities smaller in size [Yeh and Liu, 1982], additional refraction from which leads to the illumination of the caustic shadow region.

5.3. Random Ionospheric Irregularities

In view of the fact that the random fluctuations of dielectric permittivity are usually much smaller than the value of background permittivity we can write

$$\varepsilon(\mathbf{r}) = \bar{\varepsilon}(\mathbf{r}) + \tilde{\varepsilon}(\mathbf{r}), \tag{58}$$

where $\bar{\varepsilon}(\mathbf{r}) = \langle \varepsilon(\mathbf{r}) \rangle$ is background dielectric permittivity, and $\tilde{\varepsilon}(\mathbf{r})$ is dielectric permittivity of random irregularities, with

$$\left|\tilde{\varepsilon} / \bar{\varepsilon}\right| \sim \nu \ll 1. \tag{59}$$

We can now find the solution for the ray equations (42)-(43) in the form of series in the small parameter ν:

$$\boldsymbol{\rho} = \boldsymbol{\rho}^{(0)} + \boldsymbol{\rho}^{(1)} + \boldsymbol{\rho}^{(2)} + \ldots, \tag{60}$$

$$\mathbf{p} = \mathbf{p}^{(0)} + \mathbf{p}^{(1)} + \mathbf{p}^{(2)} + \ldots, \tag{61}$$

where

$$\boldsymbol{\rho}^{(i)} = O\left(\nu^{i}\right),\ \mathbf{p}^{(i)} = O\left(\nu^{i}\right),\ i = 0,1,2,\ldots$$

Substitution of (58) and the series from perturbation theory (60)-(61) into (38)-(39) provides [Kravtsov and Orlov, 1990] similar series for the eikonal ψ and the Jacobian of transformation D that is needed to determine the amplitude A (39). The other approach [Rytov et al, 1989] uses the perturbation method in solving not the ray equations but directly the eikonal equation (29). It can be demonstrated [Kim and Tinin, 2007], that these approaches are equivalent.

On substituting ψ and D into (28), we obtain

$$U(\mathbf{r}) = \frac{\bar{U}(\mathbf{r})}{\sqrt{1 + \tilde{D}(\mathbf{r})}} \exp\left\{ik\tilde{\psi}(\mathbf{r})\right\} \tag{62}$$

where

$$\tilde{\psi}(\mathbf{r}) = \psi^{(1)}(\mathbf{r}) + \psi^{(2)}(\mathbf{r}) + \ldots \tag{63}$$

$$\tilde{D}(\mathbf{r}) = \left[D^{(1)}(\mathbf{r}) + D^{(2)}(\mathbf{r}) + \ldots\right] / \bar{D}(\mathbf{r}) \tag{64}$$

$\bar{U}(\mathbf{r})$ is the field in the absence of an irregularity; $\psi^{(i)} = O\left(\nu^{i}\right)$, $D^{(i)} = O\left(\nu^{i}\right)$.

It is most customary to confine oneself to the first approximation of perturbation theory, where

$$\begin{cases} \tilde{\psi} \approx \psi^{(1)} \\ \tilde{D} \approx D^{(1)} / \bar{D} \end{cases} \tag{65}$$

When the background medium is homogeneous ($\bar{\varepsilon} = 1$), the formula (62) within the small-angle approximation assumes the known form [Rytov et al, 1989]:

$$U(\mathbf{r}) = \frac{A_{sp}}{|z - z_0| \sqrt{1 + \tilde{D}_{sp}(\mathbf{r})}} \exp\left\{ ik\left[z - z_0 + \frac{1}{2} \frac{(\boldsymbol{\rho} - \boldsymbol{\rho}_0)^2}{z - z_0} + \tilde{\psi}_{sp}(\mathbf{r}) \right] \right\} \tag{66}$$

where

$$\tilde{\psi}_{sp}(\mathbf{r}) = \frac{1}{2} \int_{z_0}^{z} \tilde{\varepsilon}\left[\mathbf{s}(\boldsymbol{\rho}, \boldsymbol{\rho}_0, z'), z' \right] dz', \tag{67}$$

$$\tilde{D}_{sp}(\mathbf{r}) = \frac{1}{2} \int_{z_0}^{z} \Delta_{\perp} \tilde{\varepsilon}\left[\mathbf{s}(\boldsymbol{\rho}, \boldsymbol{\rho}_0, z'), z' \right] (z - z')(z' - z_0)/(z - z_0) dz' = (z - z_0) \frac{\partial^2}{\partial \boldsymbol{\rho} \partial \boldsymbol{\rho}_0} \tilde{\psi}_{sp}(\mathbf{r}), \tag{68}$$

$$\mathbf{s}(\boldsymbol{\rho}, \boldsymbol{\rho}_0, z') = \boldsymbol{\rho} \frac{z' - z_0}{z - z_0} + \boldsymbol{\rho}_0 \frac{z - z'}{z - z_0} \tag{69}$$

is the ray in the absence of an irregularity;

$\Delta_{\perp} = \frac{\partial^2}{\partial x^2} + \frac{\partial^2}{\partial y^2}$ is the transverse Laplacian.

When $z_0 \to -\infty$ we obtain the field of the plane wave incident normal to an inhomogeneous semi-space $z > z_s$:

$$U(\mathbf{r}) = \frac{\bar{U}(z_s)}{\sqrt{1 + \tilde{D}_{pl}(\mathbf{r})}} \exp\left\{ ik \tilde{\psi}_{pl}(\mathbf{r}) \right\} \tag{70}$$

where

$$\tilde{\psi}_{pl}(\mathbf{r}) = \frac{1}{2} \int_{z_s}^{z} \tilde{\varepsilon}\left[\boldsymbol{\rho}, z' \right] dz' \tag{71}$$

$$\tilde{D}_{pl}(\mathbf{r}) = \frac{1}{2} \int_{z_0}^{z} \Delta_{\perp} \tilde{\varepsilon}\left[\boldsymbol{\rho}, z' \right] (z - z') dz' \tag{72}$$

If it is assumed in (66) and (70) that $\tilde{D}_{sp}(\mathbf{r})=0$ и $\tilde{D}_{pl}(\mathbf{r})=0$, we obtain the formulas of the phase approximation usually [Rytov et al, 1989] used in investigating the spatial and frequency coherence functions, the angular spectrum, and the mean envelope of the pulsed signal in a random inhomogeneous medium.

The conditions of validity of the GO formulas (66) and (70) imply primarily the relatively large irregularity size when compared with the characteristic size of the field. This characteristic size is usually represented by the Fresnel radius of the incident wave at the location of the irregularity. When the irregularity is located at $z \approx z_c, \boldsymbol{\rho} \approx \boldsymbol{\rho}_c = 0$ the Fresnel radius is equal to

$$l_{fs} = \sqrt{\lambda(z-z_c)(z_c-z_0)/(z-z_0)} \tag{73}$$

for the spherical wave, and

$$l_{fp} = \sqrt{\lambda(z-z_c)} \tag{74}$$

for the plane wave.

For irregularities with the size l_ε larger than the corresponding Fresnel radius

$$l_{fs,fp} < l_\varepsilon \tag{75}$$

we can then make an estimate

$$\left|\tilde{D}_{sp,pl}(\mathbf{r})\right| < k\left|\tilde{\psi}\right|. \tag{76}$$

Hence, in the case of small phase fluctuations

$$k\left|\tilde{\psi}\right| < 1 \tag{77}$$

fluctuations for the irregularity larger than the Fresnel zone, we can use the phase GO approximation. However, with increasing phase fluctuations, even for the irregularity with the size larger than the Fresnel zone, the GO approximation can become invalid because of the appearance of random multipathing and random caustics in the region, where

$$\left|\tilde{D}_{sp,pl}(\mathbf{r})\right| \geq 1 \tag{78}$$

and the amplitude in (66) and (70) can become infinite.

With the background refraction taken into account, the region in which random caustics appear, is determined by the condition

$$\left|\tilde{D}_{sp,pl}(\mathbf{r})\right| \square \left|\bar{D}\right| \tag{79}$$

The region, in which background caustics appear, is determined by the condition $\bar{D} \to 0$, and the condition (79) for the appearance of random caustics is made significantly easier. It is therefore of interest to describe the wave field with the inclusion of background caustics and random irregularities simultaneously.

6. Go Generalizations in Statistical Problems of the Ionospheric Radio Wave Propagation, and Tendencies of Their Perfection

6.1. Integral Representations of the Field

There are well-known [Brekhovskikh, 1973; Felsen and Marcuvitz, 1973] integral representations of the spherical wave field in a plane-layered medium obtained by solving the Helmholtz equation using the method of separation of variables. These integral representations constitute a spherical wave in the form of the superposition of plane waves propagating in a plane-layered medium. As has been pointed out above, the caustics of plane waves are surfaces of the turning points. Yet, the turning points are well apart from the caustic of the spherical wave, from its lower branch at least (see Fig. 10). This different placement of the caustics for plane and spherical waves allows using the expansion of the solution in terms of GO plane waves in the neighborhood of the caustic of the spherical wave. One such generalization of the method of separation of variables is the interference integral [Orlov, 1972; Kravtsov and Orlov, 1999; Avdeev et al, 1988].

In a broad sense, the interference integral (II) is any expansion in terms of any waves referred to as partial. However, when the II is used in statistical problems [Tinin, 1983; Tinin et al, 1992; Afanasiev and Tinin, 1982], it is desirable that the partial waves should be convenient for a subsequent calculation of the statistical moments. In the ray II, the solution for the wave equation (26) is represented as an integral over the GO approximations of the partial waves.

When the field in the neighborhood of the source is expanded in terms of partial waves, the partial waves are plane waves incident on an inhomogeneous medium. In the Maslov method [Maslov and Fedoryuk, 1976; Kravtsov, 1968; Kravtsov and Orlov, 1999], a somewhat different integral representation is constructed for the field in which the partial waves are represented by waves which are plane waves not in the neighborhood of the source but in the neighborhood of the observer.

That is, both in the II method and in the Maslov method, integral representations for the field have the form of Fourier-transforms in some coordinates of the source and the observer, respectively. For the Green function, in view of its symmetry, from the results of the II method we can obtain the results [Permitin and Fraiman, 1969] of the Maslov method by interchanging the location of the source and the observer. For that reason, at this stage we shall principally consider, following [Tinin, 1983; Tinin et al, 1992; Afanasiev and Tinin, 1982], the II method. Of course, in a plane-layered medium the results of both methods degenerate into the known results of the method of separation of variables [Brekhovskikh, 1973; Felsen and Marcuvitz, 1973].

Using the Fourier-transform, it is an easy matter to obtain the expansion in terms of plane waves for the field of a point source (the spherical wave)

$$U_0(\mathbf{r}) = \int_{-\infty}^{\infty} U_{s0}(\mathbf{s},\mathbf{r}) d^2 s\,, \tag{80}$$

where

$$U_{s0}(\mathbf{s},\mathbf{r}) = \frac{ikA_{sp}}{4\pi\sqrt{1-s^2}} \exp\left\{ik\left[\mathbf{s}(\boldsymbol{\rho}-\boldsymbol{\rho}_0) + \sqrt{1-s^2}\left(z-z_s\right)\right]\right\}. \tag{81}$$

We seek the solution for the field of a point source in an inhomogeneous medium in a form similar to (80):

$$U(\mathbf{r}) = \int_{-\infty}^{\infty} U_s(\mathbf{s},\mathbf{r}) d^2 s\,, \tag{82}$$

where $U_s(\mathbf{s},\mathbf{r})$ satisfies a homogeneous Helmholtz equation (26) with the boundary conditions ensuing from the form (81) when $z \to z_0$

$$U_s(\mathbf{s},\mathbf{r})\big|_{z=z_0} = U_{s0}(\mathbf{s},\mathbf{r}) = \frac{ikA_{sp}}{2\pi\sqrt{1-s^2}} \exp\left\{ik\left[\mathbf{s}(\boldsymbol{\rho}-\boldsymbol{\rho}_0)\right]\right\}. \tag{83}$$

In determining $U_s(\mathbf{s},\mathbf{r})$ we use the aforementioned technique from subsection 5.3 for obtaining the GO approximation for the plane wave with the boundary conditions

$$\psi_s\big|_{z=z_0} = \mathbf{s}(\boldsymbol{\rho}-\boldsymbol{\rho}_0)\,, \tag{84}$$

$$A_s(\mathbf{s})\big|_{z=z_0} = A_0(\mathbf{s}) = \frac{ikA_{sp}}{2\pi\sqrt{1-s^2}}\,. \tag{85}$$

We can now avail ourselves of the solution (65) for the amplitude and eikonal and use, as the rays, the solutions of the ray equations with the boundary conditions characteristic for the plane wave specified originally on the plane $z = z_0$ and received at the point $\mathbf{r} = \{\boldsymbol{\rho}, z\}$

$$\begin{cases} \overline{\mathbf{p}}(z_0) = \mathbf{s} \\ \overline{\boldsymbol{\rho}}(z) = \boldsymbol{\rho} \end{cases}. \tag{86}$$

When at the background of a plane-layered medium there are random irregularities, for the field outside of the irregularity we obtain the expression

$$U = \int_{-\infty}^{\infty} A_0(\mathbf{s}) \exp\left\{ik\left[\bar{\psi}(\mathbf{s}) + \tilde{\psi}(\mathbf{s})\right]\right\} d^2 s, \tag{87}$$

where

$$\bar{\psi}(\mathbf{s}) = (\boldsymbol{\rho} - \boldsymbol{\rho}_0)\mathbf{s} + \left(\int_{z_0}^{z_t} + \int_{z}^{z_t} \right) \sqrt{\bar{\varepsilon}\left(\boldsymbol{\rho}(z'), z'\right) - s^2}\, dz' \tag{88}$$

$$\tilde{\psi}(\mathbf{s}) = \left(\int_{z_0}^{z_t} + \int_{z}^{z_t} \right) \frac{\tilde{\varepsilon}\left(\bar{\boldsymbol{\rho}}(z'), z'\right)}{\sqrt{\bar{\varepsilon}(z') - s^2}} dz' \tag{89}$$

Integration in (88)-(89) is performed over unperturbed rays $\bar{\boldsymbol{\rho}}(z')$, that is, over the solutions of the ray equations (43)-(43) when $\varepsilon(\mathbf{r}) = \bar{\varepsilon}(z)$ with the boundary conditions (86).

The solution (87) contains a random function $\tilde{\psi}(\mathbf{s})$ under the integral sign in the exponential only. This alleviates a calculation of the field moments for a normal distribution $\tilde{\psi}(\mathbf{s})$. For the second frequency-spatial moments we obtain

$$\begin{gathered} \left\langle U(\omega_1, \boldsymbol{\rho}_1) U^*(\omega_2, \boldsymbol{\rho}_2) \right\rangle = \int_{-\infty}^{\infty}\int_{-\infty}^{\infty} d^2 s_1 d^2 s_2 A_0(\omega_1, \mathbf{s}_1) A_0^*(\omega_2, \mathbf{s}_2) \times \\ \exp\left\{ \frac{i}{c}\left[\omega_1 \bar{\psi}(\omega_1, \mathbf{s}_1, \boldsymbol{\rho}_1) - \omega_2 \bar{\psi}(\omega_2, \mathbf{s}_2, \boldsymbol{\rho}_2)\right] - \frac{1}{2c^2}\left\langle \left[\omega_1 \tilde{\psi}(\omega_1, \mathbf{s}_1, \boldsymbol{\rho}_1) - \omega_2 \tilde{\psi}(\omega_2, \mathbf{s}_2, \boldsymbol{\rho}_2)\right]^2 \right\rangle \right\}. \end{gathered} \tag{90}$$

From this, assuming that $\boldsymbol{\rho}_1 = \boldsymbol{\rho}_2 = \boldsymbol{\rho}$, $\omega_1 = \omega_2 = \omega$, we obtain the expression for the mean intensity

$$\begin{gathered} \left\langle |U|^2 \right\rangle = \int_{-\infty}^{\infty}\int_{-\infty}^{\infty} d^2 s_1 d^2 s_2 A_0(\omega_1, \mathbf{s}_1) A_0^*(\omega_2, \mathbf{s}_2) \times \\ \exp\left\{ ik\left[\bar{\psi}(\omega, \mathbf{s}_1, \boldsymbol{\rho}) - \bar{\psi}(\omega, \mathbf{s}_2, \boldsymbol{\rho})\right] - \frac{k^2}{2}\left\langle \left[\tilde{\psi}(\omega, \mathbf{s}_1, \boldsymbol{\rho}) - \tilde{\psi}(\omega, \mathbf{s}_2, \boldsymbol{\rho})\right]^2 \right\rangle \right\}. \end{gathered} \tag{91}$$

Since the partial waves in the expression under the integral sign (87) have caustics at the turning points, the formulas (90)-(91) can be used to calculate the moments in regions at a distance from the surfaces of the turning points, specifically in the neighborhood of the lower branch of the caustic of the spherical wave (see Fig.10) or the MUF, if the field behavior is considered not in the space but in the frequency range.

In calculations in the neighborhood of the upper branch of the caustic (Figs. 9-10), the observer can be in the neighborhood of the turning points , where the expressions (87)-(89) does not hold, and it is necessary to use other partial waves [Orlov, 1974].

The main contribution to the integral (91) is made by the neighborhood $\mathbf{s}_1 \approx \mathbf{s}_2$, because outside of this neighborhood the expression under the integral sign either strongly oscillates or decrease rapidly. Taking this into account, in (91) we can use the new integration variables

$$\begin{cases} \mathbf{s} = (\mathbf{s}_1 + \mathbf{s}_2)/2 \\ \mathbf{s}_- = \mathbf{s}_1 - \mathbf{s}_2 \end{cases} \tag{92}$$

and evaluate the integral over $\mathbf{s}_-$ asymptotically. As a result, we obtain

$$\left\langle |U|^2 \right\rangle = \frac{2\pi}{k^2} \int_{-\infty}^{\infty} d^2 s \frac{|A_0(\mathbf{s})|^2}{\sqrt{\Delta(\mathbf{s})}} \exp\left\{ -\frac{1}{2\Delta(\mathbf{s})} \left[\left(\frac{\partial \bar{\psi}}{\partial s_x} \right)^2 \sigma_y^{\ 2} - 2 \frac{\partial \bar{\psi}}{\partial s_x} \frac{\partial \bar{\psi}}{\partial s_y} K_{xy} + \left(\frac{\partial \bar{\psi}}{\partial s_y} \right)^2 \sigma_x^{\ 2} \right] \right\}, \tag{93}$$

where

$$\frac{\partial \bar{\psi}}{\partial \mathbf{s}} = \boldsymbol{\rho} - \boldsymbol{\rho}_0 - \bar{\boldsymbol{\rho}}(\mathbf{p}, z), \tag{94}$$

$$\sigma_x^{\ 2} = \left\langle \left(\frac{\partial \tilde{\psi}}{\partial s_x} \right)^2 \right\rangle, \quad \sigma_y^{\ 2} = \left\langle \left(\frac{\partial \tilde{\psi}}{\partial s_y} \right)^2 \right\rangle, \quad K_{xy} = \left\langle \frac{\partial \tilde{\psi}}{\partial s_x} \frac{\partial \tilde{\psi}}{\partial s_y} \right\rangle, \quad \Delta = \sigma_x^{\ 2} \sigma_y^{\ 2} - K_{xy}^{\ 2}. \tag{95}$$

We will assume that the reference frame is chosen such that $y = y_0 = 0$.

Under conditions of the ionospheric propagation of short radio waves the wave field in the horizontal plane usually has a more regular behavior than in the vertical plane, and the wave propagates in the plane $y = 0$. Based on this, after performing an asymptotic integration over s_y , from (93) we obtain a two-dimensional formula

$$\left\langle |U|^2 \right\rangle = \frac{(2\pi)^{3/2}}{k^2} \int_{-\infty}^{\infty} ds_x \frac{|A_0(s_x, 0)|^2}{\sigma_x \bar{x}(s_x)} \exp\left\{ -\frac{[x - x_0 - \bar{x}(s_x)]^2}{2\sigma_x^{\ 2}} \right\}. \tag{96}$$

This formula coincides with the solution of a two-dimensional problem [Tinin, 1983] up to the coefficient, which is associated with the difference of the linear source in [Tinin, 1983] from the point source considered here.

When $\sigma_x^2 \to 0$, that is, for the case of a small dispersion of the partial rays, the integral (96) can be evaluated asymptotically by the saddle point approximation. The main contribution to the integration in this case is made by the vicinity $s \approx s_{xc}$ of zero of the expression in the exponential (96) which is determined from the adjusting equation of the unperturbed ray

$$x - x_0 = \overline{x}(s_{xc}). \tag{97}$$

The width of this vicinity is

$$\delta_s = \sigma_x(s_{xc}) \Big/ \left| \frac{\partial \overline{x}(s_{xc})}{\partial s_{xc}} \right|. \tag{98}$$

If this interval (98) is small enough, so that in the exponential under the integral sign (96) we could change

$$\overline{x}(s_x) \approx \overline{x}(s_{xc}) + \frac{\partial x(s_{xc})}{\partial s_{xc}}(s_x - s_{xc}) \tag{99}$$

and in the other terms assume that $s_x = s_{xc}$, then we obtain

$$\left\langle |U|^2 \right\rangle = \frac{(2\pi)^2 |A_0(s_{xc},0)|^2 s_{xc}}{k^2 \overline{x}(s_{xc}) \left| \frac{\partial \overline{x}(s_{xc})}{\partial s_{xc}} \right|}. \tag{100}$$

Taking (56) into account, it is easy to see that (100) is the intensity of the GO wave in the absence of random irregularities.

Note that in the neighborhood of the caustic

$$x \approx x_m \equiv \overline{x}(s_m) + x_0, \tag{101}$$

where s_m is determined from

$$\frac{\partial \overline{x}(s_m)}{\partial s_m} = 0, \tag{102}$$

the expression (100) becomes invalid. However, the asymptotic formula (96) remains also valid in the neighborhood of the caustic provided that [Tinin, 1983; Tinin et al, 1992]

$$\sigma_x > \Lambda \equiv \left(\frac{1}{2k^2} \left| \frac{\partial^2 \overline{x}(s)}{\partial s^2} \right|_{s_m} \right)^{1/3} \tag{103}$$

which implies the case where the ray dispersion σ_x exceeds the width of the near-caustic zone Λ.

If in the neighborhood of the boundary of the dead zone $x \approx x_m$ the expansion

$$\overline{x}(s) \approx x_m + \frac{1}{2} \frac{\partial^2 \overline{x}(s)}{\partial s^2} \bigg|_{s=s_m} (s - s_m)^2 \tag{104}$$

is possible, then from (96) we can obtain a local asymptotics [Tinin, 1983; Tinin et al, 1992]:

$$\left\langle |U|^2 \right\rangle = \frac{(2\pi)^2 \left| A_0(s_m, 0) \right|^2 s_m}{k^2 x_m \sqrt{\sigma_x(s_m) \left| \frac{\partial^2 x(s_m)}{\partial s_m{}^2} \right|}} \exp\left(-\frac{\chi^2}{4} \right) D_{-1/2}(\chi) \tag{105}$$

where $\chi = (x_m - x) / \sigma_x(s_m)$, $D_{-1/2}(\chi)$ is a function of a parabolic cylinder order of -1/2 [Abramowitz and Stegun, 1964]. Comparing with the known field behavior in the neighborhood of the caustic [Kravtsov and Orlov, 1999], we must remark that at the occurrence of random irregularities in the mean intensity there disappear oscillations caused by interference of the waves corresponding to the upper and lower rays.

Similar to (93), (96), (105) asymptotic expressions were obtained in [Tinin et al, 1992; Afanasiev et al, 1982; 1983] for the spatial coherence function of a random field of the wave and its spectrum in the neighborhood of the caustic, as well as for the frequency coherence function, and for the mean form of the pulsed signal. Fig. 11 presents the results [Afanasiev et al, 1983] from calculating the mean form of the pulsed signal for a number of frequencies in the neighborhood of and above the MUF. A Gaussian pulse of width 2×10^{-2} sec is used as the initial pulse, and the path length is 2000 km; the profile of the ionospheric layer was specified thus:

$$\overline{\varepsilon}(z) = 1 - \left(\frac{f_c}{f} \right)^2 \exp\left\{ -\left(\frac{z - z_{\max}}{d} \right)^2 \right\} \tag{106}$$

In calculations, the parameters of the layer were taken to be $f_c = 7$ MHz, $z_{\max} = 300$ km, $d = 100$ km.

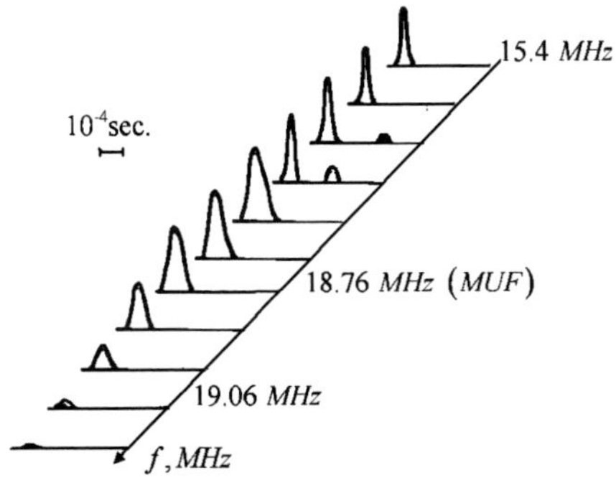

Figure 11. The results from modeling [Afanasiev et al, 1983] of the mean form of the Gaussian radio pulse at a number of frequencies in the neighborhood of and above the MUF

The correlation function was specified in the form

$$K \equiv \langle \tilde{\varepsilon}(\mathbf{r}_1)\tilde{\varepsilon}(\mathbf{r}_2)\rangle = \sigma_{\varepsilon 0}^2 \{1-\bar{\varepsilon}[(\mathbf{r}_1+\mathbf{r}_2)/2]\}^2 \exp\{-(\mathbf{r}_1-\mathbf{r}_2)^2/l_e^{\,2}\} \quad (107)$$

for $\sigma_{\varepsilon 0} = 10^{-2}$ и $l_e = 10$ km.

It is evident from Fig. 11 that, as the MUF is approached, the pulses corresponding to the lower and upper rays come closer together, and in the immediate vicinity of the MUF they merge into a single pulse. With a further increase in frequency (above the MUF), unlike calculations within the GO approximation, here we have a pulse, with a decreasing amplitude. This is just the result of illumination of the caustic shadow region by random rays.

In addition, the formulas obtained in [Tinin, 1983; Tinin et al, 1992; Tinin and Kulizhsky, 1994; Kulizhsky and Tinin, 1994] by the II method were used to calculate the relative intensity variance (the scintillation index) in the illuminated region, in the neighborhood of a regular caustic, and in the caustic shadow region.

6.2. The Validity Range of Integral Representations of the Field in a Random Inhomogeneous Medium, and Some Avenues for Their Expansion

Before proceeding to estimating the validity range of the GO generalizations considered above, we must remark that a number of assumptions were made when deriving the integral representation (87) of the field of the spherical wave in an inhomogeneous medium. First, it was assumed that a set of partial waves is complete enough for the field of the spherical wave in an inhomogeneous medium to be described by means of the superposition of these waves. Secondly, it was assumed that the source and the observer are sufficiently far from the surface of the turning points, that is, from a regular caustic of plane waves. And, thirdly, the use of the GO approximation for partial waves implies the absence of multpathing, caustics and strong focusing of these waves not only because of a regular (background) irregularity but also due to a random irregularity of the propagation medium. The combined analysis of all these factors involves difficulties. At this stage, for estimating the validity range of the integral representations we shall compare the properties that ensue from the formulas obtained, with the known properties of the waves in inhomogeneous media.

In the case of solving the problem of wave propagation in a background plane-layered medium with no random inhomogeneities, the II method changes to the method of separation of variables, and the field generated in such a medium is well understood [Brekhovskikh, 1973; Felsen and Marcuvitz, 1973]. Therefore, at this point we consider the use of the II method to describe the propagation of the spherical wave through a random inhomogeneous medium. For a better comparison with existing research results we confine ourselves to the small-angle approximation of the wave propagation in a random inhomogeneous medium with no background inhomogeneity. In this case we can obtain a corresponding integral representation either from (87)-(89) by assuming in it that $\bar{\varepsilon}(z)=1$ and $|\mathbf{s}|=s \ll 1$ or by using the small-angle approximation of plane waves (70)-(72) as partial waves in (82). As a result we obtain

$$U=\frac{ikA_{sp}}{2\pi}\int_{-\infty}^{\infty}\frac{d^2s}{\sqrt{1+D^I_{pl}}}\exp\left\{ik\left[z-z_0+\mathbf{s}(\boldsymbol{\rho}-\boldsymbol{\rho}_0)-\frac{s^2}{2}(z-z_0)+\tilde{\psi}^I_{pl}\right]\right\}, \quad (108)$$

where

$$\tilde{\psi}^I_{pl}=\frac{1}{2}\int_{z_0}^{z}\tilde{\varepsilon}\left[\boldsymbol{\rho}-\mathbf{s}(z-z'),z'\right]dz', \quad (109)$$

$$D^I_{pl}=\frac{1}{2}\int_{z_0}^{z}\Delta_{\perp}\tilde{\varepsilon}\left[\boldsymbol{\rho}-\mathbf{s}(z-z'),z'\right](z-z')dz'. \quad (110)$$

To estimate the validity range of the phase GO approximation of particle waves we can make use of the analysis of the validity of the phase GO approximation of plane waves made in subsection 5.3 of this Chapter. If the irregularity with the size l_ε is located at $z' \approx z_c$, $\boldsymbol{\rho}\approx\boldsymbol{\rho}_c=0$, then the condition of validity of the GO approximation for partial waves can be obtained from the formulas (74)-(75)

$$l_\varepsilon>\sqrt{\lambda(z-z_c)}. \quad (111)$$

It is of interest that for the spherical wave the Fresnel radius l_{fs} is smaller than the Fresnel radius of the plane wave l_{fp}

$$l_{fs} = \sqrt{\frac{\lambda(z - z_c)(z_c - z_0)}{z - z_0}} = \sqrt{\lambda(z - z_c)}\sqrt{\frac{z_c - z_0}{z - z_0}} < l_{fp} . \quad (112)$$

Hence it follows that the condition of validity of the GO approximation for partial waves is more stringent than that for the initial spherical wave.

The condition of validity of the phase GO approximation of partial waves, that is, the condition of smallness of the amplitude fluctuations of these waves (108) in this case has the form

$$(z - z_c)\max\left|\tilde{\varepsilon}\right| < l_\varepsilon \quad (113)$$

It is interesting that for the initial spherical wave, a similar condition of weakness of the amplitude fluctuations has the form

$$\frac{z_c - z_0}{z - z_0}(z - z_c)\max\left|\tilde{\varepsilon}\right| < l_\varepsilon \quad (114)$$

As has been shown above (see (76)), the condition of weak phase fluctuations (77) suffice to satisfy these conditions when (111) holds. At the occurrence of strong phase fluctuations the condition of validity of the phase GO approximation of partial waves can be violated even for irregularities with the size larger than the Fresnel radius. It follows from the comparison of the conditions (113), (114) that, with distance away from the source, the observer initially reaches the region of occurrence of strong fluctuations (random caustics) for plane partial waves. After that, at large distances there appears a region of random focusing for the spherical wave. Such a sequence of appearance of caustics of the partial waves and of the spherical wave can change in the neighborhood of a background caustic of the spherical wave, where additional background focusing compensates the divergence of the spherical wave, and statistical focusing of the spherical wave can occur before focusing of the partial waves. In this case the II with the use of the phase GO approximation of partial waves can be used in the region of regular and statistical focusing of the spherical wave.

Note that the region that is essential for integration in (108) is defined by the relation

$$\left|\mathbf{s} - \mathbf{s}_c\right| \sim 1/\sqrt{k(z - z_0)} , \quad (115)$$

where $\mathbf{s}_c$ is the stationary phase point (at $\boldsymbol{\rho} = \boldsymbol{\rho}_0 = 0,\ \mathbf{s}_c = 0$).

In view of this, for the irregularity with the size

$$l_\varepsilon > (z - z_c)/\sqrt{k(z - z_0)} \quad (116)$$

the II is calculated by the stationary phase method to give the GO approximation. Hence it follows that for the plane wave (that is, when $z_0 \to -\infty$) the GO approximation is valid for the irregularity with any size, which is inconsistent with the known results of investigation into the propagation of plane waves in inhomogeneous media [Rytov et ai, 1989]. This is explained by the use of the incomplete set of partial waves when $z_0 \to -\infty$, which corresponds in this case to an attempt to represent the plane wave as a set of plane waves.

The above reasoning can all be repeated for the Maslov method by interchanging the point of radiation and the point of reception [Tinin, 1998]. More specifically, it can be demonstrated [Tinin, 1998] that at a large distance of the observer from the inhomogeneous medium, the Maslov method changes to the GO approximation for all irregularities and instead of an enhancement of the intensity fluctuations with subsequent saturation, following from the known research results [Rytov et ai, 1989; Yakushkin, 1974], their decrease is observed. The fact that the Maslov method and the II method have limitations in the statistical problems is suggested by the absence of a symmetry of the Green functions obtained by these methods (with the phase GO approximation of partial waves), which is inconsistent with the reciprocity theorem.

The cited paper [Tinin, 1998] provides a comparison of the results from these methods with the known results obtained by the method of smooth perturbations, and by the method of phase screens. It follows from this comparison that the results from the II method and the Maslov method are in qualitative agreement with the known results on the condition of symmetry of the problem in, for example, a statistically homogeneous medium. A quantitative agreement is observed in the case of a symmetric position of the source and the observer in range relative to the inhomogeneous region of a small extent. Such a picture is rather typical for the oblique propagation of radio waves where the source and the observer are on the ground. Also, the path of the wave in the ionosphere with irregularities often does not exceed several hundred kilometers, whereas the distance between the correspondents reaches 3000-4000 km. In this case, integral representations can be used to calculate statistical characteristics of ionospheric radio waves in the neighborhood of and above the MUF, that is, in the neighborhood of a caustic, and in the caustic shadow region.

The development of the methods to describe fields in a multiscale inhomogeneous medium involves switching over to the new systems of partial waves which, on the one hand, describe sufficiently accurately the incident wave, and, on the other, random irregularities have a smaller influence on their GO amplitudes. It is worth noting at this point that the partial wave in the II method are plane waves in the neighborhood of the source, and in the Maslov method they are plane waves in the neighborhood of the observer. That is, the partial rays that are convergent in the Maslov method are divergent in the II method, and vice versa. One would therefore expect that the combined use of the two methods would decrease the focusing effect of the partial waves.

Let us consider the two-point propagator in the form of a product of two Green functions

$$g(\boldsymbol{\rho}_1, \boldsymbol{\rho}_2, \boldsymbol{\rho}_{01}, \boldsymbol{\rho}_{02}) = U(\boldsymbol{\rho}_1, \boldsymbol{\rho}_{01}) U^*(\boldsymbol{\rho}_2, \boldsymbol{\rho}_{02}) \tag{117}$$

We introduce the summary-difference coordinates

$$\boldsymbol{\rho}_+ = (\boldsymbol{\rho}_1 + \boldsymbol{\rho}_2)/2, \quad \boldsymbol{\rho}_- = \boldsymbol{\rho}_1 - \boldsymbol{\rho}_2 \tag{118}$$

We can then obtain within the small-angle approximation the parabolic equation for $g = g(\boldsymbol{\rho}_1, \boldsymbol{\rho}_2, \boldsymbol{\rho}_{01}, \boldsymbol{\rho}_{02})$ in a random inhomogeneous medium with no background inhomogeneity:

$$\frac{\partial g}{\partial z} = \frac{i}{k}\frac{\partial^2 g}{\partial \boldsymbol{\rho}_+ \partial \boldsymbol{\rho}_-} + \frac{ik}{2}\left[\tilde{\varepsilon}\left(\boldsymbol{\rho}_+ + \frac{\boldsymbol{\rho}_-}{2}, z\right) - \tilde{\varepsilon}\left(\boldsymbol{\rho}_+ - \frac{\boldsymbol{\rho}_-}{2}, z\right)\right] g = 0 \qquad (119)$$

Using the method of two-scale expansions [Frankenthal et al, 1984; Mazar et al, 1985; Whitman and Beran, 1985] or a combination of the II and the Maslov method [Tinin, 1998], we can obtain a mixed integral representation for the two-point propagator which agrees in the regime of weak scattering with the results from method of smooth perturbations, and in the regime of strong fluctuations with the results from the phase screen method. The results of this approach were generalized to the case of an inhomogeneous background in the region illuminated by regular rays [Mazar and Felsen, 1987] and in the neighborhood of a background caustic [Tinin et al, 2000].

However, often (for instance, in the case of diagnostics) it is necessary to know not the two-point propagator but the field itself or the one-point-propagator. Furthermore, a mixed integral representation for the two-point propagator has retained the absence of symmetry of the solution for the points of the sources and the observers. That this representation is inaccurate is also suggested by some results of calculations [Martin and Flatte, 1990].

Therefore, a double (in the coordinates of the source and the observer) weighted Fourier-transform was used in [Kravtsov and Tinin, 2000] in determining the wave field in an inhomogeneous medium. An integral representation was obtained in [Kravtsov and Tinin, 2000] for the field within the small-angle representation for a two-dimensional problem, and a generalization to a three-dimensional case was carried out in [Kravtsov and Tinin, 2003]. The method of the fifth parameter was used in [Tinin, 2004; Tinin et al, 2002] to obtain integral representations for the field without assuming the small-angle approximation. The resulting integral representation makes it possible to describe not only the weak and strong forward scattering but also to take into account the back scattering. As was shown in [Tinin, 2004], taking into account the following approximation for the rays when constructing the partial waves in this integral representation will make it possible to simultaneously allow for the effects of different-scale irregularities and, in particular, obtain asymptotic expressions for the scintillation index which were derived previously using the method of path integrals [Wang and Dashen, 1993]. In this manner, after regular refraction is taken into account, this method will provide a means of correctly taking into account the various irregularities in the neighborhood of a regular caustic.

7. On the Possibility of Using the II Method in Investigating the Propagation of Radio Emission in the Solar Corona

A combination of the effects of random irregularities and a background (regular) irregularity occurs not only in the case of the ionospheric propagation of radio waves but also in radio diagnostics of laboratory plasma (for instance, fusion plasma in devices of the tokomak type

[Mazzucato, 2006]), and in investigations of stellar and planetary atmospheres in, for example, the case of probing the solar corona using natural and artificial radiation sources [Yakovlev, 2002]. Recently there appears a number of publications [Afanasiev, 2005; Afanasiev, 2006a; Afanasiev, 2006b; Afanasiev and Altyntsev, 2006] using the II method to estimate the effects of random irregularities in radio probing of the solar corona. They consider a two-dimensional problem in cylindrical coordinates centered on the Sun's center which is represented in this case as an infinite cylinder. Fig. 12 presents the geometry of the problem considered in [Afanasiev, 2005].

Since the Sun is opaque to any radio frequencies, the situation is identical to the one in the ionosphere (see Fig. 9) but differs by the geometry. The point source that is located in $\mathbf{r} = \mathbf{r}_0$ produces a field of rays with a caustic separating the region illuminated by two rays from the shadow region. As the observer moves across the surface, he traverses the caustic surface, as is the case in the aforementioned ionospheric problem. It is this similarity of the problems that prompted the authors [Afanasiev, 2005; Afanasiev, 2006b; Afanasiev and Altyntsev, 2006] to use here the results of [Tinin, 1983; Tinin et al, 1992] by representing them in cylindrical coordinates.

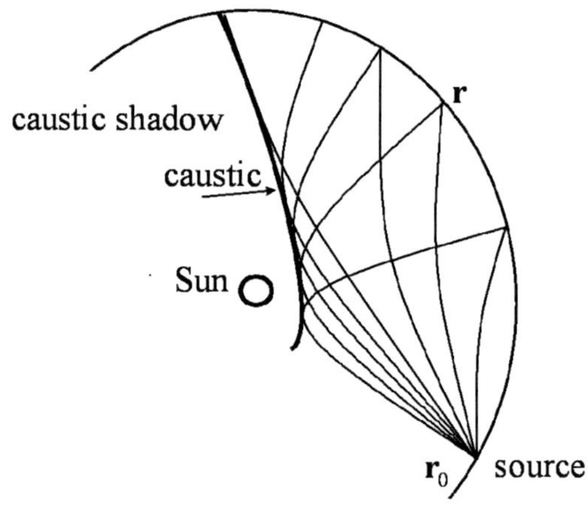

Figure 12. Schematic of the rays and caustic in the solar corona.

In a cylindrical layered medium, unlike the plane-layered medium considered above, a constant of separation is equal to the quantity $a = \sqrt{\varepsilon(\mathbf{r}_0)}r_0 \sin\theta_0$ called the impact parameter. Therefore, now integration in II is performed [Avdeev et al, 1988] over the impact parameter a rather than over $s_x = \sqrt{\varepsilon(\mathbf{r}_0)} \sin\theta_0$. Unfortunately, the authors of [Afanasiev, 2005; Afanasiev, 2006b; Afanasiev and Altyntsev, 2006] do not provide the boundary conditions under which the eikonal and transfer equations, and the corresponding ray equations for partial waves are solved. However, as was shown above, the specification of these conditions determines the system of partial waves used. For instance, when the II method is used, it is necessary to specify the initial conditions for the eikonal on some surface in the neighborhood of the source. Then the ray problem must be solved as a boundary problem with one condition for the exit angle θ_0 determined from $a = \sqrt{\varepsilon(\mathbf{r}_0)}r_0 \sin\theta_0$, and with the other condition for the ray hitting the point of observation.

The main difference of the problems of radio probing of the solar corona from ionospheric problems implies a substantial (for radio astronomy) absence of the symmetry of the solar corona. In particular, the inhomogeneous model of a random field of inhomogeneities of the solar corona is used in [Afanasiev, 2005], and the coordinates of the source and the observer differ drastically. As was shown in subsection 6.2, however, the II method is unsuitable for the description of the effects of random irregularities in such situations. Moreover, the applicability of the II method is violated in the situation considered in [Afanasiev, 2006b; Afanasiev and Altyntsev, 2006], where the properties of radio bursts are explained by the complicated non-layered structure of background plasma in the solar corona. Besides, switching over to integration in terms of partial waves is unlikely to alleviate the problem of determining the field here, because the structure of partial waves in a background medium is most likely more complicated than the field structure of the initial source.

The II method is used in [Afanasiev, 2006a] to describe the field of the plane wave (the pulsar wave) propagating through the solar corona. The initial conditions are also not specified here, and hence the system of partial waves is not concretized. Since the parameter of expansion in terms of partial waves here is a coordinate at the front of the incident wave, the representation used correspond in fact, to a greater extent, to the phase approximation of the Huygens-Kirchhoff method [Mironov, 1981] rather than to the II method. Furthermore, the partial waves must have the form of cylindrical waves emitted by points on the surface of the phase front of the incident wave and calculated within the phase GO approximation. The difference from the II method implies mainly a difference of the partial wave amplitude (not defined in the cited reference). Note that in the supplement to [Afanasiev, 2006a] the author gives (without making a reference to the source [Afanasiev and Tinin, 1982]) the results of analysis (using the II method) of the frequency coherence function in three-dimensional Cartesian coordinates, which have no any relation to the problem discussed in the basic text of [Afanasiev, 2006a].

Thus the extension of the methods of interference investigations to other problems requires a careful analysis of the limitations of the II method under conditions of the new problem, and an accurate construction of the partial waves including the solution of the ray equations with corresponding boundary conditions.

8. Conclusion

In this paper we have considered mainly the effects of two kinds of multiscaleness, one of which is associated with the anisomery of the ionospheric irregularity, and the other is due to the formation of the background caustic in the ionosphere with random irregularities.

The anisomery effects were investigated for the weak scattering within the Born approximation. As analysis and modeling show, for the elongated irregularity, with an increase in the longitudinal size, there appears first the aspect character of the scattering with a maximum on the conical surface. Furthermore, with an increase in the longitudinal size, there is taking place a narrowing of the illuminated ring. When anisomery is strong, the longitudinal size of the irregularity exceeds the corresponding Fresnel radius. In this case the dependence of the spatial distribution of the scattered intensity on the irregularity scale is reversed – with an increase in the longitudinal size, the width of the illuminated region increases.If there is a difference from the cylindrical character of the irregularity (for instance, when one of the transverse sizes increases), here appears an intensity redistribution across the illuminated ring. With a further increase in this transverse size, that is, when the cylindrical irregularity turns into a disk-shaped irregularity, the intensity distribution in the transverse plane tends to two points which are specular relative to the plane of the disk. This corresponds to a change of the conical surface, in the neighborhood of which the scattered energy from the elongated irregularity is concentrated, to two rays: the direct ray and the specularly scattered ray from the plane of the disk. However, when the two major axes of the disk-shaped irregularity become larger than the Fresnel radius, the dependence of the size of the illuminated region on the size of the disk changes – with an increase in the size of the disk, there is an increase in the size of the illuminated region. Unlike the scattering from the elongated irregularity, the size of the region on the terrestrial surface illuminated by the scattered wave from a thin flattened irregularity whose plane is parallel to the terrestrial surface depends weakly on the location of the transmitter relative to this irregularity. These properties of the scattering from anisomeric irregularities must be taken into account in meteor radio communication, in probing of the upper atmosphere using meteor radars, and in investigating the polar ionosphere using the SuperDARN radar system [Vallie`res et al, 2003] that is currently being deployed.

The effects of irregularities with different sizes larger than the Fresnel radius can be examined using the GO approximation or its integral generalizations representing the superposition of the phase GO approximation of partial waves having the form of plane waves, either in the neighborhood of the source (in the II method) or in the neighborhood of the observer (in the Maslov method). Whilst in a number of problems of ionospheric radio wave propagation (for instance, in the neighborhood of a regular caustic in the ionosphere which is statistically homogeneous in range) these methods describe rather well the combined effects of different-scale irregularities, their further development is necessary for a wider use. As is apparent from the research results [Tinin, 1998], the use of mixed integral representations holds the greatest promise for such purposes. An especially good agreement with existing results is provided by the use of a double weighted mixed integral representation for the field without the small-angle approximation [Tinin, 2004]. With such an approach, it is possible to investigate the scattering from anisomeric irregularities not only for the conditions of weak scattering but also for the conditions of strong scattering and backscattering. With the progression of mixed integral representation, one can anticipate the advent of methods that correctly describe the complicated mechanisms of radio wave propagation in such strongly regular and random inhomogeneous media as the ionosphere, solar plasma, and fusion plasma. This, in turn, would provide a means of developing the methods of diagnostics of such media with improved resolution.

This work was done with support from the Russian Foundation for Basic Research, grant No. 06-02-16357.

References

Abramowitz, M.; Stegun, I. A. *Handbook of Mathematical Functions*; National Bureau of Standards: Washington, 1972.

Afanasiev, A. N. *J. Atmos. Solar-Terrest. Phys.* 2005, 67, 1002 -1013.

Afanasiev, A. N. *Astron. Rep.*, 2006, 50, 762–772.

Afanasiev, A. N.; Altyntsev, A.T. *Solar Phys.* 2006, 234, 151-167.

Afanasiev, A. N. *Solar Phys.* 2006, 238, 87–104.

Afanasiev, N. T.; Tinin, M. V. *Issled. Geomagn., Aeron. Fiz. Solntsa.* 1982, 61, 236-241.

Afanasiev, N. T.; Grozov, V. P.; Krasikov, A. A.; Nosov, V. E.; Tinin, M. V. *Issled. Geomagn., Aeron. Fiz. Solntsa.* 1983, 63, 180-196.

Avdeev, V. B.; Demin, A. V.; Kravtsov, Yu. A.; Tinin, M. V.; Yarygin, A. P. *Izv.Vyssh. Uchebn. Zaved. Radiofiz.* 1988, 31, 1279-1294.

Brekhovskikh, L. M. *Waves in layered media;* Nauka: Moscow, 1973 (in Russian).

Budden, K. G. *The propagation of radio waves;* Cambridge University Press: Cambridge, 1985.

Felsen, L. B.; Marcuvitz, N. *Radiation and Scattering of Waves;* Prentice-Hall: Englewood Cliffs, NJ, 1973.

Frankenthal, S ; Whitman, A. M.; Beran, M. J. *J. Opt. Soc. Am.* 1984, *A6* , 585-594.

Ginzburg, V. L. *The Propagation of Electromagnetic Waves in Plasma*; Fiz.-mat. Literatura: Moscow, 1960.

Ishimaru, A. *Wave Propagation and Scattering in Random Media. Volume 1. Single Scattering and Transport Theory*; Academic Press: New York, 1978.

Kim, B. C.; Tinin, M. V. *J. Geod.* 2007, 81, 189–199.

Kravtsov, Yu. A. *Sov. Phys. Acoust.* 1968, 14, 1-17.

Kravtsov, Yu. A. *Rep. Prog. Phys.* 1992, 55, 39-112.

Kravtsov, Yu. A.; Fejzulin, Z. I.; Vinogradov, A. G. *Transmission of radio-waves through an Earth's atmosphere*; Radio i svyaz: Moscow, 1983 (in Russian).

Kravtsov, Yu. A.; Orlov Yu. I. *Geometrical Optics of Inhomogeneous Media*; Springer-Verlag: New York, 1990.

Kravtsov, Yu. A.; Orlov, Yu. I. *Caustics, Catastrophes and Wave Fields;* Springer-Verlag: New York, 1999.

Kravtsov, Yu. A.; Tinin M. V. *Radio Sci.* 2000, 35, 1315-1322.

Kravtsov, Yu. A.; Tinin M.V, *Cosmic Research.* 2003, 41, 357-358.

Kulizhsky, A. V.; Tinin, M. V. *Waves Random Media.* 1994, 4, 149-165.

Martin, J. M.; Flatte, S. M. *J. Opt. Soc. Am.*1990, *A7*, 838-847.

Maslov, V. P.; Fedoryuk, M. V. *Quasi-Classical Approximation for Quantum Mechanics Equations;* Nauka: Moscow, 1976 (in Russian).

Mazzucato, E. *Plasma Phys. Control. Fusion.* 2006, 48, 1749–1763.

Mazar, R; Gozani, J.; Tur, M. *J. Opt. Soc. Am.* 1985, *A2*, 2152-2160.

Mazar, R.; Felsen, L. B. *J. Acoust. Soc. Am.* 1987, 81, 925-937.

McKinley, D. W. R. *Meteor Science and Engineering*; McGraw Hill: New York, 1961.

Mironov, V. L. *Propagation of a Laser beam in a Turbulent Atmosphere*; Nauka: Novosibirsk, 1981 (in Russian).

Orlov, Yu. I. *Trudy MEI* . 1972, 119, 82-91.

Orlov, Yu. I. *Izv. Vyssh. Uchebn. Zaved. Radiofiz.* 1974, 17, 1035–1041.

Permitin, D. V.; Fraiman, A. A. *Izv. Vyssh. Uchebn. Zaved. Radiofiz.* 1969, 12, 1836-1841.

Rytov, S. M.; Kravtsov, Yu. A.; Tatarskii, V. I. *Introduction to Statistical Radiophysics, vol. 4, Wave Propagation through Random Media*; Springer: New York, 1989

Tatarskii, V. I. *The Effect of a Turbulent Atmosphere on Wave Propagation*; National Technical Information Service: Springfield, Va, 1971.

Tinin, M. V. *Issled. Geomagn., Aeron. Fiz. Solntsa.* 1973, 29, 157-166.

Tinin, M. V. *Izv. Vyssh. Uchebn. Zaved. Radiofiz.* 1983, 26, 36–43.

Tinin, M. V. *Waves Random Media.* 1998, 8, 329-338.

Tinin, M. V. *Waves Random Media.* 2004, 14, 97-108.

Tinin, M. V.; Afanasyev, N. T.; Mikheev, S. M.; Pobedina, A. P.; Fridman, O. V. *Radio Sci.* 1992, 27, 245-255.

Tinin, M. V.; Afanasyev, N. T.; Kulizhsky, A. V. *Izv. Vyssh. Uchebn. Zaved. Radiofiz.* 2000, 43, 17–27.

Tinin, M. V.; Kim, B. C. *Izv. Vyssh. Uchebn. Zaved. Radiofiz.* 2004, 47, 1057-1065.

Tinin, M. V.; Kim, B. C. *J. Atmos. Solar-Terrest. Phys.* 2006, 68, 977–988.

Tinin, M. V.; Kim, B. C.; Kolesnik, S. N. *Waves Random Complex Media.* 2005, 15, 61–69 .

Tinin, M. V.; Kolesnik, S. N.; Kravtsov, Yu. A. *Proc. SPIE.* 2002, 4678, 605-609.

Tinin, M. V.; Kulizhsky, A. V. *Waves Random Media.* 1994, 4, 83-95.

Vallie`res, X.; Villain, J.-P.; Andre´ R. *Radio Sci.* 2003, 38, 1003, doi:10.1029/2001RS002550.

Whitman, A. M.; Beran, M. J. *J. Opt. Soc. Am.* 1985, *A*2, 2133-2143.

Wang, G.; Dashen, R. *J. Opt. Soc. Am.* 1993, *A*10, 1226–1232.

Yakovlev, O. I.. *Space Radio Science;* Taylor and Francis Inc.: New York, 2002.

Yakushkin, I. G. *Izv. Vyssh. Uchebn. Zaved. Radiofiz.* 1974, 17, 1350-1355.

Yeh, K. C.; Liu, C. H. *Proc. IEEE.* 1982, 70, 324-360.

In: Solar Physics Research Trends
Editor: Pingzhi Wang, pp. 233-258

ISBN 978-1-60021-987-0
© 2008 Nova Science Publishers, Inc.

Chapter 6

A Fresh Insight into the Energy Source and Transport for Large Solar Flares

*G. M. Simnett**
School of Physics and Astronomy, University of Birmingham, B15 2TT, UK

Abstract

The physical processes involved in the total energy release and transport in a large solar flare are not well understood. The energy is derived from magnetic fields and the magnetic free energy arises through convective motions of plasma at levels below the photosphere. In studying flare energetics the goal is to establish a comprehensive and self-consistent picture of the sources and transport of energy. The first objective must be to define the relevant volume encompassing the important energy sources and sinks and it is believed that most previous studies of flare energetics have grossly underestimated this. A second objective is to identify the principal energy transport mechanisms. Much of the work to date has concentrated on mass motions, Joule heating, radiative transport and thermal conduction. Energy transport by non-thermal ions has been given a secondary, often minor role, although transport by non-thermal electrons has featured in an unsatisfactory way in some models. In this paper we re-examine flare energetics by postulating the relevant volume for the energy source to be more like the volume of the corona involved in the initiation of the coronal mass ejection and eruption of coronal magnetic fields that almost invariably accompany a major flare. We then examine how the energy transport might proceed if the initial result of the release of magnetic energy is to create a population of non-thermal charged particles in which the ions contain the bulk of the energy released. An inevitable outcome of this approach is that the active region flare becomes primarily a secondary phenomenon to energy release elsewhere. The processes occurring in the volume of the corona containing the magnetic structures that erupt into the interplanetary medium are the primary driver of the energetic transient, which includes the flare. Often these erupting structures contain excess mass over the ambient corona, in which case they are identified as coronal mass ejections. The photospheric/chromospheric active region does play an important role, including provision of some energy, in forming the spectacular dynamic structures that for over a century have been the epitome of the optical flare, and we incorporate this into the global picture.

*E-mail address: gms@star.sr.bham.ac.uk

1. Historical Development of Solar Flare Models

A typical large solar flare occurs in an active region of the photosphere which contains sunspots with strong magnetic fields. The clearest manifestations of the flare are a rapid optical brightening of a part of the existing active region plage and a burst of non-thermal hard X-rays and microwave radio emission. The X-rays clearly indicate the presence of near-relativistic electrons, as the only plausible physical X-ray production mechanism is via collisional bremsstrahlung as the electrons encounter dense regions in the chromosphere. The microwaves are mainly produced by the same population of high energy electrons moving in the magnetic field and gamma ray emission is occasionally observed when protons above about 20-30 MeV interact in the dense atmosphere. It is common for energetic protons to be emitted into space along with relativistic electrons.

The intensity-time histories of the electromagnetic emissions tend to become less impulsive as the photon energy goes down (radio wavelengths excepted), so that in the H α spectral line the flare may last 1-2 hours. Most large flares are accompanied by a coronal mass ejection (CME), often with a speed in excess of 1000 km/s. (A big CME is actually made up of several, if not many, discrete structures which erupt within minutes of each other. However, we shall ignore this subtle point and refer to the event simply as a CME.) There is evidence in a few flares for electrons up to around 100 MeV and protons up to around 25 GeV. Justification for the latter comes from the observation of events at ground level on the Earth near the geomagnetic equator. Estimates for the total flare energy have been put at a few times 10^{25} J for the largest events (see Svestka, 1976 for a review) but the energy involved in a large CME may be an order of magnitude higher than this. The challenge is to formulate a model which accounts for the principal features of flares in a consistent way, which we interpret here as including the CME and energetic particle emission plus a plausible source for the energy and mass.

It is appropriate to review briefly our understanding of solar flare models around the beginning of the space era, as it was only then that the emissions over the full spectral region from radio waves to hard X-rays (and subsequently gamma rays) began to be appreciated. Also, it was in the early 1960s that the full scope of energetic particle emissions started to be revealed. The reader is referred to the AAS-NASA Symposium on "The Physics of Solar Flares", Ed W.N. Hess, NASA SP-50, 1964, for an excellent account of the observations, concepts and theory existing at that time. Two things are clear from this early work: (1) the energy is derived from the magnetic field and (2) the energy released in collapsing magnetic fields goes into energetic particles.

Elliot (1964) suggested that energetic particles, gradually accelerated in the corona over a period of the order of a day before a major flare, could suddenly be dumped into the lower solar atmosphere to provide the energy for the optical flare. Part of the motivation for this suggestion was to provide a relatively slow coronal acceleration process for producing the > 1 GeV protons emitted at the time of some large flares. The highest energy protons very occasionally reach energies up to 25 GeV. Such a gradual acceleration to these high energies should be visible via gamma ray production. The lack of such gamma ray emission prior to large flares was unsupportive of this model and it was largely abandoned. The problems that Elliot was trying to address were the energy supply for flares and the acceleration mechanism for highly relativistic protons. These have not yet been solved. It is interesting to note that Carmichael (1964) emphasised that the magnetic field which was important for the flare was not that in the active region but the coronal field "high above the active region" perhaps extending to 2 $R_{\odot}$. However, Carmichael's field concept was rather unphysical, in that it was mushroom-shaped, with a narrow connection to the active region. Elliot assumed the storage field was that coming from a bipolar sunspot group, and only extended about 0.2 solar radii above the photosphere. Warwick (1962) on the other hand had suggested that the acceleration was in or near the photosphere. These authors all considered that the energy for powering the flares came from energetic protons. Such models became unpopular mainly due to the difficulty of accelerating the high energy protons coupled with the invisibility of such protons. All flare emissions were ascribed to the energy loss of energetic charged particles.

A major observational advance came in 1973/74 with the three Skylab space missions, which had a variety of instruments dedicated to observing the Sun. The X-ray region of the spectrum became accessible from space, and from such data the emphasis shifted to explaining the intense X-ray bursts seen at the start of flares, rather than the energetic protons responsible for the ground-level events seen at Earth. Since then there have been other dedicated solar space missions, notably the Solar Maximum Mission (SMM), Hinotori, Yohkoh and SOHO, all of which have in their own way advanced our knowledge of the solar flare.

It is now over 30 years since Švestka (1976) published his book on "Solar Flares", but in many ways his review of flare models is still an extremely useful starting point for those trying to understand the fundamental issues of the flare problem. In particular Svestka summarised no fewer than 37 basic properties of flares which a flare theory should explain. Inevitably some are more important than others and we concentrate here on those which are more global in their significance rather than addressing details, interesting as these might be. The reason we adopt this approach is that there may be a variety of ways of providing an answer to, say, the problem of producing homologous flares, whereas others which have causality or timing constraints, or mass or energy supply constraints are more critical and these clearly need to be given prominence. The relative timing of various emissions is very important and has frequently been overlooked. For example, if the primary energy transfer mechanism is into non-thermal electrons in the corona, then we would expect the signatures of such electrons, such as hard X-ray and microwave radio emission, to be present at the start of the flare, probably with the radio emission slightly ahead in time. As this is not the case, then the postulate should be dismissed as untenable.

In an attempt to focus on the energy supply, it was recognised that emerging magnetic flux might well provide a suitable energy source in approximately the right place if a magnetic bipole emerged within an active region. This was the essence of the flare model proposed by Heyvaerts et al, (1977). Around the same time, the energy supply for the solar flare hard X-ray burst was investigated, and Lin and Hudson (1976) pointed out that if the hard X-rays were produced via thick target bremsstrahlung from an electron population with a power law distribution in energy, then perhaps half the flare energy in the impulsive phase resided in the electrons. This left very little energy available for ions. In flares without gamma rays, ions are almost invisible, so it was understandable that the focus remained on accounting for the electrons, which were certainly present to produce the hard X-rays and the microwave radio emission. Thus the early concept of powering flares with energetic protons remained out of favour.

It is an accident of history that we had essentially a century of increasingly sophisticated studies of solar flares, mainly using optical telescopes, before the development of spaceborne coronagraphs by Tousey (1965), which eventually led to the discovery of coronal mass ejections when a coronagraph was flown on OSO-7 (Koomen et al, 1975). CMEs did not play a significant role in Svestka's review of flares, and although the CME phenomenon was addressed by the Skylab and SMM workshops it remained on the periphery of the flare problem. Both Svestka and Webb et al (1980) were aware of the huge amount of magnetic energy convected through the corona by a large CME. Estimates of this energy were in excess of the estimated energy for even the largest flares. Hypothetically, had solar eclipses been commonplace and of long duration we might well have discovered CMEs before flares. In this case the thrust of solar research may have been to account for the optical flare as a by-product of the CME. As we know, events proceeded from the other direction.

In fact, mass ejections from the Sun at the times of large flares were known from attempts to explain the Forbush decrease phenomenon. In 1960, using Pioneer V instrumentation, Coleman et al (1960) measured the arrival at around 1 AU of a strong magnetic cloud, with the field strength of 50 nT. Carmichael (1962) in his review of high energy solar particle events, noted that plasma clouds were emitted from the Sun around the time of large flares and carried strong magnetic fields which were ultimately responsible for scattering galactic cosmic rays, thus resulting in the intensity (Forbush) decrease when the clouds enveloped the Earth. This concept was first advanced by Gold (1959). Emission of a plasma cloud through coronal magnetic field closed out to around 4 $R_\odot$ was suggested in order to explain subtle details relating to the release of energetic particles from the Sun (Simnett, 1971). We note in passing that the ejection of significant mass into the interplanetary medium was actually observed by Hansen (1961) following a flare behind the solar limb on 20 November, 1960. We now know that Hansen's observations were of a prominence eruption, which is often the brightest part of a large CME. When the Culgoora solar radio telescope began observing the Sun routinely in 1968, moving type IV radio bursts were seen to accompany mass ejections out to the edge of the field-of-view at 6 $R_\odot$ (Sheridan, 1970). Such bursts probably result from emission from energetic trapped electrons and this strengthened the shift in emphasis from protons to electrons.

It would be remarkable if the two most dramatic and energetic solar transients, the large coronal mass ejection and large flare, were truly independent. The search for a causal connection is on-going, but so far has not been established. A separate, but longstanding, problem is the source of the energy released in a large flare as in many, if not all, cases this does not appear to come from the active region magnetic field. We discuss here a model whereby non-thermal energetic particles provide both the link between CMEs and flares, and a significant part of the flare energy. From an energetics viewpoint ions dominate, and to avoid gamma ray production which would be observable by sensitive current instrumentation, the ions should be below a few MeV/nucleon. Plausible ways of generating an adequate ion population within the CME structure, which for large events extends over 90 degrees in sun-centred coordinates, are indicated. Evidence for the release into space of some of this population is presented. The non-thermal electrons responsible for the hard X-ray and microwave signatures are generated near the chromospheric flare site, together with the highly relativistic ions which often accompany large flares.

It is not always evident what the key evidence is regarding the problems with the conventional explanations for solar flares. In the next section we discuss briefly some of the more relevant ones.

2. Current Status of Solar Flare Models

Later in this paper we concentrate on the search for a causal link between CMEs and large flares, which we emphasise would have been a natural way for the subject to proceed had CMEs been discovered first. A key feature of our model is the provision of the majority of the flare energy via non-thermal ions (protons) so after reviewing first the important features of a major flare, we then review the case for powering flares with protons rather than electrons.

2.1. Selected Important Flare Properties from Svestka (1976)

Svestka (1976) did not give much prominence to the recently-discovered CMEs in his discussion of the basic properties of flares. Nevertheless, these properties are still relevant, and we select the following as being extremely pertinent to the present work. Items marked " " are direct quotations from Svestka's work.

(1) The flare starts several minutes before the onset of the impulsive phase, which is generally taken as the onset of the rapidly-time-varying hard X-ray burst. "Therefore the electrons responsible for the hard X-ray burst cannot be the primary process giving rise to the flare."

(2) "When a quiescent filament is present close to the position of the flare, it becomes activated minutes or tens of minutes before the flare occurrence. This gives evidence that in many (and possibly all) cases changes in the magnetic field start tens of minutes before the flare onset."

(3) The most important flares occur in active regions that are magnetically complex. Flares that occur in old, spotless regions only accelerate protons, at best, to low energies (eg $\sim$10 MeV).

(4) In terms of energetic particles emitted into space, on an energy basis, the electron flux is orders of magnitude smaller than the proton flux.

(5) "The earliest manifestation of a flare is an increase in the soft X-ray and EUV flux, reflecting a heating in the corona or transition layer" This cannot be achieved by input from non-thermal electrons as they would give a bremsstrahlung spectrum yielding photons up to the highest electron energy; which is not observed.

(6) The stationary type IV radio burst is probably produced by relativistic electrons which are trapped by corona magnetic fields. This demonstrates that coronal magnetic loops can trap energetic particles for extended periods (see also (7)).

(7) "The very broad spectral lines seen in the loop prominences " (following a flare) "appear to be a visible manifestation of non-thermal particles trapped in flare-associated magnetic loops."

Svestka argued that no single flare model can explain all aspects of the complex flare phenomena. We would like to challenge this assertion by virtue of having the apparent differences or complexity in the flare produced by variations in the magnetic topology and intensity and spectrum of the accelerated particles; and the size of the erupting coronal magnetic structure.

Some of Svestka's more speculative assertions may ultimately prove to be correct. For example, he suggested that 1-15% of accelerated particles go downwards, while most are trapped in coronal arches. It is difficult to prove this, but if true, then it would suggest that the heating mechanism for post flare loops could well be Coulomb collisions of energetic particles.

2.2. More Recent Developments

In the decade following Svestka's work, two publications may be singled out for special mention. The first is the Proceedings of the Second Skylab Workshop on Solar Flares (Ed: P. A. Sturrock, 1980) and the second is the Solar Maximum Mission (SMM) Flare Workshop Proceedings (Ed: M. R. Kundu and B.E. Woodgate, 1986). More recently the topic has been addressed by workshops on High Energy Solar Physics (Ed: R. Ramaty , N. Mandzhavidze and X.-M.Hua, 1995) and High Energy Solar Physics: Anticipating HESSI (Ed: R. Ramaty and N. Mandzhavidze, 2000). It has been recognised over the years by many sources that coronal phenomena may cause the flare, although Wagner (1984) stressed that CMEs may occur without a flare. This is also very evident from the incidence of CMEs at solar minimum. Earlier Sheeley et al, (1975) pointed out that magnetic reconnection involving large amounts of energy does not necessarily lead to a flare. Therefore unless we complicate matters by invoking a variety of physical processes for CMEs, flares do not produce CMEs. This is a cornerstone of our thesis.

When trying to assess the energy balance, it is difficult to know which components are primary and which are secondary. For example, it is most probable that the energetic protons derive their energy via magnetic reconnection directly from the magnetic field. Therefore they represent a primary component in the flare energy budget. When they interact in the dense chromosphere/photosphere they transfer their energy to a variety of secondary components, such as heat, mass motions, radiant energy. Wu et al (1986, Figure 5.1.1) endeavoured to partition the energy in various categories as a function of time, and in retrospect this may be misleading. It may well be true that essentially all the primary energy release occurs early in the flare, say before the peak in the non-thermal hard X-ray burst, and that this, the thermal X-rays, the optical and mechanical energies, are all secondary. It is very easy to confuse energy release with energy deposition.

2.3. The Role of Energetic Particles

In one of the first studies of the energy balance in flares by Lin and Hudson (1976) they concluded from an analysis of the energy content of the electrons responsible for the non-thermal X-ray emission that for the event they studied particle acceleration, and in particular the electrons, was the dominant output of the energy release mechanism. The energy estimate was based on the assumption that the hard X-ray emission was via non thermal bremsstrahlung of electrons with a power law spectrum (Brown, 1971) and it has been common practice to obtain the total electron energy by integrating the derived electron energy spectrum down to around 20 keV. The form of the electron spectrum is derived from the observed X-ray spectrum. The theory underpinning this analysis is quite robust. Nevertheless an electron power law spectrum such as dN/dE $\propto E^{-\gamma}$ has the bulk of the energy at the low energy end of the spectrum for γ above 2, and unfortunately this is typically the case. Hence the need for an artificial low energy cut off around 20 keV to avoid having too much energy in the electrons. Even so, there are some large flares where even a cut off at 20 keV still leads to a huge energy in the electron spectrum.. For example, the flare of 1 June, 1991 (Kane et al, 1995) had a calculated electron energy content above 20 keV of 10^{27} J requiring around 10^{41} electrons. Kane et al noted that the energy supply for such large flares could not be accommodated either from the active region or by current flare models.

The problem with producing the flare energy from non-thermal electrons had been noted already by Colgate (1978) and Simnett (1986). Simnett recognised that the flare energy budget derived from the hard X-ray analysis neglected the energy in ions, which almost certainly exceeds that in the electrons, probably by orders of magnitude. More recently it has been recognised from gamma ray observations (Ramaty et al, 1995) that the energy content of the ions above ~1 MeV is comparable to, and sometimes in excess of, that in the electrons above ~20 keV. Henoux (2000) has supported Simnett's (1986) conclusion that the energy flux carried by protons could be orders of magnitude greater than that carried by electrons. Also, in his review on stochastic particle acceleration, Miller (2000) discusses electron and proton acceleration by cascading low-amplitude Alfv́en and fast mode waves. He noted that "At equilibrium, the total energy contained in the protons dwarfs that in the electrons". He argues that electron dominated flares are a consequence of relatively small acceleration regions, ~1000 km, as he shows that the longer the acceleration region, the greater the number of protons. Miller summarises the energy and total number requirements for both electrons and protons, and notes that if the flare electrons are to come from the corona, then the acceleration region will be entirely depleted of electrons in about 1 second. This is of course unrealistic.

Simnett(1986) pointed out that extending the lower energy cut-off for the ion spectrum down from 1 MeV to 100 keV would increase the energy content by typically 2-3 orders of magnitude. He suggested that protons in the 100 keV - 1 MeV range were the dominant energy carrier. The electrons producing the hard X-ray burst were in fact all secondary to the primary energy release and accelerated in the chromosphere by processes related to the impact of intense neutral beams of such protons.

2.4. Location of the Energy Supply and Duration of the Release

Let us now return to the energy source. The total energy in the system, including the magnetic free energy, is very large compared with the flare energy (Wu et al, 1986). It has been known since the Skylab era that the most difficult task is to know in what form the energy released from magnetic field annihilation gets transferred into something we can recognise (Kahler et al, 1980). Since that time other important tasks are to identify the location in the corona where the magnetic annihilation occurs, and the timescale over which it happens. There are strict physical limits on the speed with which magnetic energy can be dissipated, and these appear to be incompatible with the rapid evolution of solar flares if the energy release, as distinct from energy deposition, is concurrent with the principal flare emissions. The only place we can make any measurements of the magnetic field strength is in the photosphere, and it is clear that in general attempts to measure the decrease in magnetic energy in the active region encompassing the flare fail to account for the estimated flare energy.

2.4.1. Location

It has been generally accepted that the energy for transient solar activity comes from magnetic fields in the region above the photosphere. How high into the corona is relevant for flare energetics is unclear, but the thrust of much of the work to date has only considered magnetic fields associated with the active region, and therefore the relevant coronal height is just of the order of the scale of the active regions, namely around 50000 km. We are proposing that the energy release site is distributed, at several $R_\odot$, within the extended low latitude magnetic fields that erupt as a CME. All coronal magnetic field lines come through the photosphere, so any point in the corona where there are closed field lines must map, magnetically, into the photosphere. Thus reconnection in the high corona may be driven by photospheric motions.

There has recently been direct evidence for reconnection occurring in the high corona around 3-4 $R_\odot$ (Simnett, 2004). The events he studied were detected by the LASCO/C2 coronagraph (Brueckner et al, 1995) and had excess mass travelling both outwards and inwards along the same (projected) radius vector. The outward moving mass was always faster than the inward moving mass by an amount consistent with twice the solar wind speed at that coronal location. This supports our model where such events are occurring in the high corona in the absence of flares or other non-thermal signatures. In a study of bulk flow velocities in the corona over a whole year near solar minimum, Lewis and Simnett (2000) found a marked outward acceleration at mid-latitudes around 4-5 $R_\odot$ which can only be explained by having significant energy release at this altitude.

The physical energy release mechanism has been widely studied employing magnetohydrodynamic (MHD) theory, which predicts that the coronal magnetic field is sheared, or twisted, as this implies a storage of free energy. In a recent review of flare models Priest (2000) considers only the magnetic loops in an active region; however the analysis is equally applicable to the much larger scale fields involved in the CME eruption. Regarding particle acceleration, the strong electric field at the reconnection site may provide direct electric field acceleration, although many more sophisticated models, often involving MHD shocks, have also been proposed (see below).

Kahler et al (1980) considered that the primary output from the magnetic energy annihilation would be in particle acceleration, energy in thermal plasmas, mechanical energy in mass motions, and shock waves. We would agree with this. However, if the energy release takes place in a very low-density region, such as the high corona, then there can be little energy in either mass motions or the enthalpy of the plasma. In this case the energy must go into shock waves and energetic particles. Kahler et al. also took the view that "the magnetic field of the active region is the source of the flare energy". Here we are disagreeing with this.

2.4.2. Timescale

It is well known that there is a huge discrepancy between the timescale for magnetic energy release via diffusion and the timescale of a flare. The diffusion timescale is gven by $\tau_D = D^2/\eta$, where D is the dimension of the diffusion region and η is the magnetic diffusivity in the corona, which is ~1 m^2/s. If we have thicknesses of the diffusion region of the order of 1000 km then τ_D is 10^{12}s, whereas the risetime of a major flare is ~10^2s and certainly not more than 10^3s. Therefore the magnetic energy release process must be driven and even then it has not been possible to come up with a satisfactory theoretical model. Thus it is a huge advantage if we can invoke a temporary storage mechanism for the energy output of the reconnection.

Therefore we wish to revive one of the original concepts (Elliot, 1964) that the majority of energy release occurs over a considerable period of time, say 10^5s, and that the output is stored in high coronal magnetic fields. In discussing relevant timescales it is important to separate primary from secondary phenomena. It is possible that we have been misled by assuming that the timescale of the optical or X-ray flare represents the timescale of the primary process.

2.4.3. Conclusion

The conclusions we reach from these considerations are (a) that the magnetic energy release timescale may be considerably longer than th at inferred from studying the flare impulsive phase duration and (b) that the location of the magnetic fields from which the energy is derived may be in a region of the corona where measurements of field strength are currently not possible. The first of these is simply a consequence of the laws of physics, which would require the plasma resistivity to be anomalously high if the energy dissipation time is short. The second may be summarised thus: if the annihilating fields were in a region where field strength measurements were possible, then after many decades of observations, we would surely by now have solved the problem.

Therefore without prejudicing the ability of future theorists to devise a means of dissipating magnetic energy over the short timescale needed for flares, and the ability of observers to measure more accurately the coronal magnetic field strength, we take the following approach in this paper. The energy release occurs in the high corona over timescales long compared to that of the flare.

2.5. Problems with the Electron Hypothesis: The Case for Powering the Flare with Protons

The largest solar flares dissipate, in the photosphere, chromosphere and low corona, energy in excess of 10^{25} J within a timescale of the order 10^3 s. At some time during this process a significant fraction of the energy is in energetic particles, which for our purpose we define as protons above 100 keV and electrons above 20 keV. The optical identification of major flares shows that typically the bulk of the radiation comes from well within an active region. The ability of the active region to supply the total energy content for the accelerated particles is wanting, so that the only alternative is to involve a much larger volume, which can only be the high corona. The corona immediately above the active region cannot supply the sheer number of energetic electrons deduced from the X-ray emissions. For large flares the electron beam strength (under the conventional beam models) above $\sim$20 keV is 10^{36} - 10^{37} electrons/s (see Miller, 2000 for a summary of the requirements for both electrons and protons). For large flares this corresponds to a total of 2 10^{40} electrons (Emslie and Brown, 1985) which at a density of $\sim 10^9$ cm^{-3} is equal to all the electrons in a volume of the order of (3 10^8 m)^3 or almost (0.5 $R_\odot$)^3. Clearly the corona cannot supply these electrons, which leaves the chromosphere as the most probable source. Unlike the energy supply, we cannot increase the volume from which we extract the energetic electrons into the high corona, otherwise we would see intense radio emission from these hypothetical electrons as they became energised.

This is not the only serious problem with the primary electron hypothesis. Priest (2000) has given a review of the understanding of the magnetohydrodynamics of solar flares. One of the practical difficulties in MHD modelling is that it is quite difficult to accommodate energetic particle production, especially if it is used as a major external energy input. The theorists may also not take into account important but subtle observational facts. For example, Priest (2000) cites the classical flare model for hard X-ray production, in which he states that "electrons are accelerated to, say, 100 keV in the corona and then spiral downwards, creating microwaves" (and later the X-rays). Unfortunately, the microwave emission occurs after the X-ray emission, not before (Cornell et al, 1984, Costa et al, 1984). This is a natural consequence if the electron acceleration is in the chromosphere, as the electrons lose energy immediately via bremsstrahlung, whereas they have to escape into the corona to produce the radio emission. Delays of up to one second would be expected, this being the time taken for near-relativistic electrons to escape into a loop of dimension $\sim 10^4$ km. This point was discussed in detail by Simnett (1995) who concluded: "Starting with an electron beam in the corona, it is a triumph of ingenuity to always delay the radio emission from the X-rays, when simplistically it should be the converse!"

The total flare energy in the electrons responsible for the hard X-ray burst severely constrains the flare energy budget. As if this were not a problem in itself, Ramaty et al (1995) have estimated the total ion energy responsible for the gamma ray production in some flares and have concluded that the energy in ions above 1 MeV may equal or exceed the energy in the electrons. However, as the spectrum is believed to be a steep power law, extending it down even to 100 keV would require the ion energy to exceed the electron energy by some orders of magnitude. Given the implausibility of having equipartition in energy between protons and electrons, we are reduced to having the protons be the dominant energy carrier. The role of protons in flares has been reviewed by Simnett (1995).

Many soft X-ray events do not have any significant hard X-ray burst. This is naturally explained if the production of the electrons, say above 20 keV, is a secondary process which does not always materialise. The energy for the heating, and thermal X-ray emission, comes from energy deposition by the protons.

Spectroscopic observations also provide important constraints on models. Winglee et al, (1991) made a study of the interrelation of soft and hard X-ray emission during flares. They compared the line broadening and blueshifts for limb and disc flares and concluded that explosive heating could not explain the data. They argued for electric field acceleration of electrons in the chromosphere. Plunkett and Simnett (1994) showed that the blue-shifted component can start before the onset of the hard X-ray burst which is not possible if the energy input is from non-thermal electrons. Observations of red/blue shifts are very important to the understanding of the energy deposition in the atmosphere. Red shifted H α emission is typically seen at the onset of flares (Svestka, 1976). The simplest explanation for this is momentum deposition from energetic ions moving downwards.

Simnett and Benz (1986) made a study of major hard X-ray flares observed by SMM and found that around 15% of large flares were quiet in the metric/decimetric spectral region at the sensitivity of the radio patrol observatories. Benz et al, (2005) in a more recent study found that 17% of X-ray flares were radio-quiet. This precludes the flare energy coming from beams of electrons accelerated in the corona. Rieger et al (1999) have discussed a major gamma ray flare that was radio-silent at the start. The gamma ray line emission was present at the flare onset, which indicates that energetic protons were present at that time. However, the gamma ray continuum which evolved later was dominated by high energy electron bremsstrahlung, which indicated that for this flare energetic electron production was very efficient vis-à-vis energetic proton acceleration (eg >50 MeV). The lack of metric/decimetric radio emission indicated that the electrons were confined to high density regions. Rieger and Marschauser (2000) concluded that the gamma ray line emission and the electron-dominated bursts came from different areas on the Sun and that they were accelerated low in the atmosphere.

The observation of Hα polarization in flares has been thought due to the impact of low energy protons (Henoux et al, 1990, Vogt and Henoux, 1996). Theoretically H α polarization may be produced optimally by electrons around 200 eV or proton around 200 keV. As 200 eV electrons cannot penetrate down the the Hα formation level in the photosphere, then any polarization is likely to be a signature of 200 keV protons. Hanaka (2003) has reported Hα polarization and red-shifted emission in a gradual flare without any evidence of non-thermal hard X-ray emission. As momentum transfer by protons may produce the redshift, then both signatures are consistent with the presence of low energy proton beams. This has also been discussed by Emslie et al, (2000), who concluded from the analysis of a well-observed flare that their results were consistent with a high proton energy content. Linear polarization has been detected in cool dwarf stars, and Saar et al (1994) concluded that proton beams in the hecto-keV region were the most probable cause. Bianda et al (2005) have recently questioned the validity of the polarization results, based on a negative result from an analysis of 30 flares, where they found no evidence for Hα polarization. In recognition of this we therefore might cast doubt on the ability of the polarization observations to identify low-energy protons. However, it does not alter our basic premise.

3. Important but Subtle Points from the Particle Data

Observations of solar energetic particles have been made with increasing sensitivity, energy and time resolution since the dawn of the space age. From the simultaneity (to within about 1 second) of the hard X-ray and gamma ray emissions from flares observed by SMM it was concluded that the production of the electrons and the >30 MeV protons was coincident (to within 1 second). The phenomenon of coronal acceleration of electrons, first identified by Lin (1970) has been studied in depth (Robinson and Simnett, 2002, Simnett, 2005). The differential energy spectrum of such events is a power law with an exponent typically between 3 and 5, which faced with the quiet time background energetic particle levels in the heliosphere means that many of the smaller events are only observed in the lower energy channels of the detectors. By contrast the energy spectrum from large flare events has an exponent less than around 2.5 (Simnett, 2006a). This indicates that there are at least two acceleration processes occurring at the time of flares.

If we invoke two accelerations, one in the high corona and the other in the active region, then the latter might be due to electric field acceleration (Holman, 2000). This could possibly account for the displacement of the hard X-ray and gamma ray locations in the 23 July 2002 flare (Hurford et al, 2003). Most type III radio bursts are not associated with flares, which indicates that there is an acceleration in the corona that produces beams of electrons moving upwards without any significant energy deposition below the transition region. Therefore the electron energisation process for the type III bursts must be largely independent of the processes occurring in the active region at the time of a flare.

The proton acceleration occasionally extends to GeV energies, with the highest being around 25 GeV, which can be detected at the Earth by neutron monitors on the ground. The duration of the solar emission for these events is difficult to determine from the Earth, as trapping and scattering in the inner heliosphere, say out to 2 AU, may extend the period over which very high energy particles may be detected at the Earth. A recent event on 20 January 2005 was observed as a short pulse with a duration down to 1/10 maximum, at the neutron monitor with the highest response, of only around 25 minutes, which given the inevitable propagation path differences suggests an impulsive acceleration at the flare site of perhaps 10 minutes or less. This is consistent with the duration of the gamma ray lines seen by RHESSI (Lin et al, 2002). Analysis of the 20 January 2005 (Simnett, 2006a) showed that the GeV protons were accelerated promptly, around the time of the impulsive phase as defined by the hard X-ray burst. This demonstrates conclusively that (for this event at least) the high energy protons are not accelerated by the CME-driven shock. Hurford et al. (2006) have studied the gamma ray emission from three large GOES X-class events in October/November 2003 and came to the same conclusion. For some large flares RHESSI provides evidence that the high energy protons are accelerated locally in the active region where the flare occurs. It is frequently assumed that if mildly energetic particles are seen at the time of a CME-driven shock, then they must have been accelerated by this shock. Simnett (2005) in a detailed analysis of energetic particles from the major flare of 28 October, 2003, showed that although the shock produced some acceleration, most of the electrons were trapped behind the front of the CME.

Even in 1976 it was known that heavy ion abundances and charge states were highly variable in flare plasmas. More recent measurements have been much more accurate than the early ones and confirm the variability. Thus it is very clear that the charged states are not thermal in origin. Highly variable charge states naturally follow from our model, via the collisional ionisation caused by the impact of the protons in the chromosphere (Brosius et al, 1995). Frequently the brighter leg of a CME is the leg away from the active region containing the associated flare. Qualitatively we explain this by chromospheric evaporation due to energy deposit at the remote site, such that the upward moving material has a relatively unimpeded passage into the structure which is erupting. It is unlikely that this site will produce a flare with non-thermal emissions as a complex active region is required to provide the extra energy and acceleration for the particles responsible for the emissions. There is frequently a small brightening from the remote site (Simnett and Harrison, 1985).

The equivalent process in the leg near the active region results in most of the material going into what become the bright, post-flare loops (cf point 7 in section 2.1). This asymmetry serves to emphasise the importance of the active region in supplementing the energy input from the high corona.

4. Acceleration of the Energetic Particles

Unless one advocates direct electric field acceleration for the energetic particle production, then the energy release process may well proceed in the following way: magnetic reconnection – shock waves – particle acceleration. A description of how this process might work was first formulated by Petschek (1964). Kahler et al (1980) argued that the reconnection was collision dominated, namely that the reconnection was driven by a steady plasma flow. The active region may contribute energy if the neutral sheet is "flare active" (Kahler et al, 1980). When twisted flux tubes are brought into contact the neutral sheet so formed can accelerate particles as envisioned by Speiser (1965), if the process is non-explosive, or via the formation of shocks, if the process is like Petschek (1964). There are other possibilites and Miller (2000) has reviewed stochastic acceleration in flares.

In our model we consider an initial acceleration for the moderate energy electrons and ions (say up to a few keV and a few hundred keV respectively), and separate secondary accelerations of the highly relativistic electrons and the relativistic ions. This may seem unduly complicated but the electromagnetic signatures and the observations of ions and electrons in the vicinity of the Earth seem to preclude a simpler scenario. The initial acceleration takes place in the high corona and the secondary acceleration takes place in the active region.

5. A Flare Model Where the CME Powers the Flare

5.1. Introduction

Our goal is to focus on having the erupting magnetic structure, that is often observed as the CME, play a central role in the development of any associated flare (Simnett, 2006c). It would be remarkable if the two most dramatic and energetic solar transients, the large CME and large flare, were truly independent. We concentrate on the search for a causal link between CMEs and large flares, which would have been a natural way for the subject to have proceeded had CMEs been discovered first. The production of non-thermal energetic particles is given a prominence not normally found in recent flare models. In particular we address where the particles might come from, the source of the energy, the relative energy content in the electron and proton populations and the timescale for the acceleration. We require that the CME and the active region both provide energy but that the highest energy particles are all accelerated in the active region.

It is surprising that the thrust of solar flare theoretical research did not change following the discovery of CMEs in 1970. It was as though the flare community noted the arrival of the new phenomenon, but largely thought it was irrelevant. Faced with all the problems with "conventional" flare theory, the community should have been eager to embrace the newly-discovered phenomenon as a possible saviour for their frustration. This did not happen. For the next decade at least, until the results from SMM were assimilated, it was thought that the CME was effectively the result of a blast wave produced by the flare itself (Dryer, 1982). It was recognised over two decades ago (Simnett and Harrison, 1985) that a flare associated with a major CME was located near one footpoint of the CME, and not near the centre of the CME, as would be expected in a blast model. It was this fact that was fatal to the blast model. We have been careful to refer to coronal transients as "erupting coronal magnetic structures", rather than CMEs, as there is no fundamental reason why an erupting magnetic loop should have any excess mass over the ambient corona. There have been white-light coronagraph observations of CMEs which have so little excess mass that they are barely visible above the ambient corona (Lyons and Simnett, 2001) and therefore by extrapolation there are presumably some eruptions that are invisible. For convenience we shall refer in future to any major erupting magnetic structure as a CME, regardless of whether it actually has significant excess mass.

The main objective in all theoretical approaches should be to account for the most extreme facets of the flare, which are the acceleration of ions and electrons to highly relativistic energies and the associated eruption of a major, fast CME. Once we have the energetic particle population, then by allowing it to interact in the dense solar atmosphere in the vicinity of strong bipolar sunspot magnetic fields, it is surely relatively easy to account for the remaining flare phenomena. Even now flares are being treated as though they are a single, albeit complex, process occurring in an active region. We are advocating that it is a tandem process, with the first stage linked to the eruption of a CME, with the important physical processes occurring in the high corona; and the second relating to specific conditions within an active region and obtaining significant energy from the output of the first stage. This circumvents most, if not all, of the problems identified above with trying to account for all big flares as though they are confined to the active region with a single particle acceleration mechanism.

The salient features are as follows. Non-thermal energetic particles provide both the link between CMEs and flares, and a significant part of the flare energy. The excess pressure within the coronal structure due to the particles may be a factor causing the ultimate CME eruption, although it is beyond the scope of the present work to discuss this further. From an energetics viewpoint ions dominate, and to avoid gamma ray production which would be observable by current instrumentation, the ions should be below around 1 MeV/nucleon. They certainly must be below $\sim$30 MeV, which corresponds to the binding energy of the ^{4}He nucleus. Once protons have enough energy to break up ^{4}He then free neutrons will be produced, resulting within $\sim$100s in the formation of the neutron capture gamma ray line at 2.223 MeV. References to plausible ways of generating an adequate ion population within the CME structure, which for large events covers over 90 degrees in a sun-centred reference frame, were indicated in section 4. The non-thermal electrons responsible for the hard X-ray and microwave signatures are generated near the chromospheric flare site, together with the highly relativistic ions which often accompany large flares.

The concept is based on gradually populating a large coronal structure, of typical length scale around 2 $R_\odot$, with energetic ions (referred to hereinafter as protons) up to energies around 1 MeV. A schematic representation of how this structure might look in relation to the photosphere and an active region is shown in Fig. 1. We have drawn two rectangles in Fig. 1. The larger represents the primary volume, from which most of the energy is

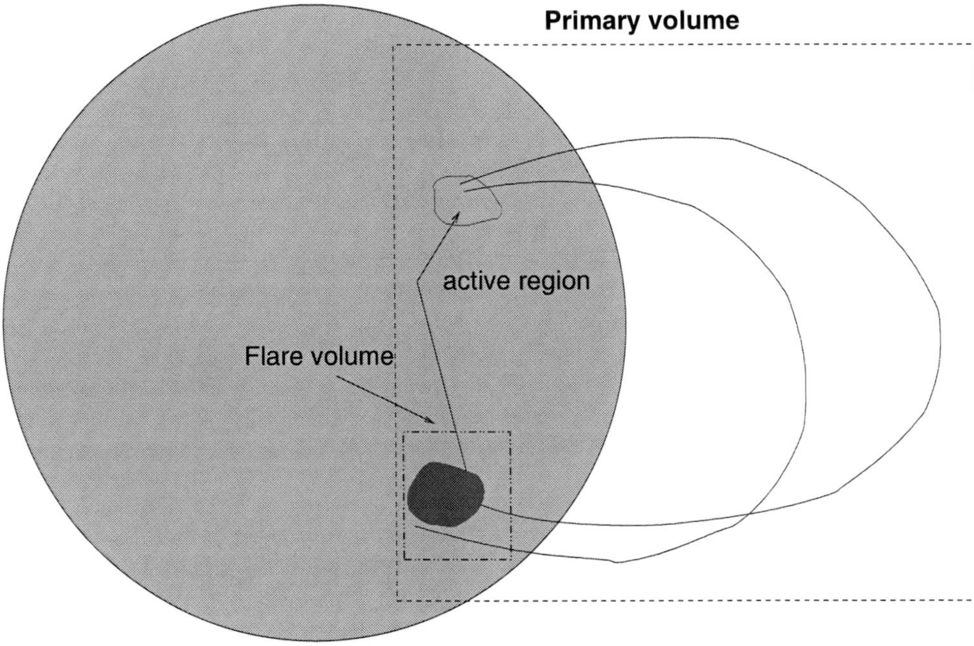

Figure 1. A possible configuration of the magnetic fields in the erupting structure and their footpoints in the photosphere. The solar disc is indicated by the lightly shaded circle, and sample magnetic field lines are shown linking two active regions, one of which (heavy shading) is complex enough to produce a flare.

extracted, and the smaller one represents the flare volume.

The energy release is by magnetic reconnection high in the corona which results in the acceleration of protons and electrons to modest energies, say up to 10-20 keV for electrons and up to $\sim$1 MeV for the protons. Some of this output is released into the interplanetary medium and it may be detected if the connection is favourable by instruments near Earth. Some of the output remains trapped in the coronal structure, but the electrons, being faster than the protons and more easily scattered, tend to be lost fairly quickly in the lower solar atmosphere. Note that the speed of a 1 MeV proton is around c/25, while that of a 20 keV electron is around c/4, where c is the velocity of light. The coronal trap would be leaky so that particles would be lost slowly both to space and into the loss cone.

The coronal structure would gradually accumulate the protons up to the equipartition energy. Possibly through simplification of the magnetic fields near the photosphere the structure erupts as a CME and the trapped particles precipitate to the active region to provide *most* of the energy to power the optical flare. The active region magnetic field is dynamic and manifestations of this are the motions of active region filaments prior to flares (cf point 2 in section 2.1).

The electrons for the X-ray burst are accelerated in or near the chromosphere by a secondary process associated with the dumping of the protons. Note that not all erupting magnetic structures would be expected to be triggered by an excess of trapped particles, but such particles would certainly contribute to the onset of instability. Low (2001) has pointed out that the coronal magnetic field is naturally sheared by footpoint motion, and

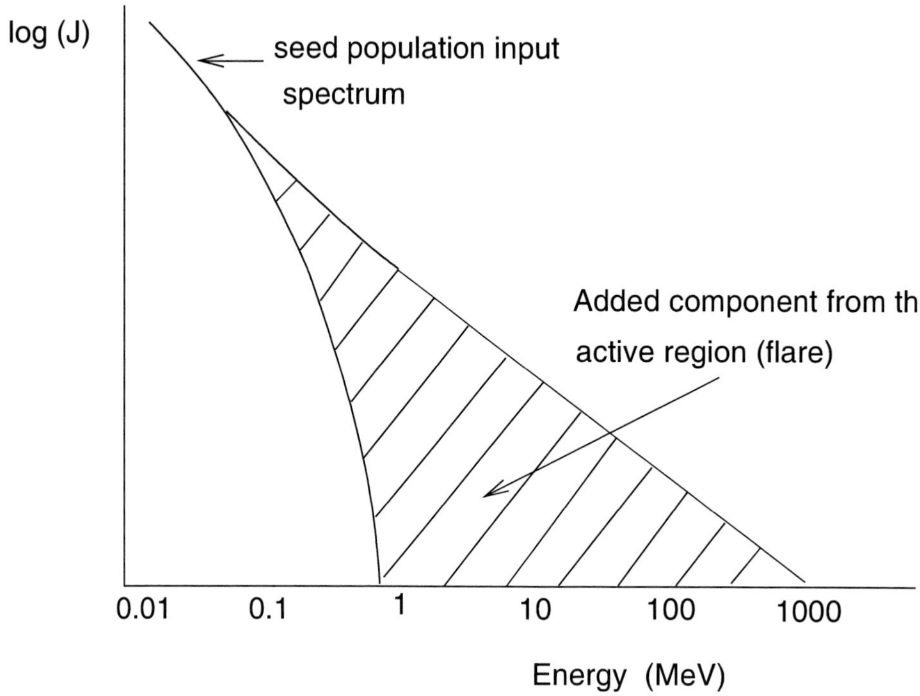

Figure 2. The energy spectrum of the seed population, shown as the left part extending to around 1 MeV, and the additional flare population, shown hatched. The ordinate scale is deliberately left blank to allow for variations in spectral index from actual events.

that an eruption is simply the way the corona releases the shear. The mass in a CME is not essential to the eruption, but it is normally present when there is a major flare, so that in addition to the process which releases the shear, energy must be deposited at the base of the corona to drive the excess mass observed in a CME into the corona. It is frequently observed that there is more mass in the side of the CME remote from the flare, and this is driven from the chromosphere via the part of the proton population which is dumped at the remote site. If the remote site is not a magnetically-complex active region it does not develop the signatures of a typical major flare. Coulomb collisions by the precipitating protons provide the high charge states of the material evaporated from the chromosphere. At the footpoint in the active region, this material tends to populate the active region loops, and does not necessarily contribute much to the CME; hence the observed mass asymmetry in some CMEs. We have yet to understand how many erupting structures there are without excess mass, and consequently invisible to a coronagraph. A comprehensive study near Earth of magnetic features in the solar wind which are not associated with a visible CME would help to clarify this point.

The active region, which has strong magnetic fields in a metastable state now receives a large seed population of energetic protons. There is no doubt considerable turbulence generated by the incoming particles which we envisage coming in as a neutral beam. However, from an energetics viewpoint the energy in the electrons is 1/1836 times the energy in the protons and so may be neglected. The active region now accelerates both ions and electrons to high energies. The maximum energy gained by the various particle species is clearly a function of the active region morphology and will very from flare to flare, as will the relative importance of the ions and electrons. Therefore in some large flares electron acceleration may be favourable and this will lead to the occasional electron dominated flare (Trottet et al, 1998). Fig. 2 shows how the proton energy spectrum might appear. The spectrum from the seed population is shown going out to ~1 MeV. Acceleration at the flare site by the mechanism of choice adds the hatched region. The active region magnetic fields provide the means of achieving this, including the provision of some energy, although the precise details are not known, nor do we attempt to discuss them here.

It should be emphasised that the "flare accelerator" does not have to accelerate ions from the ambient thermal plasma to GeV energies, but from the non-thermal seed population. In Fig. 2 we show this going to 1 GeV, but this is clearly flare dependent (see figure caption). The key point here is that the acceleration to high energies is in or just above the chromosphere, where there is an ample supply of material, especially to provide enough electrons for the hard X-ray burst.

5.2. Energetics Considerations

Estimates of the total magnetic energy in the active region surrounding a major flare typically do not show an adequate diminution following the flare, which proves that the active region magnetic field is not the main energy source. We now estimate the potential for the high corona to supply the energy. Following Simnett (2003), consider a large coronal loop system of radius 1.5 $R_\odot$, cross section of 10^{17} m^2 with a field strength of $2\ 10^{-4}$T. The total magnetic energy is ~$5\ 10^{24}$ J. Thus in principle such a structure could contain an energetic particle population of this energy without disruption; and if the energy could be transferred to the active region it could power the major part of the flare. The limit to the stability of the large structure may well occur when the total energy of the trapped particles approaches the total energy of the magnetic field. Therefore, the trigger for the eruption of the CME seen at the time of a major flare could be the final input of energy from, say, a small-scale reconnection within the overall configuration, ultimately driven probably by photospheric motion. As the active region producing the flare is an integral part of the system, then motions within the region, such as emerging flux, could suffice. Note we do not advocate this process, namely the high pressure of trapped particles, to account for the eruption of *all* CMEs.

Another cornerstone of this model is the acceleration of non-thermal particles in the high corona. Evidence for high coronal particle acceleration has been available since ~1970 (see Lin, 1985 for a review of the early observations). Typically electrons are accelerated impulsively up to around 10 keV, released into the interplanetary medium, where they are adiabatically focused and arrive at 1 AU as a collimated beam. From analysis of the electron spectrum, Potter et al (1980) established that the events originated in the corona above ~1.5 $R_\odot$(sun centred). It was suggested by Cliver and Kahler (1991) that the particle acceleration occurred as a result of reconnection in the neutral current sheets of coronal streamers.

Such reconnection events are difficult to observe, but recent measurements (Simnett, 2004) of bi-directional flows detected by the LASCO C2 coronagraph (Brueckner et al, 1995) have indicated that the typical projected altitude of such events is between 3 and 4 $R_\odot$. Simnett stressed that the marked absence of events above the LASCO threshold originating below 3 $R_\odot$ was real and not a limitation of the observations. This shows that the 1.5 $R_\odot$ loops suggested above are not an overestimate, and may in fact be too small. The original theory for such reconnection advanced by Petschek (1964), involving the formation of oppositely-directed shocks, is a plausible explanation.

Particle acceleration in such events is difficult to estimate accurately. The impulsive electron events reviewed by Lin (1985) were typically low energy, below 20 keV. However, observations by Robinson and Simnett (2002) and Simnett (2005) have shown that the electron spectrum in the high coronal events may extend very occasionally up to 200 keV. It is inconceivable that the process does not also accelerate protons. The Sun is a quasi-continuous source of non-thermal protons (McDonald and Desai, 1971, Simnett, 2001) which extend up to MeV energies, even at times of quiet solar activity. MeV proton counterparts of impulsive electron events are difficult to observe at 1 AU, as the velocity of such protons is $\sim$c/25 and they therefore take several hours to cover 1 AU and are easily scattered by magnetic inhomogeneities in the interplanetary medium. Thus the frequent low energy solar proton emission could be a manifestation of the ionic part of the coronal accelerator. Fig. 3 shows the 337 - 594 keV ion intensity from the EPAM instrument (Gold et al, 1998) on the ACE spacecraft for 60 days towards the end of 2005, when the Sun was quiet, near solar minimum. It is clear that there are many distinct, and long, events during this period. Also shown in Fig. 3 is the 38-53 keV electron intensity, which is virtually uncorrelated with the protons. The data are plotted as hourly averages. We believe the data shown in Fig. 3 are supportive of our contention that protons up to $\sim$1 MeV are accelerated in the corona in the absence of both chromospheric flares and accompanying electrons above $\sim$10-20 keV. There were only three GOES C1 class X-ray flares from day 267 - day 315. It may be no accident that the most active period in terms of chromospheric activity was from day 316 - day 322, which was the minimum in the ion intensity seen at ACE. One hypothesis is that coronal energy release is a quasi-continuous process, and when the output goes primarily to the solar atmosphere, it therefore does not go into the interplanetary medium.

Lin (1985) has estimated the electron energy/event for the coronal accelerations as 10^{18} - 10^{19} J. However, because of the uncertainty in the geometry of the release into the interplanetary medium, energy losses and energy in the low energy part of the spectrum, this may be underestimated by a factor of 10-100 (Lin 1985). We adopt as a working hypothesis that the energy in the ions may be $\sim 10^3$ times the energy in the electrons (cf m_p/m_e). Then the energy /event may plausibly be in the region 10^{22} - 10^{24} J. Lin has emphasized that such events are the most common of all solar electron events, and occur many times/day over the whole Sun. The output of a number of such events may be trapped in the coronal magnetic structure.

Therefore based on the above discussion there is not an energy problem in providing the additional energy input for flares, of say 5 10^{24}J, via energetic particles accelerated in the high corona.

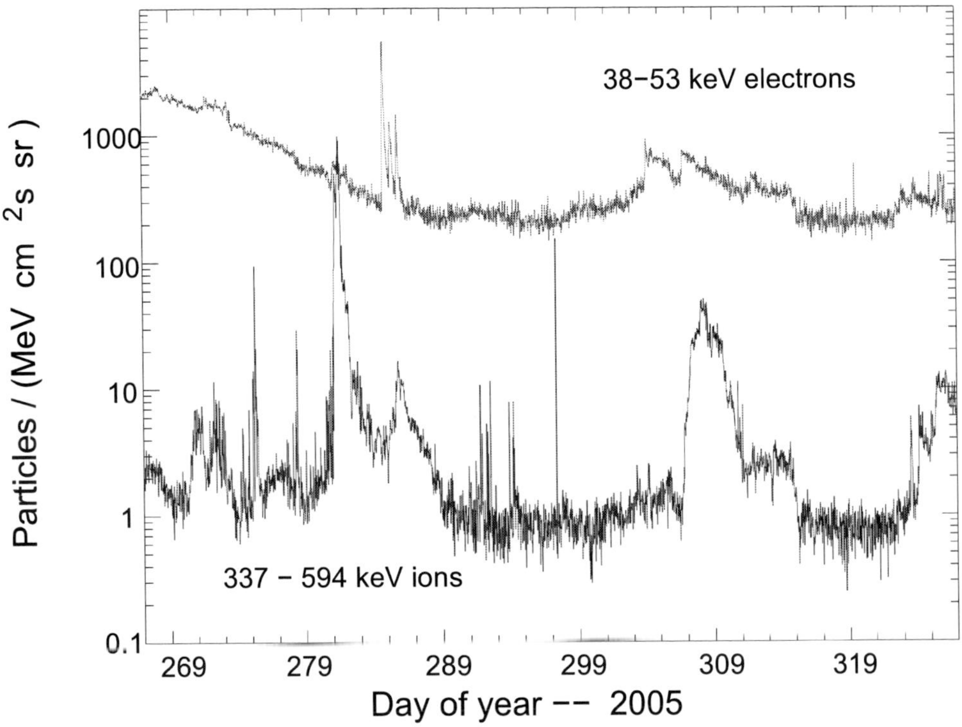

Figure 3. The intensity-time history of 38-53 keV electrons (upper trace) and 337 - 594 keV ions (sector 1; lower trace) measured by the ACE/EPAM detector from day 267 - day 327 2005. The data are plotted as 30 minute averages. The spike in the ion intensity on day 297 is an instrument calibration.

5.3. Characteristic Timescales

How rapidly can a large coronal structure empty of particles following ejection? Suppose that the ejection of the magnetic field structure causes the mirror points above the flare site to move into the chromosphere. This would be a natural consequence of a reducing magnetic field strength at a footpoint. The length of the coronal loops is around ($1.5\ \pi\ R_\odot$) = $3.5\ 10^9$ m. The protons we are advocating travel at c/50 ($\sim$200 keV) - c/25 ($\sim$1 MeV). Thus the timescale to dump the population is around 100s. (Some particles have to go to the other mirror point and back.) The trigger is caused by flux cancellation of the active region fields with the fields in the coronal structure. The rising phase of flares can last longer than 100s. Elliot (1973) pointed out that ion-cyclotron noise may help scatter particles into the loss cone.

6. Discussion

We now list some of the key elements of flares which are addressed by this model. Our starting premise was to have a causal link between the flare and the CME. As there are many CMEs at solar minimum which occur in the absence of flares, then the only direction of the causality is that CMEs are key to the flare process, rather than vice versa. The points raised in section 2.1 should be recalled also.

(1) The excess mass in the CME must be accounted for, especially in the side remote from the flare. In a large CME the 10^{13-14} kg of excess matter was not present in the corona before the event started. This magnitude of matter may be present even when the CME is not followed by a prominence eruption, so the matter is not necessarily prominence material. A further important point is that often the leg of the CME with the most material is not the leg closest to the flare (Simnett, 2006d). Therefore this material does not in general all come from the flare region. Such a large amount of excess matter can only come from the chromosphere.

(2) The energy for the flare must be available for dissipation within the timescale of the risetime of the principal flare emissions, such as the hard and soft X-ray bursts. This fact alone, notwithstanding support from magnetic field measurements, precludes the energy source coming directly and totally from the magnetic fields in the active region.

(3) The energy content in the non-thermal charged particle populations must be able to be produced by the model. In our view, the very large flares, such as 1 June, 1991 (discussed by Kane et al, 1995), cannot obtain the energetic electrons required to produce the hard X-ray burst from an electron acceleration process in the corona. There are simply not enough electrons in the corona, and there is no plausible way of energising them there even if there were.

(4) The pre-flare emission must be satisfactorily accounted for, especially the filament activation and the gradual heating. This energy certainly cannot be provided by energetic electrons. Svestka (1976) notes that prominence activations may start up to an hour before the flare starts. The onset of the precipitation of trapped ions as the CME erupts provides the energy for the early flare emissions (cf point (1) in section 2.1) This is clearly an indication that the high coronal fields are also evolving at this time. Svestka speculated that these “magnetic field changes which lead to the pre-flare activation essentially precede all flares.” At the time Svestka wrote this CMEs had only recently been discovered and had not yet featured in flare models.

(5) The wide variety of heavy ion abundances and charge states should be explained. To our knowledge there have been no observations of this matter moving into the CME structure. However, observations of upflowing material can only be made if the plasma is emitting spectral lines which can be monitored for blueshifts. The spectral line must be one which is bright enough to be visible. The analysis of Brosius et al (1995) and Brosius and Woodgate (1999) of the ionisation produced by the impact of a proton beam onto a stellar atmosphere concluded that for a strong beam, the bulk of the ambient plasma became ionised, or for the heavy elements ionised to a high charge state, within seconds of the beam impact. This makes it difficult to observe either upflows or downflows for very long, except those on the periphery of the action. This makes the observations difficult to interpret, as some red/blue shifts will be seen, but they will not reflect the bulk of the material in motion.

(6) The timing of the emission of energetic charged particles into the interplanetary medium must always be compatible with the model, propagation effects notwithstanding. Our model would have most of the high energy accelerated particles from the active region trapped behind the outward-propagating CME. There would be leakage, especially at high energy and probably some the flare emission would be onto open magnetic field lines.

Simnett (2005) showed that the bulk of the energetic particles from the major flare on 28 october, 2003 were trapped behind the outward-propagating CME and that the escaping part of the seed population was emitted into space before the flare onset.

(7) The prompt arrival of energetic flare particles at the Earth when generally the flare location is separated by many tens of degrees, in both solar latitude and longitude, from the base of the nominal Parker spiral magetic field line linking the Earth to the Sun, suggests that the particle injection is from a large scale, but leaky, magnetic structure. This would naturally be the erupting magnetic structure which is the CME.

(8) Spectroscopic observations of soft X-ray and EUV lines can reveal mass motions of the hot plasma. We discussed this in detail in section 2.6. Brosius and Phillips (2004) found that is some flares the onset of the low temperature lines preceded the onset of high temperature lines by several minutes, and concluded that this was consistent with energy transport by non-thermal ion beams.

(9) One of the surprising results from the RHESSI mission (Hurford et al,, 2003, 2006) was that the gamma ray images of the neutron capture line at 2.223 MeV were displaced from the core of the hard X-ray emission. The neutrons travel only a short distance before they thermalise and therefore the site of the gamma ray emission is close (to within 1 $''$) of the site of the original nuclear interactions. Our model of the flare has the bulk of the energy deposited in ions in the 0.1-1 MeV region in the active region where the footpoint of the coronal field is the same as one of the footpoints of the low-lying flare loops in the active region producing the flare. Dynamic effects from this process are responsible for accelerating the electrons producing the hard X-ray emission, either by a mechanism similar to that discussed by Simnett and Haines (1990) or possibly by small-scale, strong shocks produced by the large energy input. The protons are further accelerated to high energy in the strong active region field perhaps by a process similar to that suggested by Martens (1988), or by Speiser, (1965). Provided any electrons accelerated by these processes are of significantly lower intensity than those accelerated due to the energy deposition of the ions from the corona, then the centroids of the hard X-ray and gamma ray emissions will be different. The radio-quiet flares confirm this.

(10) Point (3) in section 2.1 is highly relevant. The key role of the active region is to accelerate the seed population to high energies. The strong magnetic fields associated with the sunspots are vital. When this component is missing, as it is for flares in old spotless regions, then the secondary acceleration is missing and such flares are lacking energetic protons, as noted by Svestka (1976). The 10 MeV limit quoted by Svestka may be the upper limit of the coronal acceleration. However, it may represent a modest acceleration of the seed population by processes associated with the energy deposition.

The key unknowns in our model are the precise mechanisms for accelerating the particles that populate the large coronal structure, the mechanism for dumping the trapped particles, and the chromospheric acceleration of the GeV protons and deka-MeV electrons. In our view none of these is critical and there are numerous theoretical treatments of these problems in the literature.

7. Summary

In closing we make the following points:

- Magnetic reconnection in the high corona, up to 4 $R_\odot$ accelerates the ambient coronal plasma to modest non-thermal energies, typically up to $\sim$10 keV for electrons and $\sim$1 MeV for protons. Closed coronal magnetic fields provide a leaky trap for these particles. For large flares energy from this reconnection contributes the dominant part of the flare energy, with the charged particles as the intermediate energy transfer mechanism.
- The trapped population builds up until the coronal magnetic field can no longer contain it and the structure erupts. The trigger may very well come from "re-arrangement" of one footpoint of the CME structure due to motion within the active region where the flare occurs. If energy deposition from the leaky trap has carried to the chromosphere, then "evaporated" plasma provides the excess mass for the eruption to be visible as a CME and also provides some of the anomalous charge states seen from measurements of heavy ions at 1 AU (Moebius et al, 2000). Some of the population will escape into the interplanetary medium, but it will normally not be identified as such at 1 AU due to the dominance of particles from the flare itself, and/or those accelerated by the CME-driven shock.
- The flare region may accelerate the seed (trapped) population to much higher energy, thereby producing on occasion highly relativistic protons and electrons.
- The CME-driven shock may provide some additional acceleration, but not to GeV energies.
- This model provides a causal link between large CMEs and large flares, with significant energy transfer via non-thermal protons up to $\sim$1 MeV.
- In a thorough analysis of an electron dominated gamma ray flare, Trottet et al (1998) argued that the energetic electrons, up to 56 MeV, were an upward moving beam. They also estimated the energy in ions above 1 MeV as comparable to the energy in the electrons above 20 keV. This is yet another instance where, even when there are no very energetic protons, the energy content of the protons dwarfs that of the electrons (Miller, 2000).

We have discussed many aspects of flares and have shown how our concept may apply. However, the reader must not lose sight of the three most important points, namely the total energy supply, the causality between the CME and the flare, and the origin of the mass (with its non-equilibrium charge states) in the CME. The many details which we have used to illustrate our model may not be uniquely explained, but at least they are consistent.

References

[1] Benz, A.O.; Grigis, P.C.; Csillaghy, A.; Saint-Hilaire, P. *Solar Phys.,* 2005, 226, pp121-142

[2] Bianda, M.; Benz, A.O.; Stenflo, J.O.; Kveler, G.; Ramelli, R. *Astronomy and Astrophysics,* 2005, 434, pp1183-1189

[3] Brosius, J.W.; Phillips, K.J.H. *Ap. J.*, 2004, 613, pp580-591

[4] Brosius, J.W.; Woodgate, B.E. *Ap. J.*, 1999, 514, pp430-440

[5] Brosius, J.W.; Robinson, R.D.; Maran, S.P. *Ap. J.*, 1995, 441, pp385-399

[6] Brown, J.C. *Solar Phys.,* 1971, 18, pp489-502

[7] Brueckner, G.E.; Howard, R.A.; Koomen, M.J.; Korendyke, C.M.; Michels, D.J.; Moses, J.D.; Socker, D.G.; Dere, K.P.; Lamy, P.L.; Llebaria, A.; Bout, M.V.; Schwenn, R.; Simnett, G.M.; Bedford, D.K.; Eyles, C.J. *Solar Phys.,* 1995, 162, pp357-402

[8] Carmichael, H. *Space Science Rev.,* 1962, 1, pp28-61

[9] Carmichael, H., In *The Physics of Solar Flares;* Hess, W.N. Ed.; 1964, NASA SP-50, 1964, pp451-456

[10] Cliver, E.W.; Kahler, S.W. *Ap. J. Lett.,* 1991, 366, pp L91-94

[11] Coleman, P.J., Jr.; Sonnett, C.P.; Judge,D.L.; Smith, E.J. *J. Geophys. Res.*, 1960, 65, pp1856-1868

[12] Colgate, S.A. *Ap. J.,* 1978, 221, pp1068-1083

[13] Cornell, M.E.; Hurford, G.J.; Kiplinger, A.L.; Dennis, B.R. *Ap. J.*, 1984, 279, pp875-881

[14] Costa, J.E.R.; Kaufmann, P.; Takakura, T. *Solar Phys.,* 1984, 94, pp369-378

[15] Dryer, M. *Space Sc. Rev.*, 1982, 33, pp233-264

[16] Elliot, H. In *High Energy Phenomena on the Sun*; Ramaty R., Stone R.G. Ed.; NASA SP-342, 1973; pp12-18

[17] Elliot, H. *Plan. Sp. Science.,* 1964, 12, pp657-660

[18] Emslie, A.G.; Brown, J.C. *Ap. J.,* 1985, 295, pp648-653

[19] Emslie, A.G.; Miller, J.A.; Vogt, E.; Henoux, J.-C.; Sahal-Brechot, S., *Ap. J.*, 2000, 542, pp513-520

[20] Gold, T. *J.Geophys. Res.*, 1959, 64, pp1665-1670

[21] Gold, R.E.; Krimigis, S.M.; Hawkins, S.E.,III; Haggerty, D.K.; Lohr, D.A.; Fiore, E.; Armstrong, T.P.; Holland, G.; Lanzerotti, L.J. *Space Sci. Rev.,* 1998, 86, pp541-562

[22] Hanaka, Y. *Ap. J.*, 2003, 596, pp1347-1355

[23] Hansen, R.I. *Phys. Rev. Lett.*, 1961, 6, pp260-262

[24] Harrison, R.A.; Waggett, P.W.; Bentley, R.D.; Phillips, K.J.H.; Bruner, M.; Dryer, M.; Simnett, G.M. *Solar Phys.,* 1985, 97, pp387-400

[25] Henoux, J.-C. In *High Energy Solar Physics: Anticipating HESSI*; Ramaty, R., Mandzhavidze, N. Ed.; *ASP Conf. Series,* 2000; 206, pp27-36

[26] Heyvaerts, J.; Priest, E.R.; Rust, D.M. *Ap. J.*, 1977, 216, pp123-137

[27] Holman, G.D., In High Energy Solar Physics: Anticipating HESSI; Ramaty, R., Mandzhavidze, N. Ed.; *ASP Conf. Series*, 2000; 206, pp135-144

[28] Hurford, G.J.; Schwartz, R.A.; Krucker, S.; Lin, R.P.; Smith, D.M., Vilmer, N. *Ap. J.,* 2003, 595, ppL77-80

[29] Hurford, G.J.; Krucker, S.; Lin, R.P.; Schwartz, R.A.; Share, G.H.; Smith, D.M. *Ap. J.*, 2006, 644, ppL93-96

[30] Kahler, S.W.; Spicer, D.; Uchida, Y.; Zirin, H. In Solar Flares; Sturrock,P.A. Ed.; Colorado Associated University Press, Boulder, CO, 1980, pp83-113

[31] Kane, S.R.; Hurley, K.; McTiernan, J.M.; Sommer, M.; Boer, M.; Niel, M. *Ap. J. Lett.,* 1995, 446, ppL47-50

[32] Koomen, M.; Detwiler, C.R.; Brueckner, G.E.; Cooper, H.W.; Tousey, R. *Appl. Opt.*, 1975, 14, pp743-751

[33] Kundu, M.R.; Woodgate, B.W. *Energetic Phenomena on the Sun*; NASA Conf Pub. 2439, 1986

[34] Lewis, D.J.; Simnett, G.M. *MNRAS*, 2000, 317, pp1005-1013

[35] Lin, R.P. *Solar Phys.,* 1985, 100, pp537-561

[36] Lin, R.P.; Hudson, H.S. *Solar Phys*., 1976, 50, pp153-187

[37] Lin, R.P. and 65 co-authors *Solar Phys*., 2002, 210, pp3-32

[38] Low, B.C. *J. Geophys. Res.,* 2001, 106, pp25141-25163

[39] Lyons, M.A.; Simnett, G.M. *Solar Phys.*, 2001, 200, pp203-211

[40] Martens, P.C.H. *Ap. J.*, 1988, 330, ppL131-134

[41] McDonald, F.B.; Desai, U.D. *J. Geophys. Res.*, 1971, 76, pp808-827

[42] Miller, J.A. In High Energy Solar Physics: Anticipating HESSI; Ramaty, R.; Mandzhavidze, N. Ed.; ASP Conf Series, 2000, 206, pp145-161

[43] Moebius, E.; Klecker, B.; Popecki, M.; Morris, D.; Mason, G.M.; Stone, E.C.; Bogdanov, A.T.; Dwyer, J.R.; Galvin, A.B.; Heirtzler, D.; Hovestadt, D.; Kistler, L.M.; Siren, C. In *Acceleration and Transport of Energetic Particles Observed in the Heliosphere;* Mewaldt, R.A.; Jokipii, J.R.; Lee, M.A.; Moebius, E.; Zurbuchen, T.H., Ed.; AIP Conf. Proc., Melville, NY, 2000; 528, pp131-134

[44] Petschek, H.E. In *The Physics of Solar Flares;* Hess, W.N. Ed.; 1964; NASA SP-50, pp425-439

[45] Plunkett, S.P.; Simnett, G.M. *Solar Phys.*, 1994, 155, pp351-371

[46] Potter, D.W.; Lin, R.P.; Anderson, K.A. *Ap. J. Lett.*, 1980, 235, ppL97-100

[47] Priest, E.R. In *High Energy Solar Physics: Anticipating HESSI* ; R. Ramaty, R.; Mandzhavidze, N. Ed.; ASP Conf Series, San Francisco, CA, 2000; 206, pp13-26

[48] Ramaty, R.; Mandzhavidze, N.; Kozlovsky, B.; Murphy, R. *Ap. J. Lett.*, 1995, 455, ppL193-196

[49] Ramaty, R.; Mandzhavidze, N. *High Energy Solar Physics: Anticipating HESSI;* AIP Conf Series, San Francisco, CA, 2000; 206.

[50] Ramaty, R.; Mandzhavidze, N.; Hua, X.-M. *High Energy Solar Physics;* AIP Conf Series, Woodbury, NY, 1995; 374

[51] Rieger, E.; Treumann, R.A.; Karlicky, M. *Solar Phys.*, 1999, 187, pp59-75

[52] Rieger, E.; Marschauser, H., in *High Energy Solar Physics: Anticipating HESSI;* Ramaty, R.; Mandzhavidze, N. Ed.; 2000; ASP Conf Series, San Francisco, CA, 206, pp407-412

[53] Robinson, I.M.; Simnett, G.M. *J. Geophys. Res.*, 2002, 107, 1191, DOI10.1029/2001JA000305

[54] Saar, S.H.; Martens, P.C.H.; Huovelin, J.; Linnaluoto, S. *Astronomy and Astrophysics*, 1994, 286, pp194-202

[55] Sheeley, N.R., Jr.; Bohlin, J.D.; Brueckner, G.E.; Purcell, J.D.; Scherrer, V.E.; Tousey, R.; Smith, J.B., Jr.; Speich, D.M.; Tandberg-Hanssen, E.; Wilson, R.M. *Solar Phys.*, 1975, 45, pp377-392

[56] Sheridan, K. *Proc. Astronom. Soc. Australia,* 1970, 1, pp376-378

[57] Simnett, G.M. *Solar Phys.,* 1971, 20, pp448-461

[58] Simnett, G.M., Harrison, R.A. *Solar Phys.*, 1985, 99, pp291-311

[59] Simnett, G.M. *Solar Phys.*, 1986, 106, pp165-183

[60] Simnett, G.M., Benz, A.O. Astronomy and Astrophysics, 1986, 165, pp227-234

[61] Simnett, G.M., Haines, M.G. *Solar Phys.,* 1990, 130, pp253-263

[62] Simnett, G.M. *Space Sci. Rev.,* 1995, 73, pp387-432

[63] Simnett, G.M. *Space Sci. Rev.*, 2001, 97, pp231-242

[64] Simnett, G.M. In *Solar Variability as an Input to the Earth's Environment* ; Wilson, A. Ed.; ESA SP-535, 2003; pp613-618

[65] Simnett, G.M. *Astronomy and Astrophysics,* 2004, 416, pp759-764

[66] Simnett, G.M. *J. Geophys. Res.*, 2005, 110, A090S01

[67] Simnett, G.M. *Solar Phys.*, 2006a, 237, pp383-395

[68] Simnett, G.M. *Astronomy and Astrophysics*, 2006b, 445, pp715-724

[69] Simnett, G.M. In *Solar Influence on the Heliosphere and Earth's Environment: Recent Progress and Prospects*; Gopalswamy, N., Bhattacharyya, A. Ed.; Quest Publications for ILWS and Indian Institute of Geomagnetism, 2006c; pp85-88

[70] Simnett, G.M. In *SOHO-17 10 Years of SOHO and Beyond*; Lacoste H. Ouwehand,L. Ed.; 2006d; ESA SP-617, CD-ROM 22.1

[71] Speiser, T.W. *J. Geophys. Res.*, 1965, 70, pp4219-4229

[72] Sturrock, P.A. *Solar Flares*; Colorado Associated University Press, Boulder, CO, 1980

[73] Svestka, Z. *Solar Flares*; D. Reidel Publishing Company, Dordrecht, Holland/Boston, MA., 1976

[74] Tousey, R. *Ann. Astrophys.,* 1965, 28, pp600-604

[75] Trottet, G.; Vilmer, N.; Barat, C.; Benz, A.; Magun, A.; Kuznetsov, A. *Astronomy and Astrophysics,* 1998, 334, pp1099-1111

[76] Vogt, E.; Henoux, *J.-C. Solar Phys.*, 1996, 164, pp345-359

[77] Wagner, W.J. *Ann. Rev Astron. Astrophys.*, 1984, 22, pp267-289

[78] Warwick, J.W. *Publ. Astronom. Soc. Pacific,* 1962, 74, pp302-307

[79] Winglee, R.M.; Kiplinger, A.L.; Zarro, D.M.; Dulk, G.A.; Lemen, J.R. *Ap. J.*, 1991, 375, pp366-381

[80] Webb, D.F.; Cheng, C.-C.; Dulk, G.A.; Edberg, S.J.; Martin, S.F.; McKenna-Lawlor, S.; McLean, D.J. In Solar Flares; Colorado Associated University Press, Boulder, CO, P.A.Sturrock Ed.; 1980; pp471-499

[81] Wu, S.-T.; de Jager, C.; Dennis, B.R.; Hudson, H.S.; Simnett, G.M.; Strong, K.T.; Bentley, R.D.; Bornmann, P.L.; Bruner, M.E.; Cargill, P.J.; Crannell, C.J.; Doyle, J.G.; Hyder, C.L.; Kopp, R.A.; Lemen, J.R.; Martin, S.F.; Pallavicini, R.; Peres, G.; Serio, S.; Sylvester, J.; Veck, N.J. In Energetic Phenomena On The Sun; Kundu, M.R.; Woodgate, B.W. Ed.; 1986; *NASA Conf. Pub*. 2439, pp5.1-5.73

In: Solar Physics Research Trends
Editor: Pingzhi Wang, pp. 259-286

ISBN: 978-1-60021-987-0
© 2008 Nova Science Publishers, Inc.

Chapter 7

APPLICATION OF MACHINE LEARNING IN SOLAR PHYSICS

Frank Y. Shih[1*], Ju Jing[2], Ming Qu[1] and Haimin Wang[2]

[1] College of Computing Sciences, New Jersey Institute of Technology Newark, NJ 07102, U.S.A.

[2] Center for Solar-Terrestrial Research, New Jersey Institute of Technology Newark, NJ 07102, U.S.A. and Big Bear Solar Observatory, New Jersey Institute of Technology, 40386 North Shore Lane, Big Bear City, CA 92314, U.S.A.

Abstract

We present the application of machine learning techniques in solar feature detection and solar event classification. The focus of automatic solar feature detection is on the development of efficient feature-based classifiers. We have conducted experiments using different classifiers for the detection of solar flare on the solar H-α (Hydrogen-Alpha) images. These images were obtained from the Big Bear Solar Observatory in California, USA. Series of H-α full-disk images are taken in regular time intervals to observe the changes of solar disk features. In addition, we have proposed the classifier for detection of Coronal Mass Ejections (CMEs) in Large Angle Spectrometric Coronagraph (LASCO) C2 and C3 images, which were obtained from the Solar and Heliospheric Observatory (SOHO). We use image filtering and enhancement techniques to preprocess the images and extract a set of critical features for event classification. The proposed method can detect and track solar events as well as can measure their properties. This automatic process is valuable for the forecasting and studies of solar events since it dramatically improves efficiency and accuracy.

1. Introduction

The sun is the source of space weather – the disturbances whose radiative, field, and particle energy directly impact earth. Solar activity changes the radiative and particle output of the

* E-mail address: shih@njit.edu

sun, producing corresponding changes in the near-earth space environment as well as at the earth surface. The most dramatic events on the sun, insofar as space weather effects are concerned, are solar flares, filament, and coronal mass ejections (CMEs).

A flare is defined as a sudden, rapid, and intense variation in brightness. A solar flare occurs when magnetic energy that has built up in the solar atmosphere is suddenly released. Radiation is emitted across virtually the entire electromagnetic spectrum, from radio waves at the long wavelength end, through optical emission to X-rays and gamma rays at the short wavelength end (Zirin, 1988). The amount of energy released is equivalent to millions of 100-megaton hydrogen bombs exploding at the same time. Flares are often associated with CMEs. Figure 1 shows a full disk Hα (Hydrogen-Alpha) image with a flare detected on the low-left corner.

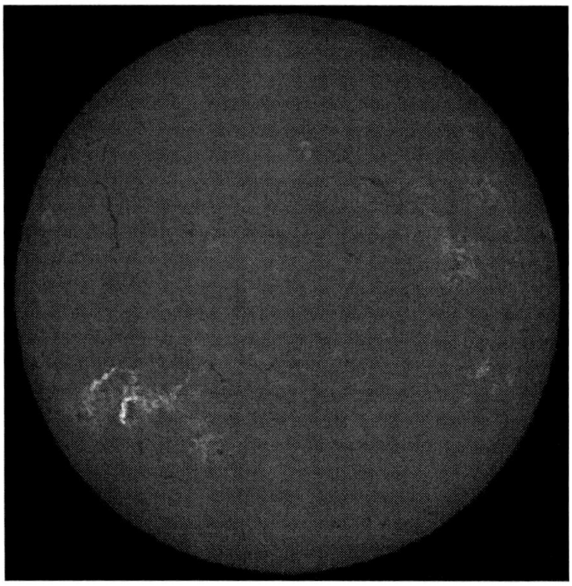

Figure 1. Solar flares on the left side of the image.

A chromospheric filament usually marks a boundary between two opposite magnetic regions. Filaments may last for multiple days, changing their manifestation, but eventually they vanish. Their disappearance may end up with large CME associated with geomagnetic storms that affect the quality of terrestrial communication, reliability of power systems and safety of space missions (Martens and Zwaan, 2001); therefore the filament tracking is an important task of solar observation. In solar images, filaments appear as dark elongated objects of different shape and size, and may be broken into separate parts, as shown in Figure 2. Qu *et al.* (2005) developed an automatic solar filament detection algorithm based on image enhancement, segmentation, pattern recognition, and mathematical morphology (Shih and Mitchell, 1989).

A CME is due to a large scale rearrangement of the solar magnetic field which can induce a geo-effective solar wind structure. Each CME may carry away a mass of up to 1013 kg and release up to 1025 J of energy from coronal magnetic fields (Harrison, 1995). Following their discovery, CMEs were soon found to be correlated with the occurrence of geomagnetic storms. CMEs are easily observed at the solar limb, where they are seen against a dark background. However, the earth-directed disk CMEs are much harder to detect – although they are the ones that have the most important geomagnetic effects. Therefore, detecting the early manifestation of CMEs on the solar disk is essential to understanding earth-directed CMEs themselves. Establishing the correlation among the occurrence of CMEs and some other solar surface phenomena, such as filament eruptions and flares, is a long-standing problem in solar physics. After firm relationships are established, the correlated solar surface phenomena can provide an early warning of the occurrence of earth-directed CMEs and geo-effectiveness of CMEs. A key challenge to move this research area forward is to have real-time, automatic and accurate detection and characterization of solar events, and to use the results for real time space weather forecasts.

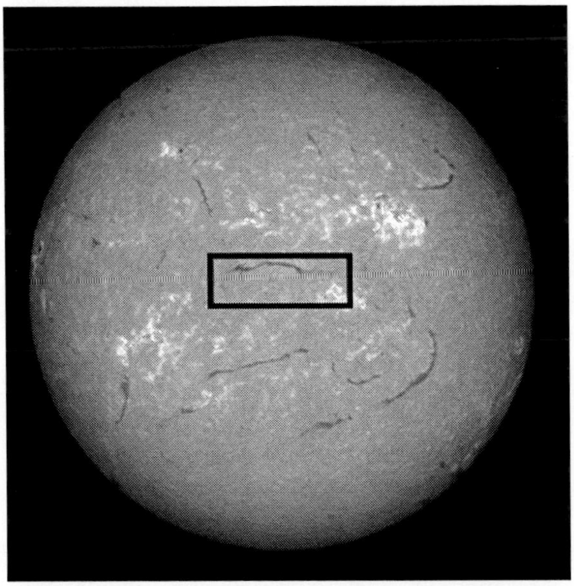

Figure 2. A fragment with a solar filament is marked by a rectangle.

CMEs have been observed from Large Angle and Spectrometric Coronagraph (LASCO) images by Solar and Heliospheric Observatory (SOHO). LASCO images include three different coronagraphs: C1, C2, and C3, whose ranges are from 1.1 to 3, 1.5 to 6, and 3.7 to 30 solar radii, respectively. Coronagraph C1 has been inoperable since 1999. A detailed description of LASCO can be found in Brueckner *et al.* (1995).

2. Solar Flare Detection

There are many ways to detect solar flares. When a solar flare first appears on the Sun's surface, there is often a radiation covering the entire range of electromagnetic waves. For example, light curves of radio waves and X-rays could be used to indicate a set of flares. However, location and properties of flares cannot be demonstrated in light curves. In the mean time, solar flares also emit high velocity charged particles that take one or two days to reach Earth. Our objective is to detect and characterize solar flares automatically in real time. Finally, most solar flares can also be detected using Hα. We start with solar images observed in Hα using an 8-inch telescope in the Big Bear Solar Observatory (BBSO) in California, operated by New Jersey Institute of Technology.

Automatic solar flare detection is the key to space weather monitoring. It is very challenging since the features of solar flares are complicated. We need a system that can handle the solar flares' complexity and scalability. Veronig *et al.* (2000) proposed a method for automatic flare detection by applying a combination of region-based and edge-based segmentation methods. Simple region-based methods are applied for a tentative assignment of flare activity, making use of one of the decisive flare characteristics: the high intensities. Borda *et al.* (2001) presented a method for automatic detection of solar flares using the neural network technique. The network used is multi-layer perceptron (MLP) with back-propagation training rule. They used a supervised learning technique and required a lot of iterations to train the network.

Qu *et al.* (2003) compared three principal techniques: multi-layer perceptron (MLP), radial basis function (RBF), and support vector machines (SVM) classifiers, for solar-flare detection. The preprocessing step is to obtain nine principal features of solar flares for the classifiers. They show that by using SVM, the best classification rate of solar flares can be obtained. Qu *et al.* (2004) further extended their work to automatically detect flares and applying image segmentation techniques to compute their properties. A solution is obtained for automatically tracking the apparent separation motion of two-ribbon flares and measuring their moving direction and speed in the magnetic fields. From these measurements with certain assumptions, the reconnection of the electric field as a measure of the rate of the magnetic reconnection in the corona can be derived.

2.1. Neural Networks

A neural network is a massively parallel distributed processor that has a natural propensity for storing experiential knowledge and making it available for classification (Haykin, 1994). Multi-layer perceptron (MLP) is a type of feed-forward networks. The learning rule is back-propagation (Ripley, 1996). The feed-forward network has three layers: input layer, hidden layer, and output layer. The hidden layer is between the input and the output layers. For example, a simple feed-forward network is shown in Figure 3, where there are two nodes in the input layer, two nodes in the hidden layer, and one node in the output layer. Weights are incorporated into the connections from input nodes to hidden nodes and from hidden nodes to the output node.

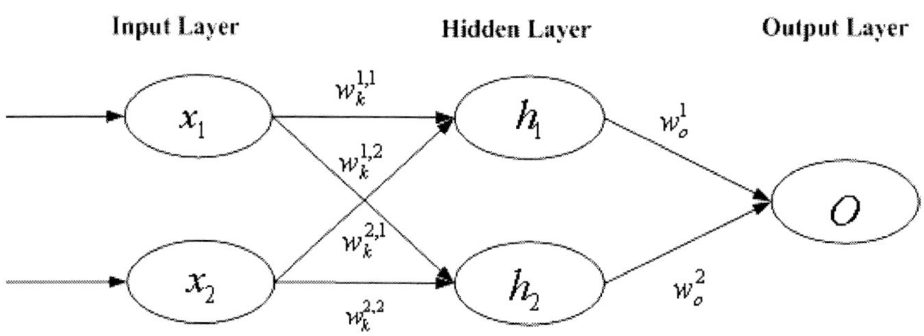

Figure 3. The architecture of MLP.

For linear processes, we can use binary thresholding in the hidden and output units. For nonlinear processes, activation functions are used rather than thresholding. We adopt the sigmoid function as the activation function

$$h_j = sigmoid(\sum_{i=1}^{N_I} w_k^{i,j} x_i) \tag{1}$$

$$O = sigmoid(\sum_{j=1}^{N_H} w_o^j h_j) \tag{2}$$

where x_i is the input feature at the input node, $w_h^{i,j}$ is the weight connecting the input node with the hidden node, w_o^j is the weight connecting the jth hidden node with the output node, N_I is the number of nodes in the input layer, N_H is the number of nodes in the hidden layer, and O is the output value. The $sigmoid$ function is defined as

$$sigmoid(x) = \frac{1}{1+e^{-x}} \tag{3}$$

The back-propogation training is conducted to obtain the correct weights. We can compute the mean square error (MSE) between the actual output and the desired output for the given input in the training set. The error function E can be obtained by

$$E = (T - O)^2 \tag{4}$$

where T is the value of the desired target. To reduce the mean square error, it is necessary to calculate the gradient of the error function with respect to each weight. One may then move each weight slightly in the opposite direction to the gradient. The gradient function for the weights in the output layer is shown below.

$$\delta_j = \frac{\partial E}{\partial w_0^j} = \frac{\partial (T-O)^2}{\partial w_0^j} \tag{5}$$

$$w_0^{j+1} = w_0^j - \eta' \delta_j \tag{6}$$

From Equation (6), the new values for the network weights are calculated by multiplying the negative gradient with a step size of parameter η' (called the learning rate). The weights in the hidden layer are updated using the same procedure. After we calculate all the correct weights, the neural network is completely constructed.

2.2. Radial Basis Function (RBF)

We can view the design of a neural network as a curve-fitting (approximation) problem in a high-dimensional space. From this viewpoint, learning is accomplished by finding a surface in a multi-dimensional space. This surface is used to interpolate the test data. Radial basis function (RBF) network is a fully connected network and generally is used as a classification tool (Broomhead and Lowe, 1988). In a RBF model, the layer from input nodes to hidden neurons is unsupervised, and the layer from hidden neurons to output nodes is supervised (Bishop, 1995). The architecture of a radial basis function neural network is shown in Figure 4. The transformation from the input to the hidden space is nonlinear, and the transformation from the hidden to the output space is linear. The hidden neurons provide a set of "functions" that constitute an arbitrary "basis" for the input patterns. These are the functions known as *radial basis functions*. Through a careful design, it is possible to reduce a pattern in a high-dimensional space at input units to a low-dimensional space at hidden units.

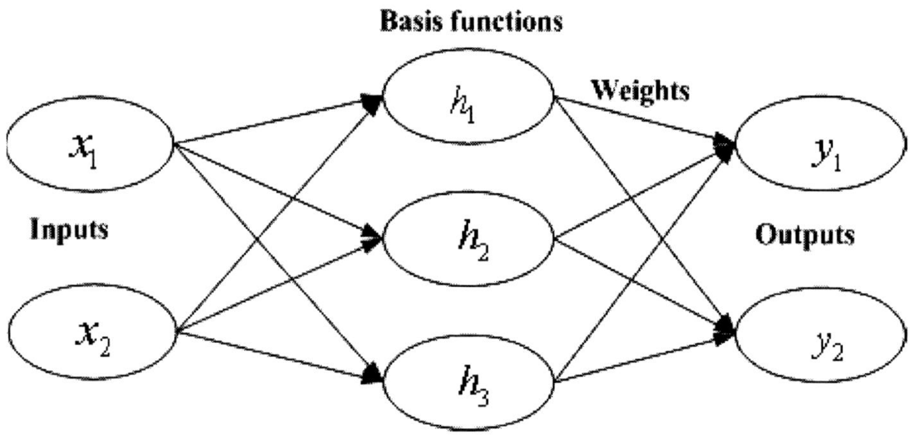

Figure 4. The architecture of a radial basis function neural network.

In Figure 4, the network forms the following equation:

$$y_k(\mathbf{x}) = \sum \omega_{kj} h_j(\mathbf{x}) + \omega_{k0} \tag{7}$$

where $h_j(\mathbf{x})$ is a Gaussian function typically, and ω_{k0} is the bias or threshold. The Gaussian basis function is used as an activation function. That is

$$h_j(\mathbf{x}) = \exp\left(-\left\|\mathbf{x} - \boldsymbol{\mu}_j\right\|^2 / 2\sigma_j^2\right) \tag{8}$$

where $\mathbf{x}$ is a d-dimensional input vector with elements x_i, and $\boldsymbol{\mu}_j$ and σ_j are respectively the center and the standard deviation of the Gaussian basis function. Since the first and the second layers of RBF network are unsupervised and supervised respectively, a two-stage training procedure is used to train the RBF model. In the first stage, the input data set is used to obtain parameters of activation functions (like $\boldsymbol{\mu}_j$ and σ_j). In the second stage, the optimal weights between hidden neurons and output nodes are obtained by minimizing a sum-of-square error function. The following procedure describes the steps of obtaining optimal weights.

By absorbing the bias parameter, ω_{k0}, into the weights, Equation (7) can be revised as

$$y_k(\mathbf{x}) = \sum \omega_{kj} h_j(\mathbf{x}) \tag{9}$$

where $h_j(\mathbf{x})$ is an extra basis function with the activation value 1. Equation (9) can be rewritten using the matrix notation as

$$Y(\mathbf{x}) = W\Phi \tag{10}$$

where $W = [\omega_{kj}]$ and $\Phi = [h_j(\mathbf{x})]$.

The sum-of-square error function, E, can be described as

$$E = \frac{1}{2}\sum\sum\left[y_k\left(x^n\right) - t_k^n\right]^2 \tag{11}$$

where $\mathbf{x}^n$ is the input data set, and t_k^n is the target value for the output unit k.

By differentiating E with respect to ω_{kj} and setting derivative to zero, the optimal weights can be obtained. The solutions of weights can be expressed using the matrix notation as

$$\left(\Phi^T\Phi\right) W^T = \Phi^T T \tag{12}$$

where $T = [t_k^n]$ and $\Phi = [h_j(x^n)]$.

By multiplying $\left(\Phi^T\Phi\right)^{-1}$ to Equation (12), the solution for the weights is given as

$$W^T = \left(\Phi^T\Phi\right)^{-1}\Phi^T T \tag{13}$$

$$W^T = \Phi^+ T \tag{14}$$

where $\Phi^+ = \left(\Phi^T\Phi\right)^{-1}\Phi^T$ being the pseudo-inverse of Φ. After computing the optimal weights, the RBF network can be used as a classifier to segment the test data into corresponding classes, with "-1" indicating a non-flare state and "1" indicating a flare state.

RBF is strongly dependent on the quality of the employed learning strategy and the quantity of training images. The aim of an adaptive learning RBF network is to reduce the required knowledge of system parameters with a minimum amount of performance loss. The RBF network requires knowledge of three different parameters per neuron:

- The center vector $\boldsymbol{\mu}_j$
- The weights ω_{kj}
- The radius σ_j

One unsupervised learning strategy is the self-organizing feature map. When the algorithm has converged, prototype vectors which correspond to nearby points on the feature map grid have nearby locations in input space. However, the imposition of the topographic property, particularly if the data is not intrinsically two-dimensional, may lead to a sub-optimal placement of vectors. Here we use the K-means unsupervised learning strategy as follows. Let $\boldsymbol{\mu}_j$ be the mean of the data points in set S_j given by

$$\boldsymbol{\mu}_j = \frac{1}{N_j} \sum_{n \in S_j} \mathbf{x}^n \tag{15}$$

The initial centers are randomly chosen from data points, and the nearest $\boldsymbol{\mu}_j$ is updated using

$$\Delta \boldsymbol{\mu}_j = \eta (\mathbf{x}^n - \boldsymbol{\mu}_j) \tag{16}$$

where η is the learning rate parameter.

The second parameter of the RBF network is the weight ω_k of the output layer. It can be done by using the LMS algorithm. The LMS algorithm was originally developed by Widrow and Hoff in 1960 and is also known as the Widrow-Hoff rule (Mitra and Poor, 1994). The weights can be summarized as

$$\Delta W = 2\eta (y_k(\mathbf{x}^n) - t_k^n) h(\mathbf{x}^n) \tag{17}$$

where η is the learning rate of LMS, and $y_k(\mathbf{x}^n)$ and t_k^n are the responses of the RBF network and the desired response, respectively. The vector $h(\mathbf{x}^n)$ contains the unweighted responses of all neurons in the hidden layer.

The last parameter of the RBF network is the radius or spread of the radial function. We may use an average of the center spread of all RBFs to calculate the radius. After the centers $\boldsymbol{\mu}_j$ are established, σ^2 can be derived from the center as

$$\sigma^2 = \frac{1}{M}\sum_{j=1}^{M} \| \mathbf{x}^n - \mu_j \|^2 \tag{18}$$

where M is the number of hidden nodes.

2.3. Support Vector Machine (SVM)

Support Vector Machine (SVM) is a learning system based on advances in statistical learning theory. It has been successfully applied in text categorization, hand-written character recognition, image classification, bio-sequences analysis, etc.

Suppose we are given l observations. Each observation consists of a pair: a vector $\mathbf{x}_i \in R^n, i = 1,\ldots,l$ and the associated "truth" y_i, given to us by a trusted source. It is assumed that there exists a density distribution $p(\mathbf{x}, y)$ from which these data are drawn; i.e., the data are assumed independently drawn and identically distributed. Suppose that we have a machine to learn the mapping $x_i \to y_i$. The machine is actually defined by a set of possible mappings $x_i \to f(\mathbf{x},\alpha)$, where the function $f(\mathbf{x},\alpha)$ is labeled by the adjustable parameter α. For example, a neural network with a fixed architecture and with α corresponding to the weights and biases, is a learning machine in this sense. $f(\mathbf{x},\alpha)$ could represent a set of radial basis functions or MLPs with a certain number of hidden neurons (Vapnik, 1982). The expectation of the test error for a trained machine is therefore

$$R(\alpha) = \int \frac{1}{2} | y - f(\mathbf{x},\alpha) | \, p(\mathbf{x}, y) d\mathbf{x} dy \tag{19}$$

where $R(\alpha)$ is called the *expected risk*.

The empirical risk $R_{emp}(\alpha)$ is defined to be just the measured mean error rate on the training set (for a fixed, finite number of observations).

$$R_{emp}(\alpha) = \frac{1}{2l}\sum_{i=1}^{l} | y_i - f(\mathbf{x}_i,\alpha) | \tag{20}$$

The empirical risk minimization (ERM) inductive principle is to minimize the risk functional in Equation (20) on the basis of empirical data ($\mathbf{x}_i, y_i$). The ERM principle is based on the law of large numbers convergence in probability to the expected risk (strategy followed by MLP and RBF neural networks). By choosing some η such that $0 \le \eta \le 1$, the following bound holds (Vapnik,1998).

$$R(\alpha) \le R_{emp}(\alpha) + \sqrt{\left(\frac{h(\log(2l/h)+1) - \log(\eta/4)}{l}\right)} \tag{21}$$

where h is a non-negative integer called the *Vapnik Chervonenkis* (VC) dimension. The right hand side of Equation (21) is the "bound on the risk", and the second term on the right hand side is called the "*confidence interval*". When l/h is large, the confidence interval is small. The actual risk is then close to the value of the empirical risk one. In this case, a small value of the empirical risk guarantees a small value of the expected risk. When l/h is small, a small $R_{emp}(\alpha)$ does not guarantee a small value of the actual risk. In this case, to minimize the actual risk we have to minimize the confidence interval. To minimize the right-hand side of the bound risk in Equation (21), we have to make the VC dimension a controlling variable.

In the previous section, we consider a classical neural network which implements the first strategy: keep the confidence interval fixed and minimize the empirical risk. Below we consider SVM to implement the second strategy: keep the value of the empirical risk fixed and minimize the confidence interval (Vapnik, 1995). SVM is based on the structural risk minimization (SRM) inductive principle. The SRM principle is intended to minimize the risk functional with respect to the empirical risks and the confidence interval (Vapnik and Chervonenkis, 1974). Therefore, SVM is a better strategy than the classical neural network, such as the MLP neural network. Its basic idea is to separate the fixed given input pattern vectors into two classes using a hyperplane in the high dimensional space (Vapnik and Chervonenkis, 1991). This means we always increment the VC-dimension h by one if we fail to separate the input pattern vectors. We do not stop the process until the separation is satisfied. SVM intends to separate the input pattern using the minimum VC-dimension h, which minimizes the total risk by considering both empirical risk and confidence interval. Figure 5 shows the SVM with a linear hyperplane.

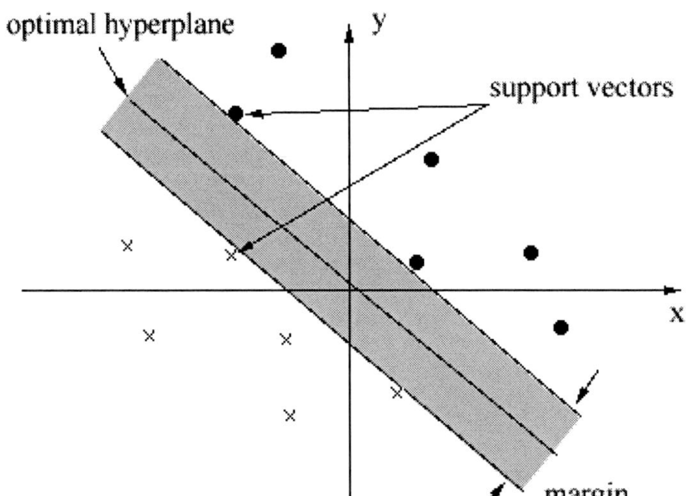

Figure 5. Support vector machine classification with a linear hyperplane that maximizes the separating margin between two classes.

Consider a two-class classification problem where patterns are represented as an-dimensional vector. For each corresponding vector, there exists a value $y \in \{-1,+1\}$.

$$\mathbf{x} = \{\mathbf{x}_1, \mathbf{x}_2, \mathbf{x}_3, \ldots, \mathbf{x}_m\} \in R^N \tag{22}$$

$$y = \{y_1, y_2, y_3, \ldots, y_m\} \in \{-1, +1\} \tag{23}$$

The problem is to find a decision function $g(\mathbf{x})$ such that the class label y of any example $\mathbf{x}$ can be accurately predicated. Using mathematical terms, we have g: $R \rightarrow \{-1, +1\}$ (Guyon and Stork, 2000).

For the linear support vector classifier

$$f(\mathbf{x}) = (\mathbf{x} \cdot \mathbf{w}) + b \tag{24}$$

where $\mathbf{w} \in R^N$ and $b \in R$.

The set of labeled training patterns

$$(y_{1,}\mathbf{x}_1), \ldots, (y_l, \mathbf{x}_l), \qquad y_i \in \{-1, 1\} \tag{25}$$

It is linearly separable if there are a vector w and a scalar b in existence such that the following inequalities are valid for all elements of the training set.

$$\mathbf{w} \cdot \mathbf{x}_i + b \geq 1 \text{ if } y_i = 1, \tag{26}$$

$$\mathbf{w} \cdot \mathbf{x}_i + b \leq -1 \text{ if } y_i = -1. \tag{27}$$

The optimal hyperplane is the unique one which separates the training data with a maximal margin. The distance $\rho(\mathbf{w}, b)$ is given by

$$\rho(\mathbf{w}, b) = \min_{\{\mathbf{x}: y=1\}} \frac{\mathbf{x} \cdot \mathbf{w}}{\|\mathbf{w}\|} - \max_{\{\mathbf{x}: y=-1\}} \frac{\mathbf{x} \cdot \mathbf{w}}{\|\mathbf{w}\|} \tag{28}$$

Because the optimal hyperplane $(\mathbf{w}_0, b_0)$ is the arguments that maximize the distance $\rho(\mathbf{w}, b)$, we have Equation (29) from (26), (27) and (28).

$$\rho(\mathbf{w}, b) = \frac{2}{\|\mathbf{w}_0\|} = \frac{2}{\sqrt{\mathbf{w}_0 \cdot \mathbf{w}_0}} \tag{29}$$

It means that the optimal hyperplane is the unique factor that minimizes $\mathbf{w} \cdot \mathbf{w}$ under the constraints (26) and (27). This problem is usually solved by means of the classical method of Lagrange multipliers.

If we denote N nonnegative Lagrange multipliers as $\boldsymbol{\alpha} = (\alpha_1, \alpha_2, ..., \alpha_N)$, the solution is equivalent to determining the saddle point of the function.

$$L(\mathbf{w}, b, \boldsymbol{\alpha}) = \frac{1}{2}\mathbf{w}\cdot\mathbf{w} - \sum_{i=1}^{N} \boldsymbol{\alpha}_i \{y_i(\mathbf{w}\cdot\mathbf{x}_i + b) - 1\} \tag{30}$$

At the saddle point, L has a minimum for $\mathbf{w} = \mathbf{w}_0$ and $b = b_0$ and a maximum for $\boldsymbol{\alpha} = \boldsymbol{\alpha}^0$. We have

$$\frac{\partial L(\mathbf{w}_0, b_0, \boldsymbol{\alpha}^0)}{\partial b} = \sum_{i=1}^{N} y_i \alpha_i^{\ 0} = 0 \tag{31}$$

$$\frac{\partial L(\mathbf{w}_0, b_0, \boldsymbol{\alpha}^0)}{\partial \mathbf{w}} = \mathbf{w}_0 - \sum_{i=1}^{N} \alpha_i^{\ 0} y_i \mathbf{x}_i = 0 \tag{32}$$

Substituting Equations (31) and (32) into the right hand side of (30), it is reduced to the maximization of the function

$$W(\boldsymbol{\alpha}) = \sum_{i=1}^{N} \alpha_i - \frac{1}{2}\sum_{i,j}^{l} \alpha_i a_j y_i y_j (\mathbf{x}_i \cdot \mathbf{x}_j) \tag{33}$$

From (32) it follows that

$$\mathbf{w}_0 = \sum_{i=1}^{N} \alpha_i^{\ 0} y_i \mathbf{x}_i \tag{34}$$

The parameter b_0 can be obtained from the Kuhn-Tucker condition.

$$b_0 = y_j - \mathbf{w}_0 \cdot \mathbf{x}_j \tag{35}$$

The problem of classifying a new data point x is now simply solved by looking at the sign of

$$\mathbf{w}_0 \cdot \mathbf{x} + b_0 \tag{36}$$

For the nonlinear case, we use the kernel functions to transfer input data to feature space. The nonlinear kernel functions are given in the following:

Polynomial support vector classifier

$$k(\mathbf{x}, \mathbf{x}') = \langle \mathbf{x}, \mathbf{x}' \rangle^d \tag{37}$$

Gaussian RBF kernel classifier

$$k(\mathbf{x}, \mathbf{x}') = \exp\left(-\|\mathbf{x} - \mathbf{x}'\|^2 / 2\sigma_j^2\right) \tag{38}$$

2.4. Feature Analysis and Preprocessing

Borda *et al.* (2001) used seven features for solar flare detection. After extensive investigations, we modify them into nine features as described below.

Feature 1: Mean brightness of the frame. Usually, when flares happen, the image becomes brighter. Let x_j denote the gray level of a pixel and N denote the number of pixels. The mean brightness $\bar{x}$ is represented as

$$\bar{x} = \sum_{j=0}^{N-1} x_j \tag{39}$$

Feature 2: Standard deviation of brightness. Usually, when flares happen, the standard deviation of the brightness becomes large. In Equation (40), std denotes the standard deviation, and V denotes the variance.

$$std = \sqrt{V} \tag{40}$$

$$V = \sum_{j=0}^{N-1} (x_j - \bar{x})^2 \tag{41}$$

Feature 3: Variation of mean brightness between consecutive images. This is the difference of mean values between the current image and the previous image.

Feature 4: Absolute brightness of a key pixel. By comparing consecutive images, we can find the key pixel that has the maximum gray level difference. Usually, when flares happen, the absolute brightness of the key pixel becomes large. The position of the key pixel k is represented as

$$k = where(\max(\mathrm{Im}_{pre} - \mathrm{Im}_{current})) \tag{42}$$

where Im_{pre} denotes the vector of the previous image, and $\mathrm{Im}_{current}$ denotes the vector of the current image.

Feature 5: Radial position of the key pixel.

Features 1-5 focus on the global view of the consecutive images. But these features are insufficient to obtain the information for small solar flares. Therefore, it is necessary to include the local view of solar images.

Feature 6: Contrast between the key pixel and the minimum value of its neighbors in a 7 × 7 window. When flares occur, this contrast becomes large.

Feature 7: Mean brightness of a 50 × 50 window, where the key pixel is at the center. Since we assume that the key pixel is one of the pixels on the flare, the 50 × 50 window will include most of the flare. This region is most effective for the solar flare detection.

Feature 8: Standard deviation of the pixels in the aforementioned 50 × 50 window.

Feature 9: Difference of the mean brightness of the 50 × 50 window between the current and the previous images.

See samples of the input nine features and the target value in Table I.

Table I. Samples of the nine features and the target value.

1	2	3	4	5	6	7	8	9	Target value
3994.25	2628.22	45	12474.00	431.23	3635.00	2011.43	1978.60	427.35	-1
3807.42	2458.64	187	16383.00	437.13	6484.00	9460.98	2430.92	7449.55	1

2.5. Classification Rates

The solar Hα images of 2032 ×2032 pixels were obtained from BBSO. We selected the images of flare events starting from January 1, 2002 to December 31, 2002. The first step taken was to carefully divide the images into two classes: flare state and non-flare state. We tried to cover all the flare events that can be recognized by human eyes. We labeled corresponding images as "flare state" and included the same number of "non-flare state" images. Because some micro-flares are difficult to be recognized even by human eyes, we regarded them as "non-flare state". Secondly, we computed the nine features for a given solar image and put it into either the training or the testing data set. We used the training data set to train the networks and used the testing data set to obtain the classification rate. Finally, we obtained flare properties using image segmentation techniques, since we are also interested in obtaining the properties of flares such as size, lifetime, and motion.

We developed the programs in Interactive Data Language (IDL) by Research Systems, Inc. The program runs on a DELL Dimension L733r with CPU time 733 Mhz and memory of 256 Mbytes under Windows 2000. There are three steps in our program:

a) Preprocessing to obtain the nine features of solar flares.
b) MLP, RBF and SVM training and testing programs used for solar flare detection.
c) Region growing and edge detection methods for obtaining the flare properties.

For the experiments of solar flare detection, we capture one image per minute and process the current image in comparison with the previous image to obtain the nine features. Firstly, we filter the noise of full-disk images using a Gaussian function, and center these two images using the IDL FIT_LIMB function. Secondly, we compute the mean brightness and standard deviation for the two images. Thirdly, we divide them to obtain the pixel with the maximum gray level difference. It is regarded as the key pixel to compute other features. After preprocessing steps, we use the nine features of the key pixel as the input to the neural network and SVM. At last, we use region growing combined with Canny edge detector to analyze the region of flares.

We tested the MLP, RBF and SVM methods based on 240 events including pre-flare, main phase and post-flare, half of them being used as the training set and the rest half as the testing set. For MLP, after about 1000 iterations for the training set of 120 events, we tested this network with 120 events. The fraction of misclassified events is about 5%. MLP uses supervised training and needs a lot of iterations to obtain the satisfied classification rate. For RBF, we obtain a better classification rate and a faster training speed. For SVM, we obtain the best performance among the three methods, such that the misclassification rate is 3.3%. The results are shown in Table II.

Table II. Classification report based on 120 training events and 120 testing events.

Methods based on events	Classification rate	Training time	Testing time
MLP with 1000 iterations	94.2%	60.20 seconds	0.01 second
RBF	95%	0.27 second	0.01 second
SVM	96.7%	0.16 second	0.03 second

The MLP architecture is to perform a full non-linear optimization of the network. Therefore, a lot of iterations to train the system to get the best result are needed. In Table III, it is demonstrated that the result performs better with more iterations. The number of hidden nodes determines the complexities of the neural networks. We can adjust the number of hidden nodes to obtain the best performance. In Table III, we show the comparison of MLP with 11 hidden nodes and MLP with other number of hidden nodes. Test correctness of MLP with 11 hidden nodes is comparable to MLP with more hidden nodes. MLP with 11 hidden nodes has better performance than MLP with more or less hidden nodes.

Table III. Comparison of different MLP iterations and hidden nodes on 120 training events and 120 testing events

Methods based on events	Classification rate	Training time
MLP with 1000 iterations and 11 hidden nodes	4.2%	960.2 seconds
MLP with 100 iterations and 11 hidden nodes	69.2%	6.2 seconds
MLP with 3000 iterations and 11 hidden nodes	94.2%	179.3 seconds
MLP with 1000 iterations and 11 hidden nodes	94.2%	60.2 seconds
MLP with 1000 iterations and 6 hidden nodes	92.5%	46.7 seconds
MLP with 1000 iterations and 20 hidden nodes	94.2%	115.6 seconds

RBF neural network, as compared with the MLP, is the possibility of choosing suitable parameters for each hidden unit without having to perform a full non-linear optimization of the network (Bishop, 1995). Usually, the number of hidden nodes equals the number of classes. For our case, we want to use the RBF classifier to separate two classes. We try RBF with different number of hidden nodes and find that RBF with two hidden nodes is good for our pattern classification. The parameters of RBF are center vector, the weight and the radius. The training architecture can be quickly constructed by K-mean algorithm.

SVM is based on the idea of "minimization of the structural risk". To obtain the best classification rate, we consider the confidence interval and empirical risk. The complex system has high confidence interval and low empirical risk. We test our flare events using a linear SVM and the non-linear SVMs respectively with polynomial kernel classifier and Gaussian RBF kernel classifier. Experimental results are shown in Table IV.

Table IV. Comparison of different SVM training strategies on 120 training events and 120 testing events

Methods based on events	Classification rate	Training time	Testing time
Linear SVM	96.7%	0.16 second	0.03 second
SVM with Polynomial kernel	95%	0.34 second	0.03 second
SVM with RBF kernel	90.83%	0.44 second	0.03 second

Table IV shows that the linear SVM is better than the non-linear kernel SVM for our events. The non-linear SVM can reduce the empirical risk using complex systems, but it has a higher confidence interval. Using the linear SVM with the nine features of the solar flare, we have both low empirical risk and confidence interval to achieve the best total risk. Therefore, we prefer the linear SVM to classify the flare patterns.

3. Solar Corona Mass Ejections Detection

The structures of CMEs can be grossly arranged into eight categories (Howard *et al.*, 1985): Halo, Curved Front, Loop, Spike, Double Spike, Multiple Spike, Streamer Blowout, Diffuse Fan, and Complex. In an effort to get an intensity classification, Dai *et al.* (2002) analyzed the velocity, span, mass and kinetic energy of CMEs and proposed three CME intensity categories, strong (including Halo, Complex), middle (including Double Spike, Multiple Spike and Loop) and weak (including Spike, Streamer Blowout and Diffuse Fan). Among these categories, the halo CMEs, which appear as expanding, circular brightenings surrounding the coronagraph's occulter, are of great concern because they may carry away a mass of up to 10^{15} kg and are moving outward along the Sun-Earth line with speeds greater than 500 km s^{-1} (Dai *et al.*, 2002). It appears that halo CMEs are an excellent indicator of increased geo-activity 3-5 days later (Brueckner *et al.*, 1998), and therefore, the automatic and accurate detection and characterization of CMEs, especially halo CMEs, in real-time can provide an early warning of the occurrence of potentially geo-effective disturbance. In Figure 6, examples of a strong and weak CMEs in LASCO C2 images are pointed by arrows.

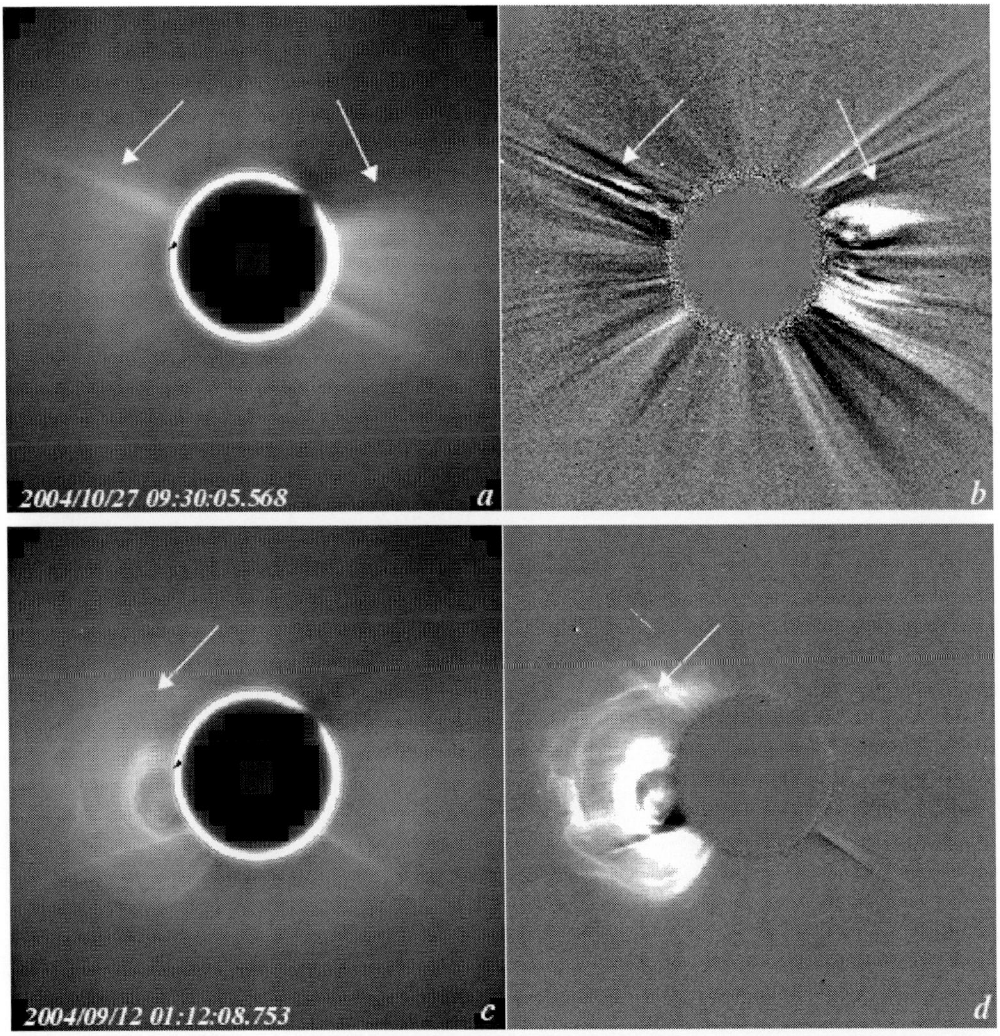

Figure 6. *a*: weak CMEs in a LASCO C2 image on October 27, 2004, *b*: the same weak CMEs in the running difference image, *c*: a strong CME in a LASCO C2 image on September 12, 2004, *d*: the same strong CME in the running difference image. CMEs are pointed by arrows.

The traditional way of detecting CMEs is based on human observation. For example, the SOHO/LASCO group provided a CME catalog based on image enhancement and human detection. However, the human observation is subjective and slow, and decision rules may vary from one operator to another. During the last few years, image processing and pattern recognition techniques have been utilized in automatic CME detection. For example, Berghmans (2002) introduced an automatic CME detection procedure using image processing techniques such as image enhancement, thresholding, and motion tracking. In 2004, Robbrecht and Berghmans developed an application, CACTus (Computer Aided CME Tracking), for automatic detection of CMEs in sequences of LASCO images which uses Hough transform to detect CMEs in the running differences. They also published their real-time CME detection results through the Internet.

We present an automatic three-step CME detection and classification algorithm. In brief, the three steps are: (1) preprocessing, in which running difference images of LASCO C2 and C3 are automatic generated, (2) detection and characterization, in which the structural properties of CMEs (such as height, span angle) are characterized using image segmentation and morphological methods, and (3) classification, in which SVM classifiers are used to classify CMEs into three groups of strong, medium and weak on the basis of characterization. Three steps are described in Section 2, 3 & 4, respectively. Finally, a comparison of LASCO, CACTus and our CME catalogs is presented.

3.1. Preprocessing

SOHO/LASCO provides C2 and C3 images in a *Flexible Image Transport System* (FITS) format which can be downloaded from the SOHO/LASCO web site. Image information such as size, center coordinates, and exposure time is written into the headers of FITS images. In our preprocessing step, LASCO images in different sizes are resized to the same size of 512×512 pixels, and aligned according to coordinates of solar centers. Like most other CME detections, brightness normalization is applied according to the ratio of exposure time of LASCO images. Furthermore, if the gray level ranges of mean brightness between two consecutive images are different after the exposure time adjustment, the images are normalized using the mean brightness ratio of the current image to the previous image. In our preprocessing, when the ratio of their gray level is greater than 1.1 or less than 0.9, Equation (43) is applied to correct the brightness of the current image.

$$M_c' = M_c \times \frac{\bar{x}_p}{\bar{x}_c} \tag{43}$$

where M_c and M_c' denote the current images before and after the normalization, and $\bar{x}_p$ and $\bar{x}_c$ denote the mean brightness of no-CME regions on the previous and current images, respectively. The mean brightness ratio is based on the regions with no CME because the mean brightness normalization may reduce CMEs' brightness. The pixel with a lower gray level than the median value of the image is considered to be the one located on the no-CME region.

In LASCO images, streamers are structured similarly to CMEs. The difference between streamers and CMEs are that streamers are steady bright structures, while CMEs are sudden brightness changes. To distinguish a CME from other stable bright structures (e.g., streamers) and stable bright noise, a running difference image, obtained by Equation (44), is commonly used. The running difference image is also useful for the measurement of CMEs' moving-front (i.e. the top moving edge) and foot-point (i.e. the lower moving bottom).

$$D = M_c - G \tag{44}$$

where G is the reference image and M_c is the current LASCO image. The reference image to use with the normal aspect is the image before the current image. This has the disadvantage that if some regions of the previous image are badly captured, the resulting running difference images are badly generated. Another issue is when a CME is evolved on a sequence of image frames, the running difference may succeed in detecting the moving front of the CME, but may fail in detecting its location of foot-point because some parts of CME regions are overlapped on a sequence of images. In this book, a reference image G' is produced recursively by using the combination of the reference and previous images, as shown in Equation (45).

$$G' = G \times c_1 + M_c \times c_2 \tag{45}$$

where c_1 and c_2 denote the percentages of effectiveness on the reference and current images. The disadvantage of using a single image as the reference image is avoided. Based on our experiments, we set $c_1 = 90\%$ and $c_1 = 10\%$. However, if a strong CME is detected in the previous image frame, $c_1 = 100\%$ and $c_2 = 0\%$ are used.

In the next section, CMEs are segmented using a threshold based on high brightness regions. When a weak CME with small brightness enhancement is located in the dark region and a strong CME is located nearby, the weak CME may not be detected because the chosen threshold is focused on the strong one. To overcome this problem, a partial division image is used to segment CME regions based on the ratio of brightness between the current LASCO image and reference image. The division image is obtained by dividing the LASCO image by the reference image. For the region with gray level close to zero, the division may enhance noise significantly and produce overflow errors. Therefore, only a partial region of the image is chosen. The partial division image V is shown in Equation (46).

$$V = \frac{M_2}{G_2} \tag{46}$$

where G_2 denotes the region on the reference image with pixels brighter than the median value of the reference image, and M_2 denotes the region corresponding to G_2 on the LASCO image. By using the relatively bright regions for G_2, the division overflow is avoided.

Some of LASCO images contain missing blocks which appear as dark or bright squares on image frames. Three criteria are proposed to find the missing blocks: (1) when a dark missing block first appears in the current image, its appearance on the current image and the running difference images are both dark; (2) its gray level of the current image is close to the minimum value of the image, and its gray level of the running difference image is less than a negative threshold t_1; (3) the number of pixels for an individual block region is set to be greater than a threshold t_2 because a missing block contains a large amount of pixels as compared to small noise. The thresholds t_1 and t_2, obtained from our experiments, are negative values of the standard deviation of the running difference in brightness and 50 pixels, respectively. A bright missing block can also be detected in the same way. Finally, those missing blocks in running difference images and division images are replaced by the minimum value of the images.

3.2. Automatic Detection of CMEs

Two consecutive LASCO images incorporated with the reference image produce two running difference images and two division images. The running difference images and division images are segmented into binary differences by applying a thresholding method. By tracking segmented CME regions in binary differences, the properties of CMEs are obtained.

3.2.1. Segmentation of CMEs

Running difference images (D) and division images (V) are constructed by using consecutive LASCO images in the previous step. In this step, a thresholding method is adopted to D and V to produce binary differences in which CME regions are separated from the background. A fixed value of the threshold would produce unstable results because contrasts of running difference images vary from time to time. In our CME segmentation, the threshold is computed using the median and standard deviation of D and V. Based on extensive experiments, the thresholds for D and V of C2 images are chosen as $m+s$, where m and s are the median and standard deviation of C2's D and V images, respectively. Similarly, we obtain the thresholds for C3's D and V images which is $m+1.5\times s$. These automatic thresholds are robust with respect to different image contrasts. The final segmentation result is the summation of the two segmentation results from the difference and the division images.

After segmentation, morphological closing (Shih and Mitchell, 1989) is applied to the binary segmentation image to eliminate small gaps. A 5×5 structuring element is used to perform the binary closing. The closing of a region A by a structure element B, denoted by $A \bullet B$, is defined as

$$A \bullet B = (A \oplus B) \ominus B \tag{47}$$

The closing of A by B is simply the dilation of A by B, followed by the erosion of the result by B. The closing method can smooth features and remove small noise. The 5×5 structuring element removes the noise whose width and length are less than 5 pixels and eliminates the gaps whose distance is less than 5 pixels.

In order to calculate the features of CMEs easily, binary segmentation images are reformed to $360\times r$ binary angular images, where r is the radius to the edge of the C2 or C3 occulting disk in the LASCO image. The results of the segmentation, morphological closing, and the angular images are shown in Figure 7. The degree of angle [0,359] is counted from north clockwise. The size of C2 angular images is 360×156 pixels, and of C3 angular images is 360×216 pixels.

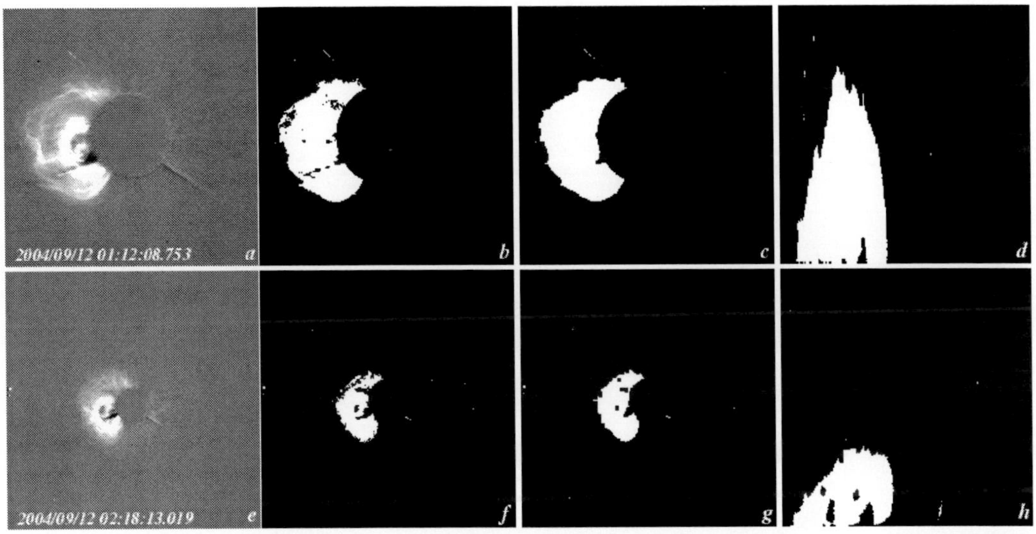

Figure 7. a: the running difference of a LASCO C2 image, observed on September 12, 2004, *b*: binary segmented result of the LASCO C2, *c*: morphological closing result of the LASCO C2, *d*: angular image of the LASCO C2. From *e* to *h*: the running difference of a LASCO C3 image, and its segmented, closing and angular results, respectively.

3.2.2. Features of CMEs

In the feature detection, all the segmented regions in a CME frame become CME candidate regions, but only continuously moving regions on consecutive images are treated as real CME regions. A CME candidate region is classified as a CME region if it occurs on the current image frame as well as on the previous image frame. The corresponding rules are given as follows: (1) the distance between the two centers of overlapping regions is less than t_3, where t_3 is the threshold for the maximum CME movement for an hour, and (2) the difference between the span widths of two overlapping regions is less than t_4, where t_4 is the threshold for the maximum degree span change for an hour. For the LASCO C2 image, t_3 is chosen as 100 pixels; for the C3 image, as 40 pixels. t_4 is chosen as a half of the span width of the CME candidate. In addition, the speed and the new increasing area for CME regions are computed by comparing the corresponding CME regions on the two binary angular images. Let us denote a CME region in the current image and its corresponding region in the previous image as A and A_p, respectively. The CME properties of A in an image frame are listed in Table V.

Table V. The properties of a CME region. The features 17-22 are obtained using the corresponding regions on the two consecutive images.

No.	Description of the CME properties
1.	The exposure time of the LASCO image.
2.	The time interval between the current and the previous image.
3.	The pixel size of the LASCO image.
4.	The mean brightness value of the reference image.
5.	The mean brightness value of the current image.
6.	The mean brightness value of the running difference.
7.	The standard deviation of the running difference.
8.	The number of pixels for A.
9.	The threshold for segmenting A from the running difference.
10.	The maximum height (arcsecs from disk center) of A.
11.	The height of the center of A.
12.	The minimum height of A.
13.	The starting angle of A. The angle is calculated from North 0 clockwise.
14.	The angle of the center of A.
15.	The ending angle of A.
16.	The angular width of A.
17.	The height difference (h_1) between the maximum height of A and A_p.
18.	The height of the new moving region (h_2) which is obtained by subtracting A_p from A.
19.	The speed which is computed using h_1 divided by the interval time cadence.
20.	The speed which is computed using h_2 divided by the interval time cadence.
21.	The span width of the new moving region.
22.	The center angle of the new moving region.

The detection reports for a CME on August 7, 2005 are shown in Figures 8 and 9. The classification of strong, medium, and weak CMEs is presented in the next section. Figures 10 and 11 show the height and velocity for the CMEs on September 1, 2002 in C2 and C3 images, respectively. The speed of the CME is around 300 km/s which is almost equal to the computed CME speed in the LASCO catalog.

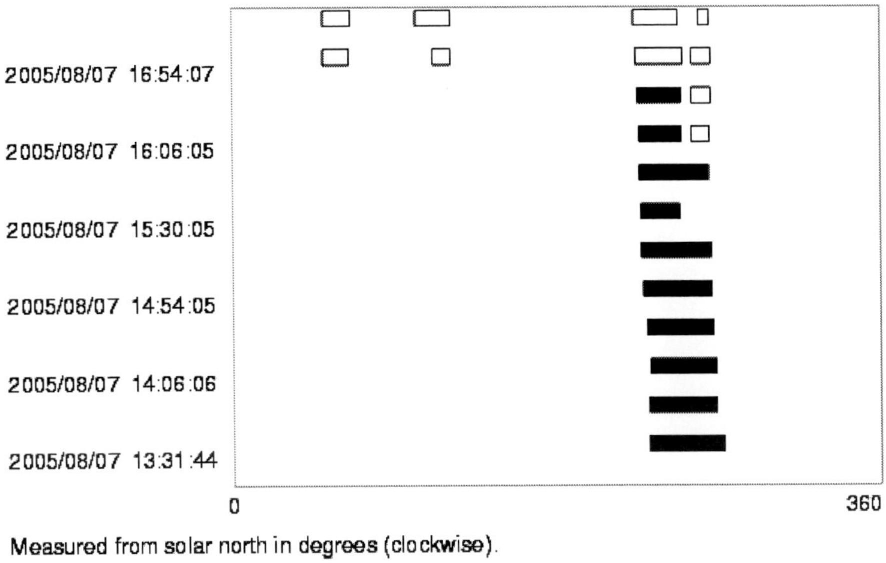

Figure 8. Detection and classification report for a detected CME using LASCO C2 images on August 7, 2005. Apparently, there is a strong CME on the West from 13:30 to 16:50UT. The solid black and the empty rectangle denote strong and weak CMEs, respectively.

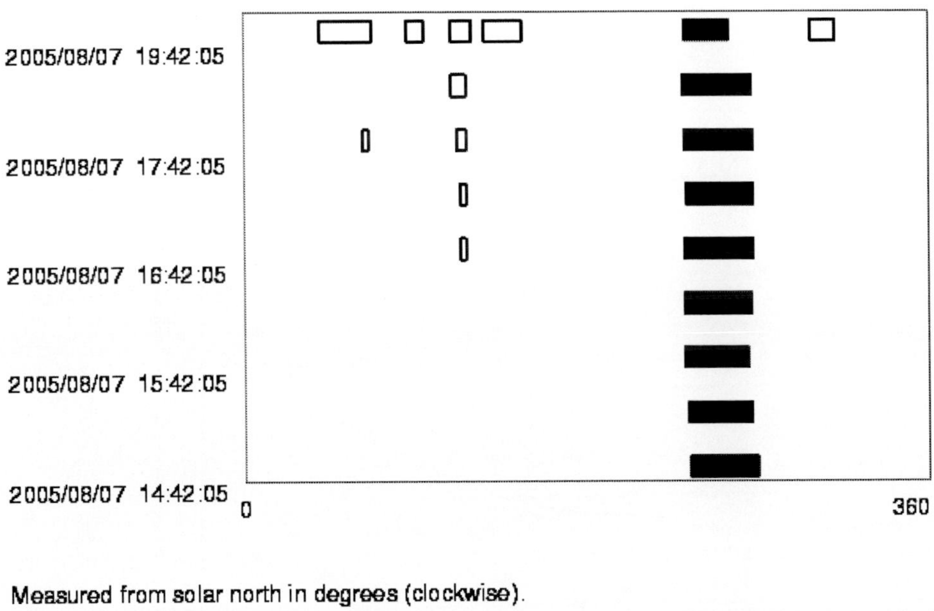

Figure 9. Detection and classification report for a detected CME using LASCO C3 images on August 7, 2005. There is a strong CME on the West from 14:40 to 19:40UT. Solid black, solid gray, and empty rectangle denote strong, medium, and weak CMEs, respectively.

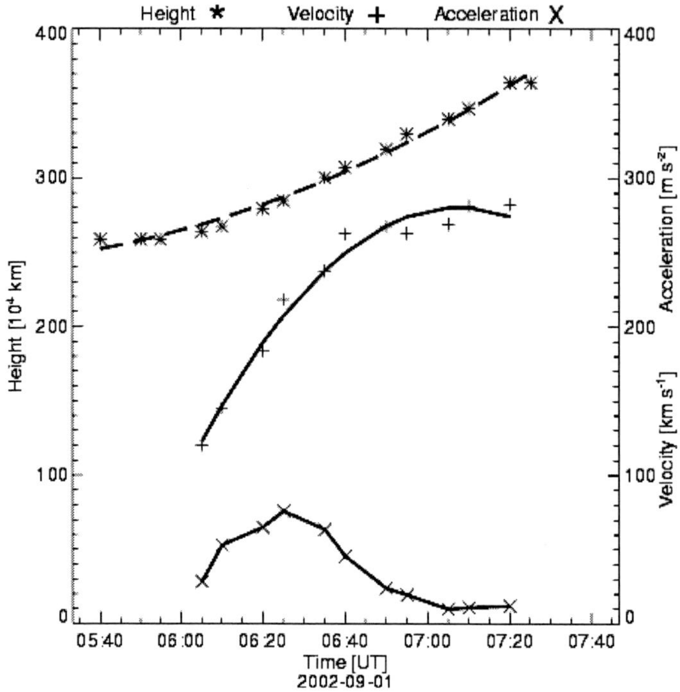

Figure 10. Height, velocity, and acceleration profile of a CME for LASCO C2 images on September 1, 2002.

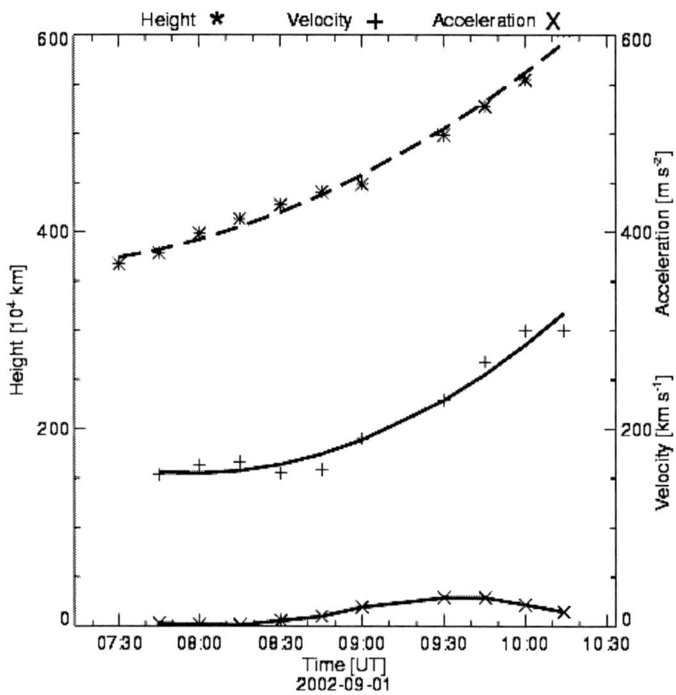

Figure 11. Height, velocity, and acceleration profile of a CME for LASCO C3 images on September 1, 2002.

3.3. Classification of Strong, Medium, and Weak CMEs

A strong CME consists of a large amount of fast moving mass. In our CME classification step, strong CMEs are selected from halo, curved front and complex CMEs according to Howard *et al.* (1985).

In this work, the Support Vector Machine (SVM) classifier with a linear kernel is used for distinguishing the strong CME from others. For the present study, a CME can be represented by twenty-two features which are obtained in the previous step. The SVM training computes the corresponding weights for each input feature for the CME classification. Six features with significant weights are selected as the inputs for classification. The six input features are proven to be robust based on our comparisons on feature combination.

The six inputs to our SVM classifier are the features with significant classification weights in Table V which are: the mean brightness in the running difference image, the number of pixels in the running difference image, the angular width of span, the height of new moving region, the span width of the new moving region, and the speed described in Feature 20 of Table V. A list of CMEs are randomly selected in 2004. Assuming that human classification for strong CMEs is 100 % accurate, we can select 50 strong (halo, curved front and most complex) and 50 other (medium and weak) CMEs by searching through a sequence of running difference images. The SVM classifier is trained by human classification results with the 100 CMEs based on the aforementioned six inputs. After training, the SVM classifier is able to classify strong CMEs from others automatically. The classification rate of the testing experiments is shown in Table VI.

Table VI. Success classification rates of the strong and non-strong CMEs based on 50 strong CMEs and 50 non-strong CMEs.

LASCO	Strong CMEs	weak and medium CMEs
C2	94%	96%
C3	96%	96%

After finding strong CMEs using our SVM classifier, further classification is applied to distinguish medium from weak CMEs using the rule proposed by Howard *et al.* (1985) and Dai, Zong and Tang (2002). The speed of medium CMEs is greater than 300 km/s, while the speed of weak CMEs is less than 300 km/s.

3.4. Comparisons of CME Detections

We have developed the software to detect and characterize CMEs. The programs were developed in Interactive Data Language (IDL) by Research Systems, Inc, and run on a DELL Dimension 4600 PC with CPU time 2.8 GHz and memory of 512 Mbytes under Linux. The computational time for detecting a CME using three LASCO images is about 5 seconds which is far less than the observational interval. The results in our catalog include a list of CMEs, a sequence of CME image frames, the classification type of each CME frame, and the properties of each CME region, such as height, velocity and angular width. The properties and classification results of CMEs are saved in our database available through our web site.

There are two, previously developed, CME detection catalogs available to the public. From the LASCO web site at http://cdaw.gsfc.nasa.gov/cme_list/, one can find the CME catalog created by visual inspection. Robbrecht and Berghmans presented their results at http://sidc.oma.be/cactus/. Our automatic detection results are currently shown at http://filament.njit.edu/detection/vso.html. It is difficult to compare the CME catalogs because there is no comprehensive catalog to be used as a reference to conduct the comparisons (Berghmans, 2002). A CME in one catalog may be considered as two CMEs in another catalog, and the beginning and ending time for a CME is hard to determine. The reasons are that the preprocessing methods and detection rules of three catalogs are different and human decision is subjective. The reference image and threshold selection may affect the final decision of a CME detection.

We select results between August 1, 2004 and August 31, 2004 to perform comparisons among the three catalogs. The LASCO catalog, used as the reference, listed 65 CMEs in this period. Our catalog missed one weak CME which happened on the West at 316 degrees on August 26, 2004 at 16:54UT. CACTus missed three weak CMEs which happened on the East at 77 degrees on August 2, 2004 at 23:06UT, on the East at 94 degrees on August 22, 2004 at 17:30UT, and on the East at 111 degrees on August 26, 2004 at 21:54UT. The missed CMEs in our catalog and in CACTus are weak CMEs. On the other hand, some CMEs missed in LASCO can be detected in CACTus and our catalog. For example, the CMEs happened on the East on August 6, 2004 at 22:30UT, on the East on August 7, 2004 at 18:54UT, and on the Southeast on August 9, 2004 at 21:30UT. By combining the results of CACTus and our catalog, the CME detection results are more complete and accurate than the LASCO catalog which is based on human eye detection. In the three catalogs, the center of principal angles, angular width and speed of the strong CMEs are given as the important CME properties. The detected properties of CMEs in the different catalogs could be varied because the region merging criteria are different. As we said previously, a CME in one catalog may count as two in another. Overall, the center of principal angles, the angular width and speed, especially for the strong CMEs, are similar in all the three catalogs.

4. Conclusion

We have presented comparisons of flare detection using three advanced techniques which are multi-layer perceptron (MLP), radial basis function (RBF) and support vector machine (SVM). Through experiments, we have found that SVM is the best for the solar-flare detection because it offers the best classification result and the training and testing speed are relatively fast. The second choice is RBF. Because its training data are presented only once, its training time is the shortest. MLP is not a well-controlled learning machine. The sigmoid function has a scaling factor that affects the quality of the approximation, and the convergence of the gradient-based method is rather slow. The result of MLP is not as good as the RBF and SVM. After the solar-flare detection, the properties of flares are analyzed using region growing and Canny edge detector. Because this work is well done by Veronig *et al.* (2000) we do not give an explanation here. For future study, we will apply SVM, RBF and MLP to more solar phenomena such as filaments, sunspots, and plages. We believe that we may create the best model to detect the solar phenomena by combining SVM and RBF classifiers.

We have also presented an automatic algorithm to detect and categorize Coronal Mass Ejections (CMEs). The first preprocessing step intends to normalize the images, remove the missing blocks, and obtain the running differences. Two consecutive images accompanying the reference image are used to produce two running differences, two division, two binary, and two angular images. The properties of CMEs such as intensity, height, span, and velocity are measured using automatic thresholding and morphology methods. The detected CMEs are saved to our database and are listed through our web site. In addition, a new advanced method, support vector machine, is used to distinguish strong CMEs from other kinds of CMEs. The strong CME classification is essential to the forecast of space weather. Compared to the previous CME detection methods proposed by SOHO/LASCO and CACTus, we have provided an alternative approach for CME detection and have proposed a new method for strong CME classification.

Acknowledgment

This work is supported by National Science Foundation (NSF) under grants IIS-0324816, ATM0548956 and ATM 0536921.

References

Berghmans, D.: 2002, The 10th European Solar Physics Meeting SP-506, 1, 85.

Bishop, C. M.: 1995, *Neural Networks for Pattern Recognition*, Oxford University Press, Oxford, p. 164.

Borda, R. A., Mininni, P. D., Mandrini, C. H., Gómez, D. O., Bauer, O. H., and Rovira, M. G.: 2001, *Solar Phys.* **206**, 347-357.

Brueckner, G. E., Howard, R. A., Koomen, M. J., Korendyke, C. M., Michels, D. J., Moses, J. D., Socker, D. G., Dere, K. P., Lamy, P. L., Llebaria, A., Bout, M. V., Schwenn, R., Simnett, G. M., Bedford, D. K., and Eyles, C. J.: 1995, *Solar Physics*, **162**, 357.

Brueckner, G. E., Delaboudiniere, J.-P., Howard, R. A., Paswaters, S. E., St. Cyr, O. C., Schwenn, R., Lamy, P., Simnett, G. M., Thompson, B., and Wang, D.: 1998, *Geophys. Res. Lett.*, **25**, 3019.

Broomhead, D. S. and Lowe, D.: 1988, *Complex Systems*, **2,** 321.

Dai, Y., Zong, W., and Tang, Y.: 2002, *Chinese Astron. Astrophys.*, **26**, 183.

Guyon, I. and Stork, D. G.: 2000, *Linear discriminant and support vector machine*, The MIT Press, Cambridge, p. 147.

Haykin, S.: 1994, in J. Griffin (ed.), *Neural Networks*, Macmillan Publishing Company, NY, USA, p. 3.

Harrison, R. A.: 1995, *A&A*, **304**, 585.

Howard, R. A., Sheeley, N. R., Jr., Michels, D. J., and Koomen, M. J.: 1985, *J. Geophys. Res.*, **90**, 8173.

Martens, P.C. and Zwaan, C.: 2001, *The Astrophysical Journal*, **558**, 872.

Mitra, U. and Poor, H.V.: 1994, *IEEE Journal on Selected Areas in Communications*, **12**(9):1460-1470.

Qu, M., Shih, F. Y., Jing, J., and Wang, H.: 2003, *Solar Phys.*, **217**, 157.

Qu, M., Shih, F. Y., Jing, J., and Wang, H.: 2004, *Solar Phys.*, **222**, 137.

Qu, M., Shih, F. Y., Jing, J., and Wang, H.: 2005, *Solar Phys.*, **228**, 119.

Ripley, B. D.: 1996, *Pattern Recognition and Neural Networks*, Cambridge University Press, Cambridge, p. 143.

Robbrecht, E. and Berghmans, D.: 2004, *Astron. Astrophys.*, **425**, 1097.

Shih, F. Y. and Mitchell, O. R.: 1989, *IEEE Trans. Pattern Analysis and Machine Intelligence*, 11, 31.

Vapnik, N. V. and Chervonenkis, A.J.: 1974, *Theory of Pattern Recognition*, Nauka, Moscow.

Vapnik, N. V.: 1982, *Estimation of Dependences Based on Empirical Data*, Addendum 1, New York, Springer-Verlag.

Vapnik, V.N. and Chervonenkis, A.:1991, *Pattern Recognition and Image Analysis*, **1**, 283.

Vapnik, N.V.: 1995, *The Nature of Statistical Learning Theory,* Springer, New York, p. 126-140.

Vapnik, N. V.: 1998, in S. Haykin (ed.), *Statistical Learning Theory*, John Wiley & Sons, Inc., NY, USA, p. 401.

Veronig, A., Steinegger, M., Otruba, W., Hanslmeier, A., Messerotti, M., Temmer, M., Brunner, G., and Gonzi, S.: 2000, *ESA*. **463**, 455.

Zirin, H.: 1988, *Astrophysics of the Sun*, Cambridge University Press, Cambridge, p. 343.

In: Solar Physics Research Trends
Editor: Pingzhi Wang, pp. 287-325

ISBN: 978-1-60021-987-0
© 2008 Nova Science Publishers, Inc.

Chapter 8

SOLAR TYPE IV BURSTS AT FREQUENCIES 10-30 MHZ

***V. N. Melnik*[1]*, H. O. Rucker*[2]*, A. A. Konovalenko*[1]*, V. V. Dorovskyy*[1]*, E. P. Abranin*[1]*, A. I. Brazhenko*[3]*, B. Thide*[4] *and A. A. Stanislavskyy*[1]**

[1] Institute of Radio Astronomy, Kharkov, Ukraine,

[2] Space Research Institute, Graz, Austria,

[3] Poltava Gravimetrical Observatory, Poltava, Ukraine

[4] Swedish Institute of Space Physics, Uppsala, Sweden

Abstract

The results of the first observations of Type IV bursts at frequencies 10-30 MHz are presented. These observations were carried out at radio telescopes UTR-2 (Kharkov, Ukraine) and URAN-2 (Poltava, Ukraine) during the period 2003-2006. Detection of Type IV bursts in wide band from 10 to 30MHz with high sensitivity and time resolution allowed to study their properties in details. These bursts have fluxes 10-2000s.f.u. at maximum phase. Their durations are about 1-2 hours and even more. Some of Type IV bursts drift from high to low frequencies with drift rates about 10kHz/s. All observed Type IV bursts have fine structures in the form of sub-bursts with durations from 2s to 20s and frequency drift rates in a majority of 1-2MHz/s. In most cases, sub-bursts with negative drift rates were registered. Sometimes sub-bursts in absorption with durations 10-200s against Type IV burst background have been observed. The Type IV burst observed on July 22, 2004 had zebra structure, in which single zebra stripes had positive, negative and infinite drift rates.

Introduction

Solar Type IV bursts were identified by Boischot [1] as continuous radiation appearing after flares connected with Coronal Mass Ejections (CME). There exist moving and stationary Type IV bursts [2]. The frequency of radio emission of moving bursts moves from high to low frequencies and the corresponding linear velocity of the source varies from 200 to

1600km/s [3]. The source velocity distribution is similar to the CME velocity distribution and Type IV bursts often follow Type II bursts caused by the shock waves. Apparently, the Type IV bursts are connected with shock waves, which are responsible for Type II bursts.

Sometimes Type II bursts are observed on the background of Type IV ones and thus are called Type II-IV bursts. The Type IV bursts are registered in the wide frequency band – from 408MHz to 18MHz [2], and more recently at lower frequencies from 14 to 1MHz [4, 5]. The duration of Type IV bursts decreases with increasing frequency: the Type IV's duration at 80MHz is in average 30min, but at much lower frequency it rises up to 2 hours [6].

The flux of a burst at fixed frequency first increases and then decreases with approximately equal rates. At the same time, the polarization degree keeps increasing during all the burst lifetime and sometimes reaches 90% [7]. The brightness temperature is higher at lower frequencies changing from 10^9K at 80MHz and 169MHz to 10^{10}K at 43MHz [8, 9]. Observations with the Nancay radio heliograph at 169MHz showed a radiation pulsation in moving Type IV bursts with the period of around 2s. This kind of pulsation was also found in Zurich at frequencies 140-250MHz during 5 min [9]. Sometimes Type IV bursts have fine structure in the form of bursts with intermediate frequency (so called fiber bursts) in emission and absorption and zebra-structure [10-12]

In the early days a synchrotron emission mechanism was used to explain the Type-IV bursts [1]. Later gyrosynchrotron mechanism was assumed [13, 14] and finally a plasma mechanism was proposed [15]. The full variety of observational data, including high emission brightness temperature, high polarization degree and its temporal variation, still cannot be completely understood within the frame of only one emission mechanism [16].

The existence of fine structure in Type IV emission is explained by different mechanisms. So, fiber bursts are supposed to appear due to coalescence of whistlers and Langmuir waves [5]. Concerning zebra-structure, the generation of Langmuir waves in double plasma resonance with their subsequent transformation into transverse waves [11, 17], or the Langmuir waves scattering on structured whistler field [11, 18] are analyzed.

In this chapter the results of observations of Type IV bursts obtained at the radio telescopes UTR-2 and URAN-2 in the frequency band 10-30MHz are considered. The observations with unprecedented sensitivity, high temporal resolution in wide instantaneous bandwidth allowed to detect Type IV bursts for the first time at such low frequencies as well as to find new properties of these bursts. These new results may appear to be the key point in understanding this kind of event.

Observations

Type IV bursts discussed in the chapter were registered at the radio telescopes UTR-2 (Fig.1) (in the years 2003-2006) and URAN-2 (Fig.2) (in 2003, 2006). Three sections of the radio telescope UTR-2 with total collecting area $30000m^2$ were used providing a beaming of $1^\circ \times 13^\circ$. All observations were carried out with the 60-channel filter bank spectrometer. Each of 60 channels has bandwidth of 3kHz and is tuned to a fixed frequency in the range between 10 and 30MHz. The channels are spaced in average by 340kHz, but not regularly in all frequency band. In practice to minimize interference, the channels are tuned out off the interference frequency. One must take into account that on dynamic spectra the gaps between adjacent channels are filled with the values of the channels, nearest to the gap. This technique allows to clarify the spectrum and makes it more smooth along the frequency axis.

Figure 1. Radio telescope UTR-2 (Kharkov, Ukraine).

The circular polarization was measured with the radio telescope URAN-2 (Fig.2), which consists of 512 wideband (10-30MHz) dipoles [19]. The effective area of the URAN-2 antenna is $28000m^2$, with a beaming of $3.7^\circ \times 7^\circ$ at a frequency 25MHz. In 2004 the polarimeter worked at two frequencies 24.5MHz and 25MHz in the frequency band of 10kHz and with a time resolution of 12.5ms.

Figure 2. Radio telescope URAN-2 (Poltava, Ukraine).

During the period 2003-2006 we registered 17 events, which could be referred to Type IV bursts. In this paper we discuss only 13 of them because each considered has some important features for the understanding of Type IV burst phenomenon. For the description of decameter Type IV bursts we consider also attendant data by WIND, SOHO and Nancay Decameter Array (NDA) if they exist.

On June 13, 2003

This Type IV burst began over the frequency range 10-30MHz nearly at 07:00 UT (Fig.3). The time profile of the burst is characterized by fast rise (approximately 1 hour) and slower decay (about 3h 30min practically till 11:30 UT) (Fig.4). The front edge of the burst drifts from high to low frequencies with the rate of 10kHz/s. The rear part of the burst, apparently, drifts with the same drift rate (observations with NDA at higher frequencies are also in favor of this (Fig.5)). On a dynamic spectrum below 20MHz one can see a shading area, the front edge of which has positive frequency drift rate and the rear edge, on the contrary, has a

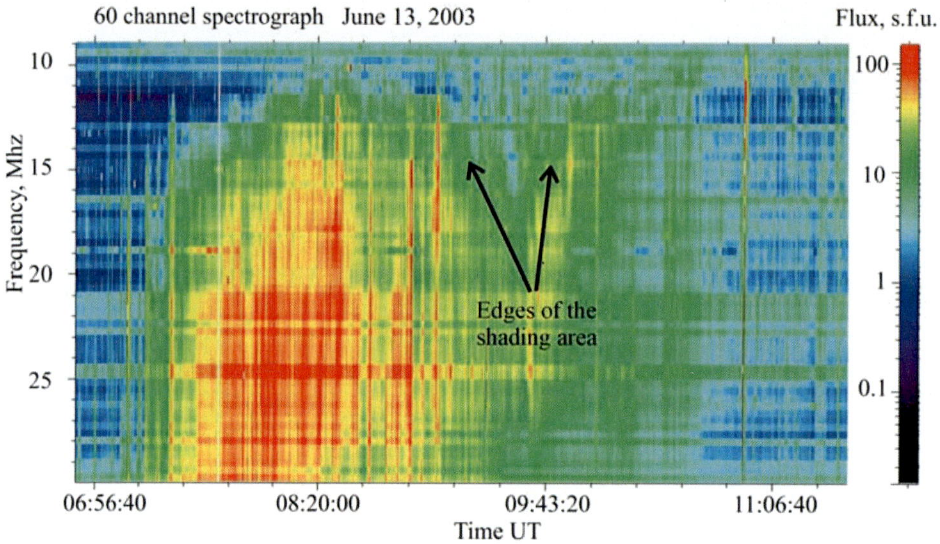

Figure 3. Type IV burst, which moves with drift rate 10kHz/s from high to low frequencies.

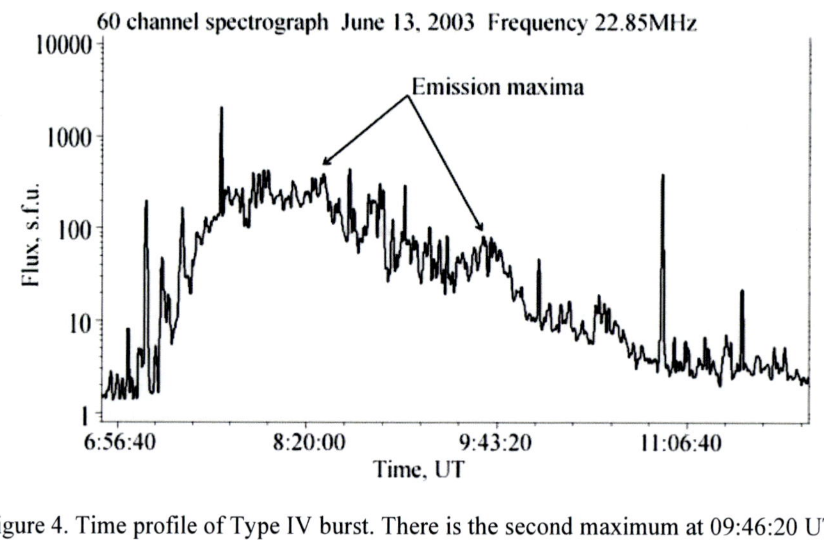

Figure 4. Time profile of Type IV burst. There is the second maximum at 09:46:20 UT.

negative one. On the time profile of the burst in addition to the main maximum (with flux reaching 1000s.f.u.) there can be seen also the second and probably the third maximum (with fluxes 20s.f.u. and 10s.f.u., respectively). The frequency profile at maximum phase of the burst (at 08:20 UT) (Fig.6) shows fast fall (from 600s.f.u. at 30 MHz to 2s.f.u. at 10 MHz) from high frequencies to low frequencies.

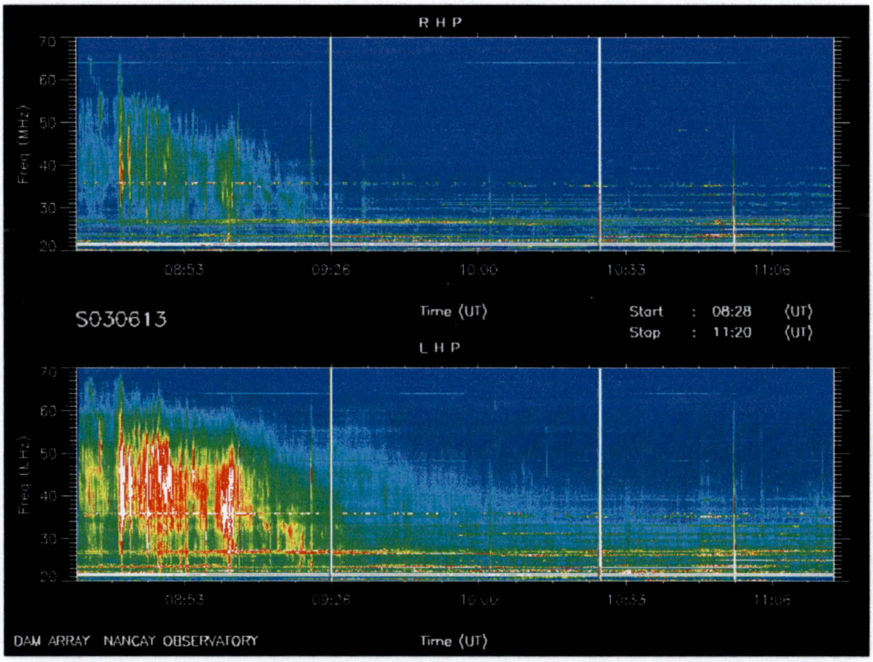

Figure 5. Right and left hand polarization of Type IV burst observed by NDA

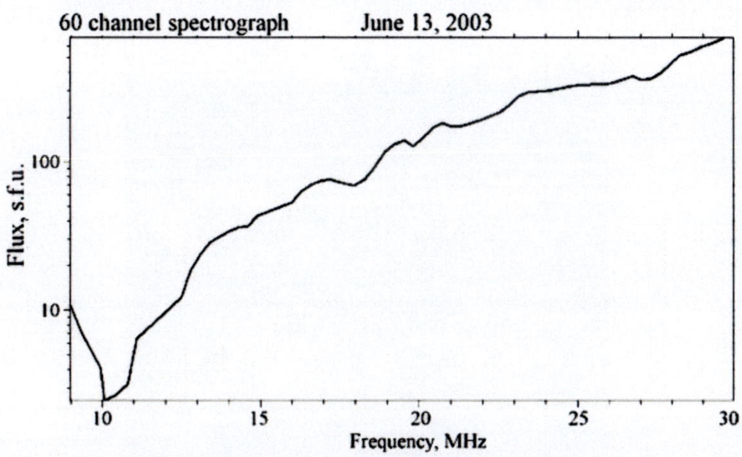

Figure 6. Frequency profile of Type IV burst maximum phase (at 08:20 UT)

As in all observed Type IV bursts, the considered burst had fine structure in the form of bursts, which in the overwhelming majority drifted from high to low frequencies with rates about 1-2MHz/s, slightly lower than the drifts of common Type III bursts for this frequency band. The duration of these sub-bursts was 10-20 s, which is more, than for common Type III bursts. It is necessary to note also, that at the beginning of the Type IV burst the sub-bursts are less observable than at the end of it. The maximum fluxes of sub-bursts with respect to the continuum flux are comprising in the majority only 10-20 % . Some fast drifting sub-bursts were also registered. Their drift rates were essentially greater, with regard to the rest of sub-bursts, and in rare cases the drift rates were positive. This Type IV burst is one of the three bursts discussed for which there are the simultaneous observations of polarization obtained by the radio telescope URAN-2 (Poltava). Fig.7 shows the evolution in time of the polarization degree, since 07:28 UT till 10:57 UT. It can be seen that the polarization degree gradually grows, reaching the maximum value of 55% at 09:40 UT, and then starts to fall practically down to 10%. Judging from the heliographic observations by Nancay (Fig.8), this burst is connected to the active regions situated on the western part of the solar disk.

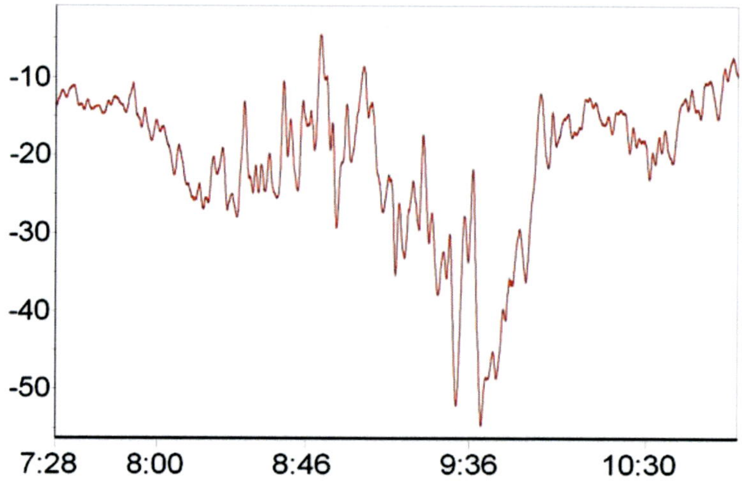

Figure 7. Polarization of Type IV burst measured by URAN-2.

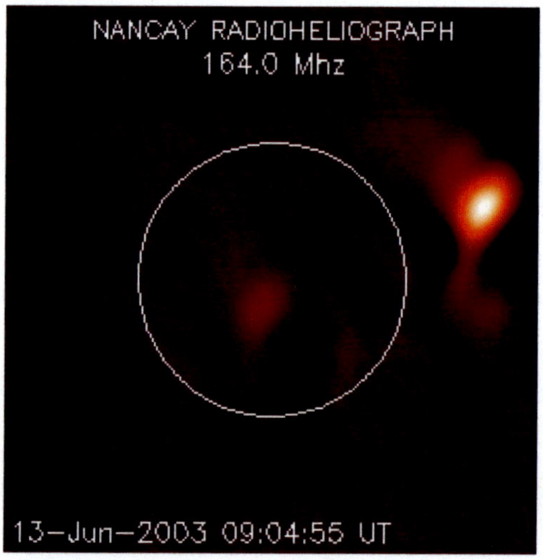

Figure 8. Position of source of 164 MHz radio emission near the maximum phase of Type IV burst.

On July 22, 2003

This Type IV burst is characterized by rather slow intensity rise. Starting at approximately 10:00 UT it reaches a maximum (100s.f.u.) only at 11.00 UT (Fig. 9). This burst also has at least two feebly marked increases of the emission intensity (Fig. 10). The internal structure of this burst has an appearance of well separated bursts similar to Type III bursts. Though the durations of these sub-bursts are 5-10s which is practically the same as for classical Type III bursts, their drift rates are in the majority less than 1MHz/s which is slower than usual Type

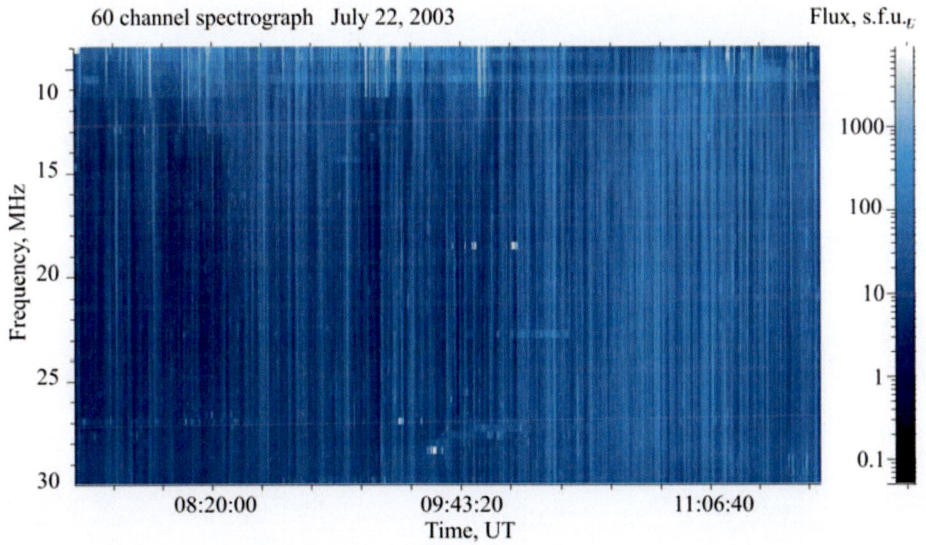

Figure 9. Dynamic spectrum of solar radio emission on July 22, 2003, observed with UTR-2.

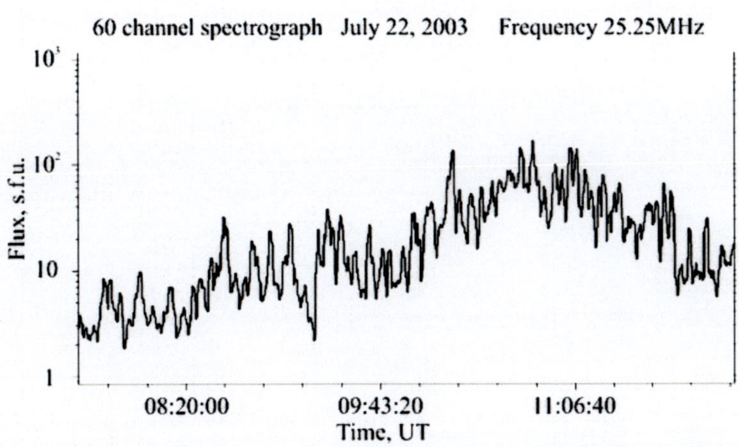

Figure 10. Time profile of Type IV burst with slow increase and maximum at 11:06 UT.

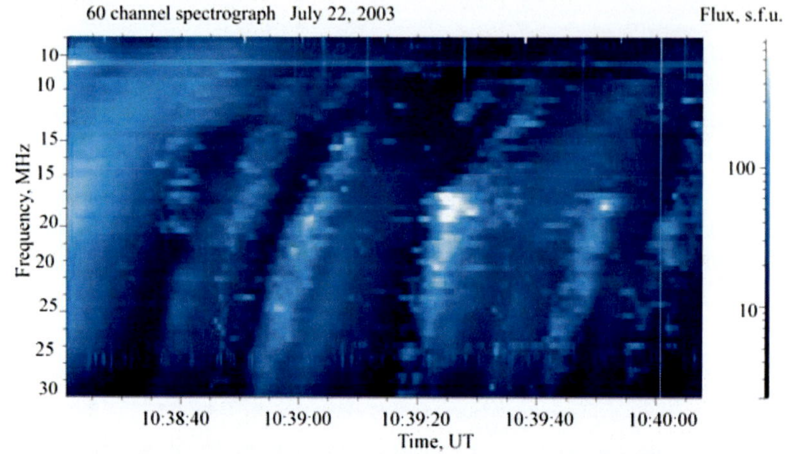

Figure 11. Fragment of Type IV burst with fine structure in the form of bursts similar to usual Type III, Type IIIb bursts, but unusual low drift rates (<1MHz/s).

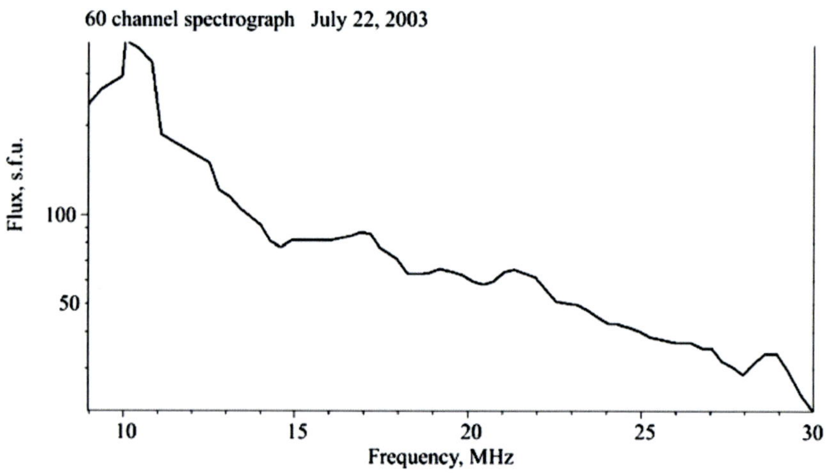

Figure 12. Frequency profile of Type IV burst at maximum phase (11:06 UT).

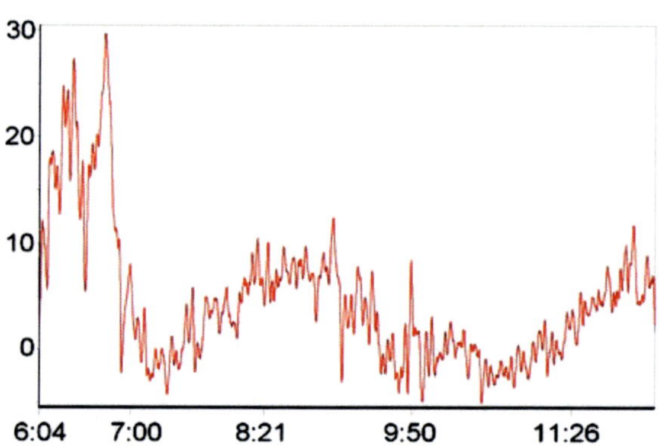

Figure 13. Nonmonotonic change of radio emission polarization at 25MHz during Type IV burst.

III bursts exhibit. As it can be seen from the Fig.11 the Type IV burst consists also of Type IIIb bursts. In general these Type IIIb bursts have the same durations and drift rates as they usually have during Type IIIb storms, but in some cases the drift rates are less than 1MHz/s. A remarkable feature of the instant spectrum of this Type IV burst (taken at maximum flux) is the flux rise towards the lower frequencies (Fig. 12). Such flux versus frequency dependence is opposite to that found for the majority of the observed Type IV bursts. The measured polarization of the burst, which is shown in Fig. 13, is rather small. This Type IV burst fills an intermediate place between Type III storm and the Type IV burst itself.

On August 3, 2003

This Type IV burst is not intensive - its maximal flux is about 20s.f.u. (Fig.14). It has a symmetric time profile with full duration of about 2 hours and a maximum intensity falling at 11:00 UT (Fig.15). The burst is observed on a background of powerful Type III bursts with fluxes reaching up to 10000s.f.u. The dependence of flux via frequency for observed Type IV burst with 100s averaging is shown in Fig.16. It is seen that the flux grows with falling frequency, as it was for the event on July 22, 2003, but conversely to the flux-frequency dependence for the 13 June 2003 event. It is interesting, that averaging the flux with 100s averaging time for all observation time on this day gives also the second maximum, which falls at 8:20 UT (Fig.16). Here the main contribution to the radiation intensity is provided by a group of intense Type III bursts. These two periods of the increased radio emission, apparently, reflect the activity of active regions. And if in case of the activity, which occurs between 08:00 UT and 09:15 UT, the increased level of radiation was provoked mainly by accelerated electron beams emitted from the active region, in the second case there was a smooth plasma heating producing the increased continuum radiation of the Type IV burst. The fine structure of the Type IV burst in the majority has an appearance of Type III bursts with their inherent durations and frequency drift rates observed on a continuous background.

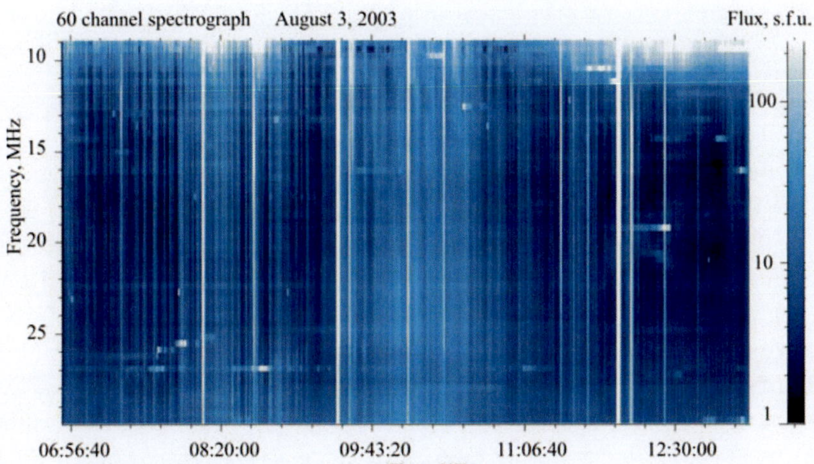

Figure 14. Weak stationary Type IV burst (from 09:30 UT to 11:20 UT).

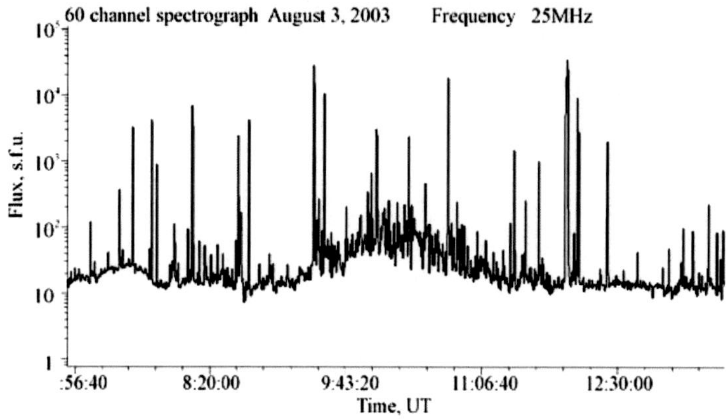

Figure 15. Time profile of solar radio emission on August 3, 2003 at 25MHz without time averaging (time resolution in these observations was 50ms)

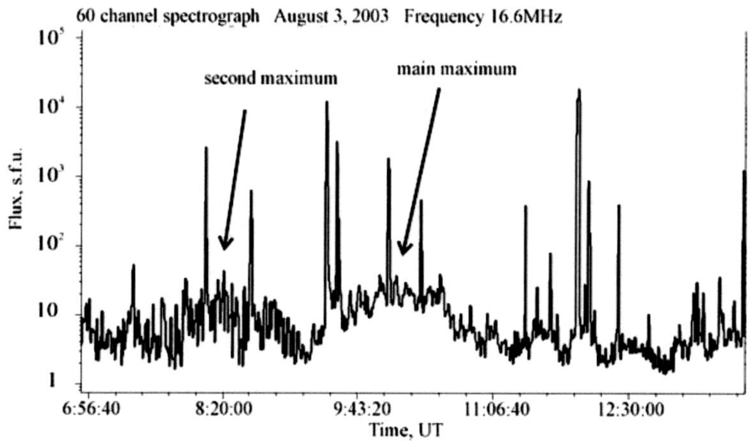

Figure 16. Time profile of solar radio emission on August 3, 2003 at 16MHz with 100s averaging.

On August 19, 2003

At 07:55 UT the group of very powerful Type III bursts with fluxes of more than 10000s.f.u. has been seen (Fig.17). Right after this a group of fiber bursts with drift rates 30kHz/s was observed (Fig.18). Their durations were up to 3s, and fluxes reach 300s.f.u. Later on the Type II burst starts drifting from high to low frequencies. This burst appeared to consist of two and probably even three harmonics with starting times 08:09, 08:16 and 08:19 UT. The first and the second harmonic had drift rates of about 40-50kHz/s and were splitted into two lanes each. Each of the Type II burst lanes consists of sub-bursts with the drift rates 1 - 4MHz/s and durations equal to 3-5s. The emission of an increased level, which is observed right after the group of intense Type III bursts steeply falls off reaching a value of 0.5s.f.u. by 08:30 UT. At low frequencies (near 10MHz) there can be seen a slight rise of emission between 09:15 UT and 10:15 UT. At 10:05 UT the lanes of Type II burst start at high frequencies (Fig.19). But in this case their drift rates equal 20kHz/s. Right after that, at 10:23 UT at 25MHz, another type II burst is registered. It has the same drift rate and is a little bit weaker. It is superimposed by two groups of Type III bursts with the peak fluxes of 500-600s.f.u. Just after them the Type IV burst begins.

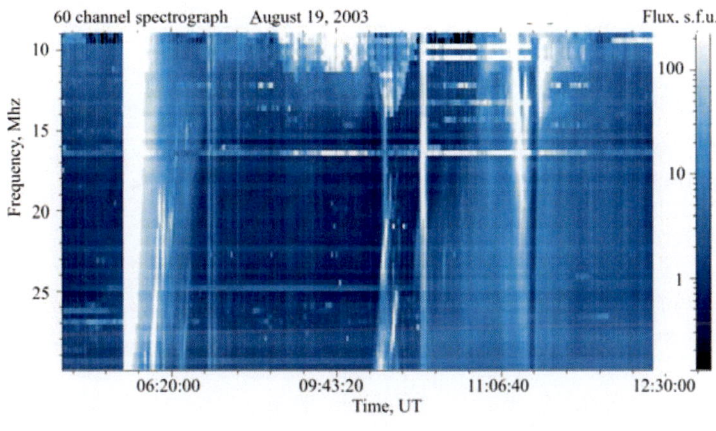

Figure 17. Type IV burst (from 10:25 UT to 12:30 UT) with attendant phenomenon group of Type III bursts, Type II bursts.

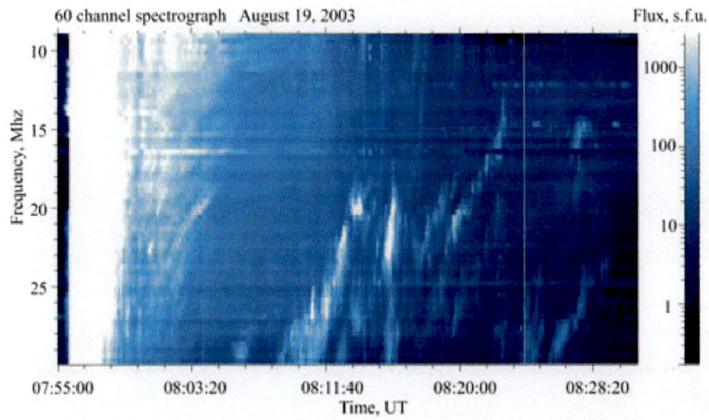

Figure 18. Group of Type III bursts at 07:55 UT, fiber bursts directly after it, fundamental and harmonic Type II burst.

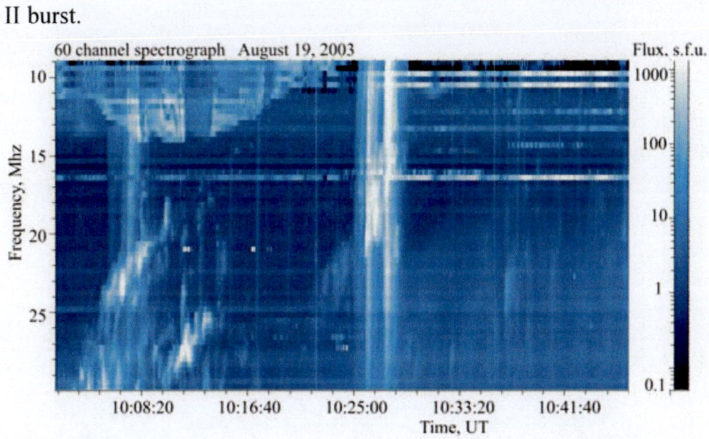

Figure 19. Type II burst (with lanes), group of Type III bursts and beginning of Type IV burst.

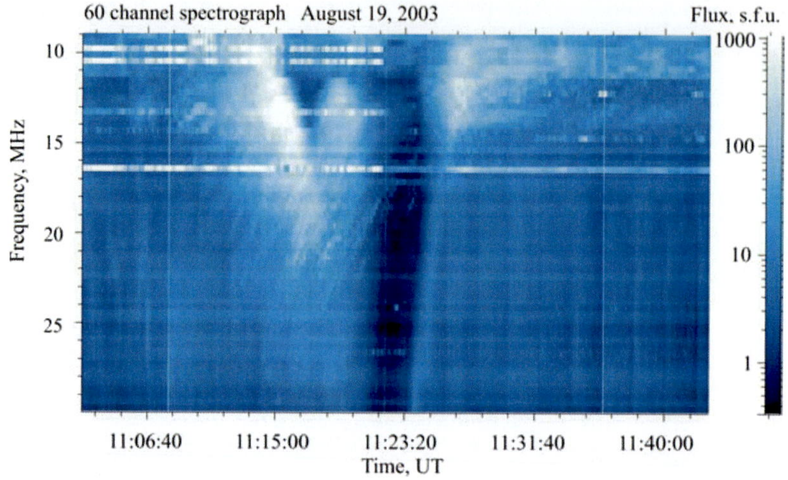

Figure 20. Absorbing area (at low frequencies at 11:17 UT) and burst in absorption against fiber bursts.

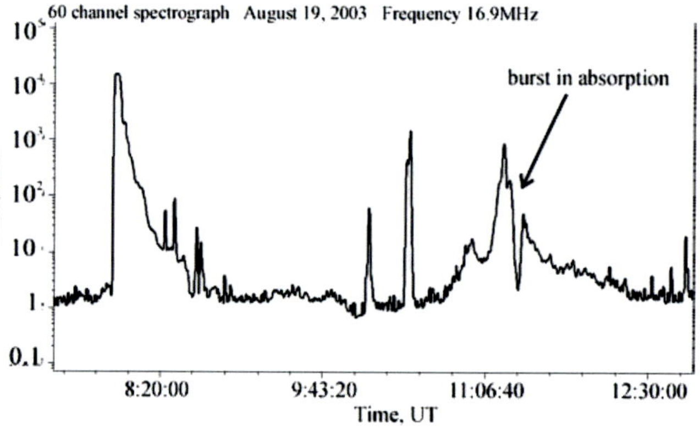

Figure 21. Time profile of solar radio emission on August 19, 2003. There is clearly seen the burst in absorption at 11:20 UT.

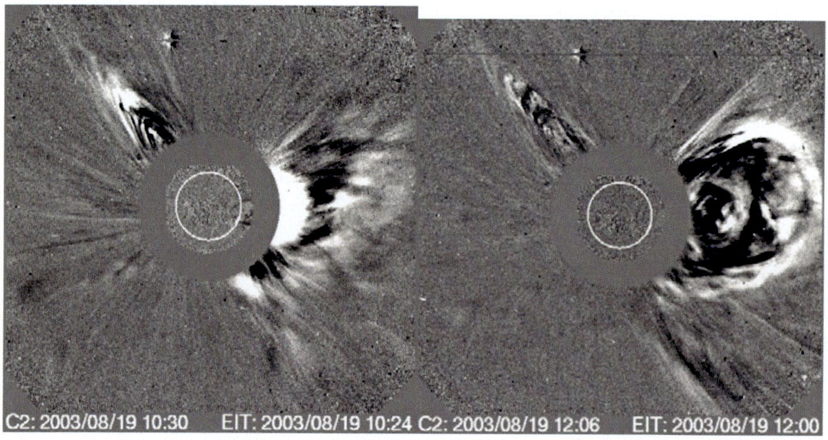

Figure 22. Movement of west CME associated with August 19, 2003 Type IV burst according to SOHO data.

The forward edge of the burst drifts slowly down to low frequencies with 10 kHz/s drift rate. The flux reaches its maximal value 30s.f.u. at 11:10 UT. Thus, rise time of this Type IV burst amounts roughly to 40min. The tail of the burst is rather long - by the time 12:45 UT (the end time of observations on that day) the emission level is still higher than 1s.f.u. The second, less pronounced emission maximum can also be found at 12:05 UT. The Type IV burst consists of very weak sub-bursts, which are hardly distinguished on a continuum background. The drift rates of these sub-bursts equal mainly 1-2MHz/s, and durations are 10-20s. At 11:17 UT the absorbing area, which widens towards low frequencies is observed below 15 MHz (Fig.20). Well visible in all frequency band from 10 to 30MHz burst in absorption starts at 11:23:20 UT (Fig.20). It drifts from high to low frequencies with drift rate of 120kHz/s. The total duration of the burst in absorption equals 5min 20s. The minimal flux at 25MHz amounts to 0.6s.f.u., which exceeds the solar emission level before Type IV burst (Fig.21). The bursts in absorption are observed on a background of fiber bursts, which drift from high to low frequencies with velocities 30kHz/s. Their durations equal 2-3s (this burst is discussed in detail in [20.21]. According to the SOHO data (Fig.22) the powerful CME has been registered at 10:24 UT, which moves with 468km/s. By the time of Type IV burst maximum (11:00 UT) the CME front was at a distance of approximately $4R_s$ from the solar surface.

On June 4, 2004

This event (Fig. 23), as well as the event on August 19, 2003, starts with a group of intense Type III bursts observed at 07:33 UT. The maximum fluxes reach 3000s.f.u. The bursts durations are about 10s, and their drift rates are 2-5MHz/s. At the end of this group one can see some fiber bursts with drift rates of 90kHz/s (Fig.24). Their durations are the same as for the event on August 19, 2003, namely 3-5s. Then the burst in absorption is observed. Its duration strongly depends on frequency varying from 2min 30s at 30MHz to 10min 50s at 10MHz. The front of the burst in absorption has practically infinite drift rate, and the tail is limited by the first harmonic of Type II burst. The latter drifts with the rate of about 40kHz/s.

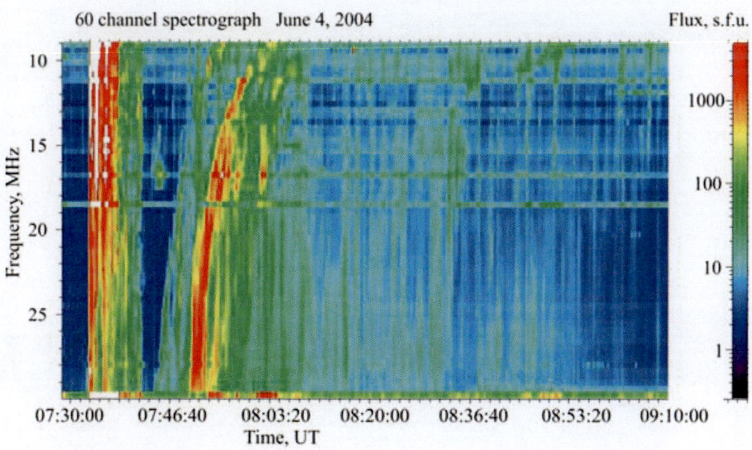

Figure 23. Type IV burst as succession of Type III bursts group, burst in absorption, and Type II burst.

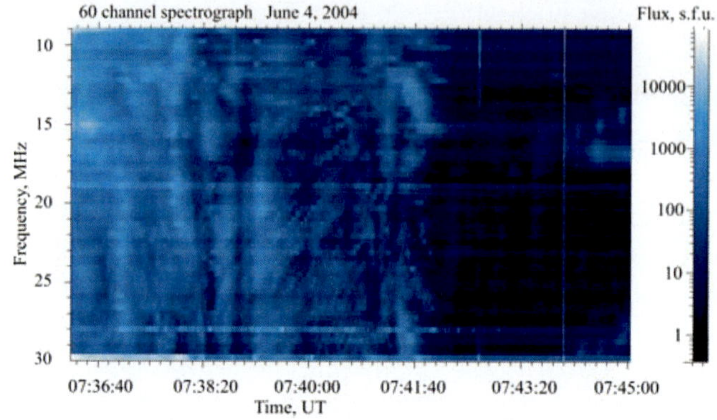

Figure 24. Fiber burst with drift rates about 90MHz/s directly after a group of intensive Type III bursts.

A remarkable feature of this Type II burst is the presence of the first, second and third harmonics (Fig.25). The second harmonic is the most intense, and the third one is the weakest. All harmonics are splitted into lanes and have fine structures in the form of sub-bursts with for decameter wavelengths standard durations (5-10s) and drift rates (1-5MHz/s). These three harmonics are also well distinguished on the dynamic spectrum obtained by the WIND spacecraft (Fig. 26). The Type II burst is followed by the Type IV burst. However we can see only a falling branch of this burst (Fig. 27). The burst lasts up to the end of observations on that day - till 9:30 UT. On the falling edge of the Type IV burst there are, at least, two increases of the emission intensity: at 08:30 UT and at 08:48 UT. The Type IV burst consists of fiber bursts (Fig. 28) with drift rates less than 1MHz/s, and sometimes more - up to 2-3MHz/s and durations of 10-20s. Practically all fiber bursts have negative drift rates although sometimes there are also bursts with positive drift rates. At the time of occurrence of the Type II burst at the decameter band the SOHO coronograph has detected a powerful CME (Fig. 29), which moved very fast in the solar corona (according to SOHO data the CME speed was 1306km/s).

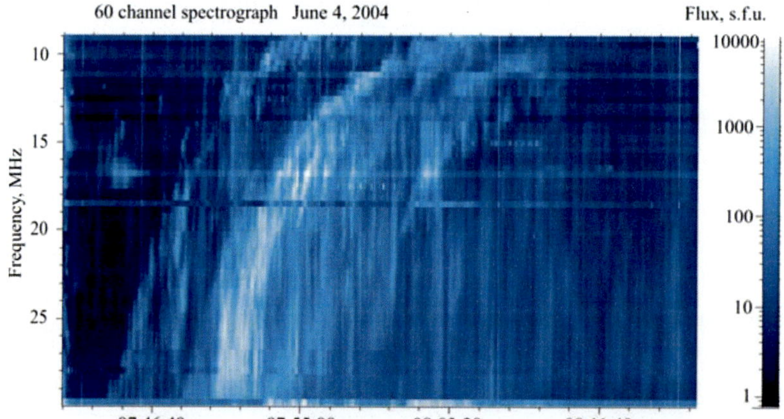

Figure 25. Type IV burst followed Type II burst with three harmonics.

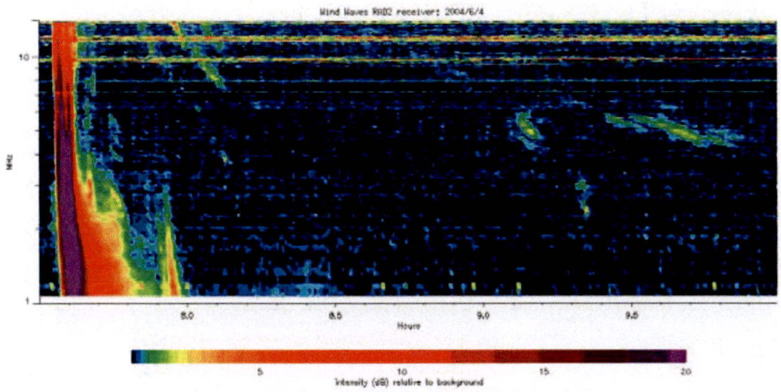

Figure 26. Type II bursts with three harmonics at frequencies 4-11MHz (WIND data).

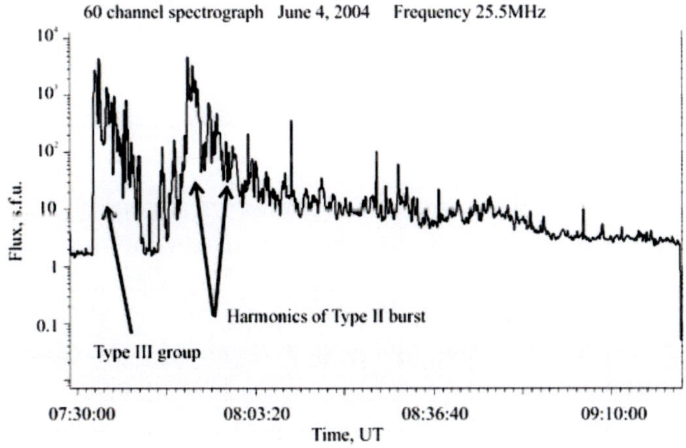

Figure 27. Time profile of June 4, 2004 event.

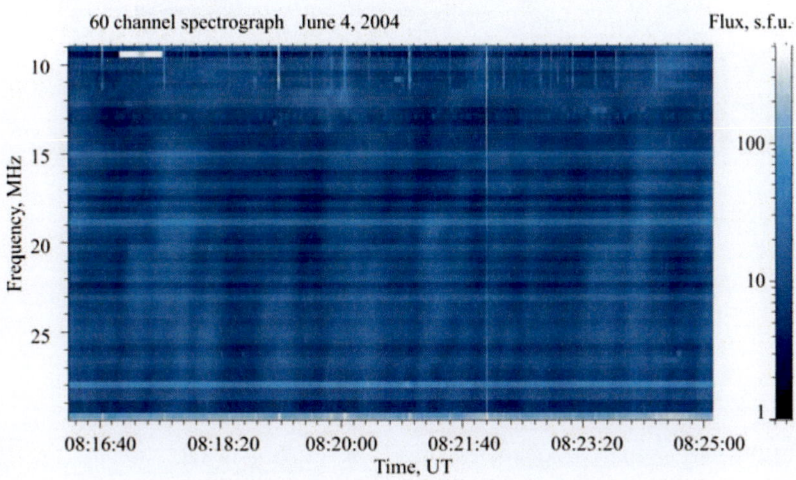

Figure 28. Fiber bursts as a fine structure of Type IV burst.

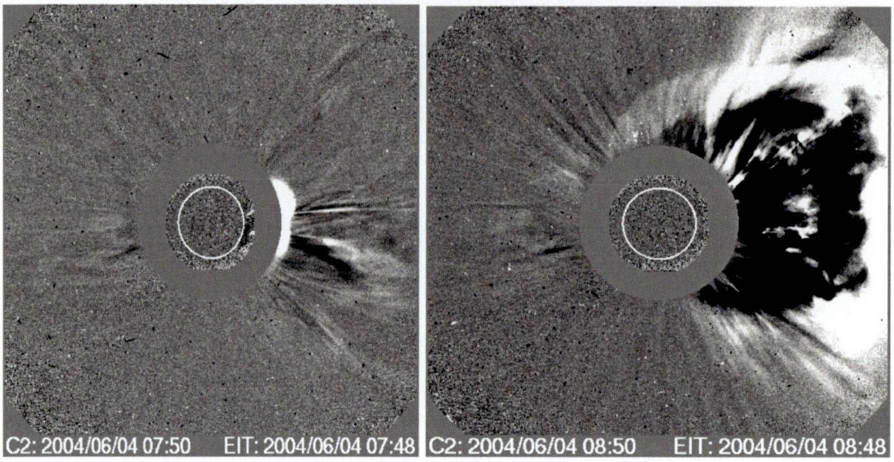

Figure 29. CME during June 4, 2004 event.

On July 13, 2004

At 08:46:50 UT (Fig.30) two intensive Type III bursts were observed. Their fluxes are about 10000s.f.u. and total duration exceeds 20s. They are followed by one more group of Type IIIb-III bursts. The Type II bursts started at 08:56:20 UT at a frequency 30MHz (Fig.31). Its drift rate was about 50kHz/s. Later on, at 09:02:20 UT the second harmonic of this burst appeared. All harmonics have fine structure in the form of drifting sub-bursts, but the second harmonic consists of the shorter sub-bursts (2-5s), with regard to the first harmonic (10-20s).

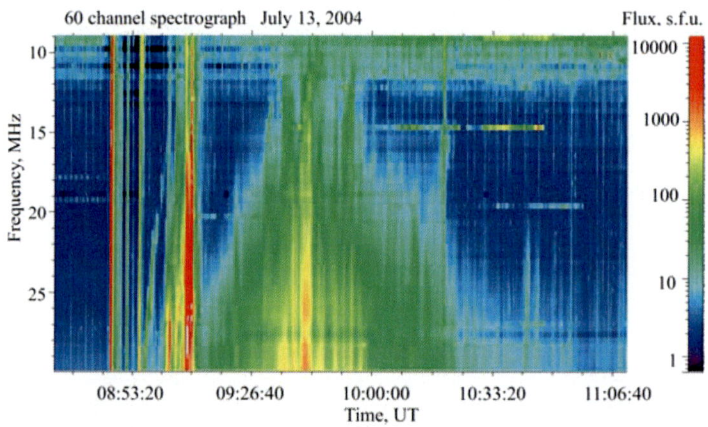

Figure 30. Low frequency part of Type IV burst with attendant Type III bursts and Type II burst.

At 09:11:00 UT at high frequencies the Type IV bursts has started (Fig.32). It also consists of sub-bursts with durations of about 5s and negative drift rates 1-3MHz/s, which are then followed by a continuum with an increased flux level of around 20s.f.u. The intensity grows rather fast and reaches 200s.f.u. at 09:45:00 UT (Fig.33).

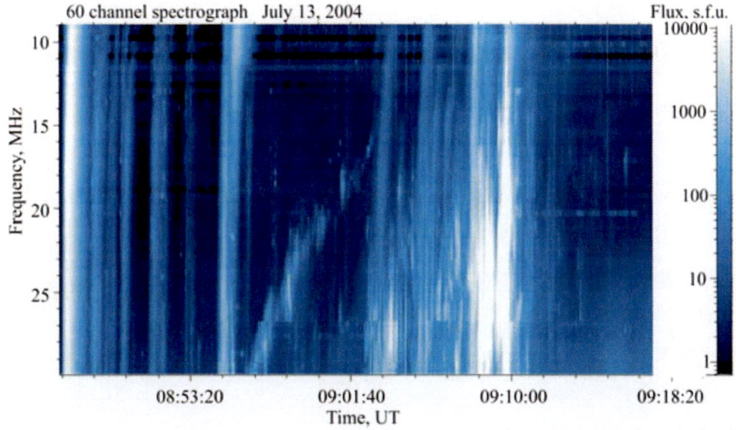

Figure 31. Group of Type III bursts and two harmonics of Type II burst intersecting with Type III bursts.

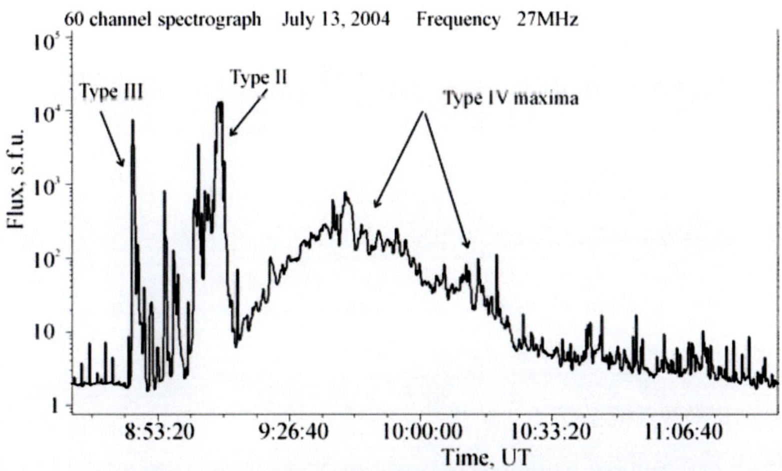

Figure 32. Time profile of Type III bursts, Type II burst and Type IV burst with two maxima.

According to our data, and data obtained from NDA (Fig.33) this Type IV burst has two components. The second component can be seen at 25MHz at 10:15:00 UT. Apparently, this burst is stationary and at frequencies 10-30MHz we observe only the lower frequency part of the burst, the main part however being at higher frequencies. The discussed burst consists of a great number of sub-bursts with various durations ranging from 4-5s up to 20s (Fig.34). The drift rates of the sub-bursts could be both negative and positive with absolute values variable from 2-3MHz/s to 10MHz/s. The sub-bursts are more distinctly seen on a falling part of the Type IV burst in spite of the fact that their intensity appreciably decreases due to increasing distance between them. As in the previous cases this burst accompanies the CME (see Fig.35) with a velocity 713km/s. The instant spectrum of the Type IV bursts at the time of the flux maximum has steep falling slope from high to low frequency changing from 300s.fu. at 30MHz down to 20s.f.u. at 10MHz.

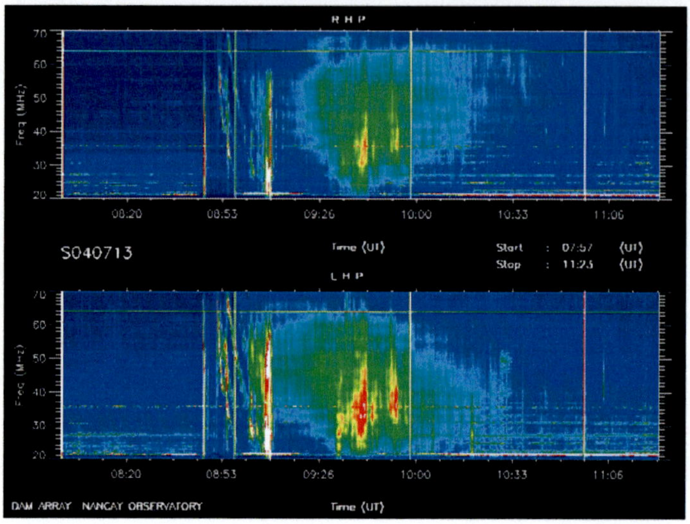

Figure 33. Right and left polarization of July 13, 2004 Type IV burst according to Nancay data.

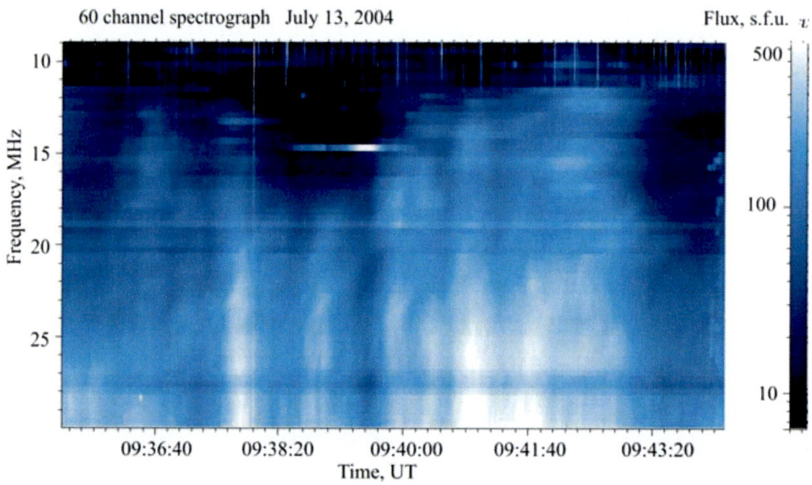

Figure 34. Fiber bursts as a fine structure of July 13, 2004 Type IV burst.

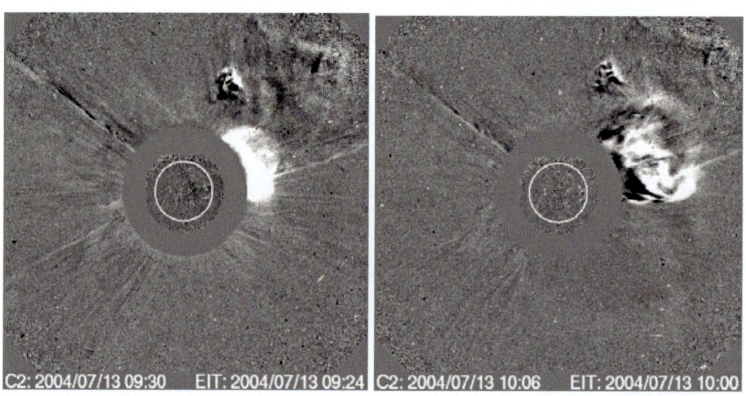

Figure 35. Evolution of a CME associated with the July 13, 2004 Type IV burst observed by SOHO.

On July 21, 2004

Data obtained by SOHO shows the CME, which began at 06:24 UT (Fig. 36). This CME is rather slow with a velocity of 419km/s. Judging from the kinetic energy estimations this CME appears to be the weakest of all here discussed events. As in the previous cases, we observe only the low-frequency continuation of the whole Type IV burst (Fig. 37 and Fig. 38). Moreover, in the last case it is already the ending phase of the burst. Nevertheless the flux is very high, reaching the value about 1000s.f.u. At frequency 25MHz it consists of two, and probably even three components separated from each other by 1-2 minutes. The burst terminates at frequency 15-17MHz. At lower frequencies the emission is practically constant and equals 80s.f.u., which is slightly higher than the level of the quiet Sun. On the background of the Type IV burst the lane of Type II burst is seen (Fig. 37). Its drift rate is near 10kHz/s. Judging from NDA data, the drift rate of the Type IV burst is the same. During these bursts a number of Type III-like bursts is observed (Fig.39). They have drift rates from 7 to 12MHz/s and durations 1-2s that is usual for decameter wavelengths. After the Type II

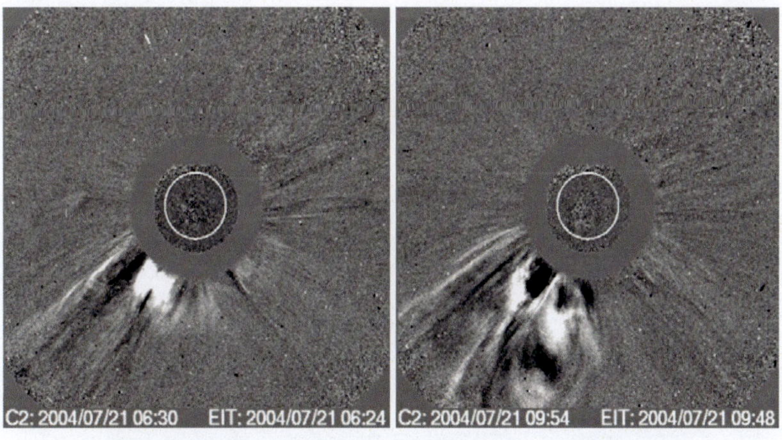

Figure 36. The CME (SOHO data), which initiates Type IV burst on July 21, 2004.

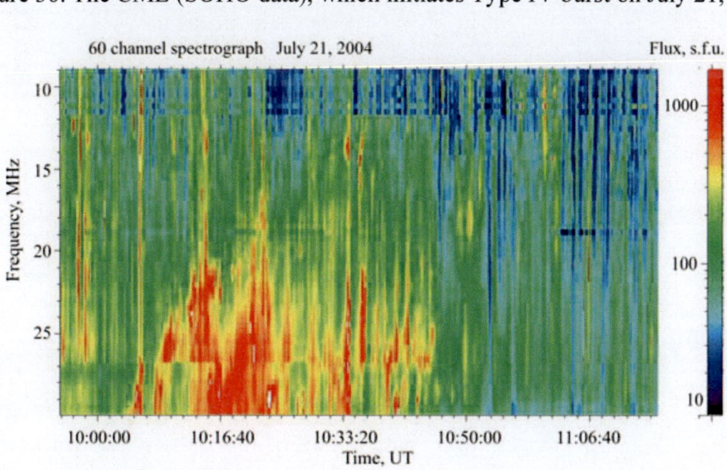

Figure 37. Decameter Type IV bursts on July 21, 2004 observed with UTR-2. Three harmonics of Type II burst are seen at frequencies 20-25MHz.

bursts there were further observed drift pairs, S-bursts and then spikes. The fast drifting bursts, apparently, are connected with the Type II burst. In addition the Nancay data show a high polarization degree.

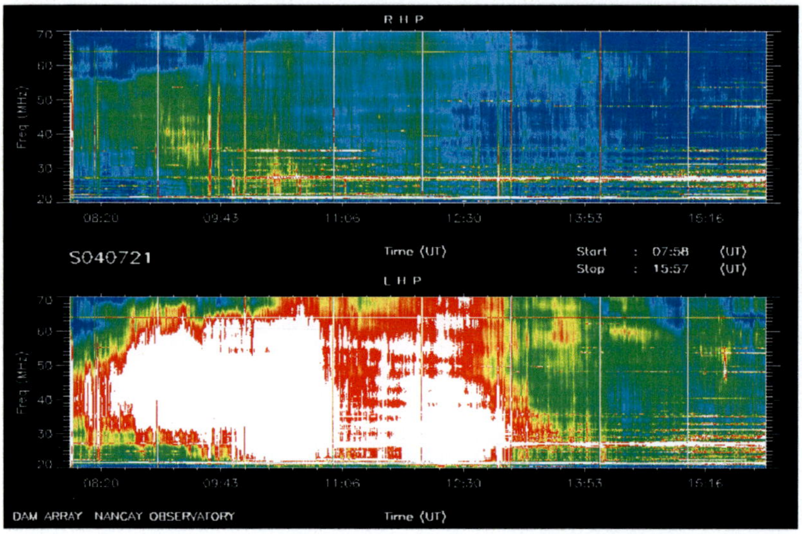

Figure 38. Type IV bursts observed on July, 21, 2004 at higher frequencies (NDA data)

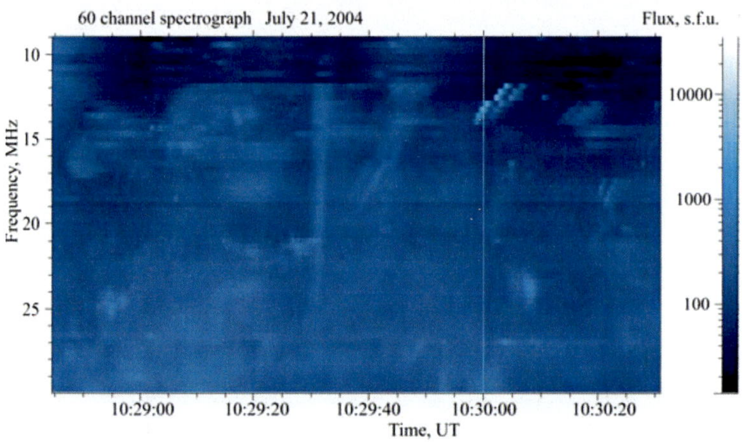

Figure 39. Fine structure of Type IV burst in the form of Type III-like bursts (example at 10:29:30) and drift pairs.

On July 22, 2004

An extremely interesting event (Fig. 40) was observed over the frequency range 10-30MHz on July 22, 2004. The increased radiation (with fluxes of more than 100s.f.u.) already existed by the start time of observations. Apparently, the Type IV burst has begun shortly before 08:20 UT. This assumption is supported by data from SOHO (the CME with the speed of

900km/s is visible at the southern part of the solar disk at 08:24 UT) (Fig. 41) and also by NDA data (Fig. 42).

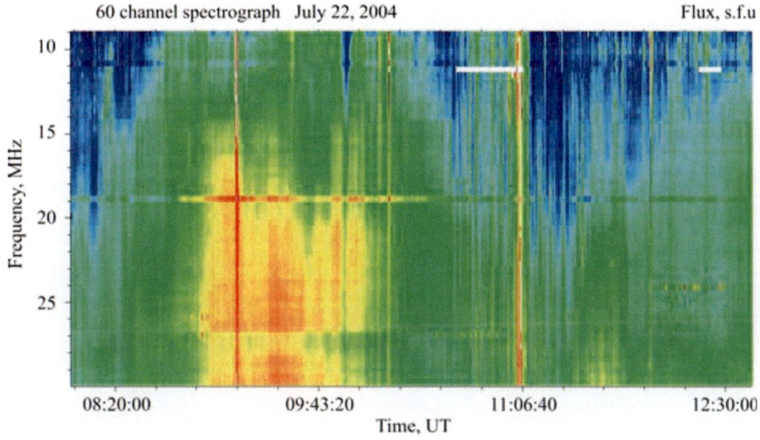

Figure 40. Decameter Type IV burst (UTR-2 data).

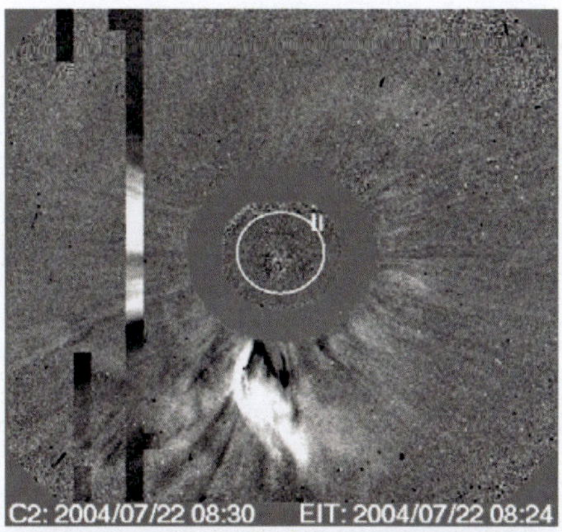

Figure 41. CME, associated with Type IV burst.

At frequencies 10-30MHz the emission intensity is smoothly rising from 08:20 UT till 09:30 UT reaching the maximum value of about 2000s.f.u. (Fig.43). After the maximum point there is an intensity decrease lasting approximately till 10.30 UT. Then one can see two weaker (400 and 500s.f.u.)and shorter (45 and 30min respectively) intensity "hills". By the end of our observations on this day the Type IV bursts still exists (according to the NDA data it lasts till 16:00 UT). At the time of the main maximum of the burst intensity the instant spectrum shows a steep decrease in flux from 2000s.f.u. at 27MHz down to 100s.f.u. at 10MHz (Fig. 44). Nearby the peak intensity point the emission is practically continuous with no fiber bursts visible, although their tracks can be seen. Far aside from a maximum fiber bursts can be detected more reliable. In this case their drift rates are in the majority -2MHz/s. Sometimes there are observed fast drifting sub-bursts, which can be attributed to the class of Type III-like bursts [22]. And, if Type III-like bursts during a storm have durations 1-2s, these bursts inside the Type IV burst have significantly greater durations - up to 10-15s. After

10.00 UT the fiber bursts in emission and absorption and with positive and negative drift rates are distinctly observed. The fiber bursts have a duration of about 10-20s.

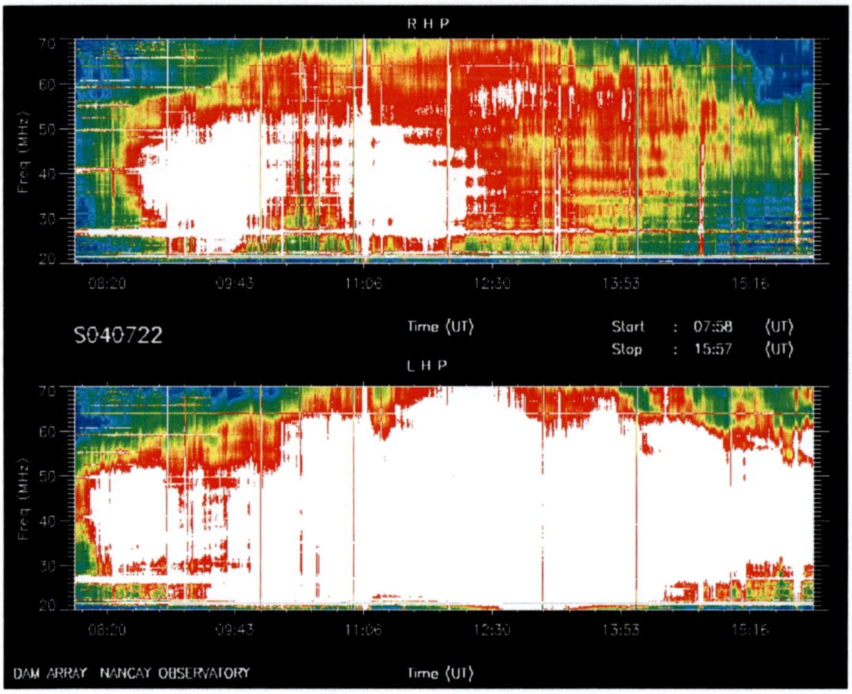

Figure 42. Type IV burst at frequencies 20-70MHz (NDA data).

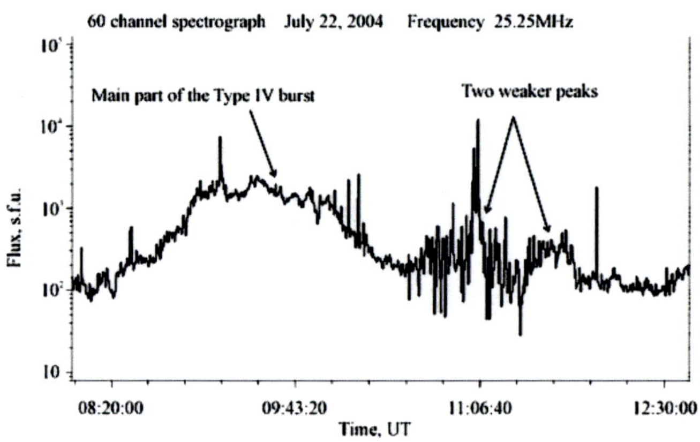

Figure 43. The time profile of Type IV burst with two increased regions (at 11:05 UT and 11:40 UT).

An interesting burst is observed at 09:53:53 UT (Fig. 45). This is an absorption burst simultaneously visible in the whole frequency band from 10 to 30MHz. At higher frequencies its duration is 1min. 20s, and at lower frequencies (9MHz), it becomes equal to 6min 20s. While the continuum emission is in the order of 1000s.f.u., the flux of absorption burst falls from 800s.f.u. at 25MHz down to 10s.f.u. at 9MHz.

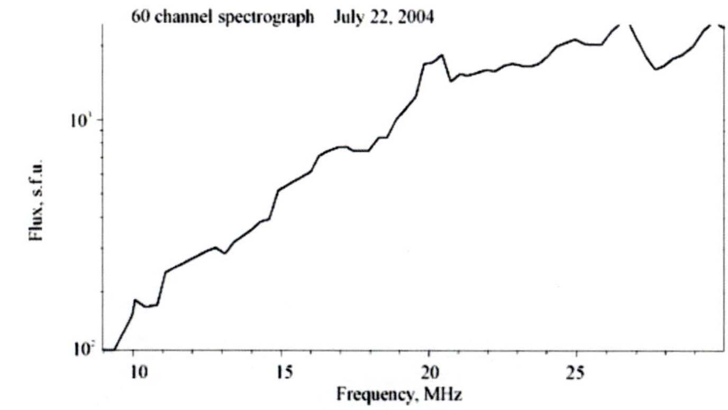

Figure 44. Frequency profile of radio emission at maximum phase of Type IV burst.

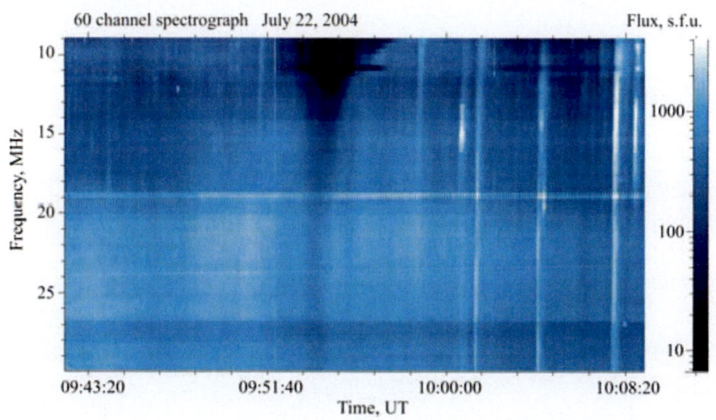

Figure 45. Burst in absorption against continuum radio emission of Type IV burst.

An extremely interesting property of this Type IV burst is a highly populated zebra-structure (unlike the higher-frequency zebra-structures in our case it is similar to finger-prints) (Fig.46). It begins at 08:10 UT and lasts with interruptions till 11:40 UT. The zebra-structures are observed in groups of 5-6 elements and more. The maximum number of elements in a group reaches 38. The stripes of zebra-structure are observed in a frequency range of 14-30MHz and may occur both in emission and absorption. For separate stripes the frequency band makes nearby 5MHz, and in rare cases it can reach up to 10MHz. Drift rates of zebra stripes can be both positive and negative, and sometimes infinite (Fig. 46). There are cases when the same stripe at higher frequency has negative drift rate, and at lower frequency positive one. The absolute average values of zebra structure drift rates are 0.5MHz/s. The distance between stripes in emission (or absorption) may range from 1.2s up to 1.3s in the beginning of the burst and remains practically constant for the given group. For the group of zebra stripes the distance between maxima is 2s at 09:03 UT, and 2.5s at 11:36 UT. Our records distinctly show, that stripes in emission are alternating with stripes in absorption (Fig. 47). The flux during emission phase equals 284s.f.u and during absorption falls down to 245s.f.u. Thus the mean flux of the Type IV bursts during this stripes group (08:38 UT) makes 252s.f.u. We must note that the fluxes in emission and in absorption for a given group remain approximately constant. The same properties are observed also in other groups of zebra-structures of this Type IV burst.

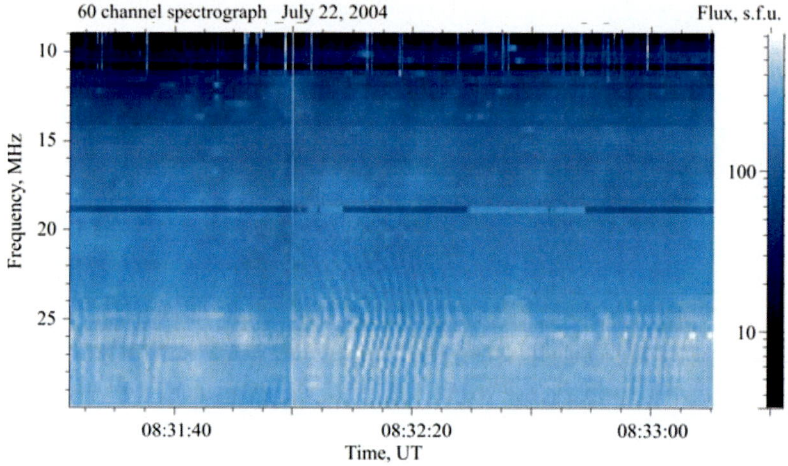

Figure 46. Three groups of zebra patterns.

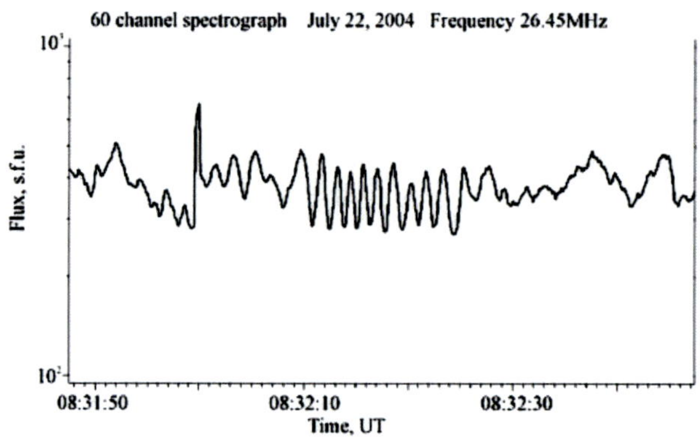

Figure 47. Succession of stripes in zebra structure in emission and in absorption.

On July 23, 2005

This Type IV burst (Fig. 48) is interesting first of all by the fact that it doesn't follow the apparently connected with a Type II burst, as it usually happens. The associated Type II bursts are observed against a background of the Type IV burst. In other words, two bursts are superimposed. Also unlike the previous cases there is no group of intensive Type III bursts before the Type IV burst. Taking into account low flux and short duration of both bursts it may be supposed that the initial event was also not so powerful. Data from SOHO (Fig. 49) show that both the bursts, apparently, are connected with the CME, which took place at 09:30 UT and had a velocity of 671km/s. The Type II burst drift rate is 20-30kHz/s. It consists of several lanes. Also this Type II burst has fine structure in the form of sub-bursts with durations 1-2s and drift rates 4-6MHz/s (sometimes sub-bursts with infinite drift rates can be detected). The Type IV burst is most likely a moving burst. It propagates as a whole from high towards low frequencies with a drift rate of 10kHz/s.

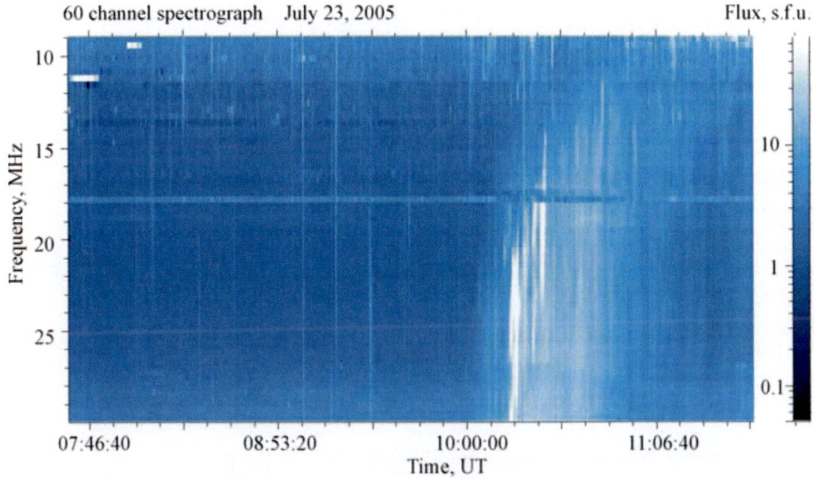

Figure 48. Decameter Type II-IV burst.

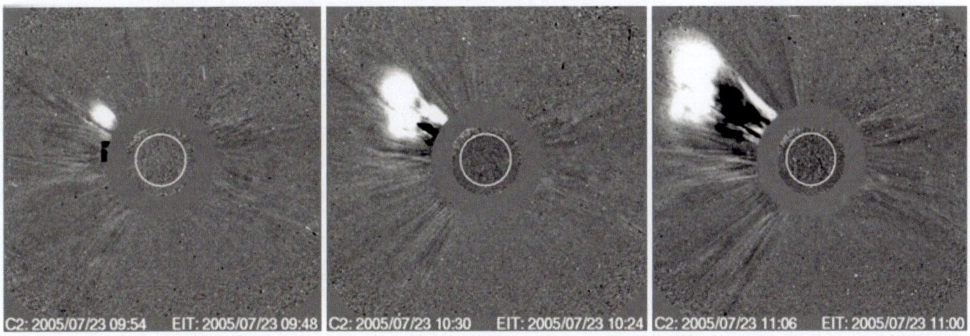

Figure 49. CME activity on July 23, 2005.

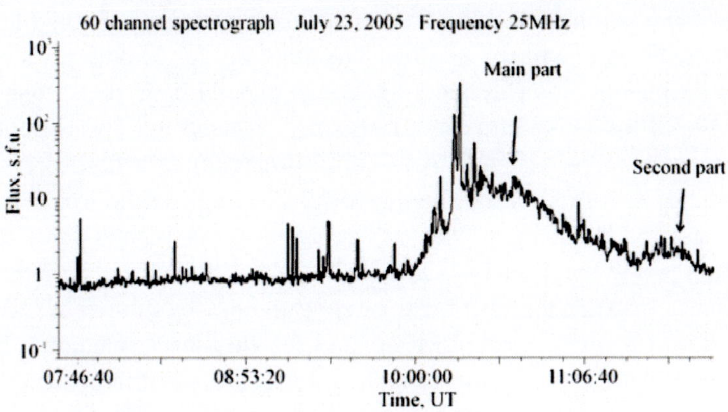

Figure 50. Type IV burst profile with two enhanced regions.

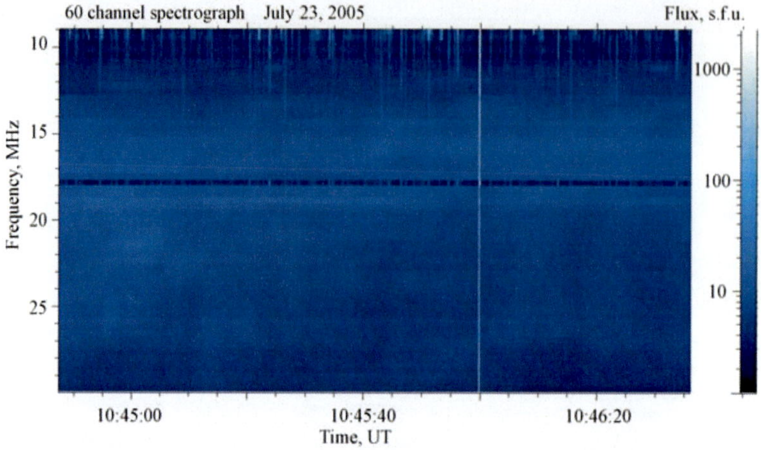

Figure 51. Fragment of continuum radio emission of Type IV burst.

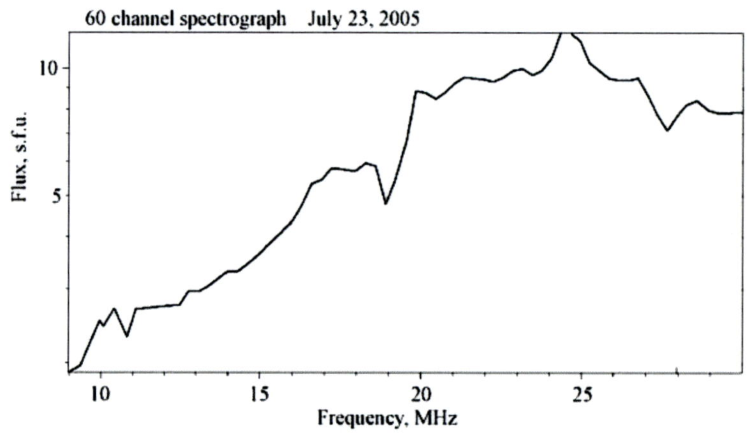

Figure 52. Frequency profile at maximum phase of Type IV burst.

In addition to the main part (maximum flux is 20s.f.u. at 25MHz) there can be distinguished also a less intensive part (maximum flux is 3s.f.u.) (Fig.50), which is observed at 11:40 UT and exhibits practically no drift. The time profiles of both parts of the burst are symmetrical, and while the first part has a duration of about 1 hour, the second part lasts only 20min. The slopes of instant spectra of two parts of the bursts are also different. In the first case the flux decreases (approximately by an order of magnitude) with decreasing frequency, while in the second case the flux remains practically constant in the whole frequency band. The Type IV burst is limited by a "bottom" frequency 11MHz, and splits in two lanes at 14MHz. The main part of the Type IV burst consists of numerous faint fiber-bursts (Fig. 51). (The flux difference between the fiber burst maxima and nearby minima is factor 1.5). Their durations are about 10s, drift rates 1-2MHz/s and the flux at its maximum is 10s.f.u. It is interesting, that at its maximum the flux remains almost constant at frequencies from 30 down to 20MHz and then steeply falls towards lower frequencies (Fig. 52). On the contrary, at the end of our observations on that particular day July 23, 2005, the flux is practically constant at low frequencies and falls off towards higher frequencies.

On July 27, 2005

The event July 27, 2005 (Fig. 53) started at 07:46 UT. As many other Type II and Type IV bursts it follows a group of Type III bursts having durations of about 10-20s and a peak flux up to 700s.f.u. On the background of these Type III bursts one can see a plenty of fiber bursts (Fig. 54) with durations up to 10s and drift rates 10-20kHz/s. This drift rate is inherent to the Type II bursts at the decameter wavelengths though apparently there were no Type II bursts visible on that day.

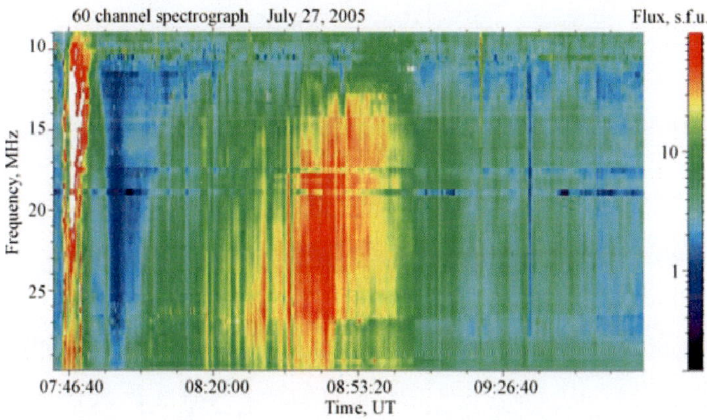

Figure 53. Type IV burst on July 27, 2005.

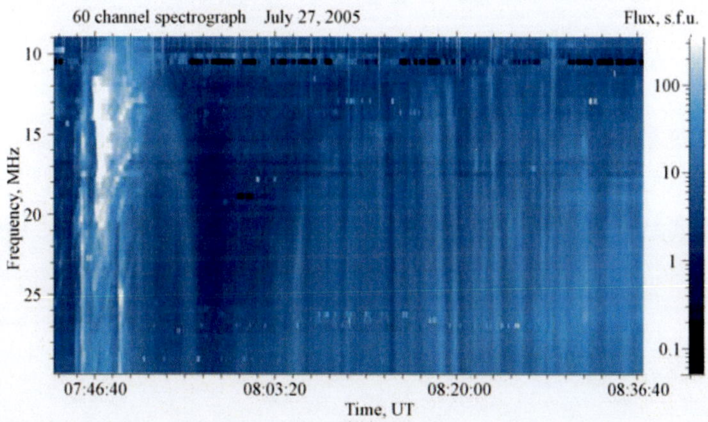

Figure 54. The beginning Type IV burst followed by a group of Type III bursts, fiber bursts and absorption region.

There are no tracks of their presence also at other frequencies, though a CME with velocity 533km/s was detected by SOHO at 07:54 UT (Fig. 55). After the group of Type III bursts and fiber bursts the intensity of radiation decreases and a shading area appears. Over the whole frequency band from 10 to 30MHz the flux in this area is constant and equal to 0.9s.f.u. The width of the shading area at 30MHz comprises 2min 30s, and at 13MHz already 7min 30s. After the shading area at 08:00 UT the Type IV emission starts rising and lasts till 08:50 UT (Fig. 56). At this time the flux has its peak value of 60s.f.u, which is practically constant over the whole bandwidth of the burst. On a falling branch of the Type IV bursts, which lasts till 11:00 UT, it is possible to distinguish some large-scale local increases in radiation with durations of about 30min.

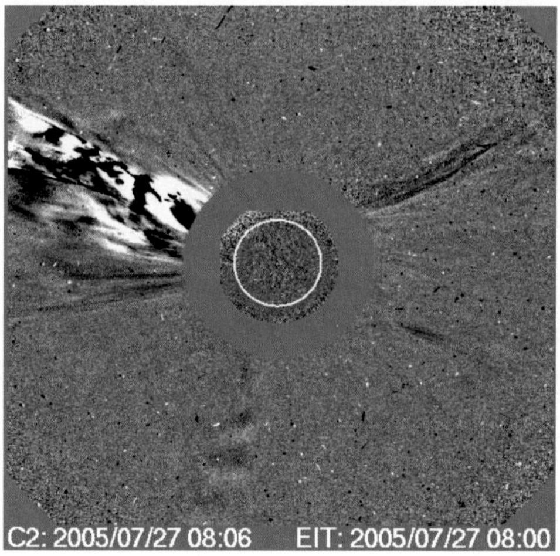

Figure 55. CME associated with July 27, 2005 event (SOHO data).

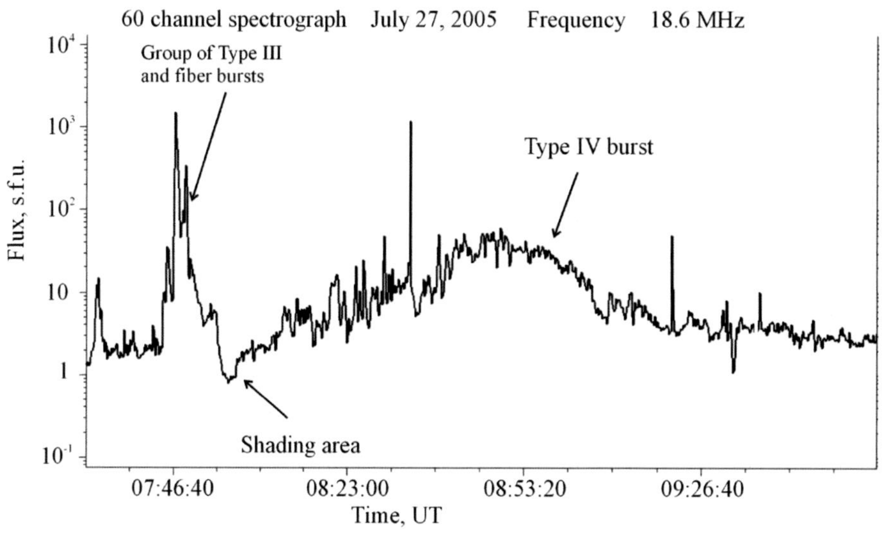

Figure 56. Time profile of the Type IV burst.

More fine structure of the Type IV burst represents faint fiber bursts with durations from 10 to 50s and both positive and negative drift rates. Moreover, a considerable number of such bursts in absorption is observed including those with the positive drift rate (so the burst at 08:14 UT had a drift rate of +8.58MHz/s and duration around 1min). On a descending branch of the Type IV burst there can be seen more contrast sub-bursts in emission and absorption with reverse drift rate (from low to high frequency). A corresponding example of such bursts is shown in Fig. 57. The first sub-burst (about 09:31 UT) has a drift rate of +1.2 MHz/s and duration 13s. This sub-burst has the knee-bend at frequency 15-16MHz and at low frequencies it has already a forward drift rate -0.36MHz/s. The drift rate of the subsequent burst in absorption is 3.0MHz/s and it lasts about 10s. The next sub-burst has a drift rate of 1.62MHz/s and a duration of 14s. And finally the last burst in absorption drifts with 9.47MHz/s and has a duration changing almost linearly from 40s at high frequencies to about 1min 30s at low frequencies. The time profile (Fig.58) is nearly symmetrical although there is a tendency of a more flat descending branch.

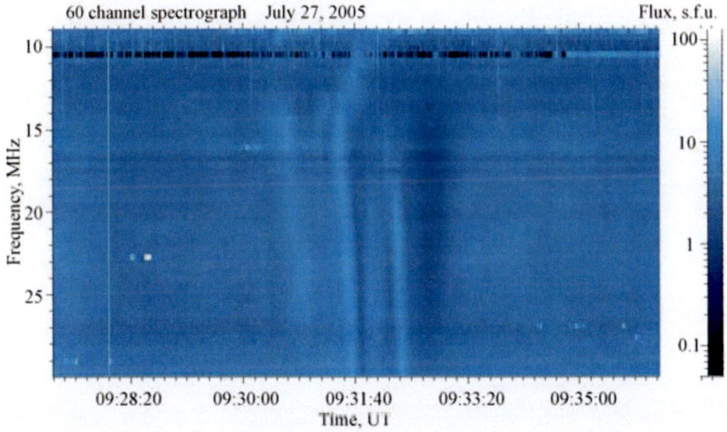

Figure 57. Bursts in emission and absorption with positive drift rates.

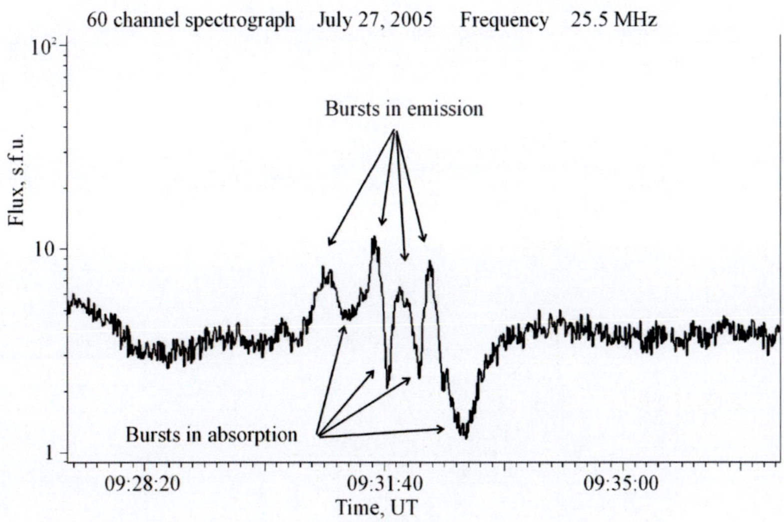

Figure 58. Time profiles of bursts in emission and in absorption.

On July 28, 2005

This Type IV burst (Fig. 59) is characterized by the absence of any associated Type III and Type II bursts. Nevertheless SOHO shows the CME started a little bit earlier, at 07:00 UT (Fig. 60) with a velocity of 573km/s. The front edge of this burst drifts with the rate close to

10kHz/s from high to low frequencies. The lowest frequency, reached by the Type IV burst is - 12-13MHz.

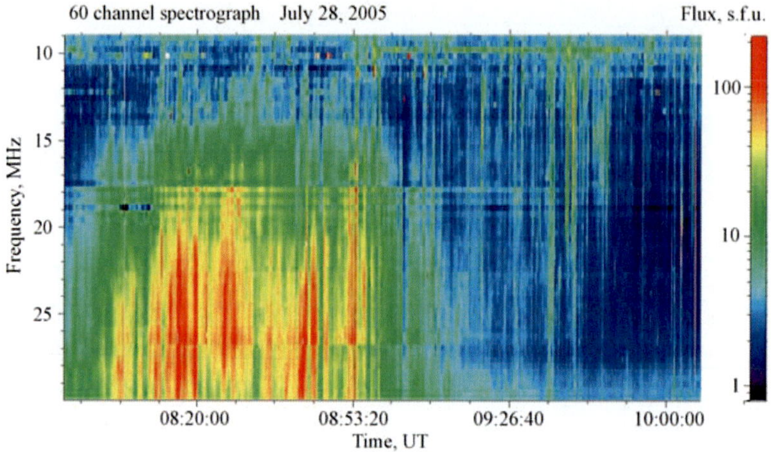

Figure 59. Type IV burst on July 28, 2005.

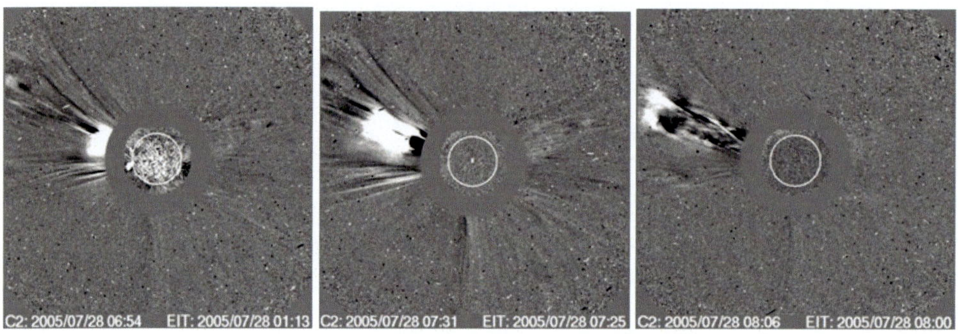

Figure 60. Evolution of CME from SOHO.

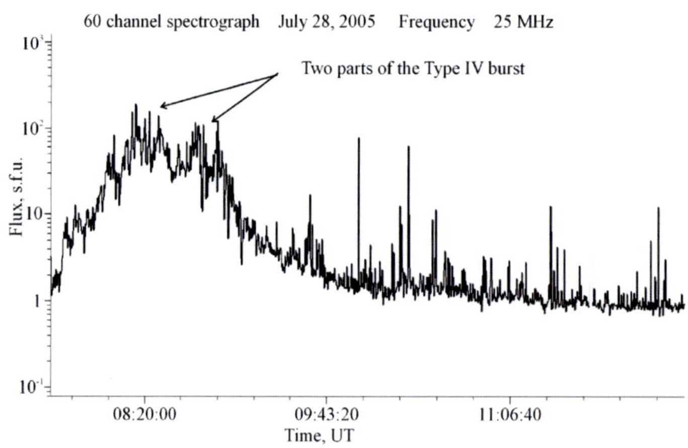

Figure 61. Time profile of Type IV burst observed on July 28, 2005 at 25MHz.

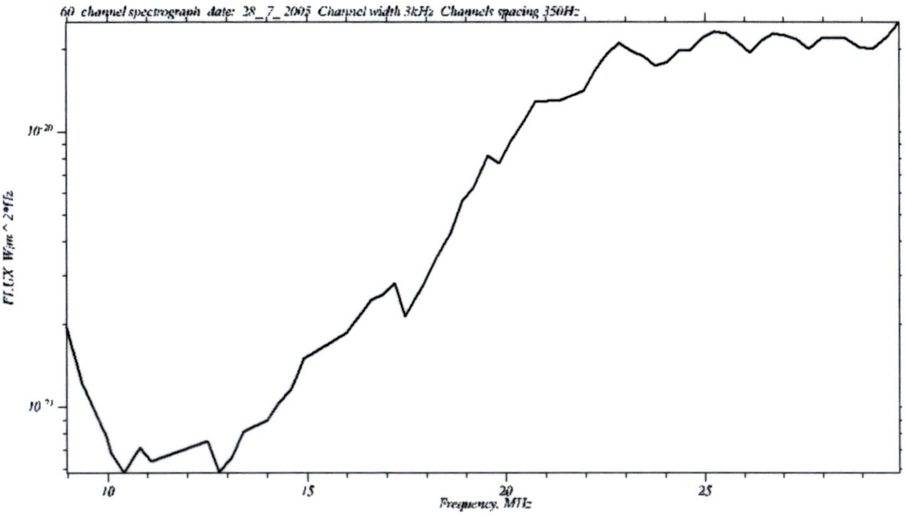

Figure 62. Frequency profile in the maximum phase of Type IV burst.

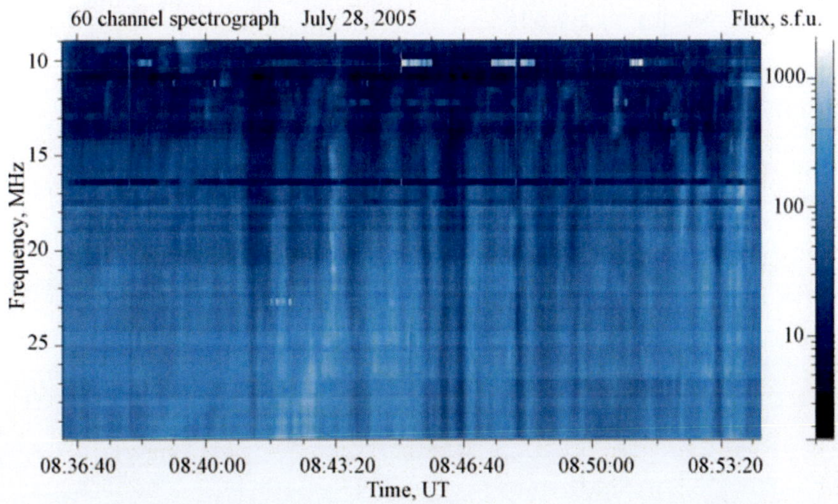

Figure 63. Fine structure of Type IV burst in the form of fiber bursts with positive and negative drift rates in emission and absorption.

At frequencies above 18-19 MHz the burst consists of two practically identical parts separated in time from each other by approximately 30min (Fig. 61). The instant spectrum at maximum flux point shows almost a constant flux from 30 down to 23MHz, which is equal to 200s.f.u. (Fig. 62). And from 23 to 13MHz the flux steeply falls down to 6s.f.u. Like the previous Type IV burst (on July 27, 2005) this burst has a fine structure in the form of fiber bursts, however they are less distinct. The duration of these fiber bursts is 5-10s, and their drift rates in the majority are typical for common Type III bursts in the decameter wavelengths, i.e. 2-4MHz/s. At the same time sub-bursts with the reverse drift rates are sometimes observed (Fig.63).

In the majority sub-bursts have a restricted frequency band of existence, given by 3-8MHz. These bursts are more clearly seen on the tail of the Type IV burst, and in addition there are also bursts in absorption with positive and negative drift rates (Fig. 63).

On July 31, 2005

This event we also file to the class of Type IV bursts since an increase of continuous emission is observed from the quiet Sun background up to a maximum value of 20 s.f.u. at 10:00-10:15 UT (Fig. 64). After that the flux decreases. The total estimated duration of the Type IV burst comprises not less than 7 hours. At the peak phase of the burst the flux practically does not

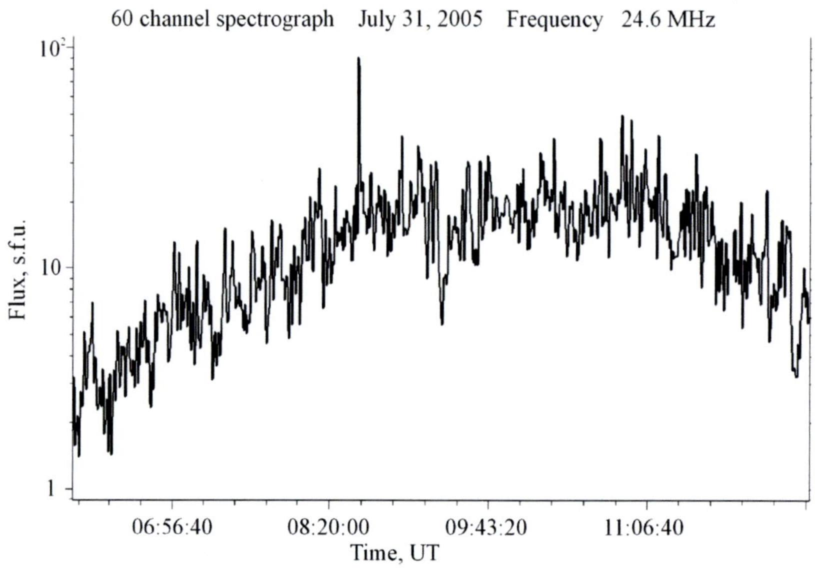

Figure 64. Time profile of prolonged Type IV burst.

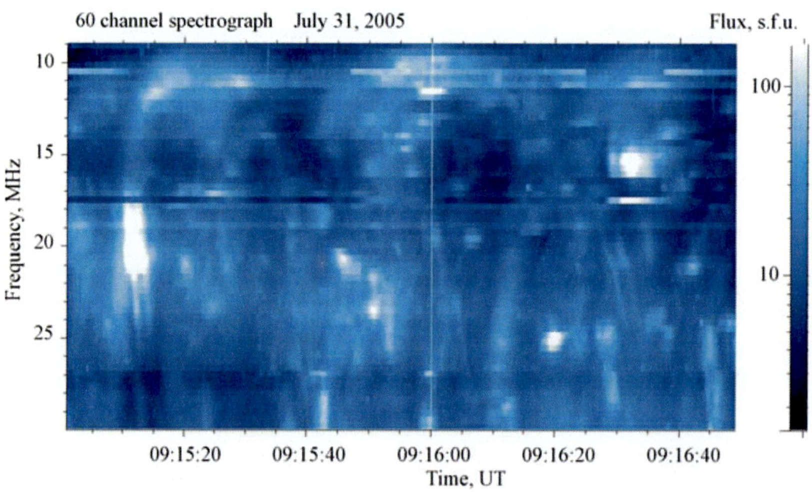

Figure 65. Fine structure of Type IV bursts in the form of fiber bursts and drift pairs.

vary with frequency. The prominent feature of this burst is that it consists of both the fiber bursts, and a number of forward and reverse drift pair bursts. Fiber bursts have various drift rates (Fig. 65) (positive, negative and infinite), durations (from 2 s up to 10-20s), and frequency bands (from a few MHz up to 20MHz). In some cases drift pairs appear to be brighter with fluxes several times the continuous emission flux. This Type IV burst is associated with a CME (Fig. 66), which was registered at 06:30 UT and is also connected with the active region in the eastern part of the solar disk.

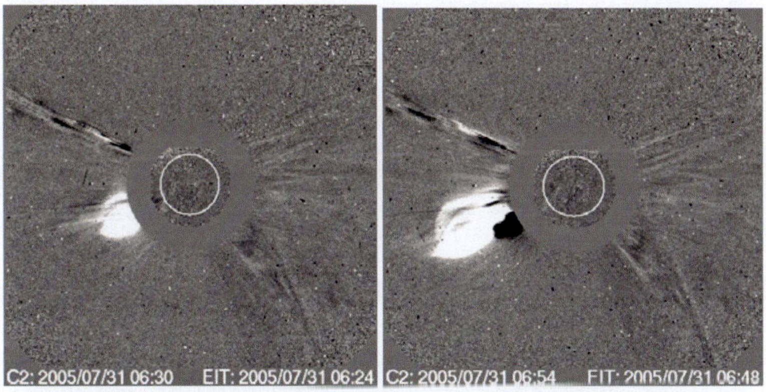

Figure 66. Dynamics of CME at 06.24 UT on July 31, 2005.

On July 6, 2006

This event (Fig. 67), apparently, is initiated by CMEs, which took place in series at 06:54 UT and 08:54 UT (Fig. 68). The group of Type IIIb and III bursts and subsequent Type II burst are connected with the first CME.

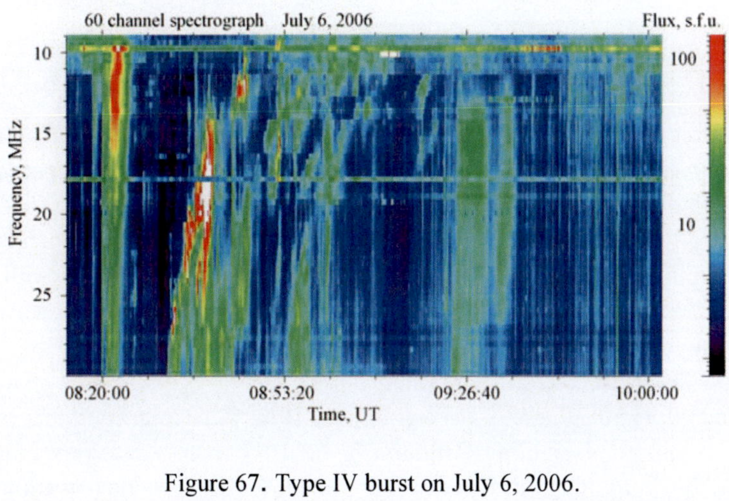

Figure 67. Type IV burst on July 6, 2006.

The Type III and IIIb bursts (at 08:20 UT) are very intensive with fluxes up to 1000 s.f.u. (Fig.69). The total duration of the group is about 2min 30s, and drift rates are 2-4MHz/s. After this group the emission level drops down to a level practically equal to the quiet Sun background (1 s.f.u.). Then, at 08:31:40 UT the Type II burst starts at 30MHz (Fig. 70). A short time later its second harmonic appears. The drift rates of both harmonics are 20-30kHz/s. Sub-bursts, which form both the first and second harmonics, have various drift rates, ranging from - 0.5 to -3MHz/s. Their durations are 2-5s, and sometimes more.

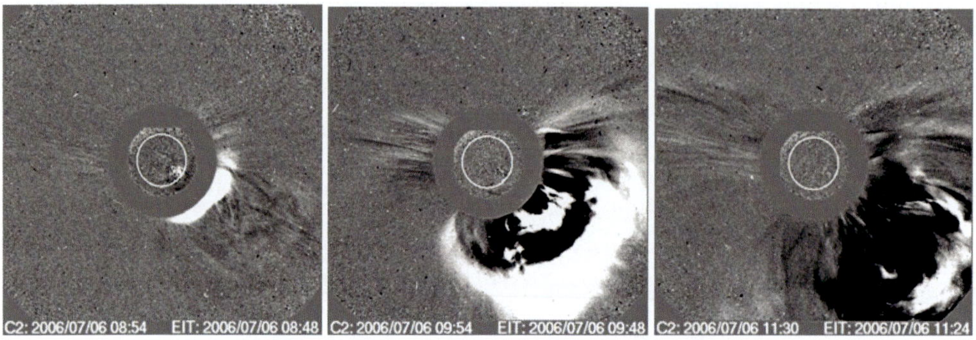

Figure 68. CME associated with July 6 Type IV burst.

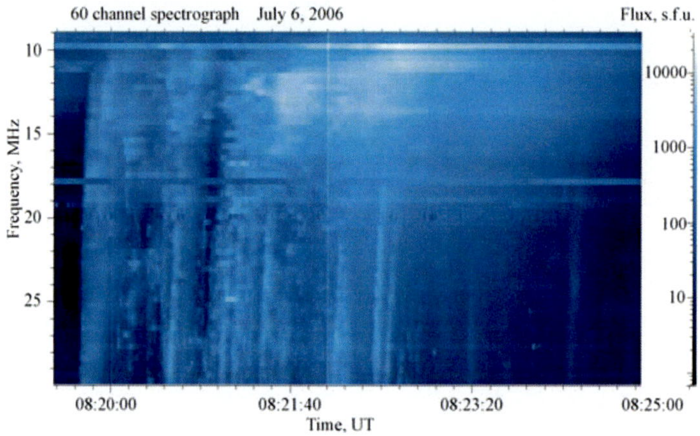

Figure 69. Group of Type III bursts as a precursor of Type IV burst.

Right after this an increased level of emission is observed (Fig. 71). It consists of 5-6 long-lived broadband flux increases reaching sometimes 100s.f.u. and spaced in time by approximately 30min. In general these increases do not consist of very intensive fiber bursts, which in the majority have durations 3-10s and drift rates of 1-3MHz/s, which is slightly less with regard to common Type III bursts. We also observed some sub-bursts with drift rates of 6-10MHz/s and durations of 1-2s. They resemble fast Type III-like bursts, which usually occur during common Type III bursts storms when the associated active region is close to the central meridian [22]. Among all mentioned 6 flux increases two cases (with maxima at 09:27:30 UT and 09:34:10UT) are of special interest (Fig. 72). Contrary to the other maxima, which consist of bursts resembling common Type III bursts, these two emission "hills" represent continuum emission with fluxes about 100s.f.u. The first one has a total duration of 7 min and splits in two elements above 18MHz. The second one has a duration near 3min. Both features drift with rates of 0.3MHz/s.

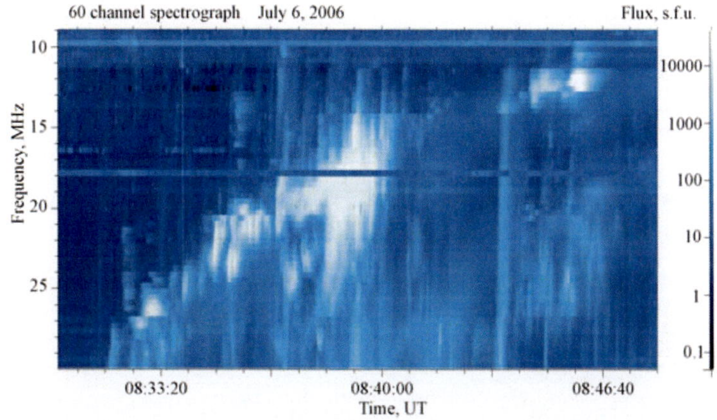

Figure 70. Fundamental and harmonic emission of Type II burst before Type IV burst.

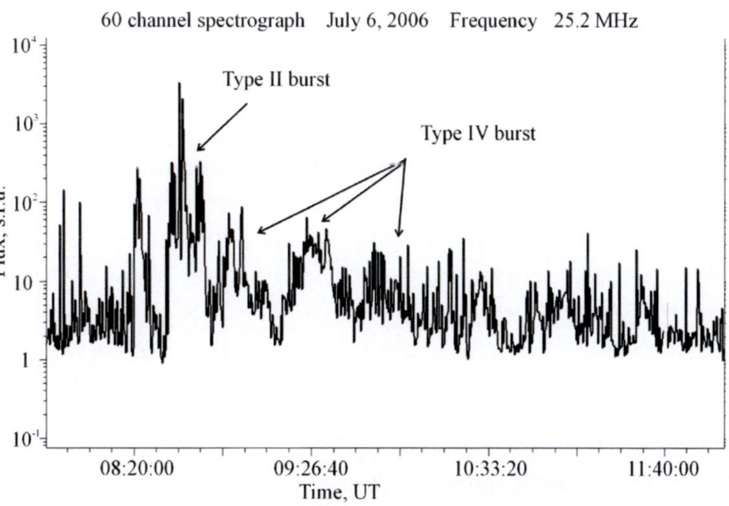

Figure 71. Time profile of Type IV burst.

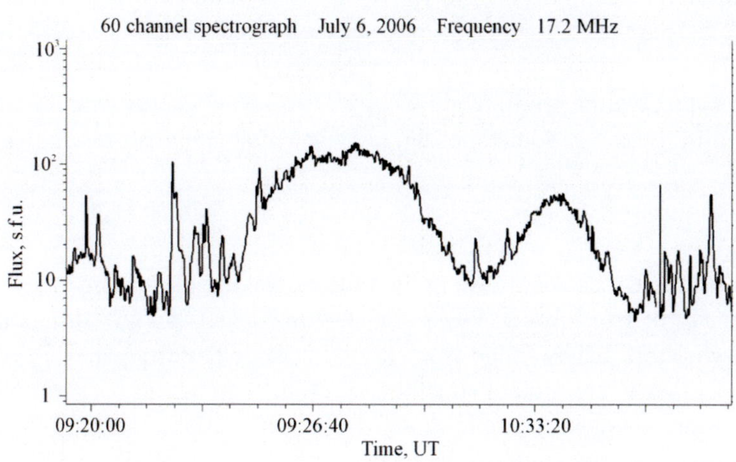

Figure 72. Two continuum "hills" in Type IV burst.

Discussion

All discussed events in this chapter can be categorized to the class of Type IV bursts due to the following reasons:

- these bursts have a noticeable continuum component, which, as a rule, rises at the beginning of the bursts and falls at the end of it;
- in the majority of cases these bursts follow the Type II bursts or the latter are seen on the background of the first;
- in all cases these bursts are associated with CMEs.

The succession of the events accompanying the Type IV bursts in the decameter wavelengths.

Preceding a Type IV burst by 10 – 30 minutes, a group of very intensive Type III bursts is observed. For decameter wavelength band they have standard durations and frequency drift rates. Just at the end of such a group and/or on its background, fiber bursts are seen with drift rates 20-30kHz/s and durations of 3-5s. (Mentioned drift rates are inherent to Type II bursts at the decameter waves). The group of Type III bursts and fiber bursts are followed by a shading area with a duration of some minutes, which uses to increases in several times towards the lower frequencies. Their drift rates can be either positive or negative and the fluxes at the minimum correspond to the background, as observed prior to the whole event. Just after the bursts shading area a Type II burst begins. The Type II bursts usually consist of fundamental, second and sometimes third harmonic emissions. A Type IV burst starts just after Type II burst or slightly earlier.

Properties of the decametric Type IV bursts.

- durations of Type IV bursts at decameter wavelengths vary from some tens of minutes to some hours;
- moving as well as stationary Type IV bursts are observed. In the case of moving ones their drift rates are close to the drift rates of the preceding and/or accompanying Type II bursts;
- as a rule, Type IV burst are formed by two or more components with the first component been the most intensive. Sometimes the sequence of components resembles an oscillation process;
- radio emission fluxes of Type IV bursts at maximum reach 2000s.f.u., however there are also weak Type IV bursts with fluxes of several tens of s.f.u. only;
- the Type IV burst spectrum taken at the time of maximum intensity shows that the flux can either fall or rise towards lower frequencies. Sometimes the flux remains constant over the whole frequency range from 10 to 30MHz. The first case happens when fine structure sub-bursts are more distinctly seen on the continuum background and the drift rates of the sub-bursts are close to those of Type III bursts in storms. In the second case, the less distinct sub-bursts have drift rates smaller than 1MHz/s, i.e. 2-3 times slower than common Type III bursts;

- in rare cases the Type IV burst fine structure represents drift pair bursts, S-bursts, spikes and Type III-like bursts;
- the fine structure known as a zebra-structure is an extremely rare event in the decameter wavelengths range. It is observed during the overall burst extent in groups, with from 5-6 up to 30-40 stripes in each of the groups;
- fiber bursts have different drift rates, both positive and negative. The fiber bursts in absorption are not uncommon;
- large scale absorption bursts are observed against Type IV burst background. These fast drifting bursts have durations of several minutes.

Fine Structure Properties of Type IV Bursts

The most frequently seen fine structure of Type IV bursts are fiber bursts, which have parameters comparable to common Type III bursts. In particular, their durations are 5-20s and drift rates from less than 1MHz/s up to 5MHz/s. Contrary to the Type III bursts, the fiber bursts may exist both in emission and absorption, may have both positive and negative drift rates and are observed in limited frequency band (from 5-6MHz up to 20MHz).

At higher frequencies (at meter wavelengths) Type IV bursts appear to have fine structure in the form of so-called zebra-structure. We found only one Type IV burst with zebra-structure (July 22, 2004). The groups of zebra-structure stripes last during the full Type IV burst duration, slowly drifting with time towards lower frequencies. In total more than 30 groups of zebra-structure stripes have been observed. At the beginning of the Type IV burst each stripe of zebra-structure has negative drift rate at higher frequencies and positive rate at lower frequencies. And at the end of the Type IV burst the stripes have mainly positive drift rates. Stripes inside a group are spaced by constant time intervals, which equal from 1.5-1.6s, at the beginning of the Type IV burst, to 2.5s by the end of it. Inside one group the stripes in emission and absorption are alternating. An interesting feature of zebra structures in the decameter band is that the fluxes at maxima of stripes in emission (as well as at minima of stripes in absorption) keep constant along a group.

Polarization of Type IV Bursts

Because of lack of numerous polarization measurements, we cannot provide any definite conclusions on the polarization of Type IV bursts at the decameter wavelengths. The only fact we can state is that although the polarization is rather high for some Type IV bursts (up to 60%), the polarization is smaller with regard to those observed at higher frequencies (e.g. up to 100% at meter wavelengths). We also do not see any monotonous increase of polarization from the beginning to the end of the Type IV burst, as it was found at higher frequencies. We can rather state that the polarization change in time during Type IV bursts may reflect some changes of inner structures inside the Type IV bursts sources.

Conclusion

Observations of the decameter Type IV bursts with unique sensitivity, high time and frequency resolutions allowed to define the Type IV bursts properties at lowest frequencies, with regard to ground-based observations. Zebra structure at those lowest frequencies was detected. The properties of zebra-structure at decameter wavelengths and at higher frequencies are found to be different. Large-scale bursts in absorption were observed on the background of Type IV bursts.

The observations showed that Type IV bursts at decameter wavelengths on the one hand are characterized by a great variety of properties, such as duration, flux, large-scale structure, and on the other hand have some similar properties – fine structure consisting mainly of fiber bursts, which resemble common Type III bursts.

Further observations of Type IV bursts, including polarization observations in a wide frequency band will support a deeper understanding of this phenomenon in large and small scales.

Acknowledgments

The authors are grateful to SOHO, WIND and Nancay teams for operating the instruments and performing the basic data reduction, and especially, for the open data policy. This chapter was partially supported by INTAS grant No. 03-5727.

References

[1] Boischot, A. *Ann. d'Astrophys.*1957, 21, 273.

[2] Stuart, R.T. *Solar Radio Physics*; McLean, D.J and Labrum, N.R. Ed.; Cambridge University Press: Cambridge, 1985; 361-384.

[3] Robinson, R.D. *Solar Phys.* 1978, 60, 383-390.

[4] Bougeret, J.-L.; Kaiser, M.L., Kellog, P.J. et al. *Space Sci.Rev.* 1995, 71, 231-263 .

[5] Chernov, G.P. *Solar Phys*. 2007, 241, 145-169.

[6] Smerd, S.F. and Dulk, G.A. *Solar magnetic fields;* Hovard, R. Ed.; Proc.IAU Symposium No.43; 1971 Reidel: Dordrecht, 616.

[7] Schmahl, B.J. *Australian Journal of Physics Supplement*, 1973, *29*, 1.

[8] Stewart, R.T., Duncan, R.A., Suzuki, S. Nelson, G.J. *Astronomical Society of Australia, Proceedings*, 1978, 3, 247-249.

[9] Trottet, G.; Kerdraon, A.; Benz, A. O.; Treumann, R. *Astronomy and Astrophysics*, 1981, 93, 129-135.

[10] Elgaroy, Ø. *Astrophisica Norvegica*, 1961, 7, 123 .

[11] Kuijpers, J. *Solar Phys*., 1975, 44, 173-193.

[12] Slottje, C. *Atlas of Fine Structures of Dynamic Spectra of Solar Type IV -dm and Some Type II Bursts*, 1981, Utrecht Observatory, Utrecht.

[13] Kai, K. *Proceedings of the Astronomical Society of Australia*, 1969 , 1, 189 .

[14] Dulk, G. A. *Proceedings of the Astronomical Society of Australia,* 1970, 1, 372 .

[15] Duncan, R. A. *Solar Phys*., 1981, 73, 191-204.

[16] Melrose, D. B. *Space Science Reviews*, 1980, 26, 3-38.

[17] Zheleznyakov, V. V.; Zlotnik, E. Ya. *Solar Phys.* 1975,43, 431-451.

[18] Chernov, G. P. *Sov. Astron.* 1976,20, 449- 459.

[19] Brazhenko,, A.I.; Bulatsen, V.G., Vashchishin, R.V.; Konovalenko, A.A.; Falkovich, I.S.; Abranin, E.P.; Ulyanov, O.M.; Zakharenko, V.V.; Lecacheux, A.; Rucker, H.O. *Kinematika i Fizika Nebesnykh Tel*, 2005, 5, 43-46.

[20] Konovalenko, A. A.;Stanislavsky, A. A.;Abranin, E. P.;Dorovsky, V. V.; Mel'Nik, V. N. *Kinematika i Fizika Nebesnykh Tel, Suppl*, 2005, 5, 78-81.

[21] Mel'Nik, V. N.;Konovalenko, A. A.;Abranin, E. P.;Dorovskyy, V.V.; Stanislavsky, A. A.; Rucker, H. O.; Lecacheux, A. *Astronomical and Astrophysical Transactions*, 2005, 24, 391-401.

[22] Mel'nik, V. N.; Konovalenko A. A.; Rucker, H. O.; Rutkevich, B. P.; Dorovskyy, V. V.; Abranin, E. P.; Brazhenko, A. I.; Stanislavskii, A. A.; Lecacheux, A. *Solar Phys.* 2007 (submitted).

In: Solar Physics Research Trends
Editor: Pingzhi Wang, pp. 327-344

ISBN 978-1-60021-987-0
© 2008 Nova Science Publishers, Inc.

Chapter 9

FORCING OF THE EARTH'S ATMOSPHERE BY SOLAR RADIATION AND SOLAR WIND

Klemens Hocke and Niklaus Kämpfer
Institute of Applied Physics, University of Bern, Sidlerstrasse 5
CH-3012 Bern, Switzerland

1. Introduction

If you imagine the Earth's atmosphere without insolation, you may see a quiet, dark and cold atmosphere with small atmospheric fluctuations due to the Earth's rotation and gravitational forcing by the Moon, Sun, and planets. There is no doubt that the electromagnetic radiation from the Sun is the main driver of the climate, the circulation, and the wave flux of the Earth's atmosphere.

In the first part of our study, we investigate how the annual variation of the total solar irradiance (TSI at the position of the Earth) influences the global mean temperature of the lower troposphere. The annual variation of TSI is caused by the changing Sun-Earth distance during the year, and its amplitude amounts 45.7 W/m^2. The annual variation of TSI is much larger than the other variations of TSI, e.g., the 11-year solar cycle corresponds to an amplitude of around 1 W/m^2. We estimate the impact of the annual variation of TSI on the time series of global mean temperature (from NCEP/NCAR reanalysis). Additionally we derive the influence of the inter-hemispherical asymmetry of the Earth on the time series of the global mean temperature. Since the Northern Hemisphere (NH) has more land surfaces and mountains than the Southern Hemisphere (SH), the characteristics of wind circulation, water vapor convection, atmospheric waves, concentration of greenhouse gases, oceanic currents, surface albedo, etc. are different and lead to a different thermal equilibrium for the NH than for the SH. Particularly the warmest sea surface temperatures are observed in the NH which can be explained by complex ocean-atmosphere interactions depending on surface wind, cloud cover, coastal geometry of the continents, the depth of the oceanic thermocline, and other factors (Philander et al. , 1996).

The time series of the global and hemispheric mean temperature, humidity, and geopotential height of the lower troposphere reveal a linear, positive trend which is mainly caused by the strong increase of anthropogenic emissions of greenhouse gases such as CO_2. The feedback of atmospheric water vapor to the increase of temperature is derived from the time series of the time interval 1996-2006.

In the second part of our study, the relationship between the active Sun evidenced by solar flares, solar wind speed, and geomagnetic activity, and tropospheric dynamics is investigated. Strong solar flares are often associated with interplanetary coronal mass ejections (ICME) which travel at high speeds through the heliosphere and cause severe disturbances of the Earth's magnetosphere, ionosphere, and upper atmosphere. The strongest flares of the solar cycle-23 are selected, and a superposed epoch analysis is performed by using the time series of solar X-ray flux, solar wind speed, planetary geomagnetic activity A_p, and geopotential height. The typical sequence of the processes is: 1) sun-spot darkening in TSI, 2) X-ray burst and faculae brightening in EUV/UV, 3) ICME arrival at the Earth and enhanced geomagnetic activity.

2. Forcing of the Earth's Atmosphere by the Annual Variation of the Solar Radiation

2.1. True TSI and the Global Mean Temperature of the Lower Troposphere

True TSI is the total solar irradiance at the actual position of the Earth. The elliptical orbit of the Earth around the Sun causes a geometrical dilution W of the incident solar radiation depending on the day of the year

$$W(t) = \left(\frac{r(t)}{R}\right)^2, \tag{1}$$

where r(t) is the Sun-Earth distance as function of time, and R is 1 astronomical unit which is the mean of the Sun-Earth distance. Usually the TSI series refer to a fixed Sun-Earth distance of 1 astronomical unit. The true TSI can be obtained by multiplication of the TSI series with the dilution factor $W(t)$. In case of the SORCE satellite mission (SOlar Radiation and Climate Experiment, (Kopp et al. , 2005; Lean et al. , 2005)), both parameters are provided in the data files, TSI and true TSI. The upper panel of Fig. 1 shows a composite of the true TSI from the SORCE experiment (blue line) and the NOAA series (red line, provided by C. Fröhlich and J. Lean). The mean value of TSI in Fig. 1 is 1366 W/m^2, and the annual variation is 45.7 W/m^2 correponding to a 3.34%-variation of the dilution factor W.

The time series of the global mean temperature are calculated by using daily averages of the global temperature field at the 850-hPa pressure level ($h \approx 1500$ m). The 850-hPa pressure level has been selected since it is representative for the lower troposphere, and possible problems due to the surface boundary are reduced at this altitude. The temperature fields are provided by the Climate Diagnostic Center (`http://www.cdc.noaa.gov/`) and originate from meteorological reanalyses of NCEP/NCAR. The NCEP/NCAR Reanalysis 1 project is using a state-of-the-art analysis/forecast system to perform data assimilation using past data from 1948 to the present. We emphasize that the temperature fields from reanalyses are mainly driven by the radiosonde and satellite observations. Thus the reanalyses data should be regarded as 'observations' adjusted for consistency with model physics. For

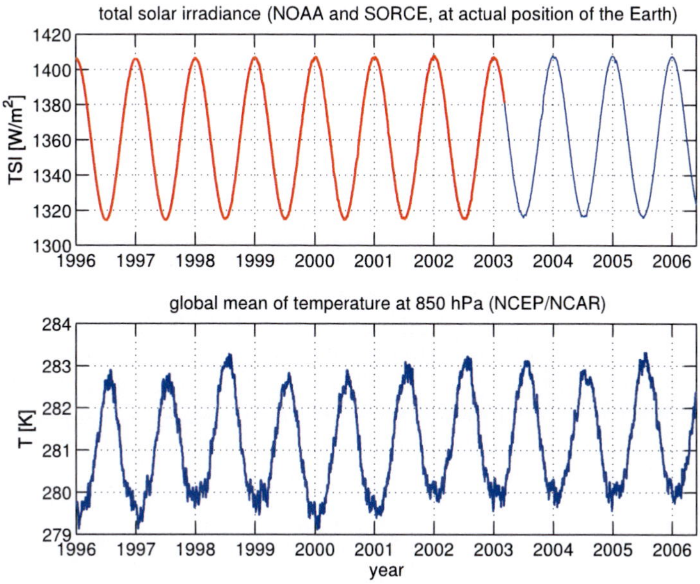

Figure 1. Upper panel: Annual variation of true total solar irradiance TSI from NOAA (red line) and SORCE (blue line) due to the annual variation of the Sun-Earth distance. Lower panel: Global mean of lower tropospheric temperature at 850 hPa (ca. 1500m altitude, from NCEP/NCAR reanalysis). T and TSI are obviously anti-correlated (tickmarks are always on January 1).

each day, the temperature values of the equally-spaced global grid (2.5×2.5 degrees) are averaged with consideration of the changing surface size of the area elements. The resulting time series of global mean temperature at 850 hPa are shown in Fig. 1. It is amazing that the global mean temperature is anti-correlated to the time series of TSI. One might expect that the temperature of the lower troposphere should be proportional to the absorption rate of solar electromagnetic radiation at optical and infrared wave lengths (main contributors to TSI).

Later we will try a decomposition of the time series of temperature in order to "solve" the puzzle of the annual variation of the global mean temperature and to estimate the effect of the inter-hemispherical asymmetry. For this purpose, we firstly determine and remove the linear regression line from the temperature series. We take the opportunity to discuss the temperature trend and its consequences for the atmospheric water cycle.

2.2. Linear Trends of Global and Hemispheric Means of Temperature, Water Vapor, and Geopotential Height

The linear trends are determined by least-square fits of straight lines to the NCEP/NCAR series from January 1, 1996 to December 31, 2005 (in the graphics, the straight lines are extended to cover the time range until June 2006). The time series and trends of the global mean (blue), the NH mean (green), and the SH mean (red) are shown in Fig. 2 for the temperature at the pressure level 850-hPa. The trend of the global mean temperature is

0.6°K per decade. This would correspond to an increase of the lower tropospheric, global mean temperature of around 3°K in the next 50 years. According to the Intergovernmental Panel on Climate Change (IPCC) the strong increase of global mean temperature is due to increasing concentrations of CO_2, CH_4, and other greenhouse gases in the Earth's atmosphere resulting from the persistent increase of anthropogenic emissions.

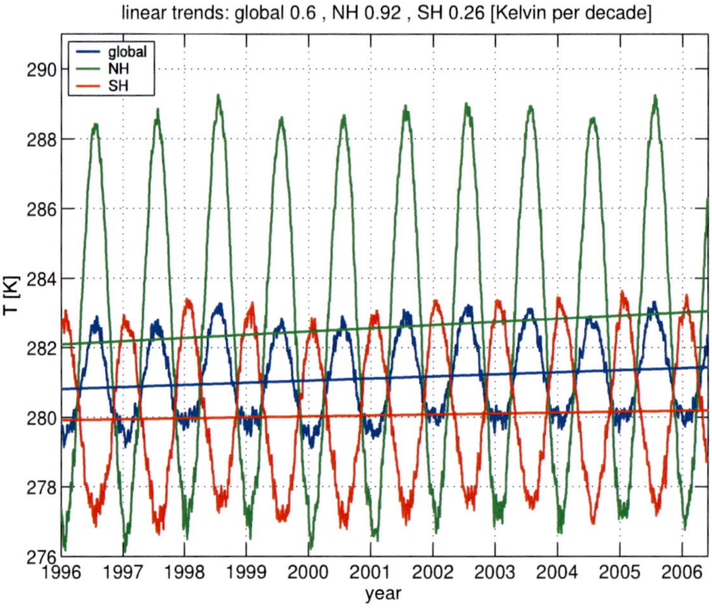

Figure 2. Least-square fits of straight lines to the time series of global mean (blue), NH mean (green), and SH mean (red) of temperature at 850-hPa (ca. 1500 m altitude). The linear trends are annotated in the title line. The increase of temperature is possibly due to the increasing concentrations of greenhouse gases such as H_2O, CO_2 and CH_4.

The trend of the NH temperature is even greater with 0.92 °K per decade while the SH temperature increases with 0.26°K per decade. The annual amplitude of the NH temperature is around 6°K, while the SH temperature only varies with an annual amplitude of around 2.5°K. How can we explain this large inter-hemispherical asymmetry? Possibly, a stronger water vapor loading of the lower atmosphere during NH summer is the reason (Trenberth and Guillemot , 1994). Since water vapor is the most important greenhouse gas of the lower atmosphere, the enhanced water vapor loading of the NH leads to a higher temperature in summer compared to the SH (Fig. 2). Solar radiation-induced convection of water vapor is the engine of tropospheric transport processes and redistributes the heat from tropical to middle and high latitudes.

These considerations can be qualitatively checked by analyzing the trends and annual variations of the specific humidity s (grams of water vapor per 1 kg wet air). Fig. 3 shows the time series and trends of specific humidity. The global trend of specific humidity at 850 hPa is 0.1 g/kg or 1.7% per decade. This trend is in agreement with the trend observed in column-integrated atmospheric water vapor by SSM/I (special sensor microwave imager) which measured a value of $1.3 \pm 0.3\%$ per decade for the ocean as a whole during the time interval from 1988 to 2003 (Trenberth et al. , 2005). As already assumed, the water vapor loading of the NH during summer is much greater than for the SH. The time series of the specific humidity of the Southern Hemisphere show a positive anomaly during the SH summer 1997/1998 which is caused by the severe 1997/1998 El Nino/Southern Oscillation (ENSO). The inter-hemispherical asymmetry of water vapor is essential for understanding the sun-climate relationships. Effects of solar radiation changes and of anthropogenic emissions can be amplified by a higher concentration of atmospheric water vapor, leading to a higher temperature which may again induce more evaporation of water and/or less rain precipitation. A possible chain of forcing and feedback processes is illustrated in Fig. 4. The adjustment processes should stop when a new equilibrium of evaporation and condensation of water is reached (Washington , 1992).

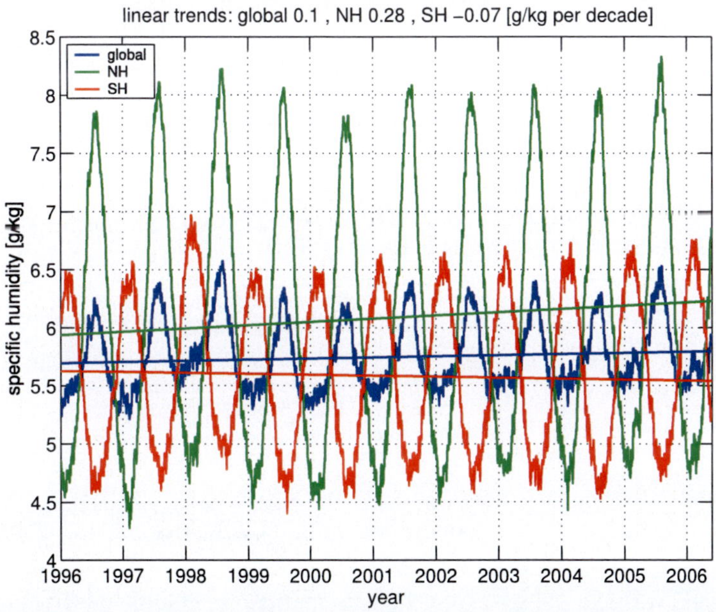

Figure 3. Least-square fits of straight lines to the time series of global mean (blue), NH mean (green), and SH mean (red) of specific humidity at 850-hPa (ca. 1500 m altitude). The linear trends are annotated in the title line. The water vapor loading during NH summer is greater than during SH summer, though the Sun-Earth distance is greater during NH summer.

The sensitiviy of water vapor to a change of temperature can be estimated from the linear trends of temperature and specific humidity in Figs. 2 and 3. According to the global mean trends at 850-hPa, an increase of 1 °K causes an increase of specific humidity of 0.17 g/kg (around 3.0%). The feedback is larger for the NH: 0.30 g/kg (or 4.9%) for a temperature change of 1 °K. Surprisingly, the feedback seems to be negative for the SH: -0.27 g/kg (or -4.8%) for a temperature increase of 1 °K. The trends of the SH are relatively small, and the extrapolation to 1 °K gives a higher uncertainty for the SH than for the NH.

Meteorological reanalyses prefer fixed pressure levels as a grid of the atmospheric parameters. In this system, the variation of the geopotential height of a pressure level is the

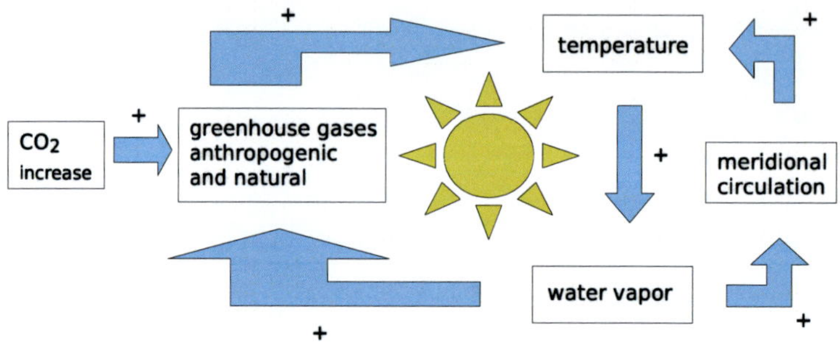

Figure 4. Scheme of a possible chain of forcing and feedback processes in the Earth's lower troposphere: anthropogenic emission of CO_2 leads to a higher concentration of greenhouse gases and to a higher global mean temperature by insolation. The increase of temperature induces more evaporation of water. The higher concentration of atmospheric water vapor causes enhanced convection and meridional exchange of tropical and polar air masses, so that the global mean temperature is expected to rise. Additionally the higher concentration of water vapor causes an increase of temperature due to the greenhouse effect.

equivalent to the pressure variation at a fixed altitude. An increase of the geopotential height of a pressure level corresponds to an increase of the atmospheric pressure at a fixed altitude. The variation of the geopotential height at $p = 850$ hPa can be roughly converted into a pressure variation at $z = 1500$m

$$\Delta p = 10.3 \, \frac{\text{Pa}}{\text{m}} \, \Delta z, \tag{2}$$

where we make the approximations of an average, vertical pressure gradient at $p = 850$ hPa and the substitution of the geopotential height by the geometric height.

Fig. 5 shows the time series and trends of the geopotential height of the 850-hPa pressure level. The NH summer shows again the largest increase. Is this increase caused by the increasing partial pressure of atmospheric water vapor? The NH trend is 4.33 m per decade. According to equation 2, the trend of the NH geopotential height corresponds to a NH pressure trend of around 44.6 Pa per decade. The trend of the specific humidity s can be inverted into a trend of the water vapor pressure e by using the approximation $\Delta e \approx p \, \Delta s / 0.622$ which is valid for $e \ll p$. With $p = 85000$ Pa, $\Delta s = 0.28/1000$ kg/kg per decade (from Fig. 3), the trend of the partial pressure of water vapor is determined by $\Delta e = 38.3$ Pa per decade for the NH. Thus, the trend of water vapor pressure contributes about $86\,\%$ to the total pressure trend of the NH.

2.3. Decomposition of the Annual Variation of the Global Mean Temperature

Firstly, it should be said that a decomposition of the annual variation of the global mean temperature into a earth-asymmetry component and a solar-radiation induced component might be impossible because of the complexity of climate feedback processes.

Assuming direct solar forcing without feedback (e.g., greenhouse effect), the temperature response to the annual variation of TSI is given by the simple radiative equilibrium of

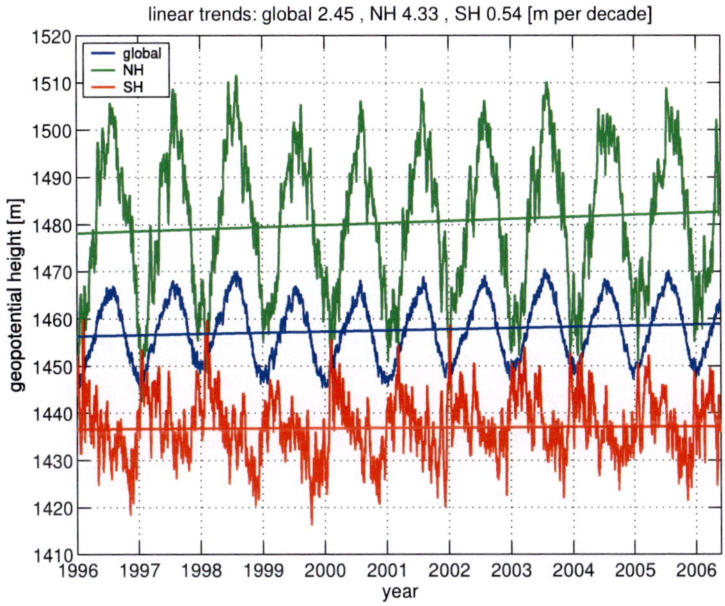

Figure 5. Least-square fits of straight lines to the time series of global mean (blue), NH mean (green), and SH mean (red) of geopotential height of the 850-hPa pressure level. The linear trends are annotated in the title line. Increase of geopotential height corresponds to an increase of the pressure at a fixed altitude (e.g., induced by more water vapor). Please note the short-term fluctuations of the NH and SH series, disappearing in the global mean series.

irradiance I and temperature T (Douglass et al. , 2004; Schneider , 1992)

$$\frac{\Delta T}{T} = \frac{\Delta I}{4I} \tag{3}$$

$$\Delta T(t) = 281.1\,^{\circ}\mathrm{K}\,\frac{\Delta \mathrm{TSI}_{true}(t)}{4 \cdot 1366\,\mathrm{W/m^2}} = 0.051\,^{\circ}\mathrm{K}\,\frac{\Delta \mathrm{TSI}_{true}(t)}{1\,\mathrm{W/m^2}} \tag{4}$$

281.1°K is the mean global mean temperature at 850 hPa. With the approximation $\Delta\mathrm{TSI}_{true}(t) \approx \mathrm{TSI} \cdot W(t) = W(t) \cdot 1366\,\mathrm{W/m^2}$, equation 4 is

$$\Delta T(t) \approx 281.1\,^{\circ}\mathrm{K}\,W(t)/4. \tag{5}$$

Inserting the annual amplitude of the geometrical dilution factor $W = 3.34\%$ or the average annual amplitude of TSI, gives a temperature amplitude of around 2.3 ^{o}K. The annual temperature variation due to solar forcing (sun component) is shown as the red line in Fig. 6). The phase lag of the temperature variation at 850-hPa is assumed to be around 1 month with respect to the solar signal. The small phase lag is neglected in Fig. 6. The total temperature oscillation (blue line in Fig. 6) is obtained by subtraction of the linear trend from the series of global mean temperature (as previously shown in Fig. 2). Now we can estimate the topography-induced component of the annual temperature variation by subtraction of the red line from the blue line in Fig. 6 yielding the green line (the earth component).

The earth component exceeds the sun component by 1.5-2 °K. The earth component is maximal around July and minimal around January. This is possibly due to the inter-hemispherical asymmetry of the Earth's topography, inducing a stronger convection of water vapor during NH summer than during SH summer.

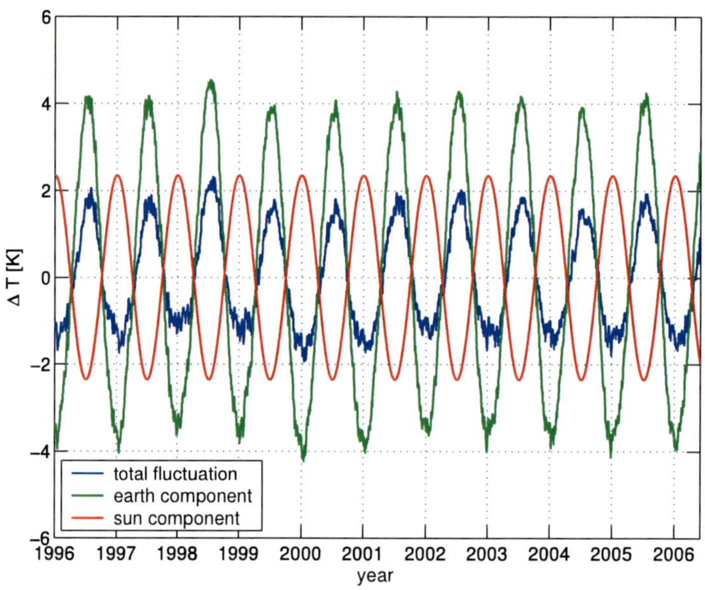

Figure 6. Decomposition of the temporal fluctuations of the global mean temperature at 850 hPa (blue line). The red line is the estimated temperature fluctuation solely due to the annual variation of the insolation (true TSI). The green line is the difference between the blue and the red line. The green line is the estimated, annual variation of the global mean temperature as caused by the inter-hemispherical asymmetry of the Earth's topography (e.g., stronger water vapor convection and meridional circulation during NH summer compared to SH summer).

Since the Climate Diagnostic Center (NOAA CDC) provides wind, humidity, sea surface temperature, and temperature maps of the Earth atmosphere, we have all informations to retrieve the details of the atmospheric processes induced by the Earth's asymmetry. The trends of the wind circulation have not been investigated here but the atmospheric water cycle influences and depends on the wind circulation. Investigation of the forcing of the middle and upper atmosphere by the Earth's asymmetry is also interesting. Other questions are: Can advanced atmospheric circulation models provide the green line of Fig. 6 if the Sun-Earth distance is assumed to be constant in the model world? Was the decomposition analysis of our study wrong?

3. Forcing of the Earth's Atmosphere by Solar Flares

3.1. Consequences of a Solar Flare

Solar flares are major explosions in the Sun's atmosphere, releasing large amounts of stored magnetic energy within time scales from seconds to tens of minutes. The energy release is usually associated with strong X-ray bursts (enhancements of the X-ray flux by a factor of > 10) and moderate enhancements of the solar EUV/UV radiation (several percents) due to faculae brightening. The flares are usually associated with dark solar spots causing sun spot darkening of the solar radiation at optical and infrared wavelengths (Lean et al. , 2005).

Besides these variations of the solar electromagnetic spectrum, the flares can cause major disturbances of the solar wind. Strong flares are commonly accompagnied by interplanetary coronal mass ejections (ICME) leaving the Sun and traveling at high speeds through the interplanetary space. If such solar wind disturbances hit the Earth's magnetosphere, the solar wind variations induce electric currents and energetic particle precipitations within the Earth's magnetosphere and ionosphere. Ground-based magnetometers measure the magnetospheric and ionospheric fluctuations as strong increase of temporal geomagnetic fluctuations which are scaled by the index of the planetary geomagnetic activity A_p.

3.2. Superposed Epoch Analysis of the X-ray Flux of Solar Flares

The superposed epoch analysis is a data analysis tool which can reveal typical characteristics, time schedules, periodicity, precursors, and consequences of a special event. The superposed epoch analysis is often used in solar-terrestrial physics, e.g., for the analysis of the response of mesospheric winds to geomagnetic storms (Ma et al. , 2001). For investigation of solar flares, the sudden burst of X-ray flux is an adequate timing mark. The time point or timing mark of the X-ray burst is clearly defined by the sudden increase of X-ray flux. This timing mark shall correspond to the epoch time 0, and all observed features which are associated with this event can be measured in days of epoch time before and after the X-ray burst happened.

The GOES satellites in geostationary orbits continuously measure the solar X-ray flux at wave lengths from 1 to 8 Å. Using the time series of GOES-8 and GOES-10, we identify 129 major X-ray bursts within the time from January 1996 to January 2006 which is roughly the solar cycle 23. We are using the hourly averages of the X-ray flux as provided by the Space Physics Interactive Data Resource (SPIDR, `http://spidr.ngdc.noaa.gov/`).

The 129 X-ray bursts are automatically selected by using the condition that the X-ray burst must exceed the long-term, background X-ray flux by a factor of 20, and this must happen from one hour to the next. The timing marks of the X-ray bursts are set to the epoch time 0, and all selected data segments of X-ray flux can be added and averaged, resulting in one typical function of X-ray flux in dependence on the epoch time. The average behaviour of the solar X-ray flux before, during, and after a major solar flare is depicted in Fig. 7a) for a sampling rate of 1 hour, while Fig. 7b) shows the moving daily average of the series. Fig. 7a) shows a strong, average increase of the X-ray flux by a factor of 100 at epoch time 0. For further details the averaged time series of Fig. 7b) is better. The major X-ray burst is followed by a smaller burst 16 days later. Of course this is not a fixed rule. The gray lines are denoting the error of the mean, and the standard deviation of the flare ensemble is greater by a factor of 11. There are further indications of enhanced solar activity 90 days before and 135 days after the major flare. It is known that solar activity has periodicities on such time scales (Bai and Sturrock , 1993). This is partly influenced by the solar rotation period, e.g., a flare and sun spot region is again active after some rotations of the Sun (one rotation takes around 27 days from the view point of the Earth). On the other hand, the typical periodicities of solar activity often differ from one solar cycle to the next.

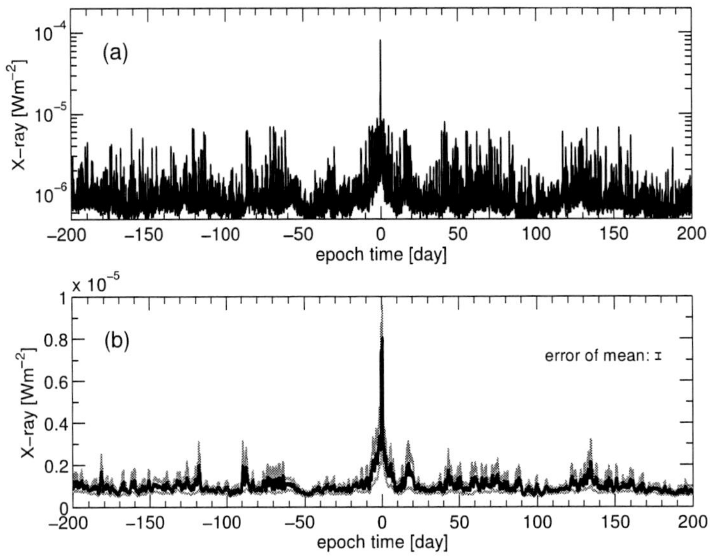

Figure 7. (a) Superposed epoch analysis of 129 X-ray bursts related to major solar flares of the solar cycle 23 (composed from hourly data of solar X-ray measurements of the GOES satellites) (b) Same as (a) but timely averaged over 1 day. The gray lines are showing the range of the mean error. Obviously, the main flare at day 0 is accompagnied by enhanced solar activity around days -90, 16, and 135.

3.3. Superposed Epoch Analysis of TSI and the Solar EUV/UV Activity

We use a merged time series of the total spectral irradiance TSI (at 1 astronomical unit) from NOAA and SORCE. The TSI data are interpolated to a series of daily averages from 1994 to 2007. Temporal variations with periods greater than 300 days are removed by high pass filtering (the digital non-recursive, finite impulse response (FIR) high pass filter performs zero-phase filtering by processing the time series in forward and reverse direction). Using the timing marks of the 129 major X-ray bursts, we select the relevant data segments of the Δ TSI series and compute their average as function of the epoch time. The typical behaviour of the TSI variation is depicted in Fig. 8. A significant depletion of TSI occurs around 5 days before the flare. The depletion is around -0.22 W/m^2 corresponding to a relative variation of -0.16 per mille. The depletion is mainly due to sun spot darkening which occurs at optical and infrared wavelengths. Lean et al. (2005) illustrate the wave length dependence of sun spot darkening and faculae brightening for the impressing case of the October 2003 solar storm.

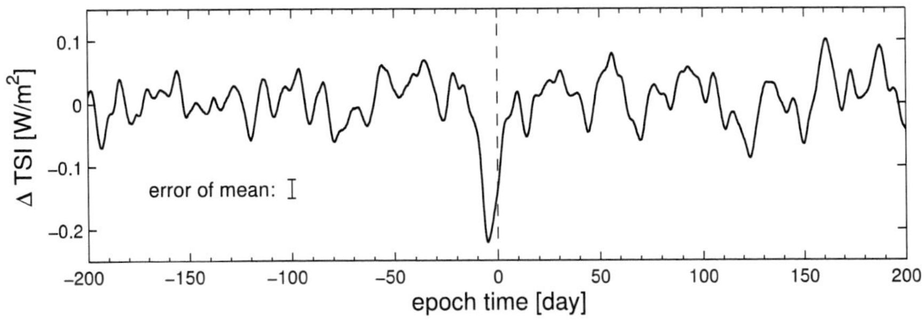

Figure 8. Superposed epoch analysis of the variation of the total solar irradiance (combined TSI series of NOAA and SORCE). The timing marks of the 129 X-ray bursts (Fig. 7) have been used for selection of the relevant data segments of the TSI series. Sun spot darkening seems to precede the flare event.

Faculae brightening takes place in the Sun's atmosphere at EUV and UV wavelengths. The Mg-II index is a measure of the solar EUV/UV activity, and NOAA provides a long-term series of daily averages (composed by R. Viereck). Using the same analysis method for the Mg-II series as for the TSI series, the relative variation of the Mg-II index can be regarded as an estimate of the relative EUV/UV emission. The averaged data segments of the relative Mg-II series are shown as function of epoch time in Fig. 9. Between 50 and 10 days before the flare event, the solar EUV/UV emission is a bit decreased. Then a brightening occurs with a maximum on Day 0, coincident with the X-ray burst but wider in time. After the flare event the solar EUV/UV emission is modulated with a period of 27 days which is the solar rotation period. The flare-induced maximum of the EUV/UV emission is around 2 per mille which is around one magnitude greater than the sun spot-induced depletion of TSI.

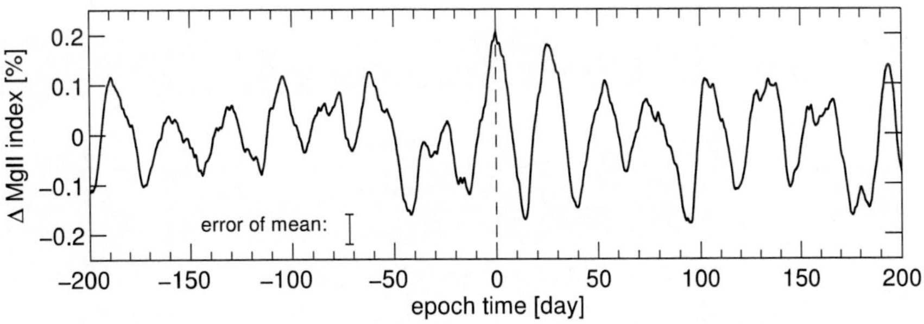

Figure 9. Variation of the Mg-II index which is a measure of the solar EUV/UV activity. The timing marks of the 129 X-ray bursts (Fig. 7) have been used for selection of the relevant data segments of the Mg-II series of NOAA. The EUV/UV emission is enhanced at the flare onset and repeats with a cycle of 27 days.

Individual, strong solar flares, faculae and sun spots are associated with relative depletions of TSI of up to 0.3% and relative variations of the Mg-II index of 2-3%. The EUV radiation ionizes atomic oxygen of the thermosphere, the UV radiation dissociates stratospheric ozone, and the optical and infrared radiation (main part of TSI) is absorbed by the Earth's surface and the lower atmosphere. This means that the flare-, faculae- and sun spot-induced perturbations vary in time, magnitude, and sign at different altitudes of the Earth's atmosphere.

3.4. Superposed Epoch Analysis of the Solar Wind Speed

The solar wind consists of protons, electrons, α-particles and other elements emitted by the Sun. The speed of the solar wind is measured by NASA's interplanetary monitoring platforms (IPM) in geocentric orbits at around 35 Earth's radii (so the satellites are often outside of the Earth's magnetosphere). Solar wind speed data and other parameters are provided by the OMNIweb (`http://omniweb.gsfc.nasa.gov/`). Here we use the daily time series of solar wind speed which we interpolate to a hourly grid. Then we select the data segments from the solar wind speed series by using the timing marks of the 129 major X-ray bursts. Again the data segments are averaged, and the typical series of solar wind speed as function of the flare epoch time is depicted in Fig. 10. Though the time resolution of the solar wind speed series is not so good (daily averages), the response of the solar wind at 1 astronomical unit is delayed by 1-3 days. Later we will precisely determine this delay by using the hourly series of geomagnetic activity. The average response of the solar wind to a flare is an enhancement of the speed by 60 km/s (around 10 %). Possibly this enhancement is higher if hourly instead of daily averages are used. A further enhancement is obvious around 154 days after the flare. This is in agreement with some other studies who reported about this periodicity in several Sun and solar wind parameters (Bai and Sturrock , 1993).

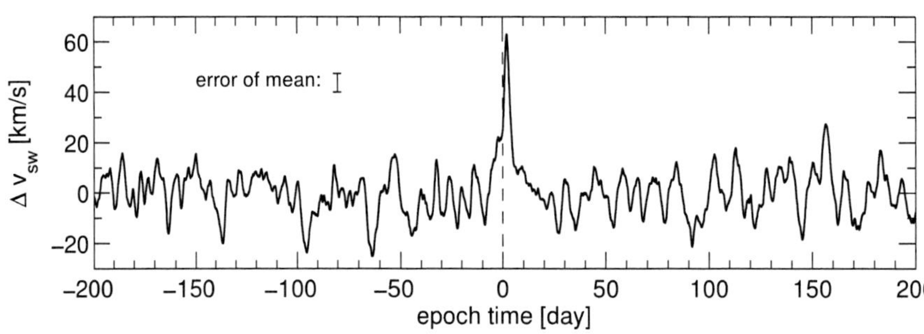

Figure 10. Superposed epoch analysis of the flare-induced behaviour of the solar wind speed at 1 astronomical unit (composed from daily averages measured by the interplanetary monitoring platform IPM). The timing marks of the 129 X-ray bursts (Fig. 7) have been used for selection of the relevant data segments of the solar wind speed series.

3.5. Superposed Epoch Analysis of the Geomagnetic Activity

The fast streams of the solar wind plasma induce electric currents and particle precipitations in the magnetosphere-ionosphere-thermosphere system of the Earth. The correlation between variations of solar wind speed and geomagnetic activity is known to be very high ($r \approx 0.7$) (Snyder et al. , 1963).

Ground-based magnetometers record time series of geomagnetic field intensity and field direction. The average intensity of the short-term fluctuations of the geomagnetic field is routinely derived from a worldwide network of magnetometers providing a time series of the planetary geomagnetic activity A_p. In the following we use the hourly A_p values from the Space Physics Interactive Data Resource (SPIDR, `http://spidr.ngdc.noaa.gov/`).

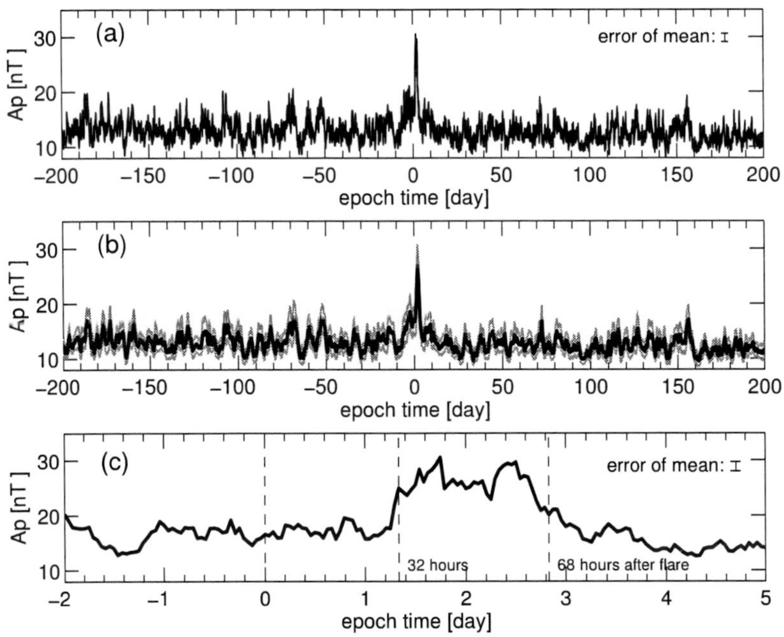

Figure 11. (a) Superposed epoch analysis of geomagnetic activity Ap using the 129 timing marks of solar X-ray bursts. (b) Same as (a) but timely averaged over 1 day. The gray lines are showing the range of the mean error. (c) The detailed view indicates an average increase of geomagnetic activity around 32-68 hours after the flare.

The timing marks of the 129 solar X-ray bursts are utilized for picking out the relevant data segments of the A_p series. These data segments are averaged as function of epoch time, and the result is depicted in Fig. 11a) with hourly sampling rate and in b) after application of a moving daily average. Shortly after the flare event the geomagnetic activity index A_p increases by more than 100%. A smaller peak is visible around 154 days after the flare which is in agreement to the peak of the solar wind in Fig. 10.

Fig. 11c) zooms the time interval from 2 days before to 5 days after the flare. The increase of geomagnetic activity starts around 32 hours (1.3 days) after the X-ray burst and stops around 68 hours (2.8 days) after the X-ray burst. The average A_p is greater than 20 nT during this time interval. Using the arrival and stop times of the enhanced geomagnetic activity we can calculate the average, expansion speed of the solar wind disturbance (ICME) on its travel from the Sun to the Earth. Generally ICMEs travel along curved paths from the

Sun to the Earth. The radial velocity v_r of the ICME (straight line propagation) is highly correlated ($r = 0.86$) with the true expansion velocity v_{exp} of the ICME (curved path), and the following relationship has been derived (Schwenn , 2006)

$$v_{exp} \approx \frac{v_r}{0.88} \approx \frac{1\text{a.u.}}{0.88\Delta t} = \frac{170 \cdot 10^6 \text{ km}}{\Delta t}. \tag{6}$$

Inserting the arrival and stop times derived from Fig. 11c), the expansion velocities v_{exp} of the flare-induced ICMEs of the solar cycle 23 are ranging from 690 to 1480 km/s. The average expansion velocity is around 1100 km/s. These values agree well with Fig. 35 of Schwenn (2006) who determined travel times and expansion velocities of ICMEs by means of detailed case studies.

3.6. Response of the Lower Troposphere to Solar Flares

Since the authors just started with the investigations of the impact of solar flares and solar wind disturbances on the lower and middle atmosphere, we cannot provide a comprehensive summary of the wide field of solar-terrestrial relationships but we will give a motivating example, showing a significant variation of the atmospheric wave activity at periods 5-10 days. Again, we use the superposed epoch analysis for relating the change of lower tropospheric wave activity to the solar flare event.

Let's go back to the time series of geopotential height of the 850-hPa pressure level in Fig. 5. The short-term fluctuations (period range: 2-40 days) of the geopotential height series of NH and SH are remarkably large. The NH and SH fluctuations must be highly anti-correlated since the short-term fluctuations of the global mean series are much smaller. Data segments of solar X-ray flux and geopotential height variations are depicted in Fig. 12 for the time interval from September 1, 2004 to March 1, 2005 with extremely high solar activity and a giant flare on January 20, 2005. The SH geopotential height increase by around 20 m, while the NH geopotential height decreases by the same amount which corresponds to a pressure variation of around 2.1 hPa (equation 2). It is special that the green and the red line come exactly together after the giant flare on January 20, 2005 (of course, it can be a conjunction without a causal relationship).

For understanding the nature of the short-term fluctuations, we recall that the time series are composed of daily averages, and the parameters are zonally averaged. Migrating tides and traveling planetary waves with zonal wave numbers greater than zero are surpressed due to this averaging processes. We suggest that the remaining short-term fluctuations of the NH and SH series indicate meridional atmospheric flow from the Northern to the Southern hemisphere and vice versa. Ebel et al. (1977) found evidence of a 7.5-day global-scale wave with zonal wave number zero in stratospheric ozone observations. Like Ebel et al. (1977), we will concentrate on the 5- to 10-day waves which might be related to the crossings of the interplanetary sector boundaries of the heliospheric magnetic field.

We start with band-pass filtering the NH and SH series of geopotential height. The period range of the band pass (non-recursive, finite impulse response, zero-phase filtering) is from 5 to 10 days. We also tried other period ranges but the results are best for the waves at a period of 7.5-day. The band-pass filtered fluctuations of the NH and the SH series are squared, yielding series of NH variance and SH variance. Now we can take the average

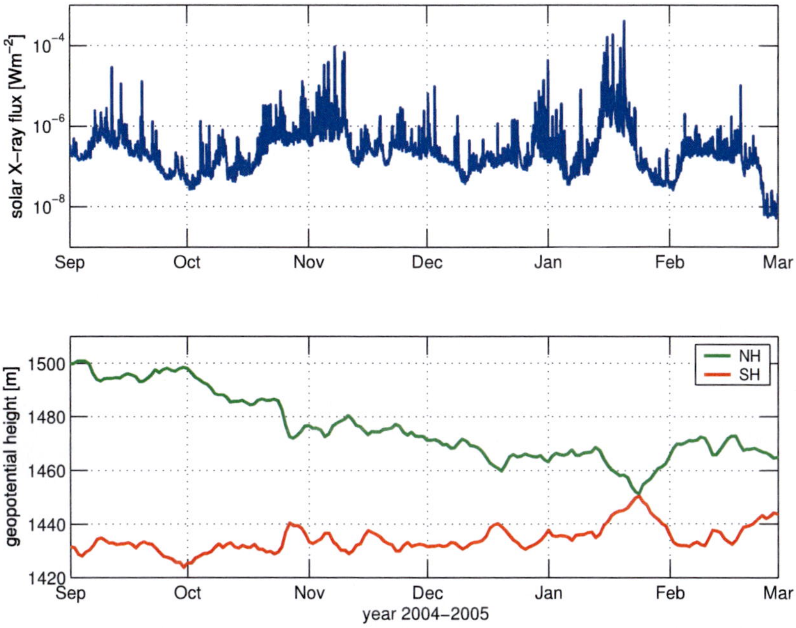

Figure 12. Time series of the solar X-ray flux as observed by the GOES-10 satellite. The time interval from September 1, 2004 to March 1, 2005 shows strong solar activity with major flares around November 2004 and a super flare on January 20, 2005. The time series of geopotential height of the 850-hPa level (from daily averages of NCEP/NCAR reanalyses) are separately shown for the Northern and the Southern Hemisphere. The temporal fluctuations of the NH and SH series are highly anti-correlated. A relationship between the variations of the solar activity and the geopotential height seems to exist.

of both series, and apply the superposed epoch analysis method to this series. Again, the timing marks of the major solar X-ray bursts are utilized for selection of the relevant data segments of the atmospheric variance series.

We should note that the variance series are a bit shorter than the A_p series. The variance series start in January 1996 and stop in June 2006 (the interval selection was influenced by minor internet problems: downloading of a huge 10-year series was possible at Climate Diagnostic Center but problems occurred for longer time intervals). The consequence of the shorter time interval is that the superposed epoch analysis is performed for 118 instead of 129 solar flares.

The flare-related behaviour of the average variance of the 7.5-day wave is depicted in Fig. 13. A similar curve can be obtained, separately for the NH and SH, or when the time interval is changed. 5 days before the X-ray burst, the 7.5-wave variance increases by around 30%. The maximum of geopotential height variance is around 0.8 m^2 corresponding to a pressure amplitude of 9 Pa (equation 2). The temporal behaviour of the atmospheric wave variance is quite similar as for the total solar irradiance TSI which showed a depletion due to sun spot darkening 5 days before the flare. Further analysis is required to verify this assumption.

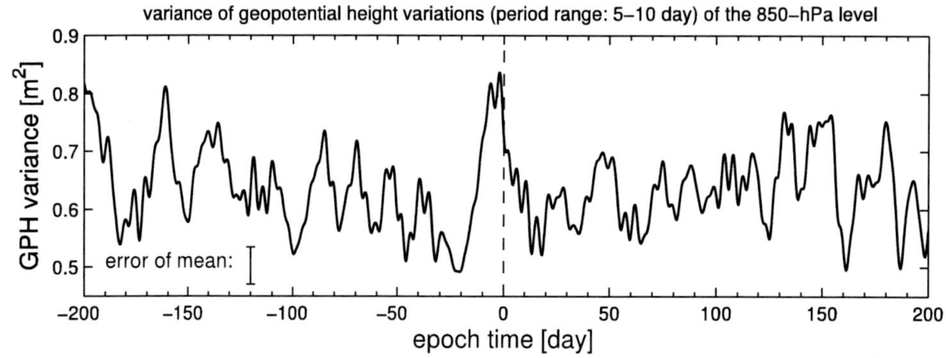

Figure 13. Superposed epoch analysis of the flare-induced behaviour of the 7.5-day wave variance of geopotential height of the 850-hPa level (from daily averages of NCEP/NCAR reanalyses). The timing marks of 118 major X-ray bursts (Fig. 7) have been used for selection of the relevant data segments of the atmospheric variance series.

On the other hand, the heliospheric magnetic field evolves in response to the photospheric field at its base. Thus the heliospheric magnetic field in vicinity of the Earth might be closely connected to the strong magnetic fields of the sun spots. The variance of the 7.5-day wave could be induced or influenced by the crossings of the interplanetary sector boundaries of the heliospheric magnetic field having a similar period. This assumption sounds a bit sophisticated but it is supported by a second feature of Fig. 13. Enhancements of the atmospheric variance appear 154 days after the flare, which seems to be related to a similar behaviour of the solar wind speed and the geomagnetic activity in Figs. 10 and 11.

It is going beyond the aims of the present study to decide which kind of solar forcing is responsible for the 7.5-day atmospheric wave variance: forcing by solar electromagnetic radiation or forcing by solar wind structures and the heliospheric magnetic field. Further analyses of solar, geophysical and meteorological data sets, considerations of additional aspects (e.g., cosmic ray flux), and modeling of the Sun-Earth system have to be performed before coming to explanations.

4. Conclusion

Our investigations have been helpful for us - and hopefully for the reader too - for understanding of the response of the Earth's atmosphere to variations of the solar radiation and the solar wind. The importance of the inter-hemispherical asymmetry has been estimated and visualized for the annual variation of the global mean temperature of the lower troposphere. It seems that enhanced water vapor loading during NH summer explains the inter-hemispherical asymmetry of the lower troposphere. The Earth's asymmetry compensates the effect of the annual variation of the Sun-Earth distance and induces higher water vapor concentration, pressure, and temperature in the NH compared to the SH.

The analysis of the NCEP/NCAR series shows a strong increase of the lower tropospheric temperature during the last decade, due to increasing concentrations of anthropogenic greenhouse gases such as CO_2 and feedback processes by the natural greenhouse gas H_2O. The trends of temperature, water vapor, and geopotential height (pressure) are greater for the NH than for the SH. The positive trend of the atmospheric pressure (at 1500 m altitude) of the NH is mainly explained by the increase of the partial pressure of water vapor. The trend of water vapor pressure contributes about 86 % to the total pressure trend of the lower troposphere of the NH.

It is likely that atmospheric anomalies/disturbances/trends 'propagate' upward into the middle and upper atmosphere. Additionally the upper atmosphere is strongly forced by solar flares, solar wind disturbances, and the heliospheric magnetic field. The upward wave flux from the lower atmosphere can be interrupted by wave-mean flow interactions in the middle and upper atmosphere. This process could have a significant impact on the circulation of the lower atmosphere. The method of superposed epoch analysis was successfully applied to solar, geophysical and meteorological data sets. The typical sequence of solar-flare related processes is: 1) sun-spot darkening in TSI, 2) X-ray burst and faculae brightening in EUV/UV, 3) ICME arrival at the Earth and enhanced geomagnetic activity. The Earth's troposphere seems to react before the solar X-ray burst because of sun-spot darkening in TSI and effects related to the structure of the heliospheric magnetic field. Further activities are planned for investigation of the response of ozone, water vapor, tidal waves, etc. to variations of the Sun's radiation, the solar wind, the heliospheric magnetic field, and the galactic cosmic ray flux.

Acknowledgments

We thank Professor Christian Mätzler for improvements and corrections. The data sets have been perfectly provided by the Space Physics Interactive Data Resource (SPIDR, http://spidr.ngdc.noaa.gov/), NASA's OMNIweb (http://omniweb.gsfc.nasa.gov/), and NOAA's Climate Diagnostic Center (http://www.cdc.noaa.gov/). Particularly we appreciate the free use of the NCEP/NCAR reanalyses, data of the satellites GOES, IMP, SORCE, and A_p-values from the network of ground-based magnetometers.

References

T. Bai and P. A. Sturrock. Evidence for a fundamental period of the sun and its relation to the 154 day complex of periodicities, *Astrophysical Journal*, **409**, 476-486, 1993.

David H. Douglass, B. David Clader, and Robert S. Knox. Climate sensitivity of Earth to solar irradiance: update, *SORCE meeting on Decade Variability in the Sun and the Climate*, Meredith, New Hampshire, October 27-29, 2004.

A. Ebel, A. Ghazi, and W. Bätz. Evidence of global-scale waves with zonal wave number zero in the stratosphere. *Pure and Applied Geophysics* **116**, 8-31, 1977.

Greg Kopp, George Lawrence, and Gary Rottman. The Total Irradiance Monitor (TIM): Science Results, *Solar Physics*, **230**, 1, 129-140, 2005.

Judith Lean, Gary Rottman, Jerald Harder, and Greg Kopp. SORCE Contributions to New Understanding of Global Change and Solar Variability, *Solar Physics*, **230**, 1, 27-53, 2005.

G. Ma, K. Igarashi, and K. Hocke. Mid-latitude winds in the mesosphere: a superposed epoch analysis over the geomagnetic storm times, *J. Atmos. Solar-Terr. Phys* **63**, 1993-2001, 2001.

S. G. H. Philander, D. Gu, D. Halpern, G. Lambert, N.-C. Lau, T. Li, and R. C. Pacanowski. Why the ITCZ is mostly north of the equator, *Journal of Climate* **9**, 2958-2972, 1996.

S. H. Schneider. Introduction to climate modeling, *in Climate system Modeling edited by K. E. Trenberth*, Cambridge University Press, 1992.

Rainer Schwenn. Space Weather: The Solar Perspective, *Living Rev. Solar Phys.* 3, http://www.livingreviews.org/lrsp-2006-2, 2006.

C. W. Snyder, M. Neugebauer, and U. R. Rao. The solar wind velocity and its correlation with cosmic ray variations and with solar and geomagnetic activity. *J. Geophys. Res.* **68**, 6361-6370, 1963.

K. E. Trenberth and C. J. Guillemot. The total mass of the atmosphere. *J. Geophys. Res.* **99**, 23079-23088, 1994.

K. E. Trenberth, J. Fasullo and L. Smith. Trends and variability in column-integrated atmospheric water vapor. *Climate Dynamics* **24**, 741-758, 2005.

W. M. Washington. Climate-model responses to increased CO_2 and other greenhouse gases, *in Climate system Modeling edited by K. E. Trenberth*, Cambridge University Press, 788 pp., 1992.

In: Solar Physics Research Trends
Editor: Pingzhi Wang, pp. 345-356 ISBN: 978-1-60021-987-0
© 2008 Nova Science Publishers, Inc.

Chapter 10

NUMERICAL STUDY OF FORMATION AND PROPAGATION OF GIANT ELECTRIC JETS BETWEEN A THUNDERCLOUD AND THE LOWER IONOSPHERE

Lizhu Tong* and Kenichi Nanbu
Institute of Fluid Science, Tohoku University, Sendai 980-8577, Japan

Abstract

The lightning from thundercloud toward the ground has been widely studied. It has been found, however, that more elusive lightning forms, *e.g.*, dubbed elves, sprites, and blue jets, flit above the thunderclouds. Recently, giant electric jets, which link the top of thunderclouds to the lower ionosphere, have been observed. These events are a new important factor in the model of the earth's electrical and chemical environment. However, the observations of giant electric jets are not enough to understand the factors causing the formation and propagation of the events because no associated cloud-to-ground (CG) lightning discharges were detected in the underlying thunderstorms. In this article, we use the numerical simulation method to study the formation and propagation of giant electric jets. The initiating conditions of streamer formation in laboratory situations are extended to construct a criterion for the formation of giant electric jets. The Monte Carlo technique is used to perform the microscopic estimation of the growth of electrons produced in giant electric jets. A randomly stepped three-dimensional (3D) model is developed to simulate the propagation of giant electric jets. The simulation presents a 3D overall picture for the well-developed giant electric jets, which agrees well with the observations.

Keywords: giant electric jet; initiation; propagation; modeling; simulation.

* E-mail address: lizhu-tong@uyemura.co.jp, Tel./Fax +81-72-834-3501

Introduction

Transient luminous events in the atmosphere were recently found in the region between thunderclouds and the ionosphere [1-4]. These phenomena have been divided into three groups based upon their altitude, width and duration. These are dubbed elves, sprites with and without sprite halos, and blue jets. The elves appear at the altitudes of ~70-90 km and extend over several 100 km horizontally. They are visible for less than 0.1 ms. Red sprites were found at the altitudes of ~55-80 km with narrow tendrils extending below 55 km. Their visible lifetime is from a few to some tens of ms. Sprite halos appear at the altitudes of ~70-85 km for a short time of ~1-2 ms, sometimes preceding the development of streamer structures of sprites. Blue jets propagate from cloud tops to an altitude of ~40 km with a lifetime of 300 ms. More recently a new discharge phenomenon, named as giant electric jets in this article, was discovered, which linked the top of thunderclouds to the ionosphere [5,6]. These events are potentially important factors in the model of the earth's electrical and chemical environment. The investigation of these events will help us to understand their contribution to the global electrical circuit [7].

Observations of giant electric jets are not enough to understand the factors causing the initiation and development of these events because no associated cloud-to-ground (CG) lightning discharges were detected in the underlying thunderstorm [6]. The initiating conditions and propagating mechanisms of giant electric jets are unknown. In this article, the initiation of giant electric jets is modeled based on the streamer discharge theory and is simulated by using the quasi-electrostatic (QE) model and the Monte Carlo technique. Using the fractal theory, a randomly stepped leader discharge model is proposed for simulating the propagation of giant electric jets from the top of thunderclouds. The simulation results are compared with the observational results.

Numerical Model

Quasi-Electrostatic (QE) Model

Since the extremely-low-frequency (ELF) radio waves associated with giant electric jets were detected in the observations [6] and a negative cloud to ionosphere (NCI) discharge could lead to the generation of these ELF waves, here we consider the initiation criteria for giant electric jets which are NCI discharges. The energy source causing giant electric jets is assumed to be due to the quasi-electrostatic (QE) field generated by the thundercloud charge, which is formed by a fast accumulation of a negative charge located at the top of thundercloud [8]. The thundercloud charge dynamics can be mathematically represented by [9]

$$Q(t) = Q_0 \frac{\tanh(t/\tau)}{\tanh(1)}, \qquad 0 < t < \tau \tag{1}$$

where Q_0 is the magnitude of thundercloud charge and τ is the duration for accumulation of thundercloud charge.

The two-dimensional cylindrical coordinate system (r,z) is used, the z axis representing altitude. The computational domain is $r \leq 60$ km and $0 \leq z \leq 90$ km. The quasi-electrostatic (QE) field is assumed as an axisymmetrical distribution. The boundaries at $z = 0$ km, $z = 90$ km, and $r = 60$ km are assumed to be perfectly conductive. The effect of the artificial boundary at $r = 60$ km on the quasi-electrostatic (QE) field is considered as small [10]. The charge conservation equation is represented as

$$\partial \rho / \partial t + \nabla \sigma \cdot \vec{E} + \rho \sigma / \varepsilon_0 = 0 \,, \tag{2}$$

where ρ is the charge density, σ is the conductivity, ε_0 is the dielectric constant, and t is the time. Observations have shown that the initiation altitude of giant electric jets is low, *i.e.*, ~18-24 km [6]. Thus, the total conductivity σ is dominated by ion conductivity, taken by $\sigma = 5 \times 10^{-14} e^{z/6\text{km}}$ S/m [11]. $\vec{E}$ is the electrostatic field governed by the Poisson equation $\nabla \cdot \vec{E} = (\rho + \rho_s)/\varepsilon_0$, where ρ_s is the thundercloud source charge density, which is assumed to be $\rho_{-}(r,z,t)$ with a Gaussian spatial distribution of $\rho_{-}(r,z,t) = \rho(t) e^{-[(z-z_{-})^2/a^2 + r^2/b^2]}$, z_{-} is the mean altitude of negative thundercloud charges, which is set 16 km from the observations of Su et al. [6]. a is set 2, 4, and 6 km as cases 1-3 and b is set 2 km. $\rho\,(t)$ is the charge density corresponding to $Q(t) = \int_V \rho_{-}(r,z,t) 2\pi r dr dz$, where V represents the whole computational domain. The ordinary finite difference method [12] is used to solve eq. (2). The distribution of the absolute values of charge densities over altitude at $t = \tau$ corresponding to cases 1-3 is shown in Fig. 1.

Monte Carlo Model

The one-dimensional Monte Carlo technique is used to study the kinetics of electrons produced in giant electric jets. The simulation starts at the time that electric field arrives at ionization threshold. Several hundred, or thousand, electrons are assumed as emitted from the top of thunderclouds. The motion for the electrons is simulated by the modified Verlet method [13]

$$x(t+\Delta t) = x(t) + v_x(t)\Delta t + F_x(t)(\Delta t)^2 / 2m \,, \tag{3}$$

$$v_x(t+\Delta t) = v_x(t) + [F_x(t+\Delta t) + F_x(t)]\Delta t / 2m \,, \tag{4}$$

where x and v_x are the position and velocity of electrons, respectively, m is the electron mass, Δt is the time step, and F_x is the force acting on electrons given by $F_x = q_e E_x$. Here q_e is electronic charge and E_x is electric field. For a mixture of N_2 and O_2, we consider 33 types of electron-molecule collisions. In addition to elastic, rotational, and vibrational excitations, we consider electronic excitations of $N_2(A^3\Sigma_u^+)$, $N_2(B^3\Pi_g)$, $N_2(W^3\Delta_u)$, $N_2(B'^3\Sigma_u^-)$, $N_2(a'^1\Sigma_u^-)$, $N_2(a^1\Pi_g)$, $N_2(w^1\Delta_u)$, $N_2(C^3\Pi_u)$, $N_2(E^3\Sigma_g^+)$, $N_2(a''^1\Sigma_g^+)$, and 13 eV, ionizations of $N_2^+(X^2\Sigma_g^+)$, $N_2^+(A^2\Pi_u)$, and $N_2^+(B^2\Sigma_u^+)$ for N_2, and electronic excitations of $O_2(a^1\Delta_g)$,

$O_2(b^1\Sigma_g^+)$, $O_2(c^1\Sigma_u^-)$, 6.0 eV, 8.4 eV, and 10 eV, ionizations of $O_2^+(X^2\Pi_g)$, $O_2^+(a^4\Pi_u)$, $O_2^+(A^2\Pi_u)$, $O_2^+(b^4\Sigma_g^-)$, and $O_2^+(^4\Sigma,^2\Sigma_s)$, dissociative attachment, and dissociative excitation for O_2. The corresponding cross sections for these reactions are taken from refs. [14-16].

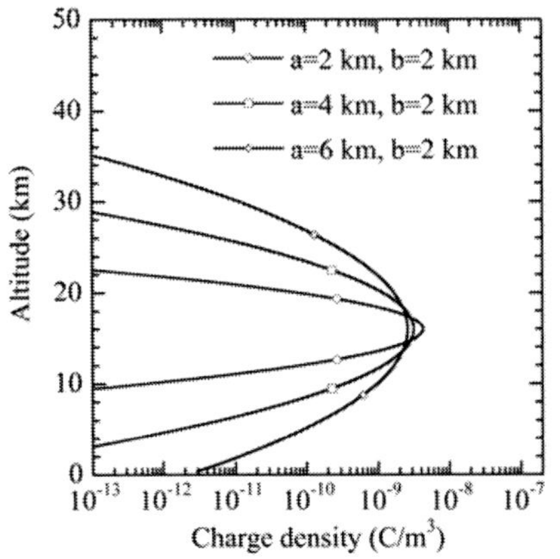

Figure 1. Distribution of the absolute values of charge densities simulating thundercloud charges.

For the estimation of electron-molecule collisions, Nanbu proposed a simple method to simultaneously determine whether an electron collides and which collisional event occurs in the case of collision [17]. By the method, we obtain the collision probability of the ith electron for the kth collisional type in $\triangle t$

$$P_k = n_n \sigma_k(\varepsilon_i) v_i \Delta t \text{ , } (k = 1,2, \ldots, K) \tag{5}$$

where, ε_i and v_i are the energy and speed of the ith electron, $\sigma_k(\varepsilon_i)$ is the cross section of the ith electron for the kth type collision, n_n is the density of neutral gas, N_2 or O_2, and K is the total number of collisional types, *i.e.*, $K = 33$ in the present work. The total probability that an electron collides with a molecule in $\triangle t$ is

$$P_T = \sum_{k=1}^{K} P_k \text{ .} \tag{6}$$

Equation (6) is written again as

$$1 = P_T + (1 - P_T) = \sum_{k=1}^{K} \left[P_k + \left(\frac{1}{K} - P_k \right) \right] . \tag{7}$$

Based on eq. (7), the method to sample a collisional event is shown in Fig. 2. The unit length is divided into K equal intervals and each interval is divided into two. We call a uniform random number U ($0 < U < 1$). The integral part of KU+1 is the number of the kth interval in which U lies. The left part of the kth interval is regarded as $1/K$-P_k and its right part is P_k. If U lies in P_k, the kth event occurs; otherwise the particle does not collide.

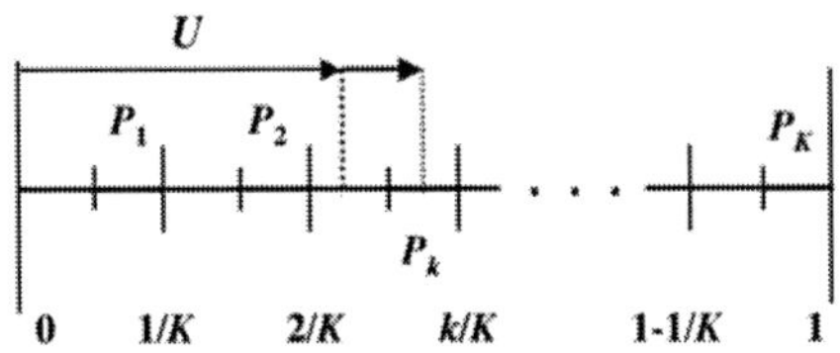

Figure 2. The method to sample a collisional event.

When a collision occurs, the post-collision velocity $\vec{v}'$ of electrons can be determined by

$$\vec{v}' = \left(\frac{m_e}{M_g} + \cos\chi\right)\vec{v} + \vec{h}\sin\chi, \tag{8}$$

where $\vec{v}$ is the pre-collisional velocity of electron, m_e is the mass of electron, M_g is the mass of gas molecule (N_2 or O_2), and χ is the scattering angle. The Cartesian components of $\vec{h}$ are

$$\begin{aligned} h_x &= v_\perp \cos\varphi \\ h_y &= -(v_x v_y \cos\varphi + v v_z \sin\varphi)/v_\perp , \\ h_z &= -(v_x v_z \cos\varphi - v v_y \sin\varphi)/v_\perp \end{aligned}$$

where $v_\perp = (v_y^2 + v_z^2)^{1/2}$ and $\varphi = 2\pi U$, where U is a uniform random number between 0 and 1. Regarding the determination of χ, please refer to ref. [18]. For inelastic collisions, the post-collision velocity $\vec{v}'$ given in eq. (8) should be replaced by $(1-\varepsilon_k/\varepsilon)^{1/2}\vec{v}'$, where ε and ε_k are the electron energy and the loss energy by excitation or ionization, respectively. It is assumed that the new produced electron by impact ionization possesses the energy being equal to its parent electron and has an isotropic velocity distribution.

Streamer Model

In order to generate a streamer in air there are three conditions to be fulfilled [19]. They are

1) The electric field must exceed the ionization threshold;
2) The initial plasma patch is sufficiently ionized so that the electric field generated by the space charges is comparable with the external field itself;
3) A seed of free electrons is required to start the ionization.

A streamer is a self-adjusting object. It produces charged particles in a high field region at the front, needed to maintain high plasma conductivity in the channel. The electric field at streamer tip is controlled by the produced space charge. This property allows streamers to develop in weak fields, much lower than the ionization threshold of air. A minimum electric field strength exists to support stable streamer propagation and it is known that the minimum field corresponding to negative streamer propagation is ~2-3 times greater than that corresponding to the positive streamers [20,21]. In the present work, we focus our research on giant electric jets which are NCI discharges. The minimum field $-E_c$ required for propagation of negative streamers in air at atmospheric pressure is about –12.5 kV/cm [21], which has been used to study blue jets [22]. It is known that the electric field of $E_0 = 3.14\times10^6\ \mathrm{V\cdot m^{-1}}$ causes a breakdown of an air gap of 1 cm at atmospheric pressure and produces an effective ionization coefficient $\alpha_{\mathrm{eff}} = 1.24\times10^3\ \mathrm{m^{-1}}$ [19]. The electric field for ionization threshold is estimated by $E_k = E_0\cdot(N/N_0)$, where $N_0 = 2.688\times10^{25}\ \mathrm{m^{-3}}$ and N is the number density of air molecules.

It is assumed that the electric field inside a plasma patch generated by electrons and ions is equivalent to that caused by the electric dipole such that positive and negative charges locate at the centre of positive ions and electrons, respectively. The electric field in the middle of the electric dipole can be simply given by

$$E_{d/2} = \frac{q}{2\pi\varepsilon_0(d/2)^2}, \tag{9}$$

where d is the distance of electric dipole, ε_0 is the dielectric constant, and q is the charge of the electric dipole. It is known that for the streamer corona, in spite of its internal structural complexity involving multiple highly branched streamer channels, its macroscopic characteristics remain relatively stable under a variety of external conditions. Therefore, a streamer can be considered to be initiated if the electric field generated by electrons and ions makes the electric field inside the plasma patch reduce from E_k to $E_c\cdot(N/N_0)$. Following similarity relations, we obtain the minimum number $N_{\min}$ of electrons generating a streamer in upper atmosphere by

$$N_{\min} = 8.68\times10^7(E_0 - E_c)\cdot(N/N_0)\cdot d^2 \tag{10}$$

The third condition to initiate a streamer can be naturally fulfilled in the domain close to the top of thundercloud if we consider giant electric jets which are negative cloud to ionosphere (NCI) discharges. The negative cloud may be regarded as a cathode that could emit electrons. It is assumed that electrons are emitted from the top of thunderclouds and the Monte Carlo calculation is terminated once the electron number exceeds $N_{\min}$. After that, a streamer will be initiated to form a gigantic get.

Leader Model

The physical processes underlying leader breakdown consist of igniting electrons, leading an avalanche to streamer corona criterion, and evolving streamer corona to leader channel. The formation and development of streamer corona is a part of the leader process. It has been indicated in streamer modeling that giant electric jets are initiated by negative streamers. Thus, the critical field E^* governing a leader in air can be considered to be equal to the field over the streamer zone in the leader. This can be deduced by such a fact that the development of a streamer discharge precedes the formation of a leader. The condition for a streamer-leader transition can be satisfied if the electric field over the length of the streamer zone in a leader exceeds the critical value [23,24]. Since the electric field decreasing with upper distance from the thundercloud is lower than the exponential decrease of the pressure with altitude, the condition for a streamer-leader transition can be substantially satisfied to initiate an upper lightning discharge. Therefore, the critical electric field for a leader is estimated by the minimum electric field required for streamer propagation [23]. E^* is set equal to E_c, which is used in streamer modeling and can be simply scaled with altitude proportionally to the neutral atmospheric density. It is assumed that as soon as the electric field around leader tip reaches the E^* field, the leader process is developed. The complex process of stream-to-leader transition is omitted in the present model.

In stepped leaders, a leader channel is formed from the strongest streamer in the corona at leader tip by a complicated transition process which involves heating and ionization of gas. It is well known that a streamer corona is formed when the field of the space charge increases to the level of the external field, which will occur when a critical number of electrons, $N_e \sim 10^8$-10^9 at atmospheric pressure, is produced in an avalanche. The critical condition can be represented by an empirical relation $\alpha x \approx 20$, where x is the length of streamer corona and α is the ionization coefficient that can be given by another empirical formula $\alpha / p = 1.17\times10^{-4}(E/p-32.2)^2\text{cm}^{-1}\cdot\text{Torr}^{-1}$ for $E/p \approx$ 44-176 V·cm^{-1}·Torr^{-1}, where p is the gas pressure [20]. Once E/p is given, *e.g.*, in the case of the critical situations to form a streamer corona, we have $\alpha \sim p$. It is known that the change in density with altitude is approximately exponential, *i.e.*, $\rho(z) = \rho(0)e^{-\delta z}$, where δ (=1.25×10^{-4}) is the scaling parameter and z is the altitude above sea level in meters [25]. The streamer corona (x) under the critical situations becomes longer when α reduces with the decrease of pressure with altitude, *i.e.*, $x \sim \alpha^{-1}$. The leader length is, however, longer than streamer zone for long leaders, so the length L of leader channel can be set $L(z) = L(0)e^{\gamma\cdot\delta z}$, where γ is the parameter that characterizes the variation of leader channel, which is determined by the electric field in leader channel discussed below. It is assumed that the change of the electric field in leader channel with leader length is linear for long leaders, *i.e.*, linearly decreasing as the increase of leader length [26]. Thus, we have $E_l(z) = E_l(0)[L(0)/L(z)] = E_l(0)e^{-\gamma\cdot\delta z}$. Assigning $\beta = \gamma\,\delta$, we obtain $E_l(z) = E_l(0)e^{-\beta z}$, where $E_l(0) = \sim 1$ kV/cm at atmospheric pressure [20] and β is determined in such a way that $E_l(z)$ located in thunderclouds is consistent with those estimated in the previous research, *e.g.*, ~0.2-0.5 kV/cm [27].

The propagation of giant electric jets can be assumed as a field controlled random growth process, which is a discrete process. The thundercloud is considered as one electrode igniting gigantic jets and the ionosphere is assumed as the other. The leader propagation is schematized as a sequence of connections between the points of a spatial Cartesian lattice [28]. The solution of electric field is divided into two stages: one is for the accumulation of the thundercloud charge and the other is for the propagation of the leader. The former is calculated until the arrival of ionization threshold. Once the initiation condition given in eq.(10) is satisfied, the point at which field value reaches ionization threshold is considered as the initiating point of a leader. The potential of this initiating point is then fixed and the discharge is propagated by adding additional links.

From the initiation of the leader, the three-dimensional electrostatic field is solved so as to satisfy random spatial trajectories of leader propagation. The equation governing three-dimensional electrostatic field is considered as Laplace equation. The field consists of moving boundaries, following leader propagation, which is determined at each growth step of the leader. At every growth step, a new grid point is linked with the leader channel, which is regarded as an imposed boundary for the recalculation of the electrostatic field. The grid points located in leader channels are assumed to retain the potential which they have acquired until the end of the simulation. The potential drop along leader channels is determined by the electric field E_l of leader channel. Next growth point of the leader is randomly chosen from the neighboring points around the leader tip. A three-dimensional isoparametric finite element method [29] is used to solve the Laplace equation.

Let us parameterize the probability p_l for the formation of a leader step as

$$p_l = \left|E_i - E^*\right|^{\alpha} / \sum_{i=1}^{I} \left|E_i - E^*\right|^{\alpha}, \tag{11}$$

where E^* is the critical field for a leader, I is the total number of the neighboring points around the leader tip, and E_i is the average field along the growth direction of the leader.

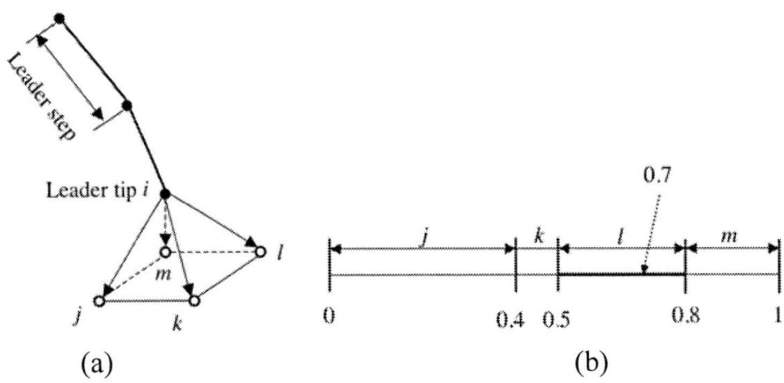

Figure 3. (a) The schematic diagram of a stepped leader and (b) The method to sample a growth point for leader propagation.

As seen in Fig. 3(a), the E_i between the points i and j is calculated by $E_i = (\phi_i - \phi_j)/d$, where ϕ_i and ϕ_j are the potentials at the points i and j, and d is the distance between i and j. α is set 1, following the previous research [30]. As shown in Fig. 3(a), there are four possible growth points at the leader tip i, *i.e.*, points j, k, l, and m. It is presumed that at each leader step only one point from the neighboring points around the leader tip can become the growth point of the leader. A uniform random number U between 0 and 1 is introduced to choose the growth point. As seen from Fig. 3(b), if the bond direction probabilities of points j, k, l, and m are 0.4, 0.1, 0.3, 0.2, respectively, and U = 0.7, the point l will be chosen as the next leader growth point.

The growth of the leader stops self-consistently when all probabilities become zero. Since the electric field E_l (= ~ 1 kV/cm at atmospheric pressure) in leader channel is an order of magnitude lower than the critical field E^* (=~12.5 kV/m). The potential at leader tip sustain a relative higher value by $\phi_{i+1} = \phi_i - E_l \cdot l_i$, where ϕ_i is the potential at leader tip, l_i is the length of leader channel, and i is the leader step.

Simulation Results

Figure 4(a) shows the distribution of electric fields on the axis ($r = 0$) for cases 1-3, in which the electric field generated by thundercloud charges just exceeds the ionization threshold E_k. The calculated initiation altitudes are ~18.6-23.6 km in agreement with the observations of Su et al. [6], which reported that an apparent emerging point of giant electric jets started from the altitudes of 18 km, 22 km, and 24 km.

In the Monte Carlo simulation, the neutral gases are composed of 80% N_2 and 20% O_2. It is assumed that about one thousand electrons are generated at the altitudes of ionization threshold. Then the electrons are accelerated and collide with neutral air molecules. The number of electrons is quickly increased due to ionizations. The electron kinetics are traced by the Monte Carlo technique until the electron number reaches N_{min} given by eq. (10). The electron energy distribution E (ε) at the time of initiation of giant electric jets is given in Fig. 4(b). As seen in Fig. 4(b), the electron energy distributions for cases 1-3 have a similar distribution, deviating from Maxwellian. The distribution has a low energy tail in which the energies of only a few electrons are over the ionization thresholds of gases, *i.e.,* 15.58 eV for N_2 and 12.1 eV for O_2 [10]. The average electron energy is ~4.97 eV, which in consistent with the previous research (~5 eV) of sprites [9].

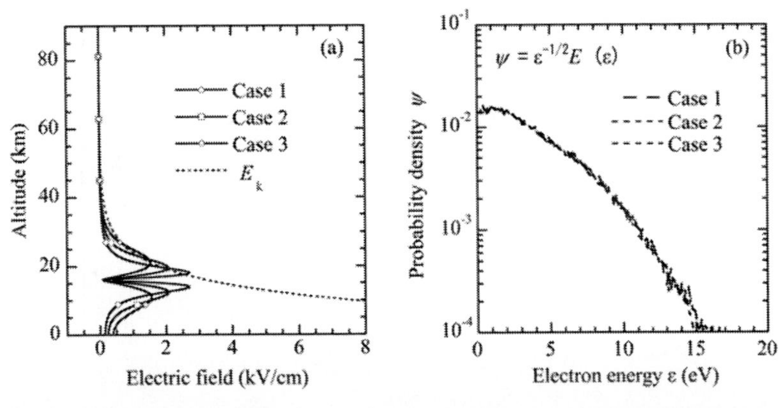

Figure 4. (a) Electric field on the axis ($r = 0$) compared with ionization threshold E_k and (b) Electron energy distribution at the time of initiation of giant electric jets.

In order to verify the developed leader model, the propagation of giant electric jets is simulated and compared with the observations. Let us consider case 1 as a typical case, *i.e.*, $a = b = 2$ km, determining the distribution of ρ_s (= ρ_-). Figure 5 shows the distribution of electric field on the axis ($r = 0$) in comparison with the critical field E^* (= E_c) for the propagation of a leader. It is obvious that the ionization threshold E_k is much larger than the critical field E^* (= E_c). The simulations show that only a few of leaders propagate over the altitude and arrive at the ionosphere. Once a leader starts its propagation, the charges will be transported upwards. The accumulation of thundercloud charge can be suppressed not enough to support the leader discharge reaching at the ionosphere. In the present work, we use a simple charge distribution of thundercloud and consider the effect of the intracloud (IC) discharge. A well-developed giant electric jet is represented in Fig. 6(a) by using the present model, in which the thundercloud charge is accumulated up to be 300 C. The prevailing leader discharges span the long distance (= ~72 km) from the initiating position around the top of thundercloud to the ionosphere. The leader propagation appears a three-dimensional overall picture in agreement with the observed giant electric jets [6], as shown in Fig. 6(b).

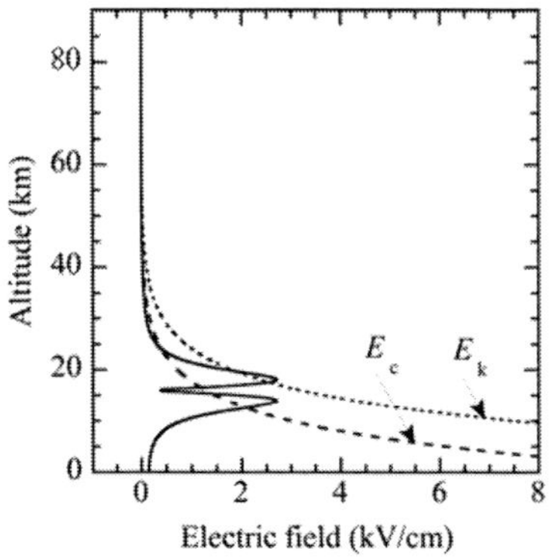

Figure 5. Electric field on the axis ($r = 0$) at the time that the leader is initiated.

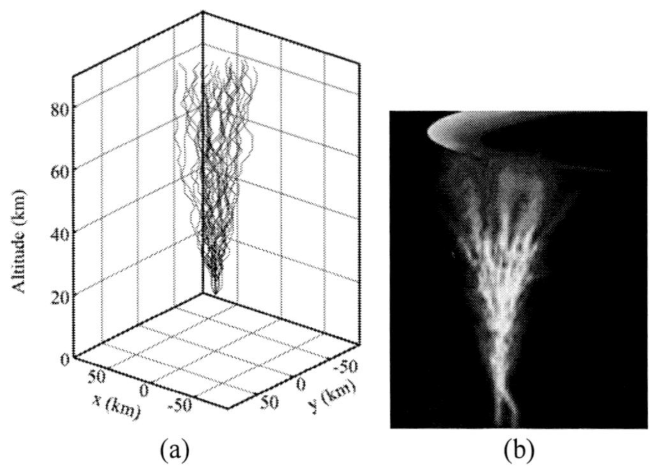

Figure 6. (a) Simulation of leader propagation and (b) Image of a giant electric jet.

Conclusion

The recently observed upper discharges above thunderclouds, *i.e.*, giant electric jets were modeled and simulated by using microscopic and macroscopic numerical methods. The Monte Carlo simulation was combined with the solution of the charge conservation equation. The streamer and leader discharge theories in laboratory situations were extended to investigate giant electric jets occurred above thunderclouds. The simulation results are consistent with the observations. The developed models can also be extended to study other upper discharge phenomena such as the streamer structure of sprite tendrils.

Acknowledgments

The vector computer SX5 at *Institute of Fluid Science of Tohoku University* in Japan is used in the present work. We would like to thank Prof. Dirk Lummerzheim of *University of Alaska* for providing us the data of cross sections for electron-molecule collisions. We also express our gratitude to Mr. Jean-Claude Vérité of *Electricité de France* for helpful discussion on leader modeling.

References

[1] Rowland, H. L. *J. Atmos. Terr. Phys.* 1998, 60, 831-844.

[2] Barrington-Leigh, C. P.; Inan, U. S. *J. Geophys. Res.* 2001, 106, 1741-1750.

[3] Miyasato, R.; Fukunishi, H.; Takahashi, Y.; Taylor, M. J. *J. Atmos. Terr. Phys.* 2003, 65, 573- 581.

[4] Tong, L. Z.; Nanbu, K.; Hiraki, Y.; Fukunishi, H. *J. Atmos. Sol. Terr. Phys.* 2005, 67, 829-838.

[5] Pasko, V. P.; Stanley, M. A.; Mathews, J. D.; Inan, U.S.; Wood, T.G. *Nature* 2002, 416, 152-154.

[6] Su, H. T.; Hsu, R. R.; Chen, A. B.; Wang, Y. C.; Hsiao, W. S.; Lai, W. C.; Lee, L. C.; Sato, M.; Fukunishi, H. *Nature* 2003, 423, 974-976.

[7] Pasko, V. P. *Nature*, 2003, 423, 927-929.

[8] Tong, L. Z.; Nanbu, K.; Fukunishi, H. *Earth Planets Space* 2004, 56, 1059-1065.

[9] Pasko, V. P.; Inan, U. S.; Bell, T. F.; Taranenko, Y. N. *J. Geophys. Res.* 1997, 102 4529-4561.

[10] Tong, L. Z.; Nanbu, K.; Hiraki, Y.; Fukunishi, H. *J. Phys. Soc. Jpn.* 2004, 73, 2438-2443.

[11] Dejnakarintra, M.; Park, C. G. *J. Geophys. Res.* 1974, 79, 1903-1910.

[12] Potter, D. *Computational Physics*, John Wiley, New York, 1973.

[13] Martyna, G. J.; Tuckerman, M. E.; Tobias, D. J.; Klein, M. L. *Mol. Phys.* 1996, 87 1117-1157.

[14] Rees, M. H. *Physics and Chemistry of the Upper Atmosphere*, Cambridge University Press, New York, 1989.

[15] Rapp, D.; Briglia, D. D. *J. Chem. Phys.* 1965, 43, 1480-1489.

[16] Phelps, A. V. *JILA Information Center Report* No. **26**, University of Colorado, Boulder CO, 1985.

[17] Nanbu, K. *Jpn. J. Appl. Phys.* 1994, 33, 4752-4753.

[18] Nanbu, K. *IEEE Trans. Plasma Sci.* 2000, 28, 971-990.

[19] Raizer, Yu. P.; Milikh, G. M.; Shneider, M. N.; Novakovski, S. V. *J. Phys. D: Appl. Phys.* 1998, 31, 3255-3264.

[20] Raizer, Yu. P. *Gas Discharge Physics*, Springer-Verlag, Berlin, 1991.

[21] Babaeva, N. Yu.; Naidis, G. V. *IEEE Trans. Plasma Sci.* 1997, 25, 375-379.

[22] Pasko, V. P.; George, J. J. *J. Geophys. Res.* 2002, 107, 1458.

[23] Petrov, N. I.; Petrova, G. N.; D'Alessandro, F. *IEEE Trans. Diele. Electr. Insul.* 2003, 10, 641-654.

[24] Tong, L. Z.; Nanbu, K.; Fukunishi, H. *Earth Planets Space* 2005, 57, 613-617.

[25] Petrov, N. I.; Petrova, G. N. *Tech. Phys.* 1999, 44, 472-475.
[26] Bazelyan, E. M.; Raizer, Yu. P. *Spark Discharge*, CRC Press, Florida, 1998.
[27] Petrov, N. I.; Petrova, G. N. *Tech. Phys.*, 1993, 38, 287-290.
[28] Tong L. Z.; Nanbu, K.; Fukunishi, H. *J. Phys. Soc. Jpn.* 2005, 74, 1093-1095.
[29] Kagawa, Y., *FEM Program 1*, Morikita Shuppan Co., Tokyo, 1994 (in Japanese).
[30] Pasko, V. P.; Inan, U. S.; Bell, T. F. *Geophys. Res. Lett.* 2000, 27, 497-500.

In: Solar Physics Research Trends
Editor: Pingzhi Wang, pp. 357-380

ISBN: 978-1-60021-987-0
© 2008 Nova Science Publishers, Inc.

Chapter 11

SUNSPOT INFLUENCE ON PHOTOMETRIC AND MORPHOLOGICAL PROPERTIES OF SOME PHOTOSPHERIC FINE STRUCTURES

Ernesto R. Rodríguez-Flores [*] ***and Ramón E. Rodríguez Taboada*** [**]

Instituto de Geofísica y Astronomía, Calle 212 No. 2906 e/ 29 y 31 Rpto. La Coronela, La Lisa, Ciudad de La Habana, C.P. 11 600, Cuba.

Abstract

Based on high resolution observations we studied the photometric behavior and morphology of solar fine structures under sunspots influence.

The pictures are white-light photographs taken with a 50-cm aperture and 120-m focal length reflector located in Tadjikistan at the Pamir Mountain Station (spatial resolution 0.3 arcsec). The photographic film characteristics are: 120-mm format, 3 DIN sensibility and a resolution of 120 lines per millimeter. From the film and filter spectra combination a bandwidth of 40 nm centered at λ 650 nm was obtained. The exposure time was 1.5 ms. The photoheliograms shows the sunspot group SD 284/78 (Solnechnye Dannye Classification). The films were scanned with an Arcus 1200 – AGFA scanner at 300 d.p.i. and 256 intensity levels. We have obtained a standardized sequence of images to characterize the spatial-temporal behavior of some umbra details of the main sunspot and surrounding granulation. Basic elements of the standardization methodology, and the morphologic and photometric results obtained are presented.

The bright granules spatial distribution and the granules surface density as function of distance from the center of the main sunspot umbra were determined. Bright granules distribute delimiting cells of dimensions in the mesogranular scale. The mean diameter of these cells does not show significant variation with the variation of the magnetic field of the sunspot. Moreover, the granules surface density does not show significant variation with distance to the sunspot umbra. Both results point to a very weak, if any, influence of the sunspot magnetic field at distances greater than 20 arcsec.

Umbral dots and nuclei were detected and their spatial distributions were analyzed. Umbral dot diameter distribution was obtained and discussed.

We conduced a threshold process at levels defined by a method based on intensity cumulative histogram to determine the areas occupied by the different photometric elements.

[*] E-mail address: erflores@iga .cu

[**] E-mail address: ramone@iga.cu

The temporal behavior of the measured quantities and their relations were obtained. We found a relation between the area of the umbral nucleus and the area of the umbra (K_{nu}) and between the area of the penumbra and area of the umbra (K). Following the statement of some researchers linking the K index to the magnetic strength; the relation between the areas of the umbral nucleus and the area of the umbra means that spots with magnetic field intensity less than a limiting value should not present umbral nucleus.

Introduction

The study of the fine structure in the solar photosphere is one of the more important aspects in modern Solar Physics. The researches of the sunspots features dynamic are directed to understand the energy and mass transfer and the characterization of the force free fields. The sunspot umbra is a region with critical energy and mass transfer conditions and a strong magnetic field. The strong magnetic field plays a main roll in any physical process taking place in the umbra and the surrounding medium and the details of the brightness distribution give information on the physical conditions in the region involved.

On the other hand, solar granulation has been studied for many years, but until now there is not a general agreement on its basic parameters. Some theoretical and observational results of the last two decades point to a non-characteristic dimension of solar granules, and their mean size or distribution are not established yet (Roudier and Muller 1987, Muller et al. 1990). To determine the probable influence of solar magnetic fields on the parameters of solar granulation is of particular interest.

Sunspots originate by the emergence of strong magnetic fields generated in the deep layers of the Sun (~ 0.2 solar radiuses). The paper of Collados et al. (1986) shows that the dimension of the granules does not change significantly with the distance to sunspots, but Macris et al. (1989) found that solar granulation depends on the distance to sunspots and until now there is not a general agreement on the subject.

Considering contrast and size, there is not a random distribution of the granules in the photosphere. Bright and big granules distribute forming a cellular pattern (Muller et al. 1990). In this way, there are clusters formed by normal granules confined by a complementary pattern of big and bright ones. Both patterns relate in a complementary way (Muller et al. 1990, Rodríguez and Gámez 2000). The pattern formed by the big and bright granules is interpreted as induced by the action of convection in the sub-photospheric layers. Look in Stein and Nordlund (1998) and Rast (2003a) for recent models

Recent models consider that the solar granulation pattern is a superficial process related to the cooling of the gas that undergoes Rayleight-Taylor instability (Hirzberger 2002, Rast 1995, 1999, 2003b). When explaining the structure of the penumbra and the origin of calcium plages around sunspots, Weiss et al. (2004) consider a magnetic field structure that surrounds the sunspot for about 25 Mm, these authors also estimate that channels of horizontal magnetic flux lay at 5 Mm depth. In this case, because the critical wave number (kc) decreases with the increase of the magnetic field under the supported gas layer, it is possible to expect an increase on the scale of granules or mesogranules in the region closer to the sunspot.

Lately, some archive material has been rescued to be digitized (Rodríguez 1995, Györi et al. 1998). Many researchers have developed methods to obtain different morphologic and geometric parameters from this digital information (Rodríguez 1995, Rodríguez et al. 1997a,

Rodríguez et al. 1997b, Pettauer and Brandt 1997, Doval et al. 1999, Rodríguez and Gámez 2000). We combine these aspects to study fine structures in the solar photosphere.

To study these fine structures is necessary to have a sequence of high resolution observations (photoheliograms) with the appropriate contrast and temporal intervals from days to minutes. The timing and quality of our photoheliograms is excellent for this purpose.

We study the spatial distribution and surface density of the solar granulation near a sunspot. We explore the pattern of the solar granulation searching for evidences of the sunspot magnetic field intensity influence on convection under the photosphere. Changes on the granulation spatial pattern with horizontal scales in agreement with the mesogranular horizontal scales, if they exist, could be considered the probable influence of sunspot magnetic field intensity at mesogranular depths. Moreover, we analyzed some indexes characterizing the photometric structures of sunspots group using the cumulative histogram method. The structures considered were the umbra, the penumbra, the umbral cores and the main spot as a whole.

Sunspot umbrae brightness inhomogeneities were known as early as 1916. These inhomogeneities have received different names from different researchers (see Sobotka et al. 1993). The more conspicuous features inside the sunspot umbrae are umbral dots and the umbral nucleus (named "voids" by Livingston 1991 and "dark nucleus" by Sobotka et al. 1993). These features differ from each other in size and brightness. We study how they change with some sunspot global properties.

Materials and Methods

Observational Data

The main material of this work is a series of photoheliograms showing at least 0.3 arcsec spatial resolution. The instrument employed to accomplish the task was a refractor telescope with a 50 cm aperture and 120 m resultant focal length. The telescope was designed specially for high precision photographic works. The observations were done in Tadjikistan at the Pamir High Mountain Station (38 N 72.5 E and 4300 m over sea level) belonging to the Main Astronomical Observatory of Russia located at Pulkovo in Saint Petersburg. (for details see Parfinenko and Miklalev (1978)).

The photoheliogram characteristics are: film sensibility 3 DIN, resolution power 120 lines/mm and 120 mm format. The combined response of the film and of the yellow filter included in the optics of the telescope gives a bandwidth of 40 nm centered in 650 nm. The time of the expositions is 1.5 ms (Parfinenko 1999). The quality of the photoheliograms is excellent as noticed in the granulation pattern sharpness and the 0.3 arcsec spatial resolution reached (figure 1). The photoheliograms show the sunspot group SD 284/78 (Solnechnye Dannye Classification) and surrounding granulation. The characteristics of this group are shown in Table I.

The outstanding conditions of diurnal seeing under which were taken the photographic records allow us to catalog our material as high resolution. This value is in the set of the smallest ones obtained observing from Earth and are the result of exceptional atmospheric conditions.

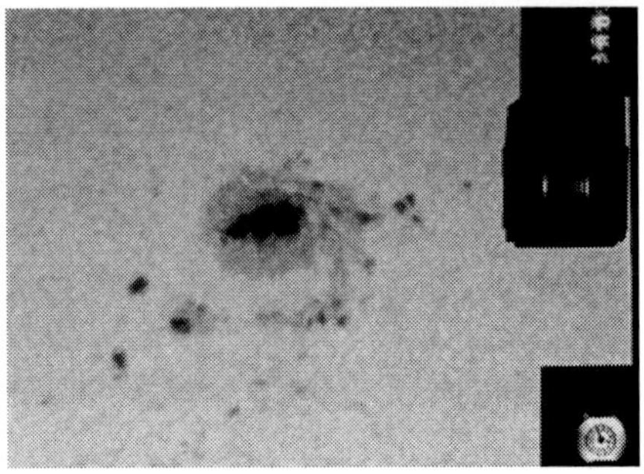

Figure 1. Photoheliogram taken on October 6th, 1978 (3^h 55^m 05^s U.T.).

The photoheliograms were digitized using an Agfa Arcus 1200 negative scanner at 300 dpi resolution (equivalent to 0.15 arcsec per pixel) and 256 intensity levels. Good enough for the planned analysis. The spatial resolution was determined by the minimum size of the structures in the images to be analyzed. The intensity level resolution was determined by the capability of the image processing programs used. From the source images were selected the regions to be analyzed: the "working images". The working images were named as shown in Table II taking into account the observation date.

Table I. Relevant data for the sunspot group SD 284 (Solnechnye Dannye classification) included in source images. dd – day, mm – month, yyyy – year, U.T. – Universal Time, m.v.s.h. – millionths of visible solar hemisphere, B –Magnetic field strength of the main sunspot in the group, r/R – distance to solar disk center in solar radius units.

dd/mm/yyyy	Time [U.T.]	Main Sunspot		
		Area [m.v.s.h.]	|B| [Gauss]	r/R
30/09/1978	3^h 59^m 26^s	240	-	0.95
02/10/1978	3^h 30^m 13^s	-	-	-
02/10/1978	4^h 34^m 02^s	154	1600	0.75
03/10/1978	2^h 58^m 20^s	-	-	-
03/10/1978	3^h 56^m 50^s	-	-	-
03/10/1978	4^h 01^m 44^s	-	-	-
03/10/1978	4^h 19^m 06^s	127	2700	0.60
03/10/1978	5^h 23^m 28^s	-	-	-
04/10/1978	2^h 22^m 20^s	-	-	-
04/10/1978	3^h 24^m 11^s	-	-	-
04/10/1978	4^h 33^m 07^s	153	3100	0.40
04/10/1978	5^h 10^m 14^s	-	-	-
05/10/1978	2^h 56^m 01^s	-	-	-
05/10/1978	4^h 37^m 08^s	191	2800	0.03
06/10/1978	3^h 42^m 23^s	-	-	-
06/10/1978	3^h 55^m 05^s	232	2400	0.36

Image processing was done using different applications. The main procedures employed were distribution of brightness levels calculation, threshold and image combination. The applications involved are DIPSY© (version 1.0) and UNESCO Bilko ® (version 2.0).

Table II. Characteristics of the working images. dd – day, mm – month, yyyy – year, *U.T. – Universal Time.*

Image	dd/mm/yyyy	Time [U.T.]
3009	30/09/1978	$3^h\ 59^m\ 26^s$
0210a	02/10/1978	$3^h\ 30^m\ 13^s$
0210b	02/10/1978	$4^h\ 34^m\ 02^s$
0310a	03/10/1978	$2^h\ 58^m\ 20^s$
0310b	03/10/1978	$3^h\ 56^m\ 50^s$
0310c	03/10/1978	$4^h\ 01^m\ 44^s$
0310d	03/10/1978	$4^h\ 19^m\ 06^s$
0310e	03/10/1978	$5^h\ 23^m\ 28^s$
0410a	04/10/1978	$2^h\ 22^m\ 20^s$
0410b	04/10/1978	$3^h\ 24^m\ 11^s$
0410c	04/10/1978	$4^h\ 33^m\ 07^s$
0410d	04/10/1978	$5^h\ 10^m\ 14^s$
0510a	05/10/1978	$2^h\ 56^m\ 01^s$
0510b	05/10/1978	$4^h\ 37^m\ 08^s$
0610a	06/10/1978	$3^h\ 42^m\ 23^s$
0610b	06/10/1978	$3^h\ 55^m\ 05^s$

Methods

Calibration and Standardization of the Images

Image calibration is fundamental in comparing different observations. We develop a process devoted to eliminate the errors and biases due to the analog-digital conversion.

Two aspects are very important dark current and non uniformity of background intensity. Following Massey (1997) we considered the expression:

Where:

I: non calibrated image

I_C: calibrated image

I_{DC}: dark current image

I_{FF}: flat field image,

$$I_c = \frac{I - \overline{I_{DC}}}{\left(\overline{I_{FF}} - \overline{I_{DC}}\right)} \times FFMV$$

The over score refers to the average images. The FFMV is the corrected flat field mean value.

Solar images have a very wide intensity spectrum because of the temperature differences between "cooler" and "hotter" features; about 4000 K in umbral darker details and 6000 K in the photosphere. The images were digitized considering only 256 intensity levels giving a big step between each brightness level. For this reason the dark current level is much less intense that the intensity resolution employed. This situation was checked with the black strips of the negative border (total absence of light) that provide a zero level. The analysis of the flat field images data shows the errors were under the intensity resolution employed. The intensity resolution selected makes unnecessary to apply the dark current and flat field corrections.

Errors due to dust grains and imperfections of the negative were avoided cleaning the negatives and selecting the areas to work with.

Characteristic Curve Analysis

This aspect of the data preparation has not been considered in detail previously (see for example Pettauer and Brandt 1997) because the photographic material was not adequate prepared for a precise photometric treatment. In our case each photoheliogram has a densitometric scale. Each step is a measure of the opacity of the element and it allows to calculate the negative density. Table III shows the opacity values of the densitometric scale steps. Decimal logarithm of this value gives the negative density (D).

Table III. Opacity values from the densitometric scale of the film

Step	1	2	3	4	5	6	7	8	9
Opacity	2.000	1.800	1.633	1.447	1.275	1.090	0.887	0.690	2.000

To obtain the intensity value of the densitometric scale steps we calculate the distribution of the values in each one. The mean value and the standard dispersion were estimated fitting a normal distribution to the data (see Table IV for details). With these values we obtained the characteristic curve of each image. An example of such curves is shown in figure 2. Notice the low intensity level values non linearity.

Considering the characteristic curve behavior, in particular:

1) Coverage of the intensity range.
2) The linear range extension of the characteristic curve (minimal number of steps in non linear region).
3) Standard deviation of the normal distribution of the step values.

The linear coverage of the intensity range is important in order to have a standardizing scale covering all the intensity intervals of the cases. Less standard deviation goes to a less error in the calculated value.

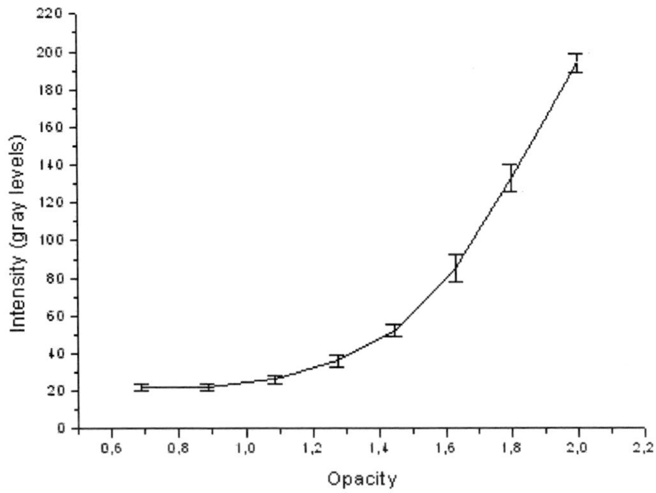

Figure 2. Characteristic curve of image 0510b.

Table IV. Intensity values (gray levels) for each of the densitometric scale step in each image. The intensity standard deviation appears too. I_i – i th step mean intensity, σ_{Ii} - intensity standard deviation of this step

Image	$I_1 \pm\sigma_{I1}$	$I_2 \pm\sigma_{I2}$	$I_3 \pm\sigma_{I3}$	$I_4 \pm\sigma_{I4}$	$I_5 \pm\sigma_{I5}$	$I_6 \pm\sigma_{I6}$	$I_7 \pm\sigma_{I7}$	$I_8 \pm\sigma_{I8}$	$I_9 \pm\sigma_{I9}$
3009	184±9	148±6	93±8	57±3	30±3	17±3	11±1	11±1	183±4
0210a	206±9	134±5	81±9	67±10	35±9	29±5	25±4	25±4	204±7
0210b	175±6	120±8	73±6	46±4	30±4	23±3	20±2	20±2	169±6
0310a	204±5	167±6	123±10	79±4	48±3	34±5	27±3	27±3	201±5
0310b	207±9	181±4	134±7	86±11	55±3	38±6	30±3	30±3	209±4
0310c	198±4	164±4	114±7	73±6	47±3	33±3	25±3	25±3	193±5
0310d	196±5	167±8	111±10	70±5	48±2	37±3	29±3	29±3	190±4
0310e	203±4	166±5	173±7	75±4	47±4	35±3	29±3	29±3	197±6
0410a	171±6	138±6	93±4	58±5	44±3	33±5	30±2	30±2	182±4
0410b	210±5	173±6	131±14	77±5	49±3	33±3	25±3	25±3	203±6
0410c	201±5	176±8	128±6	85±6	59±3	43±4	36±2	36±2	202±5
0410d	201±5	170±5	123±10	85±6	55±4	39±4	29±4	29±4	197±5
0510a	194±6	158±7	108±9	69±3	46±3	34±4	26±2	26±2	194±5
0510b	194±5	133±7	85±7	52±3	36±3	26±2	22±2	22±2	185±5
0610a	173±5	120±5	74±5	51±3	36±2	30±3	30±3	30±3	175±5
0610b	167±4	122±5	77±5	50±3	35±3	26±2	21±2	21±2	164±3

Under the explained criteria a patron image was selected and all the other reduced to that patron image. The selected image was 0510b. Table V shows the standardizing functions obtained. These functions allow to reduce the data of the different images to the patron one.

Images 0210a and 0310e were rejected because the photometric scales were not appropriate. Mean values of some steps very far from the expect values producing correction curves very different from the characteristic ones.

Table V. Standardizing functions for each image and fitness data.

Image	Function	Fitness (R)	Standard Dev. (σ)
3009	Y = 0.92X + 9.9	0.99	2.18
0210b	Y =1.11X + 0.4	0.99	0.42
0310a	Y = 0.87X – 4.2	0.98	2.78
0310b	Y = 0.84X –6.4	0.98	3.38
0310c	Y = 0.91X –3.5	0.99	2.53
0310d	Y = 0.95X –7.9	0.99	2.15
0410a	Y = 1.16X –13.3	0.99	1.30
0410b	Y = 0.84X –1.7	0.98	2.79
0410c	Y = 0.92X –14.0	0.98	3.05
0410d	Y = 0.88X – 7.5	0.98	3.36
0510a	Y = 0.95X – 5.4	0.99	2.11
0610a	Y = 1.20X –11.8	0.99	1.13
0610b	Y = 1.16 X – 3.5	0.99	0.67

Photoheliogram Exposure Uniformity

A no uniform distribution of light over the negative is an undesirable effect that could induce to consider erroneous levels of intensity when using a threshold process. To analyze the uniformity of the light distribution over the negatives we obtained the optimum threshold level for each image in granular regions in different points on the negative. The result was that they varied from one point to the other in the negative. The plot of the data showed that all the negatives fit to planes. Figure 3 shows the behavior of the background intensity in negative 0210b.

The maximum slope of background values is in the gradient direction. The standard deviation of the adjusted values is due to the granulation and intergranular spaces. When dispersion is greater that the background variation estimated by the fit we have a uniform background. Table VI shows the results.

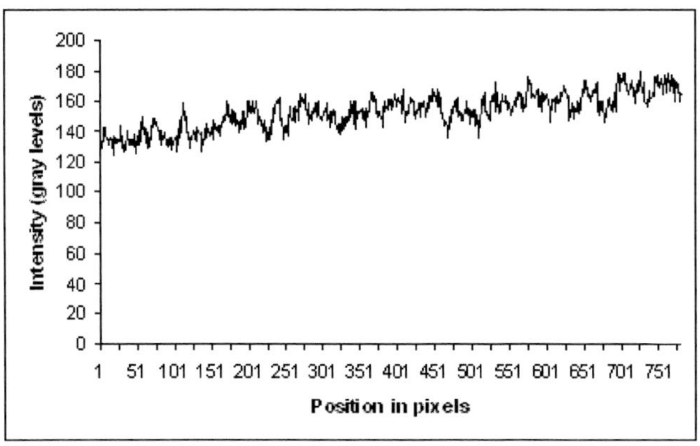

Figure 3. Intensity gradient of the image 0210b. The high frequency fluctuations are due to photospheric granulation pattern. The measurements are referred to a transec in the direction of the maximum variation (gradient) in the background values.

Table VI. Non-uniform exposure gradient values (ΔΦ), Standard deviation of the intensity values (σ) and maximal mean expected variation of intensity (ϕ) in the working image.

Image	ΔΦ [g. l./pixel]	σ [g. l.]	ϕ [g. l.]
3009	0.100	5.9	10.0
0210b	0.041	6.6	4.1
0310a	0.052	9.3	5.1
0310b	0.044	7.8	5.2
0310c	0.045	7.4	5.5
0310d	0.052	6.8	10.4
0410a	0.063	8.1	6.0
0410b	0.068	9.7	6.2
0410c	0.043	7.8	10.6
0410d	0.047	7.6	4.5
0510a	0.051	7.6	6.2
0510b	0.060	8.0	12.0
0610 a	0.083	7.6	12.4
0610b	0.071	6.6	14.2

Legend: g. l. – gray level

We correct the images subtracting the result of the filtering with a window 3 arcsec x 3 arcsec wide. In this way we eliminate the low frequency components preserving the information of the fine details. After this correction the images are ready to be measured. The corrections allow us to select the better ones to be studied. Digitizing to 256 grey levels in a wide range of intensities does not allow to notice flat field fluctuations nor the dark current values of CCD detector.

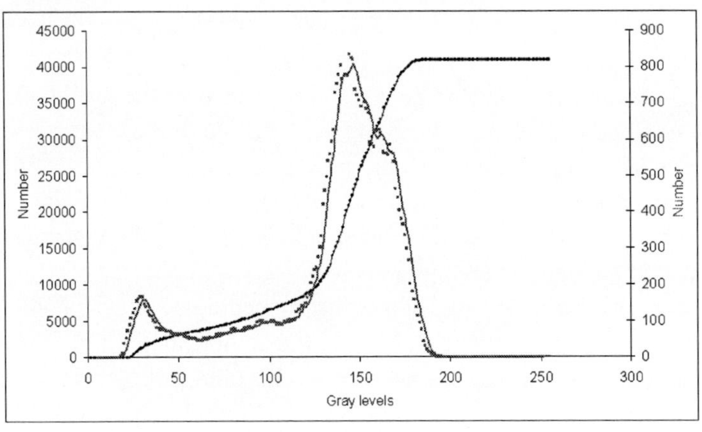

Figure 4. Histogram (in red) and Cumulative Histogram (in black) for image 0410d. The slope changes in the cumulative histogram are clearly visible.

Different authors use different methods to separate the different photometric features of sunspots group. We will use the cumulative histogram of the working images and its derivative to determine characteristic intensity values for each photometric feature to analyze. The criteria used in each case will be discussed later in the corresponding section. The figure 4 shows the cumulative histogram (in black) of the image 0410d.

Results and Discussion

The ratio between the penumbral and umbral dimension (K) is a very important parameter in the characterization of sunspots. A relationship between this parameter and the magnetic field intensity of the sunspots was reported (Antalová 1991). However Antalová (1991) point out there is not agreement in the criterion to determine the dimension of the penumbra. We employed the cumulative histogram of the images to analyze this problem.

Characterization of the Group

To characterize the sunspot group under study we calculated the areas of the main elements of sunspots and we analyzed the relationship between them. Considering the previously exposed situation we used the cumulative histogram of the main sunspot image (see an example in figure 4) to determine the intensity range of the features.

In our example (image 0410d), the visual inspection of the histogram shows a distinctive change in the slope of the curve in the 35 units intensity level range. There are other changes in the curve's slope: in the intensity level 121 and in the intensity level 175. The intensity level 175 slope change is related not to the presence of any photometric structure, but to the saturation of the cumulative histogram. No other slope change was noticed in the cumulative histogram. A more analytical determination of the slope change points was done using regressions in the three sections of the histogram data, without significant differences with the visual result. From the more simple photometric structures of sunspots we should have two frontiers, the umbra-penumbra and the penumbra-photosphere frontiers. In this first approximation analysis the penumbra-photosphere border is absent.

To associate the observed changes to the group morphology we proceed to "binarize" the image with the threshold levels determined from the histogram. The "binarizing" procedure changes all the pixels under the level to zero and the pixels over the level to 255 values.

The change of slope is evidence of a significant change of the character of the brightness distribution; that is, a border between two different photometric regions inside the image. We proceeded to compare the shape of the regions obtained from the binarizing of the image and the classical structure of the spot from its photoheliogram. The slope change in the intensity level 121 is related to the umbra penumbra frontier.

Using the intensity level 35, the binarizing procedure showed photometric structures inside the umbra. It is clear that the border of the region, determined by this method, is significantly different from the umbra as seen in the photoheliogram. We considered that these structures are umbral nucleus, as described by Sobotka et al. (1993). These features present a well defined range of brightness as seen from the maximum in the histogram (in red in figure 4).

The binarizing procedure at intensity level 121 showed structures fitting very well the umbra-penumbra frontier in the original image. The "trial and error" procedure applying binarizing to the image conduced to intensity level 155 as an estimation of the characteristic value of the penumbra-photosphere frontier. This level was determined as the limit value that gives a restricted region resembling the penumbra and does not produce any significant structure in the photosphere region. Once the photosphere values are reached the binarized regions abruptly changes allowing an accurate determination of the frontier intensity level.

The maximum in the histogram with intensity values close to the photosphere characteristic values (as is shown in section "Large scale pattern of the photospheric granules") shows a broad and noisy top. We considered this behaviour a signature of superposition of two characteristic values of photometric structures, the penumbra and the surrounding photosphere.

To explore this situation we applied an eleven points running average to the histogram and considering the uniformity of the photosphere brightness in a small area, as is our case, we proceed to adjust a normal distribution to the higher values of intensities wing of the maximum. This distribution is considered the distribution of the brightness of the photosphere and was extracted from the smoothed histogram maximum. The residual was considered the brightness distribution of the penumbra and a normal distribution fitted to the values. In figure 5 we present the smoothed histogram, the normal distributions fitted, and the superposition of both. Notice de good agreement between them.

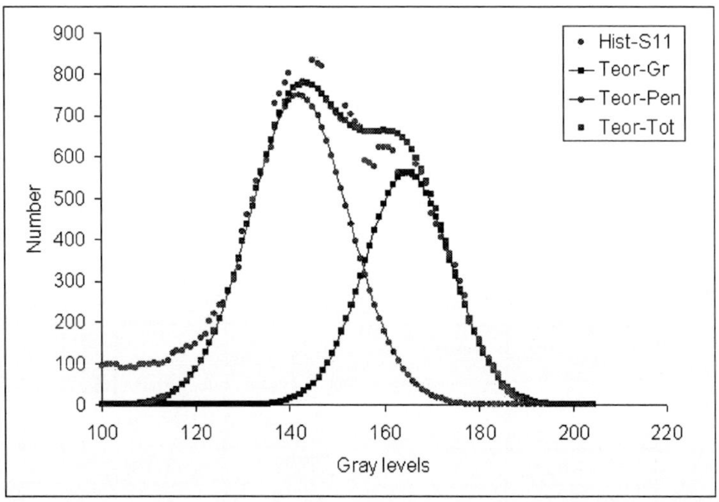

Figure 5. Notice the estimated normal distribution profiles fitted to the total main maximum of histogram profile of image 0410d. It is clear the characteristic values of the photometric features. Hist-S11: 11 points running average, Teor-Gr: normal distribution fitted to the photospheric granulation wing of the histogram, Teor-Pen: normal distribution fitted to the penumbra wing of the histogram, Teor-Tot: Sum of both normal distributions photospheric granulation and penumbra. Notice the good agreement between the total distribution calculated and the observed distribution.

The values of the parameters of the fitted normal distributions are in agreement with the previously determined ones. The 121 value is close to the 142-15=127 value estimated from the penumbra distribution fitness. The 155 value is close to the 165-13=152 estimated from the granulation distribution fitness. For the other images the values are very similar to the values of this example.

In this way we guarantee the quantitative definition of the sunspot elements frontiers employing the slope change in the histogram (used previously in Rodríguez 1995, Rodríguez et al. 1997b, Doval et al. 1999)) and the fitting of the normal distribution of the characteristic values of the features to the image histogram. The selected values are the maximum of the normal distribution plus one distribution sigma to assure the inclusion of a significant proportion of the pixels involved.

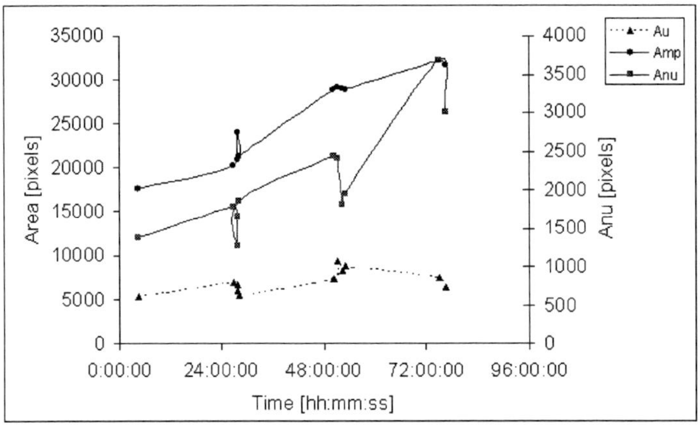

Figure 6. Temporal evolution of the area occupied by the features observed in the studied sunspot. A_{un} – Total area of the umbral nucleous, A_u – umbral area, A_{mp} – main spot area. The temporal axis is in days beginning from September 30th. The ordenate axis at right is for A_{nu}.

Table VII.- Main spot features areas. A_{un} – umbral nucleous area, A_u – umbral area, A_{mp} – Main spot area. Column 6 is the relation between the umbral nucleous and the umbral areas. Column 7 is the classic relation between the umbral and penumbral radiuses (K),Time, column 2, is in days beginning from September 30th.

Image	Time [U.T.]	A_{un} [pixels]	A_u [pixels]	A_{mp} [pixels]	(A_{nu}/A_u)	K
0210b	4:34:02	1378	5292	17599	0.260	1.82
0310a	26:58:20	1776	6929	20229	0.256	1.71
0310b	27:56:50	1643	6755	24019	0.243	1.89
0310c	28:01:44	1271	5997	20860	0.212	1.87
0310d	28:19:06	1852	5443	21243	0.340	1.98
0410a	50:22:20	2435	7387	28860	0.330	1.98
0410b	51:24:11	2402	9291	29133	0.259	1.77
0410c	52:33:07	1801	8284	28976	0.217	1.87
0410d	53:10:14	1947	8799	28865	0.221	1.81
0510a	74:56:01	3674	7474	32135	0.492	2.07
0510b	76:37:08	2996	6368	31722	0.470	2.23

Coefficient k is calculated from $(A_{mp}/A_u)^{1/2} = (r_p/r_u) = K$

Using these criteria the area of the different photometric structures was estimated in each of the working images. Figure 6 shows the general trend to growth of the different structures area. Nevertheless, significant variations of the measured values at small temporal intervals point to a noisy behavior of the analyzed areas (see table VII).

The mean value of k is 1.91± 0.11. This value differs from 2.39 ± 0.01 reported by Antalová (1991). A possible source for such a difference is the absence of a method to determine the penumbral area. But our value is very close to the 2.03 reported by Brand et al. (1990), for sunspots with dimensions similar to our case (greater than 50 m.v.s.h.). In this last paper there is not given the standard deviation, for this reason it is difficult to judge on the results. Our result has not many cases, but allows to control the behavior of our measurements and is established following a rigorous definition of the umbra and penumbra limits.

Large Scale Pattern of the Photospheric Granules (Rodríguez et al. 2005)

We explore the influence of the spot in the surrounding medium, analyzing the possible changes in the large scale granulation pattern at different locations relative to the main sunspot. To do this we selected the images with time closer to the published data (3009, 0210b, 0310d, 0410c, 0510b and 0610b, see tables I and II).

Spatial Distribution of Bright Granules

For this study we used the photoheliograms having $r/R \leq 0.6$. We selected regions of equal area (30.2 arcsec × 30.2 arcsec) with a distance to the main sunspot umbra of 65 arcsec, approximately. Selecting a high enough intensity threshold level allows us to locate the brighter granules in the images (Muller et al. 1990). Because we were searching for bright granules, we used a cumulative histogram to choose a sufficiently high level to ensure the detection (see example for image 0610b in figure 7). The level was taken selecting between 2 - 4 % of the brighter pixels in the image.

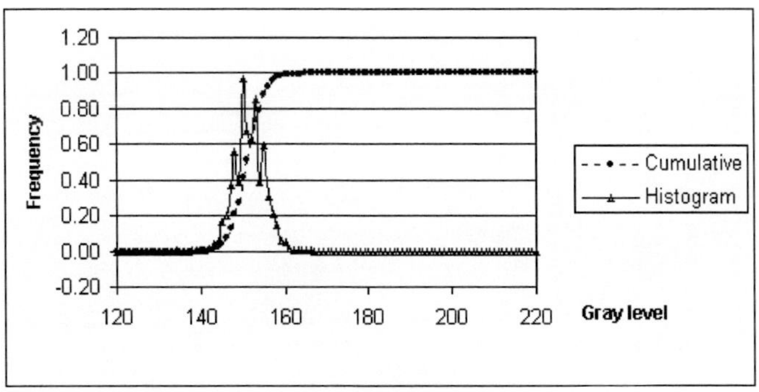

Figure 7. Cumulative histogram showing the brightness distribution of the image 0610b. The thresholding intensity level used to highlight the bright granules in this case was 157.

The brighter granules distribute surrounding regions that we call cells. Bright granules are isolated. We formed the cells connecting the bright granules by lines if they are located at less than a fixed distance; in this way we define the border of the cells. The process was a "supervised" one in which we can change the "maximum" distance to connect the bright granules (figure 8). The borders of the cells were determined individually.

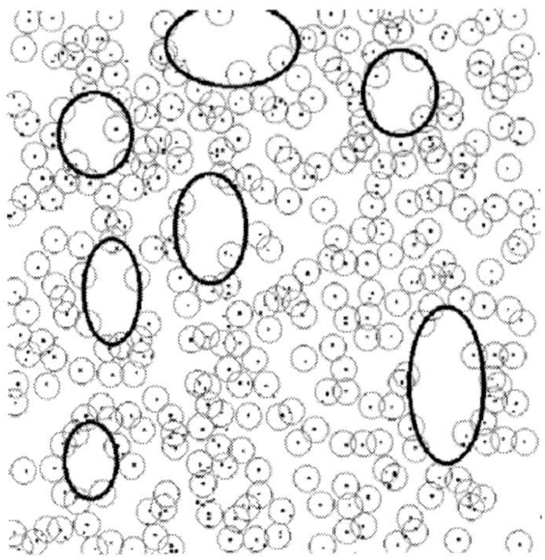

Figure 8. Points are bright pixels in the granular field, small circles are drawn at about one half of the maximum distance connecting two bright pixels. Note the complementary pattern of bright (area inside the small circles) and common granular areas. We have drawn in black some ellipses joining the bright granules and enclosing the normal granular field. These are the features we call cells.

The classical method developed by Oda (1984) to measure granule sizes was modified to measure the cell dimensions. For a full description of the method see Rodríguez and Gámez (2000), the main point of this modification to Oda's method is that the bright granule pattern enclosing the cell is used instead of the boundary of the granule. We define the diameter of a cell (*dc*) as the average between the mean value of the length (*L*) in the direction where the cell is more elongated, and the maximum of the length (*PL*) in the direction perpendicular to the previous one, see figure 9. The view angle or variation of the angular distance to the center of the Sun (θ) was taken into account to correct the cell dimensions, applying the correction factor $1/\cos\theta$. As shown in figure 10, the cell sizes distribution adjusts to a normal distribution.

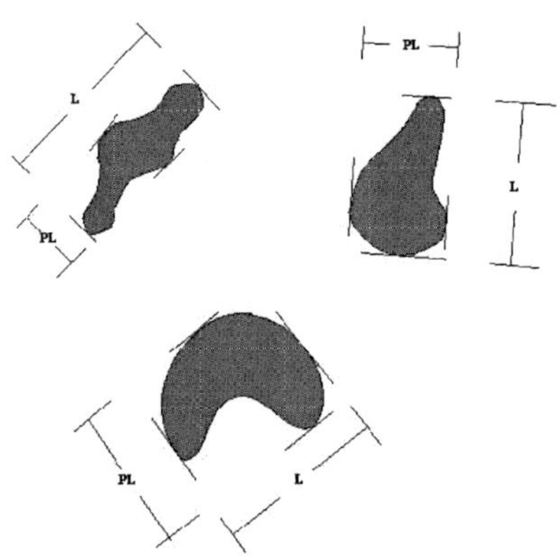

Figure 9. Illustration of the way in which the dimensions of the cells are measured, L is the largest dimension and PL the one perpendicular to L.

Table VIII shows that the mean sizes of the cells are in the mesogranular scale range, in agreement with the results of Muller et al. (1990) and Doval et al. (1999), who studied the large scale pattern of the granulation, and those of Oda (1984), considering different morphological types. For more recent papers on the subject, see Shine et al. (2000) and Müller et al. (2001).

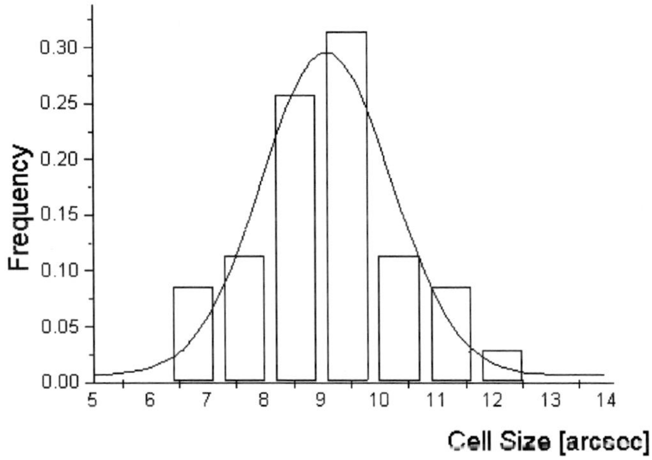

Figure 10. Cell sizes (dc) distribution in the image 0610b. A gaussian fit to the distribution is included.

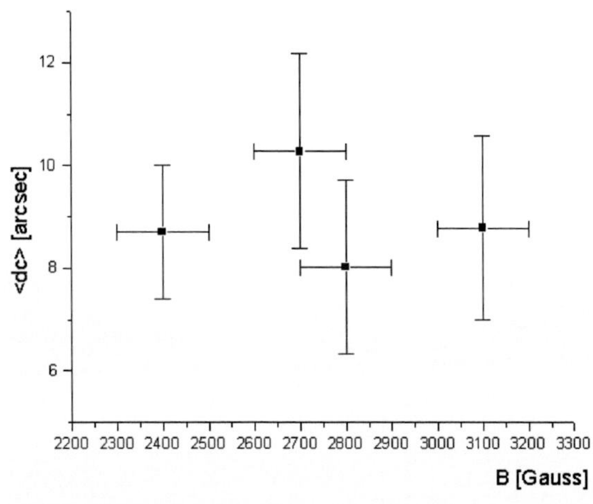

Figure 11. Behavior of the mean cell size (<dc>) and magnetic field intensity in SD 284 sunspot group during the period of observation Vertical and horizontal bars account for measurements error.

The mean distance of each selected region to the sunspot umbra is approximately constant (~ 65 arcsec). For this reason, to explore the change of the cell size due to the variation of the magnetic field strength, we considered only the changes in the magnetic field of the main sunspot umbra and the mean value of the cell diameter (Figure 11). We did not find evidence of any relationship between them. The magnetic field taken from Solnechnye Dannye (Solar Data), No. 10, 1978, changes from 2400 to 3100 Gauss, but the cell size is almost the same. The view angle changes very little, so any bias introduced by the correction done is not significant.

Table VIII.– Sizes of the cells enclosed by the bright granules. D – distance from the selected area to the sunspot center in arcsec, <*dc*> - mean value of cell diameters in arcsec, σ – standard deviation, *N* – number of cells.

Image	D [arcsec]	<dc> [arcsec]	σ [arcsec]	N
0310d	64.9	10.3	1.9	34
0410c	65.0	8.8	1.8	35
0510b	65.3	8.0	1.7	32
0610b	65.1	8.7	1.3	35

Surface Density of Solar Granulation

Some papers show that the dimension of the granules changes with the distance to the sunspot (Schmidt et al. 1988) but others have not found such a behavior (Collados et al. 1986). If the size of the granules changes with the distance to the sunspot, we expect to find a change in their density, an evidence of the action of the sunspot magnetic field on the photosphere surrounding it. The granule density in the photosphere was measured in two regions (15 arcsec × 15 arcsec each) in each of the images included in Table II. The distances to these regions, which are located between 17 to 26 arcsec and 75 to 84 arcsec, respectively, are measured radially from the center of the umbra (Table IX).

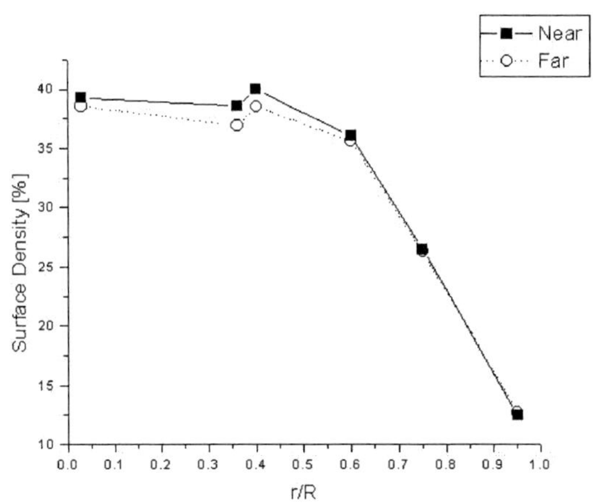

Figure 12. Percentage of the area occupied by the granules in each image vs. distance to the solar disk center (expressed in solar radius units, r/R). The solid line and square symbols correspond to the granulation near the sunspot umbra and the dashed line and dots correspond to the granulation far from the umbra.

Figure 12 shows that except in the image taken near the limb ($r/R = 0.95$), the granules surface density near the umbra is larger than in regions far from them. This behavior could be explained because of the observing conditions near the solar limb (see Jain and Muller 1989).

Near the limb, the spatial resolution becomes worse and the detection of photospheric structures decreases because of both low contrast and poor clearness. Anyway, the granular surface density distributions for "far" and "near" distances are not significantly different. In these measurements, because the sunspots magnetic field is roughly the same, we are dealing mainly with the influence of the distance to the sunspot.

Table IX. Area and distance of the selected regions where we measured the density of granules. l – distance of the granules to the main sunspot umbra center

Image	Surface Density[%]	l[arcsec]
3009	12.5	17.7
3009	12.7	84.3
0210b	26.5	24.8
0210b	26.3	78.5
0310d	36.1	22.1
0310d	35.6	79.2
0410c	40.0	24.2
0410c	38.5	77.6
0510b	38.6	23.9
0510b	36.9	75.2
0610b	39.3	25.5
0610b	38.6	76.8

Characterization of the Fine Structures in Sunspots Umbra

It is known that there are photometric fine structures inside the sunspot umbrae. Bray and Loughhead (1964) reported fine structures inside the umbrae and named it umbral granules; Sobotka et al. (1993) studying the fine structures of the umbra found what they named "umbral dots" in addition to "dark nucleus" probably the same structure reported as "voids" by Livingston (1991). In this section we explore how these features are related to sunspots characteristics.

Morphology of Umbral Dots

Based in previous experiences (Rodríguez et al. 2004) we employed the cumulative histogram of the images and its derivative to detect umbral dots.

The derivative of the cumulative histogram, calculated as the slope between the preceding and subsequent points, shows a maximum in the minimal level of the umbrae intensities (see for example figure 13). This maximum is related to the darker pixels in the umbra that locate forming the umbral nucleus or voids (following Livingston (1991) nomenclature). It is noticeable that there are little maxima in the interval of growing intensities in the umbra. To determine if they were associated to some structures in the umbra, we proceed to obtain the images corresponding to the intensity values of each of these peaks taken the intensity interval from preceding to subsequent minimum. The obtained images were rings with broaden zones (figure 14). Composing the original and the obtained images we could associate these wider zones with umbral dots. Applying the Oda's Method the dimension of the umbral dots was determined. (Oda 1984, Rodríguez 1995).

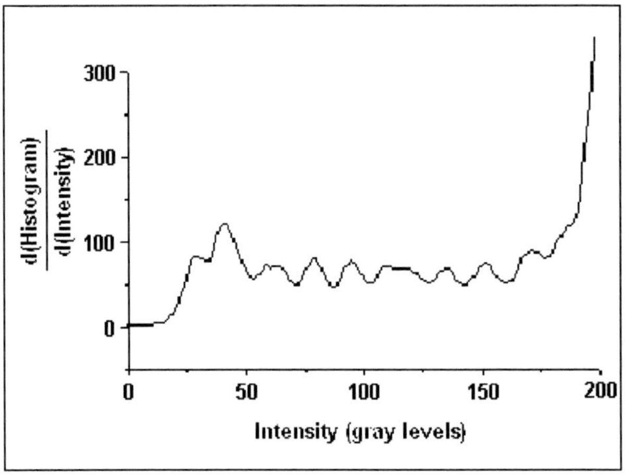

Figure 13. Cumulative histogram derivative of image 0510b.

The rings corresponding to greater values of the intensity are more distant from the center of the umbra where the umbral nuclei lay, in agreement with the results of Sobotka et al. (1993).

Figure 14. The dark region, in central part of the umbra, represent to umbral nucleus in the image 0310b. The engrossments in the ring, rounding to umbral nucleus, are the umbral dots.

Size distribution of umbral dots shows a mean values of (0.40 ± 0.08) arcsec (see image mean values in table X), in agreement with (0.42 ± 0.10) arcsec reported by Sobotka el at.

(1993) using photometric method, with a significant level of $\alpha = 0.001$. Notice the maximum in the little size region of the distribution with an approximate value of 0.3 arcsec (figure 15). Because the relative importance of this maximum respect to the main maximum changes with the evolution of the spot, been the 0.3 arcsec maximum less important in the more developed spot. We considered the difference between our observed sizes results (photometric measurements) and reports in the literature could be explained by differences in the stages of development of the spot used in the studies. Any way the hypothesis of very fine umbral dots in the limiting resolution of 0.3 arcsec of Sobotka et al. (1993) is in agreement with our measurements.

Table X.- Umbral dot sizes results. The mean values, the standard deviation (σ) and the total number of measured granules in each image.

Image	Mean value [arcsec]	σ [arcsec]	N
0210b	0.45	0.08	45
0310a	0.44	0.08	36
0310b	0.40	0.09	39
0310c	0.39	0.11	36
0310d	0.39	0.07	38
0410a	0.38	0.10	37
0410b	0.42	0.09	38
0410c	0.40	0.07	38
0410d	0.39	0.07	36
0510a	0.39	0.06	35
0510b	0.39	0.06	37

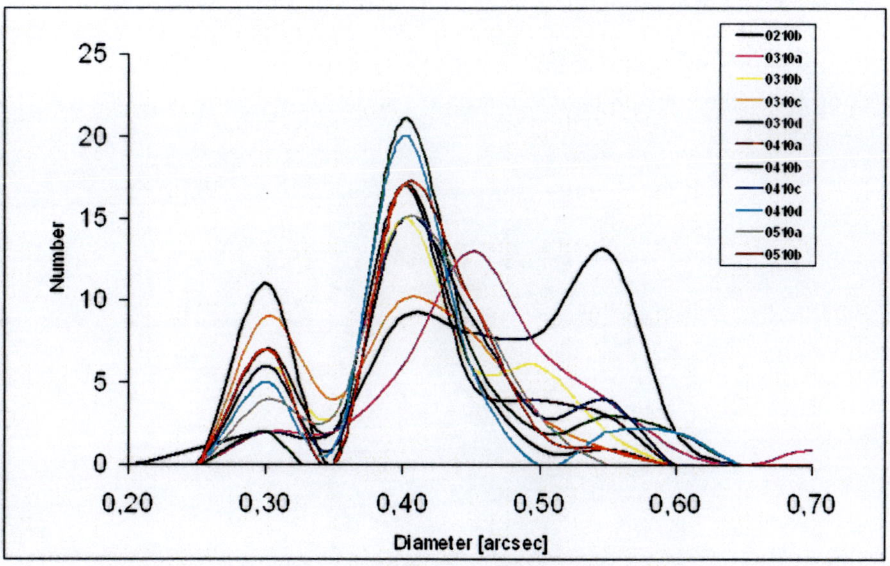

Figure 15. Continued on next page.

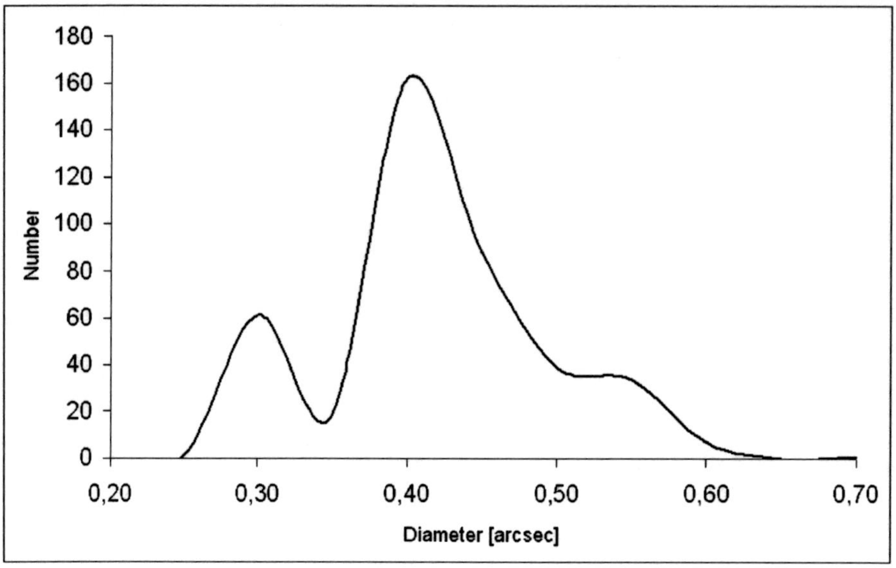

Figure 15. Size Distributions of umbral dots. Top - distributions corresponding to the umbral dot size measurements in each image. Bottom - Total distribution of umbral dot sizes

Kinematic of Umbral Dots

To analyze the kinematics of umbral dots was selected an appropriate sequence of images. Taking into account the speed and characteristic times of the evolution of umbral dots reported by other authors, we selected the images 0310b, 0310c and 0310d. The time interval between successive images is less than 20 minutes. From these images were selected the region that includes the main sunspot umbra and we analyzed the differences between the photometric features in them. To assure the equivalence of the selected regions, they were prepared considering the most stable details in the umbral nucleus. To highlight the movement of the features we proceed to generate a series of images subtracting the preceding from the considered one.

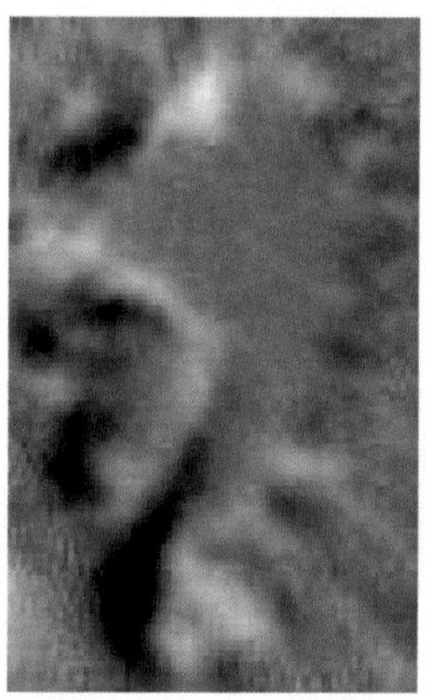

Figure 16. Results of the image processing procedure applied to the sequence of the October 3rd, 1978 (images 0310b, 0310c and 0310d).

Image in figure 16 is the result of image processing in October 3rd, 1978. The first image was subtracted from the second image; the second one was subtracted from the third one and then combined to show the movement from dark to bright regions. The umbral nucleus variations were then visible, but not the umbral dot movements. To do that we used the rings in which the changes in the position of the dots are much more noticeable.

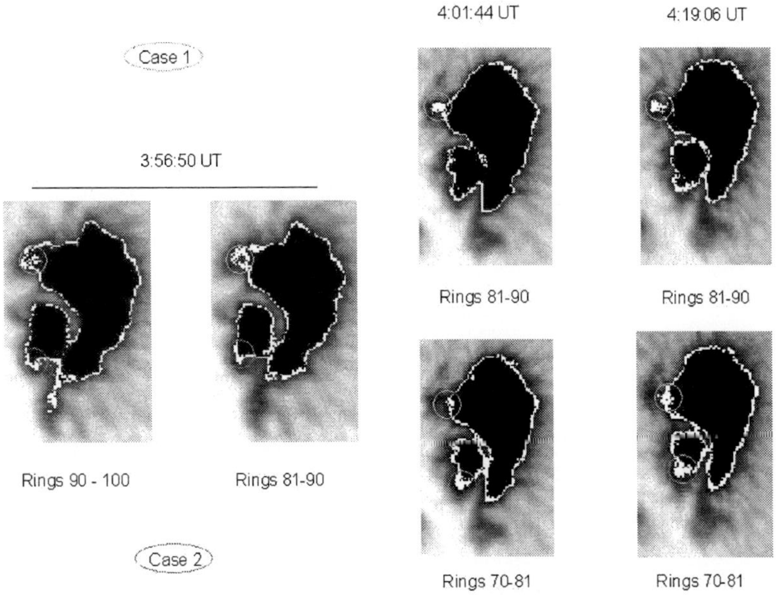

Figure 17. Displacement of the umbral dots measured. Here two cases of umbral dot movements are shown.

In figure 17 two cases of umbral dot movements are shown. The displacement is directed to the darker regions in the umbra (the umbral nucleus). The measurements of 9 umbral dots movement velocity is in the range (0.3 ± 0.1) km/s. This value is in agreement with the reports of 320 m/s (Kitai 1986), 100-200 m/s (Lites et al. 1991) and 0.5 km/s (Ewell 1992).

The umbral dots we measured are of the type classified by other authors as peripherical ones. We do not notice umbral dots inside the umbral nucleus. This could be related to the fact that we studied only one stage of development of the sunspot study case.

Characterization of Umbral Nucleus

Umbral nuclei were analyzed too. The initial exploration of the image shows that they lay close to each other, and that the external ones have intensities similar to those umbral dots close to the inner umbral nuclei.

We do not individualize the umbral nuclei in each image; we measure the total area slicing the image at the dark nucleus maximum plus one sigma value, as applied previously in the "characterization of the group" section. The values obtained appear in table VII.

Brants and Zwaan (1982) and Antalová (1991) studied the relation between the magnetic field strength and the umbral and penumbral areas ratio (coefficient K). We have not magnetic field data for the moment of the photoheliograms. But we can analyze the ratio between the umbral nucleus area and the umbral areas (K_{nu}). We determined (see figure 18) that such structures will not be present in sunspots with k less than 1.38.

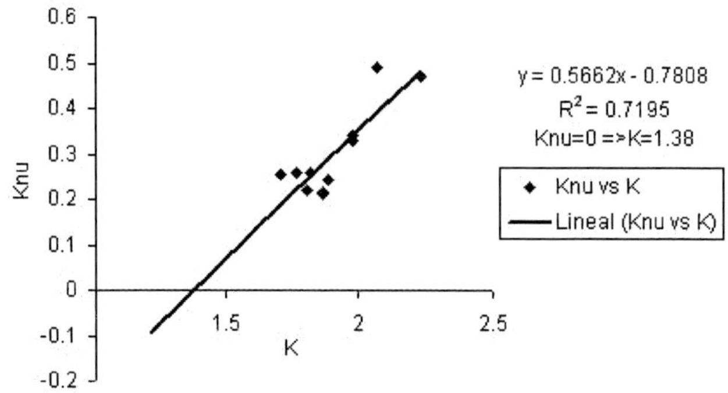

Figure 18. Diagram of the relationship between K and Knu indexes. From the linear regression it is concluded that spots with K <1.38 will not present umbral nucleus.

Conclusion

The method developed to study the morphological properties of some photospheric fine structures is applicable to digitized photographic materials. This allows to rescue old high-quality observational material to be used in solar research.

The variability of the measured areas observed in temporal scales from 24 to 0.5 hours is approximately the same. A correlation time of the measurements less than 0.5 hours points to characteristic times of the umbral changes in the tens of minutes scale.

The spatial pattern distribution and surface density of the granules showed that bright granules distribute to form the boundary of cells with a mean size in the range of the mesogranulation size (~8 arcsec), in agreement with other works found in the literature. We have not found a significant correlation between the size of the cells delimited by the bright granules and the magnetic field strength at a distance of 65 arcsec from the main sunspot center. This result is in agreement with the no significant difference in the granules surface density in regions near (17-26 arcsec) and far (75-84 arcsec) from sunspots. We conclude that the surface density of granules does not change with the distance to the strong magnetic field source (varying from 1600 to 3100 Gauss), at least farther than 17 arcsec.

At the studied distances from the sunspot umbra, we were not able to detect clear variations of the granular pattern that would imply that the convective motion of the plasma in the photosphere could be affected by the variability of the magnetic field strength. The stability of the photospheric mesogranular scale pattern, at the same distance but with different magnetic field intensities in the spot, implies no influence can be derived in the subphotospheric region in which the mesogranular pattern is assumed generated.

Using the cumulative histogram a method to determine the areas of the umbral nucleus, the umbral and sunspot areas was developed and the coefficient K was calculated for eleven images. The obtained mean value of k is 1.91 ±. 0.11, different to the value reported by Antalová (1991), but close to the value reported by Brandt et al. (1990) We considered that this difference could be explained by the absent of a precise mathematical definition of the limits of the sunspot features.

No matter we do not determined a relation of the k coefficient with the magnetic field strength; we found a limit to the appearance of umbral nucleus in sunspots depending from the K coefficient value. It means that if there is a relation between the magnetic field strength and K there is a minimal intensity of the field to the umbral nucleus to appear.

The densitometric scales in negatives allowed dynamical studies of high and low contrast structures from the same source. The cumulative histogram derivative proved to detect umbral nuclei and dots, features that by their low contrast are not clearly visible in the cumulative histogram.

The umbral nuclei are clustered in the more internal part of the umbra. Umbral dots distribute spatially forming rings resembling the border of the clustered darker umbral nuclei and are brighter as they appear farther from the umbral nuclei.

Considering the temporal evolution of the distribution of the umbral dot sizes obtained it was possible to notice that they probably came in two distinctive sizes, 0.3 arcsec and 0.4 arcsec, but the relative importance of 0.3 arcsec maximum respect to the main maximum (0.4 arcsec) changes with the evolution of the spot, been the first one less important in the more developed spot. We considered the difference between our observed sizes results (photometric measurements) and reports in the literature could be explained by differences in the stages of development of the spot used in the studies.

Any way the hypothesis of very fine umbral dots in the limiting resolution of 0.3 arcsec of Sobotka et al. (1993) is in agreement with our measurements. Following this result it possible to consider a limit value for the dimension of elementary tubes in magnetic ropes at photospheric level.

References

Adelatif, T. E., Lites, B. W., Thomas, *J.H. ApJ* 1986, Vol. 311, 1015-1024.

Antalová, A. *BAICz* 1991, Vol. 42, 316-319.

Balthasar, H., Küveler, G., Wiehr, *E. SoPh* 1987, Vol. 112, 37-48.

Brandt, P. N., Schmidt, W., *Steinegger, M. SoPh* 1990, Vol. 129, 191-194.

Brants, J. J., Zwaan, C. *SoPh* 1982, Vol. 80, 251-260.

Bray, R .J ., Loughhead R.E. Sunspot, *The International Astrophysics Series*, Vol. 7, Chapman and Hall Ltd., London, 1964, 240 p.

Collados, M., Marco, E., Del Toro, J.C., Vazquez, M. *SoPh* 1986, Vol. 105, 17 – 25.

Doval, J. P., Rodríguez, E. R., Rodríguez, R. E. *Revista de Geofísica* 1999, Vol. 50, 55 – 63 (in Spanish).

Ewell, M. W. Jr. *SoPh* 1992, Vol. 137, 215-223.

Györi, L., Baranyi, T., Csepura, G., Gerlei, O., Ludmány, A. *JAD* 1998, Vol. 4, 119-124.

Grossmann-Doerth, U., Schmidt, W., Schröter, E. H. *A&A* 1986, Vol. 156, 347-353.

Hirzberger, J. *A&A* 2002, Vol. 392, 1105 – 1118.

Jain, R., Muller, R. *SoPh* 1989, Vol. 123, 185 – 188.

Kitai, R. *SoPh* 1986, Vol. 104, 287-301.

Lites, B. W., Bida, T. A., Johannesson, A., Scharmer, G. B. *ApJ* 1991, Vol. 373, 683-694.

Livingston, W. *Nature* 1991, 350, 45.

Macris, C., Prokakis, Th., Dialetis, D., Muller, R. *SoPh* 1989, Vol. 122, 209-213.

Massey, Ph. A user´s guide to CCD reductions with IRAF, version 2.1, *National Optical Astronomy Observatories,* AZ, 1997, pp 7-19.

Muller, R., Roudier, Th., Vigneau, J. *SoPh* 1990, Vol. 126, 53 - 67.

Müller, D. A. N., Steiner, O., Schlichenmaier, R., Brandt, P. N. *SoPh* 2001, Vol. 203 211 – 232.

Oda, N. *SoPh* 1984, Vol. 93, 243 - 255.

Parfinenko, L. D., 1999, *Personal Communication.*

Parfinenko, L. D., Mikhalev, V. F. *BSolD* 1978, Vol. 8, 92 - 97 (in Russian).

Pettauer, T., Brandt, P. N. *SoPh* 1997, Vol. 175, 197-203.

Rast, M. P. *ApJ* 1995, Vol. 443, 863-868.

Rast, M. P. *ApJ* 1999, Vol. 524, 462-468.

Rast, M. P. *ApJ* 2003a, Vol. 597, 1200-1210.

Rast, M. P. Proceeding of SOHO 12/GONG+ 2002 "Local and global helioseismology: The Present and Future", Ed. H. Sawaya-Lacoste, Big Bear Lake, CA, 2003b, ESA SP-517, pp 163-172.

Rodríguez, E. R. *Análisis digital de imágenes con alta resolución de la fotosfera solar*, Thesis, Universidad de La Habana, Havana, 1995, pp 1-100 (in Spanish).

Rodríguez, E. R., Rodríguez, R. E., Doval, J. P. (1997a), *Análisis digital de imágenes con alta resolución de la granulación solar*, Memorias de Geoinfo, E-mail to request: iga@iga.cu (in Spanish)

Rodríguez, E. R., Gámez, R. (2000), "Nuevo método digital para el estudio morfológico de la granulación fotosférica", *Memorias de Geoinfo*, E-mail to request: iga@iga.cu (in Spanish)

Rodríguez, E. R., Rodríguez, R. E., Gámez, R. *Geofísica Internacional* 2004, Vol. 43, 97 – 101.

Rodríguez , E. R., Gámez, R., Rodríguez, R. E. *JASTP* 2005, Vol. 67, 1728 – 1733.

Rodríguez, R. E., Doval, J. P., Rodríguez, E. R. (1997b), *Análisis digital de imágenes con alta resolución de manchas solares*, Memorias de Geoinfo, E-mail to request: iga@iga.cu (in Spanish)

Roudier, Th., Muller, R. *SoPh* 1987, Vol. 107, 11 – 26.

Schmidt, W., Grossmann – Doerth, U., Schröter, E.H. *A&A* 1988, Vol.197, 306 – 310.

Shine, R. A., Simon, G. W., Hurlburt, N. E. *SoPh* 2000, Vol. 193, 313 – 33.

Sobotka, M., Bonet, J. A., Vázquez, A. *ApJ* 1993, Vol. 415, 832-846.

Stein, R., Nordlund, A. *ApJ* 1998, Vol. 499, 914.

Weiss, N.O., Thomas, J.H., Brummell, N.H., Tobias, S.M. *ApJ* 2004, Vol. 600, 1073-1090.

In: Solar Physics Research Trends
Editor: Pingzhi Wang, pp. 381-399
ISBN: 978-1-60021-987-0
© 2008 Nova Science Publishers, Inc.

Chapter 12

PROPAGATION OF RADIOWAVES AT FREQUENCIES BELOW THE ELECTRON GYROFREQUENCY FROM THE EARTH SURFACE TO THE HIGHER ATMOSPHERE

*V. M. Krasnov**

Institute of ionosphere, Ministry of Education and Science of Kazakhstan, Almaty, Kazakhstan

Abstract

Satellite experiments convincingly showed opportunity of penetration of radiowaves of VLF and LF- signals radiated by ground transmitters and earthquakes above F2-layer of the ionosphere (e.g. French satellite "DEMETR"). Up today the Appleton-Lassen-Hartree formula is used for calculation of dielectric permeability of ionospheric plasma. It is impossible to explain from this formula the propagation of radiowaves of a frequency range below the electron gyrofrequency from the earth surface up to heights above F2-layer of the ionosphere. The purpose of the work is to develop the theory of propagation of radiowaves in the magnetised ionospheric plasma, which explains the penetration VLF and LF- signals above F2-layer of the ionosphere. The problem was resolved in a common case, when the dependence between a vector of an electric displacement and an electric intensity is not linear (in a contrast to the solution of this problem earlier). New formula was developed for dielectric permeability of the ionospheric plasma. We also develop the formulae to calculate the critical frequencies of F2-layer of the ionosphere. The result of the frequency calculations with helping of the new formulae, practically, consists with the old ones, when a frequency of a radio wave is higher than the electron gyrofrequency. Thus, it will be difficult to show experimentally advantage of the new or old formulas. At the same time we have found conditions, when a radio wave has a frequency below than the electron gyrofrequency and can propagate upward to the altitudes above F2-layer of the ionosphere. The developed formulae can be also new theoretical basis of ray tracing, where the radio wave is used to probe the structure of the ionosphere; and in radio navigational systems and in radio geodesy, as well as in communication-system prediction and management. It is particular useful when detailed

* E-mail address: Valerie_krasnov@hotmail.com

knowledge about the interaction of the radio wave and ionosphere is required, such as in over-the-horizon radar and in single location of distant radio transmitters.

1. Introduction

Satellite experiments convincingly showed opportunity of penetration of radiowaves of VLF and LF- signals radiated by ground transmitters and earthquakes above F-layer of the ionosphere (e.g Molchanov et all, 2006). Up today for calculation of dielectric permeability (ε) of ionospheric plasma the formula (collision neglected) is used (e.g., Budden, 1961; Ginzburg, 1967)

$$\varepsilon = 1 - \frac{2w(1-w)}{2(1-w) - u\sin^2\alpha \pm \sqrt{u^2 \sin^4\alpha + 4u(1-w)^2\cos^2\alpha}}, \quad (1)$$

where $w = N/N_m$, $N_m = 1.24 f^2 10^{10}$ electron$/m^3$, $u = \Omega^2/\omega^2$, α is the angle between direction of the geomagnetic field and axis z, $\Omega = 2\pi f_H = qB/m$ is the electron gyrofrequency (we keep the sigh of Ω), q is a particle charge (negative for electron and positive for ion), B is magnetic field, m is mass. The sign "+" is for ordinary waves (O- wave) and "-"is for extraordinary waves, axis z is directed upward.

Equating the left part of equation (1) to zero we can define the critical frequencies of reflection of radiowaves for known altitude of a maximum of electronic concentration of F2-layer of the ionosphere. Reflection of an ordinary wave occurs on a frequency that equals to plasma frequency ω_0, when $w = 1$

$$f_0 = \omega_0/(2\pi) \quad (2)$$

(as though there is no the magnetic field in initial conditions of the problem). There are two opportunities of reflection for an extraordinary wave: when $w = 1 - \frac{|\Omega|}{\omega}$ and when $w = 1 + \frac{|\Omega|}{\omega}$. Hence we have

$$\omega_x = 2\pi f_x = |\Omega|/2 + \sqrt{\Omega^2/4 + \omega_0^2} \text{ and } \omega_z = 2\pi f_z = -|\Omega|/2 + \sqrt{\Omega^2/4 + \omega_0^2} \quad (3)$$

In the first case, a wave is designated as a X-wave, in the second - as a Z-wave. In practice as a rule ω_0 exceeds $|\Omega|$, and it is impossible to explain from expression (1) the propagation of radiowaves of a frequency range below module of the electron gyrofrequency from the earth surface up to heights above F2-layer of the ionosphere. The purpose of the work is to develop the theory of propagation of radiowaves in the magnetised ionospheric plasma, which explains the penetration VLF and LF- signals above F2-layer of the ionosphere.

2. Development of Equations

Up today it is known the rigorous solution of the equation of radiowave propagation in stratified inhomogeneous plasma only for the case, when the geomagnetic field is neglected (e.g., Budden, 1961; Ginzburg, 1967). It is shown that a layer reflects a radiowave, if a frequency of a radiowave is below the plasma frequency of a layer maximum. The height of radio wave reflection also depends on the value of the conductivity gradient (Krasnov, 2002). Only neglible small fraction of the energy of an incident wave can propagate upwards from a reflection point. The depth of electromagnetic wave penetration above the reflection point depends on the initial phase of the incident wave (Krasnov, 2001). The field is vanished in a narrow layer of plasma above a point of reflection. Thus, these theories can not explain propagation of VLF and LF - wave upward through F2- layer of the ionosphere. When the geomagnetic field is taken into account, the rigorous solution of the wave equation is solved only for the case of homogeneous plasma. It is supposed that dependence between a vector of an electric displacement ($\vec{\mathbf{D}}$) and an electric intensity ($\vec{\mathbf{E}}$) is linear (e.g., Budden, 1961; Ginzburg, 1967). We also shall consider the solution of the problem for the case of homogeneous plasma, but we shall not impose the assumption of linearity of dependence between $\vec{\mathbf{D}}$ and $\vec{\mathbf{E}}$.

Let the vector of induction of a magnetic field $\vec{\mathbf{B}}$ is in a plane yz (Fig.1), and α is an angle between $\vec{\mathbf{B}}$ and an axis z, directed upwards. Let a monochromatic radio wave be vertically incident on the ionosphere and electric vector $\vec{\mathbf{E}}$ is in the plane xy. Thus α is also the angle between the direction of radiowave propagation and $\vec{\mathbf{B}}$.

2.1. The Radiowave, which Has Rotating Vector of an Electric Field

Let's consider waves with electric vectors ($\vec{\mathbf{E}}_1$ and $\vec{\mathbf{E}}_2$) rotated in the opposite directions

$$\mathbf{E_x = E_1 \sin(\omega t - k_1 z) - E_2 \sin(\omega t - k_2 z)} \tag{4}$$

$$\mathbf{E_y = E_1 \cos(\omega t - k_1 z) + E_2 \cos(\omega t - k_2 z)}, \tag{5}$$

where $\mathbf{E}_1$, $\mathbf{E}_2$ are the amplitudes of waves, $\vec{\mathbf{k}}_1$ and $\vec{\mathbf{k}}_2$ are the wave vectors, directed along z-axis.

To an observer looking in the direction of positive z (Fig.1) the electric vector $\vec{\mathbf{E}}_1$ would appear to rotate anticlockwise. This direction coincides with direction of an electron rotation, when electric field is missing. The rotation of $\vec{\mathbf{E}}_2$ occurs clockwise.

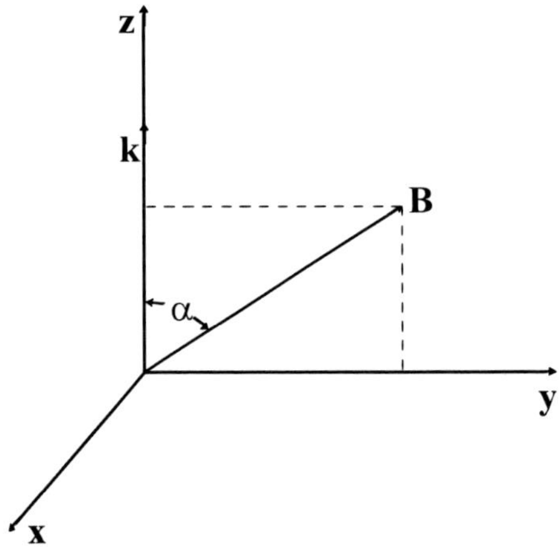

Figure 1. Geometry of the problem.

The equation of movement: $\mathbf{m}\frac{d\vec{v}}{dt} = q\left[\vec{E} + \left(\vec{v} \times \vec{B}\right)\right]$ - of the charged particles in the Cartesian coordinates for the accepted conditions of the problem becomes

$$\frac{dv_x}{dt} = \frac{q}{m}\left[E_1 \sin(\omega t - k_1 z) - E_2 \sin(\omega t - k_2 z) + v_y B_z - v_z B_y\right], \tag{6}$$

$$\frac{dv_y}{dt} = \frac{q}{m}\left[E_1 \cos(\omega t - k_1 z) + E_2 \cos(\omega t - k_2 z) - v_x B_z\right], \tag{7}$$

$$\frac{dv_z}{dt} = \frac{q}{m} v_x B_y \tag{8}$$

When α=0, the solution of the set of equations of type (6-8) for a sinusoidal wave usually is found by the approximate method (e.g. Chen, 1987). At the same time, it is possible to find the rigorous solution of this set of equations, when the angle α is arbitrary.

$$\begin{aligned} v_x = &-\frac{\Omega}{B(\omega^2 - \Omega^2)}(\omega + \Omega\cos\alpha) E_1 \cos(\omega t - k_1 z) + \\ &+ \frac{\Omega}{B(\omega^2 - \Omega^2)}(\omega - \Omega\cos\alpha) E_2 \cos(\omega t - k_2 z) \end{aligned}, \tag{9}$$

$$\begin{aligned} v_y = &\frac{\Omega}{B(\omega^2 - \Omega^2)}\left(\omega + \Omega\cos\alpha - \frac{\Omega^2}{\omega}\sin^2\alpha\right) E_1 \sin(\omega t - k_1 z) + \\ &+ \frac{\Omega}{B(\omega^2 - \Omega^2)}\left(\omega - \Omega\cos\alpha - \frac{\Omega^2}{\omega}\sin^2\alpha\right) E_2 \sin(\omega t - k_2 z) \end{aligned} \tag{10}$$

$$
\begin{aligned}
v_z = &-\frac{\Omega^2 \sin\alpha}{B(\omega^2 - \Omega^2)}\left(1 + \frac{\Omega}{\omega}\cos\alpha\right)E_1 \sin(\omega t - k_1 z) + \\
&+\frac{\Omega^2 \sin\alpha}{B(\omega^2 - \Omega^2)}\left(1 - \frac{\Omega}{\omega}\cos\alpha\right)E_2 \sin(\omega t - k_2 z)
\end{aligned} \qquad (11)
$$

Hence, we can find the trajectories of movement of charges in the assumption that path length of a charge is much less than the wave-length

$$
\begin{aligned}
X = &-\frac{\Omega}{B(\omega^2 - \Omega^2)}\left(1 + \frac{\Omega}{\omega}\cos\alpha\right)E_1 \sin(\omega t - k_1 z) + \\
&+\frac{\Omega}{B(\omega^2 - \Omega^2)}\left(1 - \frac{\Omega}{\omega}\cos\alpha\right)E_2 \sin(\omega t - k_2 z)
\end{aligned} \qquad (12)
$$

$$
\begin{aligned}
Y = &-\frac{\Omega}{B(\omega^2 - \Omega^2)}\left(1 + \frac{\Omega}{\omega}\cos\alpha - \frac{\Omega^2}{\omega^2}\sin^2\alpha\right)E_1 \cos(\omega t - k_1 z) + \\
&+\frac{\Omega}{B(\omega^2 - \Omega^2)}\left(1 - \frac{\Omega}{\omega}\cos\alpha - \frac{\Omega^2}{\omega^2}\sin^2\alpha\right)E_2 \cos(\omega t - k_2 z)
\end{aligned}, \qquad (13)
$$

$$
\begin{aligned}
Z = &\frac{\Omega^2 \sin\alpha}{B\omega(\omega^2 - \Omega^2)}\left(1 + \frac{\Omega}{\omega}\cos\alpha\right)E_1 \cos(\omega t - k_1 z) - \\
&-\frac{\Omega^2 \sin\alpha}{B\omega(\omega^2 - \Omega^2)}\left(1 - \frac{\Omega}{\omega}\cos\alpha\right)E_2 \cos(\omega t - k_2 z)
\end{aligned} \qquad (14)
$$

Figure 2 presents the trajectory of movement of an electron and ion in the different planes, when z=0, $E_1=E_2=0.1$v/m, $\mathbf{f_H = -1.2\ MHz}$, $\alpha=155°$ and $\mathbf{f = \omega/(2\pi) = 5\ MHz}$ (the case, when a radiowave frequency is more than module of the electron gyrofrequency). As a positive ion we choose the proton with charge $q=1.602\cdot10^{-19}$Q and mass $m=1.672\cdot10^{-27}$kg. So we accept the value of the proton gyrofrequency equal to Ω_p=820Hz. The wave-1 and the wave-2 rotates protons in different directions with identical radius of rotation. According to (12-14), the neglible small ratio of a proton gyrofrequency to a radiowave frequency, practically, eliminates the influence of a magnetic field on proton movement. As a result, the orbit of rotation is circular and the orbits are in the plane of incidence waves - xy. The position of particles is shown for the time t=0. The orbits of electron rotation are close to circular ones in a plane xy, but have different radius and slope in a plane zy. When $\alpha=155°$, the wave-1 rotates the electron against direction of its rotation in absence of electric field, and a wave-2 - in the opposite direction. The rotation of electrons and ions occurs synchronously: with the same angular speed ω. All changes of distances between electrons and ions occur in a plane zy. Every wave rotates an electron and an ion in the same direction, but with phase shift of 180°. These pairs of electrons and ions form the rotating electric dipoles directed counter to vectors $\vec{\mathbf{E}}_1$ and $\vec{\mathbf{E}}_2$. Thus, charged particles of plasma tend "to reduce the influence" of an external electric field. The distance between an electron and an ion is larger under action of a wave 1, than under action of a wave 2. Accordingly, in the second case the electric field of a dipole is more; and it is possible to expect the greater counteraction to an external electric field.

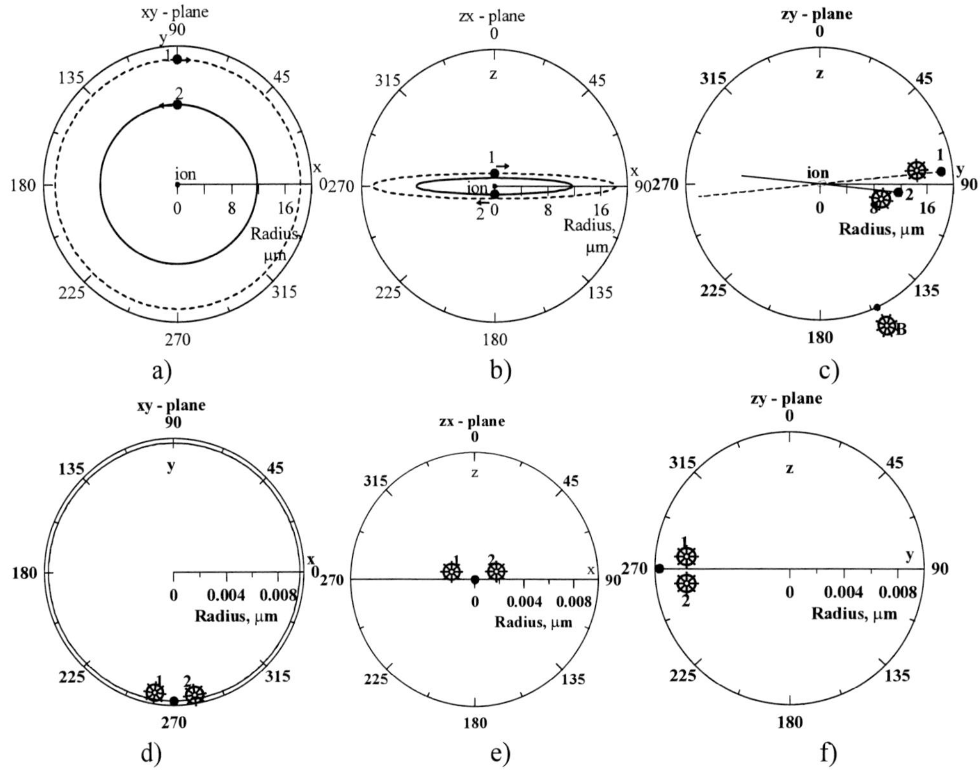

Figure 2. Trajectories of movement of an electron (a,b,c) and ion (d,e,f) under action of a wave 1 and 2 in different planes for f=5MHz. Arrows show direction of electron and ion movement, when t=0.

Figure 3 indicates trajectories of movement of an electron and ion in various planes for the same conditions, as in figure 2, but for f=20 kHz ($|\Omega/\omega| >> 1$). The comparison of figure 2 and 3 shows that movement of electrons relative to ions for radiowaves with frequency higher and lower than module of the electron gyrofrequency is essentially different. In the first case, the movement is basically defined by direction of an electric field and occurs mainly in the plane of vectors $\vec{\mathbf{E}}_1$ and $\vec{\mathbf{E}}_2$, and in the second - by direction of a geomagnetic field and occurs mainly in a plane of a vector $\vec{\mathbf{B}}$. Other important differences in figure 3 are: rotation of electrons under action of both waves occurs in one direction in a plane xy, and the radiuses of rotation differ from each other a little. The picture of rotation of protons in figure 3 has a little changed in comparison with figure 2. Little changes in orbits are connected to increase of ratio of the proton gyrofrequency to the frequency of a radiowave, and accordingly, due to the greater influence of a magnetic field. As well as in the first case the wave 2 rotates an electron and an ion in the same direction with phase shift of 180°. This pair of electrons and ions forms the rotating electric dipole directed counter to a vector $\vec{\mathbf{E}}_2$. The field of the electric dipole under action of the wave 1 becomes collinear, and it will be the maximal "compensation" of the influence on plasma of incident waves, when the electric dipoles line up along an axis y. On the other directions the "compensation" of electric field is smaller. Evidently, the qualitative distinction of the plasma response to a radiowave frequency higher and lower of module of an electron gyrofrequency should lead to essentially different conditions of radiowave propagation.

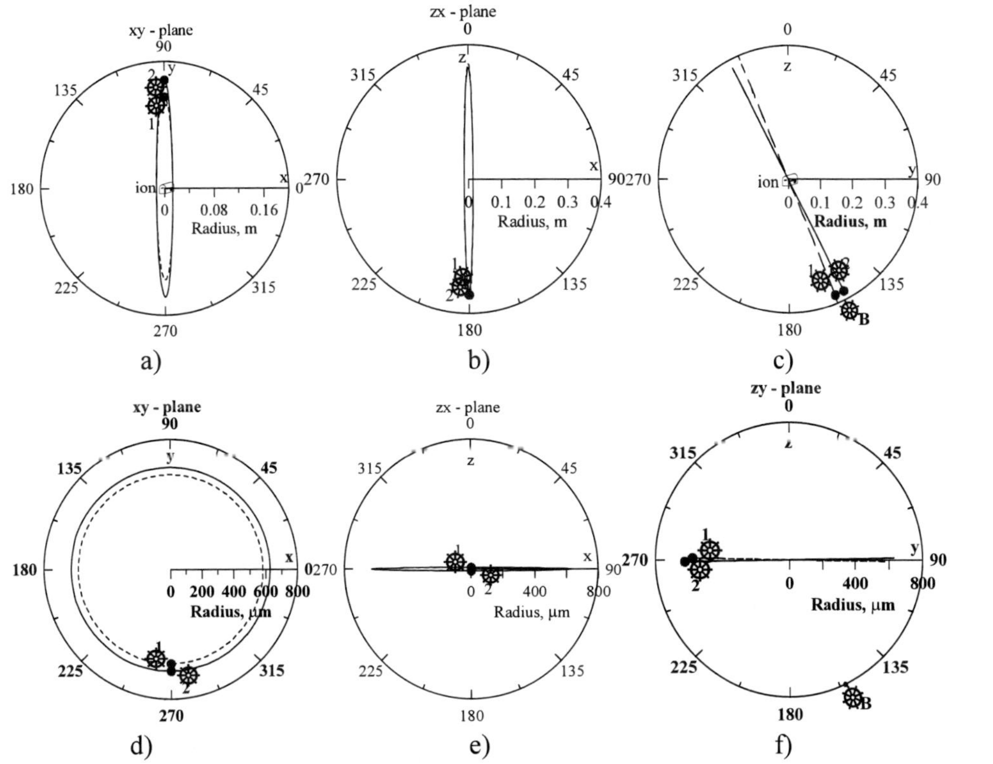

Figure 3. Trajectories of movement of an electron (a,b,c) and ion (d,e,f) under action of a wave 1 and 2 in different planes for f=20kHz. Arrows show direction of electron and ion movement, when t=0.

In the general case the vector of polarization is defined by a dipole moment of unit of volume of plasma (e.g., *Budden*, 1961; *Ginzburg*, 1967) $\vec{\mathbf{P}} = \sum_{N} \mathbf{e}\vec{\mathbf{R}}$, where N is number of electrons in the volume unit, R is a distance between an electron and positive ion. As the radius of rotation of electron around of ion is much more than radius of rotation of ion, last can be equated to zero, and then r will be radius of rotation of electron. Expressions for components of the polarization vector will become (we assume the sinusoidal wave should not lead to occurrence of a constant component of a polarization vector; otherwise the wave equation will not be satisfied)

$$\mathbf{P_x} = -\frac{\mathbf{A_1}}{4\pi}\mathbf{E_1}\sin(\omega t - \mathbf{k_1 z}) + \frac{\mathbf{A_2}}{4\pi}\mathbf{E_2}\sin(\omega t - \mathbf{k_2 z}), \tag{15}$$

$$P_y = -\frac{A_3}{4\pi}E_1\cos(\omega t - k_1 z) - \frac{A_4}{4\pi}E_2\cos(\omega t - k_2 z), \tag{16}$$

$$P_z = \frac{A_5}{4\pi}E_1\cos(\omega t - k_1 z) - \frac{A_6}{4\pi}E_2\cos(\omega t - k_2 z) \tag{17}$$

where

$$\omega_0^2 = \frac{4\pi e^2 N}{m},\ A_1 = \frac{\omega_0^2}{(\omega^2 - \Omega^2)}\left(1 + \frac{\Omega}{\omega}\cos\alpha\right),$$

$$A_2 = \frac{\omega_0^2}{(\omega^2 - \Omega^2)}\left(1 - \frac{\Omega}{\omega}\cos\alpha\right),\ A_3 = \frac{\omega_0^2}{(\omega^2 - \Omega^2)}\left(1 + \frac{\Omega}{\omega}\cos\alpha - \frac{\Omega^2}{\omega^2}\sin^2\alpha\right),$$

$$A_4 = \frac{\omega_0^2}{(\omega^2 - \Omega^2)}\left(1 - \frac{\Omega}{\omega}\cos\alpha - \frac{\Omega^2}{\omega^2}\sin^2\alpha\right), \tag{18}$$

$$A_5 = \frac{\omega_0^2\Omega\sin\alpha}{\omega(\omega^2 - \Omega^2)}\left(1 + \frac{\Omega}{\omega}\cos\alpha\right),\ A_6 = \frac{\omega_0^2\Omega\sin\alpha}{\omega(\omega^2 - \Omega^2)}\left(1 - \frac{\Omega}{\omega}\cos\alpha\right).$$

By definition the vector of an electric displacement is equal:

$$\vec{D} = \vec{E} + 4\pi\vec{P}.$$

Taking into account (4,5) and (15-17) its components are equal

$$D_x = (1 - A_1)E_1\sin(\omega t - k_1 z) - (1 - A_2)E_2\sin(\omega t - k_2 z), \tag{19}$$

$$D_y = (1 - A_3)E_1\cos(\omega t - k_1 z) + (1 - A_4)E_2\cos(\omega t - k_2 z) \tag{20}$$

$$D_z = A_5 E_1\cos(\omega t - k_1 z) - A_6 E_2\cos(\omega t - k_2 z). \tag{21}$$

If we neglect in (9-11) by terms containing Ω^2/ω^2, then V_z=0 and rotation of charged particles will have a circular orbits with a constant amplitude of the linear speed. The moment of inertia of particles relative to rotation axis remains constant with time, and accordingly their kinetic energy also is a constant. As a result the work made by the external fields, is equal to zero. When we take into account the terms containing Ω^2/ω^2, the moments of inertia of particles change in time. As a result the kinetic energy of particles also changes due to the work of external fields, and external fields lose a part of the energy. In turn, the loss of energy in plasma is possible at occurrence the oscillatory current - $\vec{\mathbf{J}}_{os}$ (e.g. Chen, 1987). In our case this current should have components J_{osz} and J_{osy}. Thus the Maxwell equations take the form

$$\mathbf{rot}\vec{\mathbf{B}} = \frac{1}{c_0}\frac{\partial \vec{\mathbf{D}}}{\partial t} + \frac{4\pi}{c_0}\vec{\mathbf{J}}_{os} \text{ и } \mathbf{rot}\vec{\mathbf{E}} = -\frac{1}{c_0}\frac{\partial \vec{\mathbf{B}}}{\partial t}.$$

We obtain for the case of a flat wave, when field changes only along z - axis:

$$-\frac{\partial \mathbf{B}_y}{\partial z} = \frac{1}{c_0}\frac{\partial \mathbf{D}_x}{\partial t}, \tag{22}$$

$$\frac{\partial \mathbf{B}_x}{\partial z} = \frac{1}{c_0}\frac{\partial \mathbf{D}_y}{\partial t} + \frac{4\pi}{c_0}\mathbf{J}_{osy}, \tag{23}$$

$$0 = \frac{\partial \mathbf{D}_z}{\partial t} + 4\pi \mathbf{J}_{osz}, \tag{24}$$

$$\frac{\partial \mathbf{E}_y}{\partial z} = \frac{1}{c_0}\frac{\partial \mathbf{B}_x}{\partial t}, \tag{25}$$

$$\frac{\partial \mathbf{E}_x}{\partial z} = -\frac{1}{c_0}\frac{\partial \mathbf{B}_y}{\partial t}. \tag{26}$$

The expression (24) shows that z - components of the displacement current and $\vec{\mathbf{J}}_{os}$ "equilibrates" each other.

$$\mathbf{J}_{osz} = -\frac{1}{4\pi}\frac{\partial \mathbf{D}_z}{\partial t} = \sigma_{z1}\mathbf{E}_1 \sin(\omega t - \mathbf{k}_1 z) - \sigma_{z2}\mathbf{E}_2 \sin(\omega t - \mathbf{k}_2 z),$$

where $\sigma_{z1`} = \dfrac{\omega A_5}{4\pi}$ and $\sigma_{z2`} = \dfrac{\omega A_6}{4\pi}$ are conductivity of plasma along axis z.

Thus, the account of a current saves us from introduction of the additional longitudinal component of field - $\mathbf{E}_z$ (e.g. Budden (1961), Ginzburg (1967)), and as a result to save the initial conditions of the problem, when we solve the movement equations (6-8).

Combining (26) and (22), it is possible to obtain

$$k_1^2 = \frac{\omega^2}{c_0^2}(1 - A_1) \text{ и } k_2^2 = \frac{\omega^2}{c_0^2}(1 - A_2),$$

from which the dielectric permeability of plasma is equal to:

$$\varepsilon_1 = (1 - A_1) = 1 - \frac{\omega_0^2}{(\omega^2 - \Omega^2)}\left(1 + \frac{\Omega}{\omega}\cos\alpha\right), \tag{27}$$

$$\varepsilon_2 = (1 - A_2) = 1 - \frac{\omega_0^2}{(\omega^2 - \Omega^2)}\left(1 - \frac{\Omega}{\omega}\cos\alpha\right). \tag{28}$$

Combining (25) and (23), it is possible to obtain y-component of a current

$$J_{osy} = \sigma_y \left[E_1 \sin(\omega t - k_1 z) + E_2 \sin(\omega t - k_2 z)\right],$$

where $\sigma_y = \frac{1}{4\pi}\frac{\Omega^2\omega_0^2 \sin^2\alpha}{\omega(\omega^2 - \Omega^2)}$ is plasma conductivity along y-axis.

As a result the expressions (19-21) become

$$D_x = \varepsilon_1 E_1 \sin(\omega t - k_1 z) - \varepsilon_2 \sin(\omega t - k_2 z),$$

$$D_y = (\varepsilon_1 + \frac{4\pi}{\omega}\sigma_y)E_1 \cos(\omega t - k_1 z) + (\varepsilon_2 + \frac{4\pi}{\omega}\sigma_y)E_2 \cos(\omega t - k_2 z),$$

$$D_z = \frac{4\pi}{\omega}\sigma_{z1} E_1 \cos(\omega t - k_1 z) - \frac{4\pi}{\omega}\sigma_{z2} E_2 \cos(\omega t - k_2 z).$$

These expressions show that in contrast to the solution of this problem earlier (e.g., *Budden*, 1961; *Ginzburg*, 1967) the dependence between vectors $\vec{D}$ and $\vec{E}$ is not linear due to a current. Components of conductivity are proportional to $\sin\alpha$. Thus, the current is absent on the magnetic poles, and the current is maximal next to the equator. Consequently the assumption of linearity is valid only, when $\alpha = 0$, and in this case it is possible to expect the coincidence of the new and old solutions. Really, when $\alpha = 0$, the expressions (27) and (28) take the form $\varepsilon_1 = 1 - \frac{\omega_0^2}{\omega(\omega + \Omega)}$ и $\varepsilon_2 = 1 - \frac{\omega_0^2}{\omega(\omega - \Omega)}$. On the other hand, when $\alpha = 0$, the expression (1) takes the form $\varepsilon = 1 - \frac{w(1 - w)}{(1 - w) \pm (1 - w)\sqrt{u}}$. By reducing the numerator and denominator of the second term by a factor $(1 - w)$, the solutions coincident with ε_1 and ε_2. At the same time we "lose" the reflection of an ordinary wave, when $\mathbf{w} = \mathbf{1}$ ($\omega = \omega_0$). This implies that an ordinary wave should "disappear" on magnetic poles. When $\alpha = 90^\circ$, the expressions (27) and (28) take the form $\varepsilon_1 = \varepsilon_2 = 1 - \frac{\omega_0^2}{\omega^2 - \Omega^2}$. At the same time the expression (1) takes the different form

$$\varepsilon_0 = 1 - \frac{\omega_0^2}{\omega^2} \text{ и}$$

$$\varepsilon_x = 1 - \frac{\omega_0^2(\omega^2 - \omega_0^2)}{\omega^2(\omega^2 - \omega_0^2 - \Omega^2)}.$$

Let us define the critical frequencies of a plasma layer for given ω_0. By equating to zero the expression (27) and (28), we can obtain two equations relative to ω: $\omega_1^3 - \omega_1(\Omega^2 + \omega_0^2) - \omega_0^2\Omega\cos\alpha = 0$; and $\omega_2^3 - \omega_2(\Omega^2 + \omega_0^2) + \omega_0^2\Omega\cos\alpha = 0$;

where $\omega_1 = 2\pi f_1$ is a variable from (27), and $\omega_2 = 2\pi f_2$ is variable from (28).

Each of these equations has three solutions

$$2\pi f_{1,1} = 2\sqrt{\frac{\Omega^2 + \omega_0^2}{3}}\cos\frac{\varphi}{3}; \tag{29}$$

$$2\pi f_{2,1} = -2\sqrt{\frac{\Omega^2 + \omega_0^2}{3}}\cos\frac{\varphi}{3};$$

$$2\pi f_{1,2} = -2\sqrt{\frac{\Omega^2 + \omega_0^2}{3}}\cos\left(60^\circ - \frac{\varphi}{3}\right);$$

$$2\pi f_{2,2} = 2\sqrt{\frac{\Omega^2 + \omega_0^2}{3}}\cos\left(60^\circ - \frac{\varphi}{3}\right); \tag{30}$$

$$2\pi f_{1,3} = -2\sqrt{\frac{\Omega^2 + \omega_0^2}{3}}\cos\left(60^\circ + \frac{\varphi}{3}\right); \tag{31}$$

$$2\pi f_{2,3} = 2\sqrt{\frac{\Omega^2 + \omega_0^2}{3}}\cos\left(60^\circ + \frac{\varphi}{3}\right), \tag{32}$$

where

$$\cos\varphi = \frac{\omega_0^2 \Omega \cos\alpha}{2\left(\frac{\Omega^2 + \omega_0^2}{3}\right)^{3/2}}.$$

In the Southern hemisphere α changes from 0^0 on a magnetic pole up to 90^0 next to magnetic equator, and in Northern hemisphere - from 90^0 next to magnetic equator up to 180^0 on a magnetic pole. Accordingly, taking into account a sign of Ω, $\cos\varphi$ is positive in the Northern hemisphere and negative in the Southern hemisphere. The angle $\varphi/3$ changes in the range from 0 up to 30^0 in the Northern hemisphere and from 30 up to 60^0- in Southern hemisphere. This implies that the positive values can accept roots of the equations: $f_{1,1}, f_{1,3}, f_{2,2}$ and $f_{2,3}$, which only have physical meaning.

2.2. Radio Wave, Which Has Constant Direction of a Vector of an Electric Field

Let the electric vector $\vec{E}$ be parallel to the x-axis in figure 1: $E_x = E_0 \cos(\omega t - kz)$, where E_0 is an amplitude and k is a wave vector directed along z-axis.

The solution of the equation of movement for this wave takes the form:

$$v_x = E_0 \frac{\Omega\omega}{B(\omega^2 - \Omega^2)} \sin(\omega t - kz),$$

$$v_y = E_0 \frac{\Omega^2 \cos\alpha}{B(\omega^2 - \Omega^2)} \cos(\omega t - kz),$$

$$v_z = -E_0 \frac{\Omega^2 \sin\alpha}{B(\omega^2 - \Omega^2)} \cos(\omega t - kz).$$

The components of a vector of an electric displacement are equal to

$$D_x = E_0 \left(1 - \frac{\omega_0^2}{\omega^2 - \Omega^2}\right) \cos(\omega t - kz),$$

$$D_y = E_0 \frac{\omega_0^2}{\omega^2 - \Omega^2} \frac{\Omega}{\omega} \cos\alpha \sin(\omega t - kz),$$

$$D_z = -E_0 \frac{\Omega\omega_0^2 \sin\alpha}{\omega(\omega^2 - \Omega^2)} \sin(\omega t - kz).$$

The Maxwell equations take the form

$$-\frac{\partial B_y}{\partial z} = \frac{1}{c_0} \frac{\partial D_x}{\partial t}, \tag{33}$$

$$0 = \frac{\partial D_y}{\partial t} + 4\pi J_{osy},$$

$$0 = \frac{\partial D_z}{\partial t} + 4\pi J_{osz}$$

$$\frac{\partial E_x}{\partial z} = -\frac{1}{c_0} \frac{\partial B_y}{\partial t}. \tag{34}$$

The expressions show that y and z - components of the displacement current and $\vec{J}_{os}$ "equilibrate" each other. Combining (33) and (34), it is possible to obtain $\varepsilon_{per} = 1 - \frac{\omega_0^2}{(\omega^2 - \Omega^2)}$, whence the critical frequency of plasma layer for given ω_0 equals to

$$2\pi f_{per} = \sqrt{\omega_0^2 + \Omega^2}. \tag{35}$$

Now, let the electric vector E be parallel to the y-axis in figure 1: $E_y = E_0 \cos(\omega t - kz)$. The solution of the equation of movement for this wave takes the form:

$$v_x = -E_0 \frac{\Omega^2 \cos\alpha}{B(\omega^2 - \Omega^2)} \cos(\omega t - kz),$$

$$v_y = E_0 \frac{\Omega(\omega^2 - \Omega^2 \sin^2\alpha)}{\omega B(\omega^2 - \Omega^2)} \sin(\omega t - kz),$$

$$v_z = -E_0 \frac{\Omega^3 \sin\alpha \cos\alpha}{\omega B(\omega^2 - \Omega^2)} \sin(\omega t - kz).$$

The components of a vector of an electric displacement are equal to

$$D_x = -E_0 \frac{\omega_0^2}{\omega^2 - \Omega^2} \frac{\Omega}{\omega} \cos\alpha \sin(\omega t - kz),$$

$$D_y = E_0 \left[1 - \frac{\omega_0^2}{\omega^2 - \Omega^2}\left(1 - \frac{\Omega^2}{\omega^2}\sin^2\alpha\right)\right]\cos(\omega t - kz),$$

$$D_z = E_0 \frac{\Omega^2 \omega_0^2 \sin\alpha \cos\alpha}{\omega^2(\omega^2 - \Omega^2)} \cos(\omega t - kz).$$

The Maxwell equations take the form

$$0 = \frac{\partial D_x}{\partial t} + 4\pi J_{osx},$$

$$\frac{\partial B_x}{\partial z} = \frac{1}{c_0} \frac{\partial D_y}{\partial t}, \tag{36}$$

$$0 = \frac{\partial D_z}{\partial t} + 4\pi J_{osz},$$

$$\frac{\partial E_y}{\partial z} = \frac{1}{c_0} \frac{\partial B_x}{\partial t}, \tag{37}$$

The expressions show that x and z - components of the displacement current and $\vec{J}_{os}$ "equilibrate" each other.

Combining (36) and (37), it is possible to obtain

$$\varepsilon_{par} = 1 - \frac{\omega_0^2}{\omega^2 - \Omega^2}\left(1 - \frac{\Omega^2}{\omega^2}\sin^2\alpha\right). \tag{38}$$

Whence the critical frequencies of plasma layer for given ω_0 equal to

$$2\pi f_{par1} = \sqrt{\frac{\omega_0^2+\Omega^2}{2} + \sqrt{\frac{\left(\omega_0^2+\Omega^2\right)^2}{4} - \Omega^2\omega_0^2\sin^2\alpha}}\,, \tag{39}$$

$$2\pi f_{par2} = \sqrt{\frac{\omega_0^2+\Omega^2}{2} - \sqrt{\frac{\left(\omega_0^2+\Omega^2\right)^2}{4} - \Omega^2\omega_0^2\sin^2\alpha}}\,. \tag{40}$$

When $\vec{\mathbf{E}}$ and $\vec{\mathbf{B}}$ are parallel vectors (α=90°) the expressions (38-40) take the form

$$\varepsilon_{par} = 1-\frac{\omega_0^2}{\omega^2}\,,\ 2\pi f_{par1} = \omega_0\,,\ 2\pi f_{par2} = \Omega\,. \tag{41}$$

Thus, dielectric permeability of plasma is different for radiowaves, when $\vec{\mathbf{E}}$ is parallel to the x-axis or y-axis in Fig.1 (in the first case $\vec{\mathbf{E}}$ is perpendicular to the yz-plane, which includes $\vec{\mathbf{B}}$; in the second case $\vec{\mathbf{E}}$ is in the yz-plane). Hence, the radiowaves have different propagation speeds and different reflection frequencies.

3. Critical Frequencies of F2-Layer

Let us compare the result of calculations of radiowave reflection frequencies on the basis of the formulae (2, 3) and (29-32), (35), (39), (40). First of all, it is necessary to note, that these expressions are rigorously valid only for a case of a homogeneous layer of plasma. Let the ionospheric profile be a step function with height, and electronic concentration is homogeneous inside each step. In this case the critical frequencies of F2-layer can be estimated as a first approximation under these formulae. For example, let us calculate f_0F2 and h_mF2 with the help of the IRI model (Bilitza, 2000) for geomagnetic latitudes in the range from -90° to 90°; a geomagnetic longitude 56.63°E; a quiet magnetic day on January, 15, 2006; t=17.00LT; A_p=6; F10.7=78.3. Figure 4a,b presents latitudinal dependence of f_0F2 and h_mF2. We used International geomagnetic Reference Field (IGRF/The 10th generation) to calculate the latitudinal dependence of B (Fig.4c) and inclination I (Fig.4d) of the geomagnetic field at the height of h_mF2. Figure 4d presents also the latitudinal dependence of α=90+I. It is shown that the angle α is changing, practically, in the all range of possible values, and thus the analysis of different situations is possible. Besides it is shown, that α can accept the same values for different latitudes. Considering also, that formulas (2,3) do not depend on α, we shall draw a graph of critical frequencies depending on geomagnetic latitude, instead of on α. Figure 5a presents the results of calculation of f_o, f_{per} and f_{par1}. The curves f_{per} and f_{par1} are practically in agreement. Additionally, these curves well agree with f_o for the most part of middle and equatorial latitudes. Visible distinctions between curves f_o and f_{per},f_{par1} are observed only at high latitudes, and especially next to geomagnetic poles, when α is close to 0 or 180°. In the general case, the expressions (35) and (39) depend on a magnetic field. The dependence on a magnetic field disappears only for f_{par1}, when α=90°

and $\vec{E}$ becomes parallel to $\vec{B}$. In this extreme case, the wave can be named by an ordinary wave, by analogy to optics. In a contrast, the expression (2) does not depend on magnetic field for arbitrary α. At the same time the equations of movement (6-8) do not depend on magnetic field only when $\vec{E}$ becomes parallel to $\vec{B}$. So it is difficult to explain why the expression (2) does not depend on magnetic field for arbitrary α. Coincidence of f_{per} and f_{parl} is caused by small distinction of a dielectric permeability ε_{per} and ε_{par}. As a result, both waves propagate almost with the same speed, and they can be perceived as one wave on a receiver (as one summarized vector of electric field).

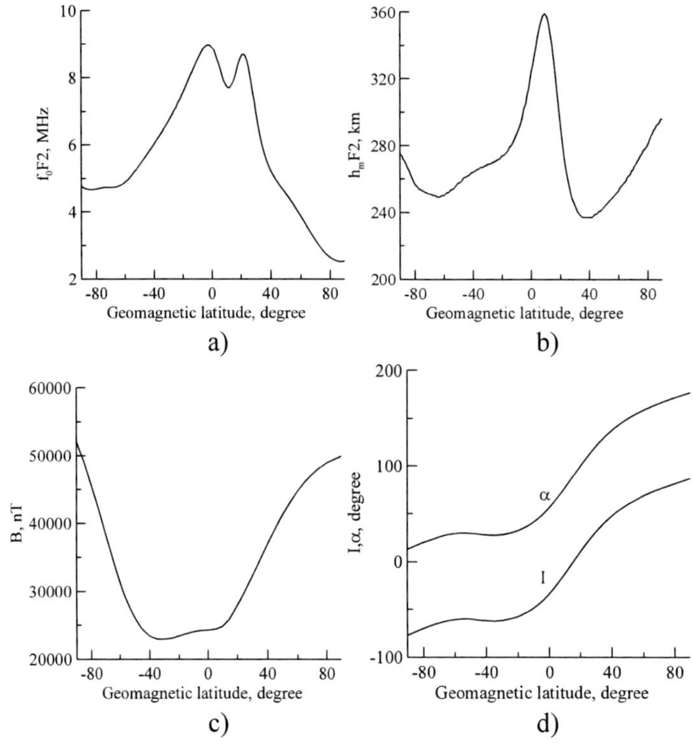

Figure 4. Latitude dependence of: a)f_0F2 , b) h_mF2 , c) amplitude of the geomagnetic field for altitude of h_mF2, d) inclination of the geomagnetic field and angle α for altitude of h_mF2.

Figure 5b presents the dependence of f_x, f_z, f_{11}, f_{22} on latitude. In the Southern hemisphere, f_{11} well agrees with f_x, and in the Northern hemisphere - with f_z. On the contrary, f_{22} well agrees with f_z in the Southern hemisphere and in the Northern - with f_x . “The transition” from one curve to another curve occurs at the latitude ~15.7°, where I=0° and α=90° (Fig. 4d). The change of a sign of a magnetic field inclination during transition through a zero leads to change of a direction of rotation of electrons and ions under action of a radiowave. Accordingly the conditions of radiowave propagation qualitatively change. At the same time the figure 5b shows that waves X and Z do not "change by the places". This result also shows advantage of formulas (29-30) in comparison with (3). It would be interesting to compare results of an experiment with calculations under formulas (29-30), (35), (39) and (2,3). However it will be difficult very much to show the advantage of those or other formulas due to little differences of the critical frequencies (Fig.5a,b). Additional difficulty is the necessity to define an angle α between a vector of a magnetic field and direction of radio wave propagation at the height of reflection. Ordinary and extraordinary waves considerably deviate from a vertical line in the region of a reflection point, and z-wave is observed in experiments only at arrival angles distinct from vertical (e.g. Ellis, 1953). Thus it is necessary to take into account tilting of the ionospheric layers. Then, a generalized solution of the problem of radio wave reflection in a linear conductive layer indicates that the height of radio wave reflection depends on the value of the conductivity gradient; and the angle of incidence is not equal to the reflection angle at the height of radio wave reflection (Krasnov, 2002). The account of plasma conductivity is uneasy problem. Additionally, calculations of radiowave trajectories are conducted in the approximation of geometrical optics which is not valid in the region of reflection point.

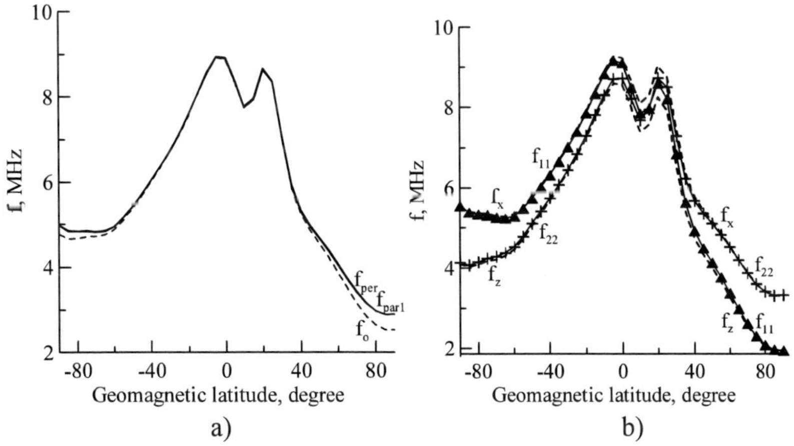

Figure 5. Latitude dependence of critical frequencies, when a frequency of radiowave higher than module of electron gyrofrequency: a) comparison of f_0 and f_{per}, f_{par1}; b) comparison of f_x, f_z and f_{11}, f_{22}.

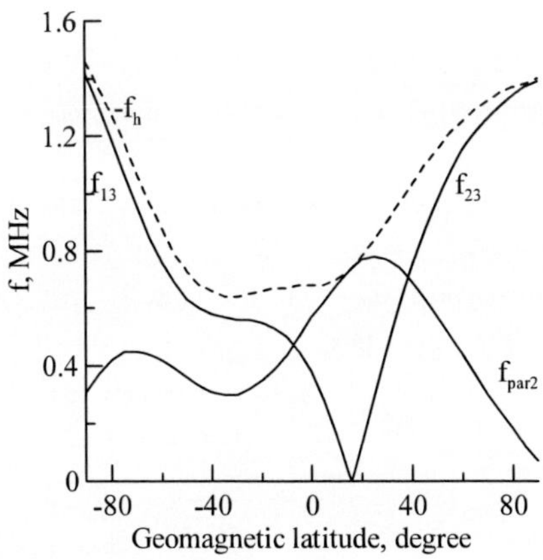

Figure 6. Latitude dependence of critical frequencies, when frequencies of radiowave lower than module of an electron gyrofrequency

Figure 6 presents variation of $|f_h|$ with latitude and variation of critical frequencies f13,f23,fpar2 with geomagnetic latitude for radiowave, when $|\Omega/\omega|>1$. Variations of f13 and f23 are constructed in the assumption that only positive values of frequencies have physical meaning. It is shown that two types of radiowaves can propagate up to the height above a maximum of F2-layer of the ionosphere. The vector of an electric field of the first type of waves should be in a plane of a vector of a magnetic field. The vector of an electric field of the second type of waves should rotate in the direction, which coincides with direction of an electron rotation, when electric field is missing. In a contrast to critical frequencies of radiowaves of the first type, the critical frequencies of radiowaves of the second type tend to zero at the latitude 15.7°, when α=90°. Thus, the radiowave of the second type has the greater "penetrating ability" for little frequencies. In principle, this result explains the regular observations of VLF and LF- signals, radiated by ground transmitters, above F2-layer of the ionosphere.

3. Conclusion

Rigorous solution of the movement equation of charged particles under action of a radiowave in the magnetized plasma allows us to show that the movement of electrons relative to ions for radiowaves with frequency higher and lower than module of an electron gyrofrequency is essentially different. For example, in the first case, the movement is basically defined by direction of an electric field and occurs mainly in the plane of vectors $\vec{E}_1$ and $\vec{E}_2$; and in the second case - by direction of a geomagnetic field and occurs mainly in a plane of a vector $\vec{B}$. As a result, there are different:

- plasma counteractions to the external electric field,
- conditions of radiowave propagation,
- formulae to calculate the critical frequency of F2-layer for radiowave with frequency higher and lower than module of the electron gyrofrequency.

The result of the critical frequency calculations, practically, consists with calculations based on the Appleton-Lassen-Hartree formula, when a frequency of a radio wave is higher than module of an electron gyrofrequency. Hence, it will be difficult to show experimentally advantage of the new or old formulae. The developed theory shows the fundamental possibility of propagation of ULF and LF radiowaves upward to the altitudes above F2-layer of the ionosphere.

Acknowledgment

International geomagnetic Reference Field model results were obtained from the web pages of World Data Center for Geomagnetism, Kyoto. This work was supported by Ministry of Education and Science of Kazakhstan under contract No. 8 - 2.03.2006.

References

[1] Bilitza D; International Reference Ionosphere 2000; *Radio Sci.* 2001, vol 36, No. 2, pp. 261-275.

[2] Budden, K. G., *Radio Waves in the Ionosphere;* Cambridge Univ. Press, New York, 1961.

[3] Chen Francis F. *Introduction to Plasma Physics;* Moscow, "Mir", 1987. (in Russia).

[4] *Ellis G.R.* F-Regio triple splitting. *Jounal of Atmospheric and Terrestrial Physics.* 1953, vol.3, pp. 263-269.

[5] Ginzburg, V.L., *Propagation of Electromagnetic Waves in Plasma*, Akad. Nauk Ross., Moscow, 1967. (in Russian)

[6] Krasnov V.M. Penetration of an Electromagnetic Wave into above a Reflection Point Layer; *Radio Sci,* 2001,vol. 36, No. 6, pp. 1505-1508.

[7] Krasnov V.M., Influence of ionospheric conductivity on parameters of a radio wave at a reflection point; *Radio Sci.,* 2002, vol 37, No. 0,XXXX, doi: 10.1029/2002RS002690.

[8] Molchanov O, A. Rozhnoi, M. Solovieva, O. Akentieva, J.J. Berthelier, M. Parrot, F. Lefeure, P.F. Biagi, L. Castellana, and M. Hayakawa. Global diagnostics of the ionospheric perturbations related to the seismic activity using the VLF radio signals collected on the DEMETER satellite; 2006, *Natural Hazards and Earth System Sciences.* No **6**, p.p. 745-753.

INDEX

A

accelerator, 250
access, viii, 137, 146
accounting, 235
accuracy, x, 43, 117, 138, 181, 259
acoustic waves, 138, 139, 143, 150, 154, 155, 157
activation, 58, 59, 75, 81, 252, 263, 264, 265
activity level, 4
actual output, 263
adjustment, 276, 331
age, 36, 88, 243
Alaska, 355
algorithm, 144, 145, 146, 147, 148, 150, 151, 153, 260, 266, 274, 276, 285
alternative, 45, 60, 63, 191, 241, 285
alters, 154
ambiguity, 189
amplitude, ix, xi, 8, 18, 19, 20, 59, 143, 144, 148, 149, 150, 156, 190, 195, 206, 207, 210, 211, 213, 214, 216, 218, 223, 225, 229, 239, 327, 330, 333, 341, 388, 392, 396
angular velocity, 142
anisomeric, ix, 195, 196, 197, 199, 201, 202, 203, 204, 205, 206, 230
annihilation, 68, 239
Antarctic polar vortex, ix, 166
anthropogenic, xi, 327, 330, 331, 332, 342
Appleton-Lassen-Hartree formula, xii, 381, 398
AR, 34, 104, 105, 106, 112, 113, 115, 116, 118, 154
Argentina, 3
assignment, 262
assimilation, 328
assumptions, 18, 143, 223, 262
asymmetry, xi, 6, 147, 152, 156, 245, 248, 327, 329, 330, 331, 332, 334, 342
asymptotics, 222
atmospheric pressure, 332, 343, 350, 351, 352
attachment, 348
attention, 54, 60, 71, 101, 124
Australia, 133, 257, 324
Austria, 287
averaging, 21, 129, 139, 140, 146, 180, 181, 295, 296, 340
azimuthal angle, 155

B

backscattering, 230
bandwidth, xii, 288, 314, 357, 359
beams, 239, 243, 244, 253, 295
behavior, vii, xii, 12, 43, 56, 87, 88, 95, 98, 99, 100, 109, 112, 118, 119, 120, 166, 170, 178, 182, 201, 220, 222, 357, 358, 362, 364, 369, 372
Beijing, 27
Bessel, 24, 54
bias, 145, 264, 265, 372
Big Bear Solar Observatory, x, 33, 35, 42, 44, 60, 95, 101, 259, 262
binding energy, 246
birth, 137
blocks, 277, 285
Born approximation, 144, 159, 196, 229
bounds, 8, 11, 120
breakdown, 190, 350
broadband, 320
burning, 120

C

calcium, 148, 358
calibration, 158, 251, 361
California, x, 259, 262
campaigns, 158
carrier, 239, 242
cast, 60, 138, 243
catastrophes, 123

categorization, 267
Catholic, 88, 89
causal relationship, 73, 82, 340
causality, 235, 252, 254
CDC, 334
cell, 370, 371, 372
certainty, 188
channels, 38, 68, 244, 288, 350, 352, 358
chaos, vii, 95, 96
charge density, 347
China, 27
Chinese, 285
chromosphere, vii, 31, 33, 38, 39, 41, 44, 45, 47, 48, 49, 54, 59, 61, 62, 74, 155, 234, 238, 239, 241, 242, 244, 247, 248, 249, 251, 252, 254
chromospheric spectral lines, vii, 31, 32
circulation, viii, ix, x, xi, 13, 165, 166, 167, 168, 170, 183, 187, 188, 189, 193, 194, 327, 334, 343
classes, 87, 265, 268, 272, 273
classification, x, 170, 259, 262, 264, 267, 268, 272, 273, 274, 276, 280, 281, 283, 284, 285, 360
cleaning, 362
cloud-to-ground (CG), xi, 345, 346
cluster model, 149
clustering, 87
clusters, 96, 101, 118, 120, 121, 122, 123, 125, 358
CO2, xi, 330, 342, 344
coherence, 120, 128, 216, 222, 229
collisions, 79, 80, 238, 248, 347, 348, 349, 355
combined effect, 8, 230
communication, xiii, 149, 196, 230, 260, 381
community, 245
compensation, 387
compilation, 11
complexity, 84, 97, 107, 119, 237, 262, 332, 350
components, 21, 23, 27, 39, 127, 141, 149, 192, 198, 208, 238, 303, 305, 322, 349, 365, 387, 388, 389, 392, 393, 394
computation, 12, 138, 143
computing, 147, 265
concentrates, 77
concentration, xi, 40, 78, 101, 102, 106, 109, 327, 331, 332, 342, 382, 395
conception, 88
concrete, 84
condensation, 45, 80, 331
conduction, 233
conductivity, 48, 67, 71, 80, 122, 129, 347, 349, 383, 389, 390, 397, 399
conductor, 64
confidence interval, 12, 268, 274
configuration, 43, 50, 63, 67, 76, 83, 85, 86, 247, 249
conflict, 51
conformity, 176, 181
conjecture, 125
consensus, 159
conservation, 65, 70, 122, 152, 347, 354
constraints, 235, 242, 269
construction, 144, 159, 210, 229
contamination, 158, 159
continuity, 41
control, 155, 369
convection, xi, 48, 71, 117, 125, 130, 139, 147, 158, 327, 330, 332, 334, 358, 359
convergence, 60, 68, 106, 267, 284
conversion, 69, 158, 159, 361
cooling, 74, 358
corona, vii, ix, x, 4, 17, 18, 19, 20, 21, 22, 23, 24, 25, 26, 31, 33, 43, 45, 47, 48, 49, 50, 54, 56, 60, 63, 65, 67, 68, 71, 72, 74, 75, 80, 82, 83, 84, 85, 86, 88, 109, 114, 117, 118, 119, 125, 129, 130, 196, 197, 228, 229, 233, 234, 235, 236, 237, 239, 240, 241, 242, 243, 244, 245, 246, 247, 248, 249, 250, 252, 253, 254, 262, 300, 350, 351
coronal mass ejection (CME), vii, x, 17, 18, 23, 25, 31, 81, 82, 83, 84, 85, 88, 106, 107, 234, 236, 240, 244, 245, 246, 247, 248, 249, 251, 252, 253, 254, 260, 274, 275, 276, 277, 278, 279, 280, 281, 282, 283, 284, 285, 287, 298, 299, 300, 302, 303, 304, 305, 306, 307, 310, 311, 313, 314, 315, 316, 319, 320
correlation(s), 4, 9, 10, 11, 12, 24, 82, 83, 88, 119, 121, 122, 123, 155, 166, 167, 188, 223, 261, 338, 344, 378
correlation analysis, 82
correlation coefficient, 4, 9, 10, 11, 12
correlation function, 24, 223
cosmic ray flux, 4
cosmic rays, 96, 236
Coulomb, 238, 248
counterbalance, 51
coverage, 362
covering, 120, 154, 261, 362
CPU, 272, 283
critical state, 122, 123, 124, 130
critical value, 56, 65, 69, 71, 86, 122, 351
criticism, 151
Cuba, 357
cumulative distribution function, 103, 104
cycles, viii, 39, 46, 165, 175, 176, 177, 179, 180, 181, 183, 184, 186, 187, 188, 189, 190, 191

D

damping, 139, 141, 158
data analysis, 144, 335

data processing, 21
data set, 100, 113, 264, 265, 272, 342, 343
database, 284, 285
death, 88
decay, 56, 100, 184, 290
decomposition, 146, 147, 329, 332, 334
deconvolution, 145
deficiency, 70
definition, 368, 369, 379, 388
deformation, 17
degenerate, 218
degrees of freedom, 11
dense matter, 47
density, xii, 18, 22, 23, 25, 27, 33, 34, 49, 59, 62, 68, 78, 79, 80, 81, 82, 83, 97, 101, 102, 107, 108, 120, 123, 125, 240, 242, 267, 347, 348, 350, 351, 357, 359, 362, 372, 373, 378
density fluctuations, 18
depolarization, 206
deposition, 238, 239, 242, 243, 244, 253, 254
depression, 80, 141, 158
destruction, ix, 166, 190
detection, x, 137, 147, 150, 152, 259, 260, 261, 262, 271, 272, 274, 275, 276, 279, 280, 284, 285, 369, 373
deviation, 40, 44, 104, 105, 110, 111, 117, 184, 196, 271, 272, 278, 362, 363, 365
diamonds, 103, 126
dielectric constant, 347, 350
dielectric permeability, xii, 381, 382, 390, 395, 396
dielectric permittivity, 24, 198, 210, 213, 214
differential equations, 95, 125
differentiation, 129
diffraction, 25, 144, 196
diffusion, 48, 105, 106, 122, 123, 241
diffusion region, 241
diffusion time, 48, 241
diffusivity, 119, 241
dilation, 278
dipole, 51, 52, 56, 65, 350, 386, 387
dipole moment, 65, 387
discharges, xi, 345, 346, 350, 354
discontinuity, 108
dispersion, 64, 154, 221, 222, 362, 364
displacement, xii, 45, 62, 72, 244, 377, 381, 383, 388, 389, 392, 393, 394
distribution, viii, xii, 11, 19, 21, 36, 38, 39, 45, 46, 47, 52, 55, 58, 59, 68, 72, 78, 80, 87, 97, 99, 100, 101, 102, 104, 105, 106, 108, 119, 122, 123, 126, 127, 128, 130, 165, 181, 182, 187, 205, 206, 230, 235, 267, 288, 347, 349, 353, 357, 358, 359, 360, 362, 364, 366, 367, 368, 369, 370, 371, 374, 376, 378, 379
distribution function, 99, 100, 101, 102, 106, 108, 126, 127, 128, 130
divergence, 140, 225
division, 277, 278, 285
dominance, 254
Doppler, 18, 22, 45, 62, 81, 101, 138, 139, 144, 157, 158, 159
draft, 160
dumping, 247, 253
duration, viii, 61, 100, 165, 166, 167, 170, 175, 181, 183, 184, 187, 188, 191, 236, 241, 244, 288, 291, 295, 299, 302, 308, 310, 312, 314, 317, 318, 319, 320, 322, 323, 324, 346
dynamic systems, 98
dynamical systems, 123

E

early warning, 261, 274
earth, xii, 12, 83, 259, 261, 332, 334, 381, 382
elaboration, 123
electric current, vii, 31, 43, 48, 61, 65, 66, 68, 71, 76, 85, 86, 122, 124, 129, 335, 338
electric field, 67, 240, 242, 244, 245, 262, 347, 349, 350, 351, 352, 353, 383, 385, 386, 396, 398
electric jets, xi, 345, 346, 347, 350, 351, 353, 354
electromagnetic, x, 32, 65, 83, 84, 85, 206, 234, 245, 260, 261, 327, 329, 335, 342, 383
electromagnetic wave(s), 206, 261, 383
electron(s), ix, xi, xii, 17, 18, 19, 21, 22, 23, 24, 25, 27, 198, 213, 233, 234, 235, 236, 237, 238, 239, 241, 242, 243, 244, 245, 246, 247, 249, 250, 251, 252, 253, 254, 295, 338, 345, 347, 348, 349, 350, 351, 353, 355, 381, 382, 383, 385, 386, 387, 396, 397, 398
electron charge, 25, 198
electron density, 17, 18, 19, 21, 22, 23, 24, 25, 27, 198, 213
electron density distribution, 22, 25
elongation, 197, 201
emission, 4, 18, 19, 21, 22, 24, 25, 27, 74, 81, 85, 100, 114, 125, 198, 209, 234, 235, 236, 238, 242, 243, 244, 250, 252, 253, 260, 287, 288, 292, 293, 294, 295, 296, 298, 299, 300, 305, 307, 308, 309, 310, 312, 313, 314, 315, 317, 318, 319, 320, 321, 322, 323, 332, 337
empirical methods, 85
energy, ix, 4, 17, 18, 19, 20, 21, 22, 23, 24, 25, 26, 27, 62, 63, 65, 66, 67, 68, 69, 71, 72, 73, 75, 79, 80, 83, 84, 85, 86, 95, 98, 99, 100, 107, 109, 110, 120, 124, 125, 127, 130, 147, 149, 157, 196, 230, 233, 234, 235, 236, 237, 238, 239, 240, 241, 242, 243, 244, 245, 246, 247, 248, 249, 250, 252, 253,

254, 259, 260, 335, 346, 348, 349, 353, 358, 383, 388
energy density, 62, 68
energy supply, 234, 235, 239, 254
energy transfer, 235, 254
environment, vii, xi, 31, 32, 83, 345, 346
equality, 55
equating, 391
equilibrium, vii, 31, 47, 48, 49, 50, 51, 52, 53, 55, 63, 64, 65, 66, 67, 68, 69, 70, 71, 72, 73, 74, 76, 79, 82, 85, 86, 88, 106, 122, 124, 239, 254, 331, 332
ERA, 193
erosion, 106, 278
estimating, 223
Europe, 13
evacuation, 80
evaporation, 331, 332
evidence, 13, 21, 40, 43, 59, 167, 194, 234, 236, 237, 240, 243, 244, 340, 366, 371, 372
evolution, vii, viii, 67, 68, 75, 76, 80, 81, 96, 98, 105, 119, 120, 124, 127, 138, 165, 167, 168, 170, 175, 176, 179, 180, 181, 182, 184, 186, 187, 188, 189, 191, 239, 292, 368, 375, 376, 379
excitation, 139, 151, 348, 349
exposure, xii, 276, 280, 357, 364, 365
external triggers, 69
extrapolation, 37, 38, 130, 246, 331
eyes, 272

F

feedback, xi, 328, 331, 332, 342
feet, 39, 61
FEM, 356
fibrils, 38, 40, 67
filament, vii, 31, 33, 34, 36, 38, 39, 40, 42, 43, 44, 45, 46, 48, 49, 50, 51, 52, 53, 54, 55, 56, 59, 60, 61, 62, 64, 65, 66, 72, 73, 74, 75, 76, 78, 79, 80, 81, 82, 83, 85, 86, 87, 88, 237, 252, 260, 261, 284
filigree structures, viii
film(s), xii, 357, 359, 362
filters, 8
finite element method, 352
fish, 39, 41
fitness, 364, 367
fixation, 71
flare, ix, x, 4, 59, 62, 66, 72, 73, 75, 84, 85, 88, 233, 234, 235, 236, 237, 238, 239, 240, 241, 242, 243, 244, 245, 246, 247, 248, 249, 251, 252, 253, 254, 259, 260, 261, 262, 265, 271, 272, 273, 274, 284
fluctuations, viii, ix, x, 12, 18, 19, 24, 25, 165, 166, 168, 184, 189, 195, 196, 214, 216, 225, 226, 227, 364, 365
fluid, 77, 78
flux rope, vii, 31, 52, 53, 54, 63, 64, 66, 69, 70, 71, 72, 73, 75, 80, 82, 83, 85, 86, 88
focusing, 25, 196, 223, 225, 226
forecasting, x, 125, 130, 259
Fourier, ix, 139, 145, 146, 196, 217, 218, 227
fractal dimension, 96, 97, 100, 117, 118, 119, 121, 123
fractal geometry, viii, 128
fractal properties, 125
fractal structure, 96, 120
fractal theory, 111, 346
fractality, vii, 95, 118, 120, 123, 127
fractional dimensions, vii
fragmentation, 99, 101, 105, 106, 113, 114, 116
France, 94, 355
free energy, ix, 83, 233, 239, 240
free fields, 358
freezing, 76
frequency distribution, 100
frequency resolution, 324
Fresnel radius, ix, 195, 197, 216, 224, 230
frustration, 245
fusion, ix, 196, 227, 230

G

gamma rays, 234, 235, 260
gases, 353
Gaussian, 8, 98, 99, 100, 102, 104, 107, 129, 139, 200, 222, 223, 264, 271, 272, 274, 347
gene, ix, 195, 196, 197, 206, 210, 223, 230
generalization, 197, 217, 227
generation, viii, 17, 18, 48, 59, 84, 165, 167, 288, 346, 395
geomagnetic field, ix, 195, 196, 198, 382, 383, 386, 395, 396, 398
geometrical optics, ix, 195, 196, 197, 206, 397
grains, 124, 362
grants, 159, 285
granules, xii, 357, 358, 367, 369, 370, 372, 373, 375, 378
graph, 395
gravitation, 47, 88
gravitational field, vii, 31, 47, 62, 85
gravitational force, 64, 127
gravity, ix, 3, 33, 47, 51, 54, 62, 70, 78, 82, 137, 138, 195, 196
grazing, 213
Great Britain, 13
greenhouse gas(es), xi, 327, 330, 332, 342, 344
grids, 11
grouping, 4, 142

groups, 33, 106, 119, 140, 151, 159, 276, 297, 309, 310, 323, 346
growth, xi, 63, 99, 111, 189, 345, 351, 352, 369
gyration radius, 122
gyrofrequency, xii, 381, 382, 385, 386, 397, 398

H

halos, 346
harmonic frequencies, 6, 7
heat(ing), 73, 74, 75, 78, 79, 80, 81, 122, 130, 147, 187, 233, 237, 238, 242, 252, 295, 330, 351
height, vii, viii, xi, 31, 33, 34, 35, 38, 39, 43, 45, 46, 47, 49, 50, 51, 52, 55, 56, 57, 58, 59, 61, 62, 65, 66, 70, 72, 76, 78, 82, 86, 87, 88, 124, 130, 165, 166, 167, 168, 169, 170, 171, 172, 173, 174, 175, 176, 178, 179, 181, 183, 184, 186, 188, 191, 192, 213, 240, 276, 280, 283, 284, 285, 327, 328, 331, 332, 333, 340, 341, 342, 343, 383, 395, 397, 398
helicity, 45, 96, 149
heliosphere, vii, xi, 95, 109, 127, 244, 328
helium, 81
hemisphere, 39, 41, 45, 46, 188, 192, 340, 360, 392, 396
histogram, xii, 103, 357, 359, 365, 366, 367, 368, 369, 373, 374, 378, 379
humidity, xi, 327, 330, 331, 332, 334
hydrogen, 4, 81, 260
hydrogen bomb, 260
hypothesis, 25, 242, 250, 375, 379

I

idealization, 12
identification, 39, 166, 188, 241
identity, 146
illumination, 213, 223
images, x, xii, 18, 26, 60, 62, 72, 74, 75, 87, 114, 119, 158, 253, 259, 260, 261, 265, 271, 272, 274, 275, 276, 277, 278, 279, 280, 281, 282, 283, 285, 357, 360, 361, 362, 363, 365, 366, 368, 369, 372, 373, 376, 378
imaging, 153
impulsive, 235, 237, 241, 244, 250
incidence, 207, 211, 238, 385, 397
inclusion, 147, 217, 368
indication, 153, 252
indicators, vii, 31, 86, 88
indices, 4, 100, 126, 192, 193
induction, 65, 67, 86, 119, 383
inelastic, 349
inequality, 56, 198
inertia, 48, 65, 388
inferences, 125, 128
infinite, x, 43, 74, 210, 216, 228, 287, 299, 309, 310, 319
inhomogeneity, 24, 25, 212, 224, 227
initial state, 96
initiation, x, 233, 345, 346, 347, 351, 353
instability, 63, 107, 125, 247, 358
instruments, 88, 138, 235, 247, 324
integration, 129, 220, 221, 225, 228, 229
intensity, vii, viii, xii, 4, 31, 43, 98, 118, 138, 152, 154, 157, 166, 167, 168, 187, 189, 200, 203, 205, 206, 219, 221, 222, 223, 226, 230, 234, 236, 238, 250, 251, 253, 274, 285, 293, 295, 300, 302, 303, 307, 313, 322, 339, 357, 359, 360, 361, 362, 363, 364, 365, 366, 367, 369, 371, 373, 374, 379, 381, 383
interaction(s), viii, xi, xiii, 3, 31, 40, 78, 83, 96, 97, 105, 107, 128, 130, 137, 138, 154, 155, 158, 159, 192, 253, 327, 343, 382
interference, 19, 196, 197, 213, 217, 222, 229, 288
internet, 275, 341
interplanetary, vii, 62, 71, 83, 84, 88, 108, 233, 236, 247, 249, 250, 252, 254, 328, 335, 338, 340, 342
interplanetary coronal mass ejections (ICME), xi, 328, 335, 339, 340, 343
interplanetary medium, x
interpretation, 83
interval, xi, 21, 36, 60, 71, 101, 112, 113, 115, 117, 167, 170, 172, 173, 174, 179, 221, 268, 274, 280, 283, 328, 331, 339, 340, 341, 348, 373, 376
inversion, vii, viii, 31, 36, 38, 39, 40, 43, 47, 48, 53, 54, 56, 58, 73, 74, 76, 78, 85, 137, 145, 146, 147, 148, 150, 151, 153, 155, 159
ionization, 81, 347, 349, 351, 353
ionosphere, ix, xi, xii, 195, 196, 197, 213, 226, 228, 229, 230, 328, 335, 338, 345, 346, 350, 351, 353, 381, 382, 383, 398
ions, ix, 198, 233, 235, 236, 237, 239, 242, 243, 245, 246, 249, 250, 251, 252, 253, 254, 350, 385, 386, 396, 398
Ireland, 119, 133, 134
isolation, 189
Italy, 90

J

Japan, 94, 345, 355
Joule heating, ix

K

Kazakhstan, 381, 398
kernel, 141, 142, 144, 145, 146, 270, 271, 274, 283
kinetic energy, 63, 67, 69, 71, 75, 79, 80, 86, 274, 305, 388
kinetics, 347, 353
Kirchhoff, 229

L

Lagrange multipliers, 269, 270
land, xi, 327
Large Angle Spectrometric Coronagraph (LASCO), x, 17, 81, 106, 240, 250, 259, 261, 274, 275, 276, 277, 278, 279, 280, 281, 282, 283, 284, 285
lattice size, 121
lattices, 120
laws, 84, 100, 109, 241
lead, xi, 10, 12, 65, 68, 113, 120, 196, 238, 249, 252, 266, 327, 346, 387
leakage, 252
learning, 262, 263, 264, 265, 266, 267, 284
lifetime, vii, 31, 106, 125, 127, 272, 288, 346
likelihood, 105
limitation, 88, 250
linear function, 105
lines of force, vii, 31
links, 351
literature, 127, 129, 253, 375, 378, 379
location, xiii, 63, 85, 124, 127, 139, 144, 170, 181, 213, 216, 217, 230, 239, 240, 241, 253, 261, 277, 382
London, 13, 92, 131, 379
long distance, 354
lying, 253

M

machine learning, x, 259
magnet, 122
magnetic field, vii, viii, ix, xii, 31, 33, 36, 37, 38, 39, 40, 43, 45, 47, 48, 49, 50, 52, 53, 54, 55, 56, 58, 59, 61, 64, 65, 67, 68, 70, 71, 72, 74, 75, 76, 77, 79, 80, 83, 84, 85, 86, 87, 88, 233, 234, 235, 236, 237, 238, 239, 240, 241, 246, 247, 248, 249, 251, 252, 253, 254, 260, 262, 324, 357, 358, 359, 366, 371, 372, 373, 377, 378, 379, 382, 383, 385, 386, 395, 396, 398
magnetic moment, 43, 122
magnetic structure, vii, x, 75, 83, 85, 95, 97, 98, 106, 114, 118, 119, 233, 238, 245, 246, 247, 253
magnetism, 123
magnetosphere, xi, 17, 96, 328, 335, 338
management, xiii, 381
mapping, viii, 137, 267
mass loss, 152
mathematical methods, 127
matrix, 144, 145, 146, 147, 265
Maxwell equations, 389, 393, 394
MDI, 37, 101, 119, 135, 138, 141, 148, 150, 151, 152, 154, 158, 159
measurement, viii, 137, 276
measures, 4, 119, 138
mechanical energy, 240
media, ix, 109, 196, 224, 226, 230, 231
median, 41, 46, 104, 276, 277, 278
memory, 119, 272, 283
meridian, 41, 320
meteor, 197, 205, 230
MHD, 54, 68, 84, 86, 93, 162, 240, 242
microwave(s), 234, 235, 236, 242, 246, 330
migration, 57, 58
Ministry of Education, 381, 398
missions, 100, 235, 260
MIT, 91, 92, 93, 134, 285
modeling, vii, 17, 32, 84, 86, 95, 120, 122, 124, 125, 203, 205, 206, 223, 229, 342, 344, 345, 350, 355
models, vii, ix, 49, 52, 53, 54, 55, 63, 64, 67, 69, 70, 71, 72, 80, 84, 85, 95, 109, 130, 147, 148, 167, 187, 233, 234, 235, 239, 240, 242, 245, 252, 334, 354, 358
molecular oxygen, 187
molecules, 350, 353
momentum, 243
monofractal, 111, 112, 119
monograph, 161
Monte Carlo, xi, 345, 346, 347, 350, 353, 354
morphology, xii, 43, 249, 260, 285, 357, 366
Moscow, 27, 31, 91, 231, 286, 399
motion, vii, 31, 45, 48, 59, 60, 62, 65, 67, 70, 72, 74, 76, 77, 78, 79, 80, 82, 83, 87, 187, 247, 249, 252, 254, 262, 272, 275, 347, 378
motivation, 234
mountains, xi, 327
movement, 27, 279, 376, 377, 384, 385, 386, 387, 389, 392, 393, 396, 398
multifractal, 97, 98, 99, 100, 111, 112, 113, 119, 123, 127, 130
multifractality, vii, 95, 98, 99, 100, 109, 111, 112, 113, 114, 115, 116, 118, 119, 120, 123, 124, 125, 128, 129, 130
multiplication, 145, 328
multiplicity, 25

N

National Bureau of Standards, 231
National Science Foundation, 285
natural evolution, 130
neglect, 388
Netherlands, 133, 162
network, 39, 40, 101, 122, 125, 262, 263, 264, 265, 266, 268, 273, 339, 343
network elements, 101
neural network(s), 262, 263, 264, 267, 268, 273
neurons, 264, 265, 266, 267
neutrons, 246, 253
New Jersey, 33, 35, 42, 44, 60, 95, 259, 262
New York, 12, 93, 132, 135, 136, 231, 232, 286, 355, 399
nodes, 39, 120, 124, 262, 263, 264, 265, 267, 273
noise, 13, 103, 118, 143, 146, 147, 251, 272, 276, 277, 278
non-magnetic, 155
normal distribution, 103, 104, 105, 106, 107, 108, 219, 362, 367, 368, 370
Northern Hemisphere (NH), xi, 193, 327, 329, 330, 331, 332, 333, 334, 340, 341, 342, 343
Norway, 96
nuclei, xii, 357, 374, 377, 379
nucleus, 246, 358, 359, 366, 373, 374, 376, 377, 378, 379

O

observations, x, xi, xii, 11, 18, 21, 26, 32, 45, 51, 62, 63, 67, 71, 81, 82, 85, 88, 100, 101, 119, 122, 123, 124, 125, 147, 151, 157, 166, 167, 188, 192, 193, 234, 236, 239, 241, 242, 243, 245, 246, 249, 250, 252, 253, 267, 287, 288, 290, 292, 296, 299, 300, 306, 307, 312, 324, 328, 340, 345, 346, 347, 353, 354, 357, 359, 361, 398
opacity, 362
operator, 275
optics, ix, 195, 359, 396
optimization, 273
orbit, 21, 23, 25, 328, 385
organization, 119
orientation, 39, 59
orthogonal functions, 167, 188
orthogonality, 76
oscillation, 3, 4, 6, 10, 11, 12, 13, 137, 138, 139, 157, 166, 184, 187, 192, 193, 194, 322, 333
oxygen, 338
ozone, viii, ix, 4, 13, 165, 166, 168, 187, 189, 190, 192, 193, 338, 340, 343
ozone hole, ix, 166, 168, 189, 190

P

Pamir Mountain Station, xii, 357
parameter, 20, 23, 57, 86, 87, 97, 106, 111, 122, 124, 146, 166, 196, 207, 209, 214, 227, 228, 229, 263, 265, 266, 267, 270, 351, 366
Paris, 72, 89, 93, 94, 131, 132, 133
particles, x, 77, 78, 79, 80, 84, 86, 127, 234, 235, 236, 237, 238, 240, 241, 243, 244, 245, 246, 247, 248, 249, 250, 251, 252, 253, 254, 261, 338, 349, 384, 385, 388, 398
partition, 238
path integrals, 227
pathways, 197
pattern recognition, 260, 275
penumbra, xii, 148, 149, 152, 155, 358, 359, 366, 367, 369
percolation, viii, 95, 100, 120, 121, 122, 123, 129, 130
percolation cluster, 120, 121, 123
percolation theory, viii, 95, 100, 120, 122, 123, 130
performance, 266, 273
periodicity, viii, 4, 7, 165, 335, 338
permeability, xii, 381
permit, 68, 88
permittivity, 214
phase shifts, 158
phase transitions, 180, 182
photographs, xii, 357
photometric measurement, 375, 379
photons, 237
physical mechanisms, 188
physical sciences, 136
physics, vii, viii, 3, 97, 100, 106, 123, 124, 127, 130, 137, 138, 140, 148, 159, 241, 255, 261, 328, 335
pitch, 34, 54, 61
plane waves, 217, 218, 223, 224, 226, 230
planets, x, 327
plasma, vii, ix, xii, 17, 18, 21, 22, 23, 24, 25, 26, 27, 31, 33, 43, 45, 47, 48, 49, 54, 62, 63, 64, 65, 67, 68, 74, 76, 77, 78, 79, 80, 81, 82, 83, 85, 88, 95, 96, 105, 108, 109, 117, 122, 125, 129, 130, 195, 197, 198, 200, 207, 227, 229, 230, 233, 236, 240, 241, 245, 249, 252, 253, 254, 288, 295, 338, 349, 350, 378, 381, 382, 383, 385, 387, 389, 390, 391, 393, 394, 395, 397, 398
PN, 122
Poisson equation, 347
Poland, 64, 70, 90, 91, 92, 93

polarity, vii, 31, 36, 38, 39, 40, 43, 47, 48, 49, 50, 51, 53, 54, 55, 56, 58, 59, 60, 73, 74, 76, 78, 85, 86, 127, 150
polarization, 243, 288, 289, 291, 292, 294, 295, 304, 306, 323, 324, 387
poor, 373
population, x, 99, 233, 234, 235, 236, 246, 248, 249, 251, 253, 254
positive correlation, 11
potential energy, 50, 66
power, 6, 21, 22, 25, 52, 96, 100, 101, 102, 103, 104, 105, 109, 111, 112, 116, 121, 122, 124, 125, 126, 130, 152, 155, 158, 159, 199, 200, 201, 202, 203, 204, 205, 206, 235, 238, 239, 242, 243, 247, 260, 359
precipitation, 252, 331
prediction, xiii, 84, 85, 86, 87, 124, 129, 188, 381
preference, 180
pressure, 33, 47, 48, 64, 78, 79, 80, 166, 167, 181, 246, 249, 328, 329, 331, 332, 333, 340, 341, 342, 343, 350, 351
probability, 3, 59, 84, 100, 103, 104, 105, 109, 120, 122, 125, 130, 182, 183, 267, 348, 352
probability distribution, 100, 103, 104, 109, 122, 125
probe, xiii, 381
production, 54, 234, 236, 242, 243, 245, 246
productivity, 113, 119
program, 68, 73, 75, 81, 272
propagation, ix, xi, xii, 17, 18, 19, 25, 147, 150, 195, 196, 197, 198, 208, 213, 220, 224, 226, 227, 230, 231, 244, 252, 262, 340, 345, 346, 349, 351, 352, 353, 354, 381, 382, 383, 387, 395, 396, 398
protons, 234, 235, 236, 237, 239, 241, 242, 243, 244, 246, 247, 248, 249, 250, 251, 253, 254, 338, 385, 386
prototype, 266
pulse(s), 107, 208, 209, 210, 213, 222, 223, 244

Q

Quasi-Biennial Oscillation (QBO), viii, 3, 4, 9, 10, 11, 12, 13, 165, 166, 167, 168, 170, 172, 174, 175, 176, 177, 178, 179, 180, 181, 182, 183, 184, 185, 186, 187, 188, 189, 190, 191, 192, 193

R

radar, xiii, 205, 206, 230, 382
radial distance, 56
radiation, x, 24, 74, 80, 83, 88, 213, 226, 228, 231, 241, 260, 261, 287, 288, 295, 306, 313, 327, 328, 329, 330, 331, 332, 333, 335, 337, 338, 339, 341, 342, 343
radiative transport, ix
radio, ix, x, xii, 4, 17, 18, 19, 21, 22, 24, 25, 27, 81, 184, 195, 196, 197, 213, 220, 223, 226, 227, 228, 229, 230, 231, 234, 235, 236, 237, 242, 243, 244, 253, 260, 261, 287, 288, 289, 292, 293, 294, 295, 296, 298, 309, 312, 322, 346, 381, 383, 397, 398, 399
radius, 19, 22, 23, 42, 64, 69, 70, 76, 78, 79, 123, 139, 195, 197, 204, 216, 224, 225, 230, 240, 249, 266, 274, 278, 360, 372, 385, 387
rain, 331
rainfall, 13
random walk, 123
range, viii, xii, 7, 8, 36, 45, 46, 47, 52, 72, 95, 97, 100, 101, 103, 104, 105, 109, 110, 111, 112, 113, 114, 117, 118, 119, 123, 130, 158, 166, 167, 168, 173, 174, 184, 188, 191, 196, 197, 220, 223, 224, 226, 230, 239, 261, 288, 290, 306, 309, 322, 323, 329, 336, 339, 340, 362, 365, 366, 371, 377, 378, 381, 382, 392, 395
reading, 126
real estate, 120
real time, 261
reality, 130, 159
reasoning, 226
recall, 96, 340
reception, 199, 200, 205, 206, 226
reciprocity, 226
recognition, 267
reconstruction, 74
recurrence, 84
redistribution, 122, 124, 125, 230
redshift, 243
reduction, 158, 324
reference frame, 220
reflection, 17, 196, 382, 383, 391, 395, 397, 399
refractive index, 196, 198
regression, 329, 378
regression line, 329
reinforcement, 113
relationship(s), xi, 11, 12, 13, 33, 47, 57, 59, 78, 82, 113, 119, 120, 159, 172, 173, 186, 189, 193, 200, 261, 328, 331, 340, 341, 366, 371, 378
relevance, viii, 137
reliability, 84, 260
reproduction, 159
residual error, 196
resolution, xii, 45, 86, 101, 103, 117, 118, 146, 150, 159, 230, 288, 357, 359, 360, 362, 373, 375, 379
retardation, 183
Reynolds number, 109

rings, 36, 373, 374, 377, 379
risk, 267, 268, 274
rotation axis, viii, 137, 388
rotations, 32, 39, 84, 336
routines, 124
Russia, 17, 31, 91, 165, 195, 359, 399

S

safety, 260
sample, 11, 12, 96, 103, 122, 138, 154, 247, 348, 349, 352
sampling, 12, 139, 335, 339
satellite, xii, 159, 189, 196, 328, 341, 381, 399
saturation, 226, 366
scalability, 262
scaling, 109, 110, 111, 112, 114, 117, 118, 125, 130, 284, 351
scaling law, 111
scatter(ing), ix, 19, 21, 25, 144, 195, 196, 197, 199, 200, 201, 202, 203, 204, 205, 206, 227, 229, 230, 236, 244, 251, 288, 349
sea level, 351, 359
search(ing), 12, 59, 236, 237, 245, 283, 359, 369
seed(s), 151, 248, 249, 253, 254, 349
segmentation, 101, 103, 260, 262, 272, 276, 278
selecting, 362, 369
self-organized criticality, viii, 123, 124
self-organizing feature map, 266
self-similarity, 96, 97
sensitivity, x, 141, 142, 143, 144, 145, 150, 159, 197, 201, 203, 243, 287, 288, 324, 343
separation, 59, 83, 87, 108, 109, 141, 205, 217, 218, 224, 228, 262, 268
series, xi, 3, 4, 6, 7, 8, 9, 10, 12, 72, 90, 114, 141, 166, 175, 177, 184, 186, 190, 199, 214, 319, 327, 328, 329, 333, 335, 336, 337, 338, 339, 340, 341, 342, 359, 376
shade, 44
shape, 32, 36, 45, 46, 67, 70, 75, 76, 77, 80, 81, 82, 87, 99, 100, 101, 102, 125, 141, 144, 158, 260, 366
shear, 53, 54, 59, 85, 167, 168, 169, 170, 171, 172, 173, 174, 176, 177, 178, 179, 181, 182, 183, 186, 187, 188, 191, 193, 248
shear deformation, 85
shock, 79, 240, 244, 245, 254, 288
shock waves, 240, 245, 288
shoot, 208
sign(s), 11, 36, 40, 41, 54, 57, 61, 98, 102, 153, 198, 199, 208, 219, 220, 221, 270, 338, 382, 392, 396
signals, xii, 18, 21, 25, 138, 139, 213, 381, 382, 383, 398, 399
signal-to-noise ratio, 139
significance level, 12, 184, 185
similarity, 97, 114, 228, 350
simulation, xi, 18, 22, 25, 119, 345, 346, 347, 352, 353, 354
Singapore, 166, 168
sites, 107
skeleton, 50
smoothing, 184, 186
SOC, 100, 114, 122, 123, 124, 125, 130
software, 283
SOI, 138, 141, 148, 150, 151, 152, 154, 158
Solar and Heliospheric Observatory (SOHO), x, 18, 37, 59, 63, 68, 73, 74, 81, 101, 106, 114, 162, 163, 235, 258, 259, 261, 275, 276, 285, 289, 298, 299, 300, 304, 305, 306, 310, 313, 314, 315, 316, 324, 380
solar disk, vii, x, 31, 32, 36, 42, 43, 87, 101, 114, 259, 261, 292, 307, 319, 360, 372
solar flares, xi, 67, 84, 236, 237, 239, 241, 242, 260, 261, 262, 272
solar plasma, ix, 95, 196
solar prominences, vii, 31, 36, 63, 92
solar wind, vii, xi, 18, 84, 107, 127, 328, 335, 338, 342, 343
Solnechnye Dannye Classification, xii, 357, 359
sound speed, 140, 142, 145, 150, 152, 155, 157, 158
sounds, 342
Southern Hemisphere (SH), xi, 327, 329, 330, 331, 333, 334, 340, 341, 342, 343
space environment, 17, 260
Spain, 92
species, 249
specific heat, 79
spectrum, 6, 7, 17, 18, 19, 20, 21, 22, 23, 24, 25, 27, 36, 98, 107, 119, 130, 196, 197, 200, 206, 216, 222, 235, 237, 238, 239, 242, 243, 244, 248, 249, 250, 260, 289, 290, 293, 295, 300, 303, 307, 317, 322, 335, 362
speed, viii, xi, 45, 47, 61, 62, 70, 71, 74, 78, 106, 107, 127, 137, 139, 140, 141, 144, 145, 150, 152, 153, 154, 157, 158, 159, 166, 168, 169, 172, 173, 174, 176, 177, 179, 182, 183, 188, 191, 234, 239, 240, 247, 262, 279, 280, 283, 284, 300, 306, 328, 338, 339, 342, 348, 376, 385, 388, 396
St. Petersburg, 165
stability, vii, 31, 45, 49, 50, 51, 55, 56, 59, 70, 72, 85, 87, 249, 378
stages, viii, 35, 83, 166, 167, 171, 175, 176, 179, 187, 351, 375, 379
stagnation stage, viii, 165, 166, 172, 174, 175, 176, 177, 178, 179, 181, 182, 183, 184, 185, 187, 188, 189, 191

standard deviation, 99, 100, 103, 180, 181, 264, 271, 272, 278, 280, 335, 362, 363, 364, 369, 372, 375
standardization, xii, 357
stars, 243
statistical analysis, 3, 186
statistics, 100, 101, 123, 124, 125, 129
storage, 63, 149, 235, 240, 241
storms, 17, 83, 84, 96, 106, 260, 295, 320, 322, 335
strategies, 274
stratosphere, viii, 3, 10, 12, 13, 165, 166, 170, 171, 176, 180, 181, 183, 184, 187, 188, 189, 190, 191, 192, 193, 194, 343
strength, xii, 38, 47, 49, 56, 61, 65, 70, 83, 86, 156, 189, 236, 239, 241, 242, 249, 251, 349, 358, 360, 371, 377, 378, 379
stress, 59
stretching, 70
structuring, 278
sub-bursts, x, 287, 291, 293, 296, 299, 300, 302, 303, 307, 310, 314, 317, 318, 320, 322
substitution, 51, 332
subsurface flow, viii
subtraction, 333, 334
summer, 180, 330, 331, 332, 334, 342
sunspot, viii, xii, 3, 4, 5, 11, 13, 33, 36, 56, 122, 137, 143, 147, 148, 149, 150, 151, 152, 153, 154, 155, 156, 157, 158, 159, 192, 235, 246, 357, 358, 359, 360, 366, 368, 369, 371, 372, 373, 376, 377, 378
superiority, 144
supply, 241, 249
suppression, 158
Sweden, 287
switching, 226, 229
Switzerland, 327
symbols, 372
symmetry, 41, 53, 54, 55, 67, 69, 87, 130, 217, 226, 227, 229
synchronization, 180
systems, vii, xiii, 60, 83, 95, 100, 106, 109, 123, 127, 128, 130, 196, 208, 226, 260, 274, 381

T

Taiwan, 138
temperature, viii, xi, 4, 33, 67, 68, 74, 78, 79, 80, 85, 107, 119, 122, 138, 165, 166, 187, 192, 193, 253, 288, 327, 328, 329, 330, 331, 332, 333, 334, 342, 343, 362
tension, 47, 54, 70, 71, 73
test data, 264, 265
textbooks, 32
theory, xii, 13, 65, 68, 84, 100, 109, 110, 114, 122, 123, 124, 125, 127, 130, 147, 193, 196, 197, 213, 214, 234, 235, 239, 240, 245, 250, 267, 346, 381, 382, 398
thermal conduction, ix
thermal energy, 86
thermal equilibrium, xi, 327
thermocline, xi, 327
threshold(s), xii, 58, 72, 87, 97, 98, 101, 102, 103, 104, 105, 114, 118, 120, 121, 122, 123, 124, 126, 250, 264, 277, 278, 279, 280, 284, 347, 349, 351, 353, 357, 360, 364, 366, 369
threshold level, 58, 364, 366, 369
thunderclouds, xi, 345, 346, 347, 350, 351, 354
tides, 340
time resolution, x, 243, 287, 289, 296, 338
time series, xi, 3, 4, 5, 6, 8, 9, 10, 11, 12, 84, 98, 125, 184, 327, 328, 329, 330, 331, 332, 333, 335, 336, 338, 339, 340, 341
timing, 167, 235, 252, 335, 336, 337, 338, 339, 341, 342, 359
Tokyo, 356
topology, 36, 43, 75, 237
torus, 63, 64, 69, 71
total energy, ix, 100, 124, 233, 239, 241, 249
tracking, 260, 262, 275, 278
training, 262, 263, 264, 265, 267, 269, 272, 273, 274, 283, 284
training speed, 273
traits, 99
trajectory, 208, 209, 385
transformation, viii, 75, 166, 168, 188, 189, 191, 208, 214, 264, 288
transition(s), 11, 118, 120, 122, 123, 182, 183, 237, 244, 351, 396
transparency, 212
transport, ix, 193, 233, 253, 330
transport processes, 330
trees, 120
trend, xi, 119, 168, 189, 190, 327, 329, 330, 332, 333, 343, 369
trial, 367
trial and error, 367
troposphere, viii, xi, 3, 165, 166, 167, 188, 327, 328, 329, 332, 342, 343
turbulence, vii, 24, 25, 95, 106, 108, 109, 110, 111, 115, 121, 127, 133, 248
turbulent, 18, 19, 21, 24, 25, 96, 105, 106, 109, 110, 114, 122, 130

U

UK, 233
Ukraine, x, 287, 289
ultraviolet, 114

umbra, xii, 148, 149, 152, 357, 358, 359, 366, 367, 369, 371, 372, 373, 374, 376, 377, 378, 379
umbral dots, xii, 357, 379
umbral nucleus, xii, 358, 374, 379
uncertainty, 80, 82, 188, 250, 331
UNESCO, 360
uniform, 58, 70, 124, 145, 158, 348, 349, 352, 364, 365
United States, 12
universality, 96
USSR, 133
UV, viii, xi, 4, 11, 12, 13, 74, 165, 166, 167, 168, 183, 184, 186, 187, 191, 192, 328, 335, 336, 337, 338, 343
UV radiation, 74, 187, 335, 338

V

validity, 144, 159, 216, 223, 224, 225, 243
values, 9, 20, 22, 23, 24, 25, 26, 43, 47, 51, 62, 70, 78, 87, 103, 104, 110, 111, 112, 146, 196, 209, 263, 271, 278, 289, 303, 309, 329, 339, 340, 343, 347, 348, 362, 363, 364, 365, 366, 367, 368, 369, 373, 374, 375, 377, 392, 395, 398
vapor, 330, 331, 332, 342, 343
variability, viii, 11, 12, 13, 83, 119, 165, 166, 167, 193, 244, 344, 378
variable(s), 12, 25, 32, 97, 99, 187, 202, 217, 218, 220, 224, 244, 268, 303, 391
variance, 8, 25, 104, 106, 223, 271, 340, 341, 342
variation, xi, xii, 3, 4, 12, 23, 152, 192, 260, 288, 327, 328, 329, 331, 332, 333, 334, 336, 337, 340, 342, 351, 357, 364, 365, 370, 371, 398
vector, xii, 19, 38, 41, 77, 109, 138, 142, 144, 145, 146, 159, 199, 200, 201, 240, 262, 264, 266, 267, 268, 269, 270, 271, 274, 284, 285, 355, 381, 383, 386, 387, 388, 392, 393, 394, 396, 397, 398
velocity, 18, 22, 24, 25, 27, 45, 47, 54, 70, 74, 76, 78, 79, 80, 83, 108, 109, 110, 117, 138, 139, 140, 141, 142, 144, 147, 148, 149, 150, 157, 158, 159, 200, 247, 250, 261, 274, 280, 282, 284, 285, 287, 303, 305, 310, 313, 315, 340, 344, 347, 349, 377
viscosity, 79, 109

W

war, 242
Washington, 88, 89, 136, 231, 331, 344
water vapor, xi, 327, 328, 330, 331, 332, 333, 334, 342, 343, 344
wave number, 20, 110, 340, 343, 358
wave power, 151
wave propagation, ix, 141, 144, 158, 196, 197, 208, 224, 230, 397
wave vector, 383, 392
wavelengths, 4, 158, 234, 300, 305, 313, 317, 322, 323, 324, 335, 336, 337
wavelet, 143
weakness, 225
wealth, 85, 138, 144
web, 276, 284, 285, 398
web pages, 398
wind, vii, viii, ix, xi, 4, 11, 12, 18, 31, 84, 107, 108, 127, 128, 165, 166, 167, 168, 169, 170, 172, 173, 174, 175, 176, 177, 179, 180, 181, 182, 183, 184, 186, 187, 188, 189, 190, 191, 192, 193, 195, 240, 248, 260, 327, 328, 334, 335, 338, 339, 340, 342, 343, 344
wind speeds, 168, 180, 181
winter, 189, 193
wintertime, 4

Y

yield, 104